POLYMER BLENDS:
PERFORMANCE

POLYMER BLENDS

Volume 2: Performance

Edited by

D. R. Paul

*Department of Chemical Engineering
and Texas Materials Institute
The University of Texas at Austin
Austin, TX 78712-1062*

C. B. Bucknall

*School of Industrial and Manufacturing Science
Cranfield University
Cranfield, Bedford MK43 0AL, United Kingdom*

A WILEY-INTERSCIENCE PUBLICATION

JOHN WILEY & SONS

New York / Chichester / Weinheim / Brisbane / Singapore / Toronto

Library of Congress Cataloging-in-Publication Data:

Polymer blends / D. R. Paul and C. B. Bucknall, editors.
    p.   cm.
    Includes index.
    Contents: 1. Formulation –
    ISBN 0-471-24825-8 (set : alk. paper). — ISBN 0-471-35279-9 (v. 1). – ISBN 0-471-35280-2 (v. 2)
    1. Polymers. 2. Plastics. I. Paul, Donald R. II. Bucknall, C. B.

TP1087.P64   1999   99–36533
668.9–dc21   CIP

Printed in the United States of America.

10 9 8 7 6 5 4 3 2 1

Contents

BLENDING FOR SPECIFIC PERFORMANCE

REINFORCED BLENDS

ELASTOMERIC BLENDS

RECYCLING

Preface

The field of polymer blends, or alloys, has experienced enormous growth in size and sophistication over the past two decades in terms of both the scientific base and technological and commercial development. It has become clear to us that an appropriate summary of this progress is needed to educate and to guide professionals working in this area into the twenty-first century. This two-volume set is a multiauthored treatise that might be viewed as an updated version of the analogous set edited by Paul and Newman and published in 1978. (See the reading list at the end of Chapter 1.) The book is intended to be a coherent entity rather than a collection of separate chapters, and a great deal of effort has been devoted to coordinating the content and style of the chapters. The editors intended each chapter to be far more than an encyclopedic summary of the literature or a review focusing only on the most recent advances in research. The authors were asked (a) to provide enough background in each chapter to enable beginners to work in the field by reading this book; (b) to sift critically through the literature and present only the most important issues (not every reference deserves mention); and (c) to write clearly but concisely, using carefully selected graphics, in order to make the important conceptual points and capture the attention of the browser.

It is the goal of these two volumes to be the authoritative source that professionals of the next decades will seek out to learn about this important field and use to set directions for future research and product development. The two volumes are roughly equal in length. Volume 1 is subtitled *Formulation* and is largely about the physics, chemistry, and processing issues associated with the formation of polymer blends and the evaluation and control of their structure. Volume 2 is subtitled *Performance* and is primarily concerned with how blends perform in practical situations. Naturally, there is a heavy emphasis on mechanical performance, but several chapters deal with a range of other properties as well. At some risk of oversimplification, it can be said that Volume 1 is about *structure*, while Volume 2 is about *properties*. Thus, the two-volume set provides a broad view of the *structure–property* relationships for polymer blends as seen by experts from around the world.

The editors have been friends and colleagues for many years. Their professional interests have been somewhat different over their careers, but there are many points of intersection. These differences and similarities have been helpful during the course of planning, which started in early 1994, and development of this book. A common view was needed in order to foster agreement on the scope, content, and choice of authors. The differences in expertise led Don Paul to have primary responsibility for Volume 1 and Clive Bucknall to oversee Volume 2.

We are thankful to many colleagues and friends who have encouraged us and given us advice on many issues.

D. R. PAUL
C. B. BUCKNALL

Contributors, Volume 2

S. ABDOU-SABET, Advanced Elastomer Systems L.P., Akron, OH 44311

R. ALEXANDER-KATZ, Depto. de Física, Universidad Autónoma Metropolitana-Iztapalapa, 09340 México, D.F., México

D. G. BAIRD, Department of Chemical Engineering and the Center for Composite Materials and Structures, Virginia Polytechnic Institute and State University, Blacksburg, VA 24061-0211

H. R. BROWN, BHP Steel Institute, University of Wollongong, Wollongong, NSW 2522, Australia

C. B. BUCKNALL, School of Industrial and Manufacturing Science, Cranfield University, Cranfield, Bedford MK43 0AL, United Kingdom

X. H. CHEN, Department of Mechanical Engineering, Center for Advanced Materials Technology, University of Sydney, Sydney, NSW 2006, Australia

C. A. CRUZ-RAMOS, Plastics Additives Research Department, Rohm and Haas Company Research Laboratories, Bristol, PA 19007

S. DATTA, Baytown Polymers Center, Exxon Chemical Co., Baytown, TX 77522-5200

J. DU, Department of Materials Science and Engineering, The University of Michigan, Ann Arbor, MI 48109

T. S. ELLIS, Delphi Automotive Systems Research and Development, Warren, MI 48090

M. EVSTATIEV, Laboratory for Structure and Properties of Polymers, University of Sofia, 1126 Sofia, Bulgaria

S. FAKIROV, Laboratory for Structure and Properties of Polymers, University of Sofia, 1126 Sofia, Bulgaria

K. FRIEDRICH, Institute for Composite Materials Ltd., University of Kaiserslautern, D-67663 Kaiserslautern, Germany

R. J. GAYMANS, Faculty of Chemical Technology, University of Twente, 7500 AE Enschede, The Netherlands

F. J. GUILD,   Department of Mechanical Engineering, Queen's Building, University of Bristol, Bristol, BS8 1TR, United Kingdom

J. KARGER-KOCSIS,   Institute for Composite Materials Ltd., University of Kaiserslautern, D-67663 Kaiserslautern, Germany

Y.-W. MAI,   Department of Mechanical Engineering, Center for Advanced Materials Technology, University of Sydney, Sydney, NSW 2006, Australia

M. A. MCLEOD,   Department of Chemical Engineering and the Center for Composite Materials and Structures, Virginia Polytechnic Institute and State University, Blacksburg, VA 24061-0211

R. A. PEARSON,   Department of Materials Science and Engineering, Lehigh University, Bethlehem, PA 18015

I. G. PLOTZKER,   Central Research and Development, Du Pont Company, Wilmington, DE 19880-0323

L. PRUITT,   Department of Mechanical Engineering, University of California at Berkeley, Berkeley, CA 94720

J. J. SCOBBO,   GE Plastics, One Noryl Avenue, Selkirk, NY 12158

P. M. SUBRAMANIAN,   S.P.M. Technologies, Hockessin, DE 19707

M. D. THOULESS,   Department of Mechanical Engineering and Applied Mechanics, The University of Michigan, Ann Arbor, MI 48109

S.-C. WONG,   Department of Mechanical Engineering, Center for Advanced Materials Technology, University of Sydney, Sydney, NSW 2006, Australia

A. F. YEE,   Department of Materials Science and Engineering, The University of Michigan, Ann Arbor, MI 48109

Contents, Volume 1

Contributors, Volume 1

V. ARRIGHI, Department of Chemistry, Heriot-Watt University, Riccarton, Edinburgh EH14 4AS United Kingdom

M. BALLAUFF, Polymer-Institut der Universität Karlsruhe, Kaiserstrasse 12, 76128 Karlsruhe, Germany

J. M. BRADY, Plastics Additives Research Department, Rohm and Haas Company Research Laboratories, Bristol, PA 19007

C. B. BUCKNALL, School of Industrial and Manufacturing Science, Cranfield University, Cranfield, Bedford MK43 0AL, United Kingdom

D. G. BUCKNALL, Department of Materials, University of Oxford, Parks Road, Oxford OX1 3PH, United Kingdom

M. M. COLEMAN, Department of Materials Science and Engineering, The Pennsylvania State University, University Park, PA 16802

C. A. CRUZ-RAMOS, Plastics Additives Research Department, Rohm and Haas Company Research Laboratories, Bristol, PA 19007

J. R. DORGAN, Chemical Engineering Department, Colorado School of Mines, Golden, CO 80401

B. D. FAVIS, Department of Chemical Engineering, University of Montréal, Ecole Polytechnique, Montréal, Québec H3C 3A7, Canada

W. W. GRAESSLEY, Department of Chemical Engineering, Princeton University, Princeton, NJ 08544

G. GROENINCKX, Department of Chemistry, Laboratory of Macromolecular Structural Chemistry, Catholic University of Leuven, B-3001 Heverlee, Belgium

S. Y. HOBBS, General Electric Company, Research and Development Center, Schenectady, NY 12301

S. D. HUDSON, Department of Macromolecular Science and Engineering, Case Western Reserve University, Cleveland, OH 44106

T. INOUE, Department of Organic and Polymeric Materials, Tokyo Institute of Technology, Ookayama, Meguro-ku, Tokyo 152-8552, Japan

A. M. JAMIESON,   Department of Macromolecular Science and Engineering, Case Western Reserve University, Cleveland, OH 44106

D. S. KALIKA,   Department of Chemical and Materials Engineering, University of Kentucky, Lexington, KY 40506

T. KYU,   Institute of Polymer Engineering, University of Akron, Akron, OH 44325

C. J. T. LANDRY,   Imaging Research Laboratories, Eastman Kodak Company, Rochester, NY 14650

D. J. LOHSE,   Corporate Research Labs, Exxon Research & Engineering Co., Annandale, NJ 08801

B. MAJUMDAR,   GE Plastics, Selkirk, NY 12158

G. D. MERFELD,   General Electric Co., Research and Development Center, Schenectady, NY 12301

P. C. PAINTER,   Department of Materials Science and Engineering, The Pennsylvania State University, University Park, PA 16802

J. P. PASCAULT,   Institut National des Sciences Appliquées, Laboratoire des Materiaux Macromoléculaires, 69621 Villeurbanne Cedex, France

D. R. PAUL,   Department of Chemical Engineering and Texas Materials Institute, The University of Texas at Austin, Austin, TX 78712

J. P. RUNT,   Department of Materials Science and Engineering, The Pennsylvania State University, University Park, PA 16802

I. C. SANCHEZ,   Department of Chemical Engineering, The University of Texas at Austin, Austin, TX 78712

M. SARKISSOVA,   Department of Chemistry, Laboratory of Macromolecular Structural Chemistry, Catholic University of Leuven, B-3001 Heverlee, Belgium

M. T. STONE,   Department of Chemical Engineering, The University of Texas at Austin, Austin, TX 78712

S. THOMAS,   Department of Chemistry, Laboratory of Macromolecular Structural Chemistry, Catholic University of Leuven, B-3001 Heverlee, Belgium

V. H. WATKINS,   General Electric Company, Research and Development Center, Schenectady, NY 12301

R. J. J. WILLIAMS,   Institute of Materials Science and Technology (INTEMA), University of Mar del Plata and National Research Council (CONICE), 7600 Mar del Plata, Argentina

POLYMER BLENDS:
PERFORMANCE

19 Quasielastic Mechanical Properties

FELICITY J. GUILD

Department of Mechanical Engineering
University of Bristol
Queen's Building
University Walk
Bristol BS8 1TR
United Kingdom

I. INTRODUCTION

The development of polymer technologies, allowing an ever-increasing range of particulate-filled (rigid or soft) polymers, is leading to increasing need for predictive modeling. The full experimental investigation of all possible materials is becoming

Polymer Blends, Volume 2: Performance. Edited by D. R. Paul and C. B. Bucknall.
ISBN 0-471-35280-2.   © 2000 John Wiley & Sons, Inc.

uneconomic, so predictive modeling is becoming more important as a cost-effective method of material investigation; with predictive modeling, elastic properties can be predicted and failure mechanisms can be investigated. This chapter is principally concerned with the prediction of quasielastic properties. Although such simulation can never entirely replace experimental investigation, its importance as an investigative tool is becoming readily apparent.

The analytical methods for simulation are first considered. It is notable that the earliest attempts to model the elastic properties of composite materials are found in the literature of the 19th century; even today, this subject is under investigation. Consideration of the methods of analytical modeling demonstrates the need for microstructural characterization of multiphase materials and for suitable particle-distribution models. Finally, the elastic properties of some filled polymers are presented, divided into overall categories based on the type of particle.

II. ANALYTICAL METHODS

Numerous analytical models have been proposed for the description of both the overall elastic properties of multiphase materials and the stress distributions within them. Some of the different methods of analysis and the main assumptions underlying them are presented in this section.

A. Classical Mechanics

The stress distributions within the matrix around an isolated spherical or cylindrical inclusion, made up of a void or a linearly elastic material, were first obtained by Goodier [1]. Later, the self-consistent approach, based on the theory of elasticity, was used to find the stress distributions around rigid spheres [2] or elastic spheres, or ellipsoids [3], embedded in an infinite elastic matrix. Eshelby's theory was generalized to include the interaction of filler particles, and expressions for elastic constants were derived by Chow [4]. Identical equations for materials containing spherical particles were obtained independently by Kerner [5]. These analyses are based on modeling the composite as an assembly of elements, each consisting of a spherical filler particle embedded in a spherical shell of the matrix, which is itself surrounded by an infinite matrix of material possessing properties of the "composite." Thus, it is inherent in this self-consistent approach that interparticle interactions are ignored.

Further analyses, producing bounds on elastic moduli, are based on the determination of the first-order moments of the random stress and strain fields in the heterogeneous solid. The exact stress fields around heterogeneities can be found as analytical solutions only in cases that allow the definition of regular unit cells. These "classical" analytical methods are based on the assumption of a defined unit cell surrounding each filler particle. The bounds generally arise from the assumption of conditions of equal stress or equal strain within the different unit cells. These approaches are well known in the analysis of continuous-fiber composites, where these simple averages are used to predict the values of longitudinal and transverse stiffness of the overall composite [6, 7]. These analyses are essentially based on parallel or series spring models in which interactions between neighboring fibers are ignored.

The bounds obtained from some of these analyses are widely spaced. Paul [8] produced bounds for the bulk modulus and shear modulus by incorporating variational principles, but his analysis involved averaging the stress fields. These bounds were improved by Hashin and Shtrikman [9] using the variational principle. This approach is based on the idea of a homogenous reference medium that allows the definition of reference values of stress and strain. It can be shown that the bounds due to Paul [8] arise from the extreme values of the trial stress field. A full derivation of the Hashin and Shtrikman bounds was later given by Willis [10].

The various models were considered by Ishai and Cohen [11], leading to the derivation of the most closely spaced bounds to apply to a particle-filled material. The (cubic) particles were assumed to be regularly arranged in perfect cubic packing. Bounds were derived only for values of tensile moduli; lateral contractions were ignored, so these bounds do not include the full elastic properties of the composite material. The interactions between neighboring particles were not fully taken into account. These bounds are compared with experimental values of tensile moduli for filled polymers containing hard and soft particles later in this chapter.

An alternative analytical approach is based on an effective medium approximation. Such approaches rely on the assumption that the second phase occupies only a small fraction of the volume of the composite material, such that the overall changes in the elastic constants between the matrix material alone and the composite are relatively small. The effective elastic constants can then be calculated using perturbation theory. As is apparent from the use of perturbation theory, these methods are applicable only to materials containing low concentrations of the second phase. Further assumptions include that the "ligament' lengths"—that is, the distance between particles—are treated as independent variables; in other words, the value of one interparticle distance is unaffected by the values of other interparticle distances. This assumption is obviously erroneous for a distribution of particles. The perturbation method has been applied to describe elastic properties of particulate-filled materials (e.g., [12, 13]).

B. Finite-Element Analysis

The limitations of the aforementioned analytical methods are apparent. Many of these limitations are overcome by the use of numerical methods. In recent years, the finite-element-analysis method has become increasingly accessible to nonspecialist users through the development of numerous graphical interfaces for both preprocessing and postprocessing and through the increasing power of "desktop" workstations for the processing of solutions. However, it is important to emphasize that increasing accessibility and increasing computer power do not necessarily correlate with increasing accuracy of the resulting solutions. It remains the responsibility of the analyst to make the correct choice of the many parameters required for the analysis, including the overall geometry to be analyzed, the element type (including formulation and integration), the mesh density, the material model, and the boundary conditions.

This list is not exhaustive, but the number of available parameters points to the care that must always be exercised when approaching this method of analysis.

Numerical models include the derivation of lateral contractions; in other words, values of Poisson's ratio ν are obtained. Assuming that the material is macroscopically isotropic, the values of shear modulus G and bulk modulus K can be calculated simply from the values of Poisson's ratio and Young's modulus E, using the following well-known expressions for isotropic elastic solids:

$$G = E/\left[2(1+\nu)\right] \tag{19.1}$$
$$K = E/\left[3(1-2\nu)\right]. \tag{19.2}$$

The importance of this full description of the elastic properties of filled polymers is becoming increasingly important as the range of their applications is growing.

The choice of overall geometry to be analyzed is the initial step in carrying out any numerical analysis. The calculations are always based on the analysis of a unit cell that is representative of the overall structure. The choice of the unit cell is dependent on the materials model chosen, as discussed in Section IV. The boundary conditions imposed on the cell reflect its interactions with its neighboring cells. Similarly, some choices regarding element type—for example, whether two-dimensional, three-dimensional, or axisymmetric elements are required—arise directly from the definition of the unit cell. Further details regarding the elements used to create the mesh, such as the integration or formulation required for a given problem, and the most suitable mesh density to be used are beyond the scope of this chapter, but may be found in several texts (e.g., [14–16]).

The models chosen to represent the properties of the different constituents must also be considered. Since this chapter is concerned with the quasielastic properties of multiphase materials, the material property models considered here are generally based on linear elasticity. All analyses carried out using "classical" methods, described in the previous section, are based on linear elastic behavior. Further finite-element simulations of these materials can now include sophisticated material property models using either the models provided in the package or new ones written specifically for the application; some of the material property models available have been reviewed recently [17].

III. MICROSTRUCTURAL CHARACTERIZATION

Quantitative descriptions of the spatial arrangement of particles can now be obtained using automatic image analysis. The quantitative methods described here generally lead to descriptions of the microstructure that may be used directly in analytical techniques, such as finite-element analysis. Other methods used for the description of particle arrangement may be described as functional methods; the microstructure is described by some parameter or function. The spatial distribution of the parameter is measured using a specialized image analyzer or software package, and the data are often transferred (for example, to a PC) for further computation. These functional methods include those described in Chapter 9; their application to fiber composites has recently been reviewed by Guild and Summerscales [18]. A major drawback in many of these analyses is the assumption that the distribution of particles can be treated as a distribution of points. In other words, the finite size of the particles is ignored. Measurements of distributions have indicated the importance of using "hard core" models, which take into account the finite particle size [19].

The analysis of microstructure, whether using direct methods or automatic image analysis, is essentially based on observations in two dimensions. The determination of three-dimensional structures, such as the microstructure of a particulate-filled polymer, via measurements in two dimensions is the subject of stereology, the fundamentals of which are contained in various texts (e.g., [20, 21]).

The development of more powerful image analyzers has opened the possibility of the more rigorous microstructural definition, as the positions of the particles themselves can be recorded. Such methods are closely related to the definition of the unit cell required for the analysis, described in Section IV. Methods of rigorous quantitative definition of the microstructure of particulate-filled materials are presently being developed, but further experimental and theoretical work is required. The need for such definition is becoming increasingly apparent as the links between microstructure and properties are investigated in greater detail.

IV. PARTICLE-DISTRIBUTION MODELS

The analysis of the mechanical properties of particulate-filled polymers requires the definition of a representative unit cell for the material, which is based on assumptions regarding the distribution of the filler. Two distinct approaches are available for describing the properties of particulate-filled polymers; these approaches may be characterized as regular models or random models. Regular models are based on the assumption of a fixed, regular distribution of particles. Alternatively, the particles can be assumed to exist in a random distribution.

A. Regular Models

The use of regular models for the analysis of particulate-filled polymers implies that the filler particles are distributed in a fixed and uniform pattern. The unit cell can then be drawn using mirror symmetry. The uniform cubic distribution is the most widely used configuration for both analytical methods based on elasticity (e.g., [11]) and those based on numerical analyses (e.g., [22]).

B. Random Models

A random distribution of particles is conceptually more likely than a regular array. Representative cells describing a random distribution in space can be derived from theoretical considerations. Davy and Guild [23] have presented a description of composite materials in terms of the interparticle distances arising from Voronoi cells. The Voronoi cell associated with a particle is the region around that particle in which all points are closer to it than to any other particle. The variability of the half-interparticle distance, or the radius of the Voronoi cells, was calculated for a random distribution. The particles were assumed to be of constant radius and were not allowed to overlap. This calculated variability can be used to describe the random distribution as a collection of spherical Voronoi cells of varying radius, each containing a filler particle at its center. The important distinction between this null hypothesis derived by Davy and Guild and the functional methods described in Section III and in Chapter 9 is that the finite size of the reinforcing domains or particles is taken into account. The approach including this effect may be described as the random hard-core model. The division of a plane into Voronoi cells is also known as Dirichlet cell tessellation [24].

The parameters of Voronoi cells derived from neglecting the finite size of the particle—that is, the distribution known as the Poisson process—and using a hard-core model have been derived and compared with measured materials parameters [25, 26]. The comparison showed the importance of the hard-core model. However, that analysis does not include the derivation of the interparticle distance, so the results presented could not be used directly for analytical models. Yang and Colton [27] attempted to define the dimensions of two-dimensional Voronoi cells by approximating measured Voronoi cells to the nearest square, in a given orientation. The application of arbitrary axes to define the orientations of these squares is hard to justify.

An alternative way of deriving interparticle distances for a random distribution is based on percolation theory (e.g., [12, 13]). However, as described previously, such a description of a random distribution includes several assumptions, including the treatment of the "ligament" lengths as independent. The ligament lengths are analogous here to the interparticle distances, and the description of random distributions of particles using Voronoi cells does not include this assumption.

The analysis of Davy and Guild [23] includes a method for deducing overall mechanical properties from analysis of unit cells representing a random distribution of particles of finite size. This materials model is based on the concept that the interactions of neighboring particles with a given particle are not directional; that is, the overall effect of the interactions is a nondirectional average of the individual interactions. It is apparent that, "on average," the shape of a unit cell is spherical. Thus, the correct overall materials model for a random distribution of particles is a collection of spherical cells of different sizes, each containing a single sphere. Ideally, analyses would be carried out for the full range of cell sizes representing the given volume fraction of particles and the overall properties deduced from the properly weighted sum. However, Davy and Guild [23] have deduced a more usable approach based on a "dispersion factor" that is applied to the value of the property (or a simple function of its value) calculated for the mean volume fraction. This dispersion factor accounts for the variation in the value of an elastic property that occurs when particle–matrix cells on either side of the mean composition have unequal effects on the value of that property. The overall results provided by this model do consist of bounds, arising from the assumption that the different cells are subjected to either equal stress or equal strain. However, these bounds are found to be close.

Recently, simulated random distributions, including distributions in particle size, have been analyzed in terms of statistical parameters and materials properties using the finite-element-analysis method [28,29]. These analyses do take into account the directionality of the interactions between neighboring particles, but they are carried out in two dimensions, assuming plane-strain (zero strain in the out-of-plane direction) conditions. Such numerical analyses require extensive computer resources.

V. ELASTIC PROPERTIES

A. Hard Particles

As intuitively expected, the addition of a particulate filler of high stiffness increases the overall modulus of the material. The size of this increased stiffness has been the subject of extensive discussion in the literature for many years, and several predictive models have been developed to describe it, as discussed in Section II.

There are various examples in the literature of comparisons between experimental data and different models. In the presentation of their model, Ishai and Cohen [11] compared experimental values for moduli with the predictions of their model and with the wider bounds arising from the Hashin analysis [2] and the analysis due to Paul [8]. The experimental results were found to exceed their predictions throughout the range of the volume fraction; the difference became larger at higher values of the volume fraction. Similarly, Young et al. [30] found that measured values of Young's modulus for epoxy resin filled with glass beads lay close to the lower bound predicted by the Ishai and Cohen model at low volume fractions of the beads, but exceeded the upper bound at higher volume fractions.

A carefully measured set of experimental data for epoxy resin filled with glass spheres is available in the literature [31]. These data are compared with the predictions of the Ishai and Cohen model in Fig. 19.1. The differences between experimental measurements and predicted bounds are similar to those reported previously [30]. For lower volume fractions of glass beads, the experimental values lie close to the lower bound; at the highest volume fraction, the experimental value exceeds the higher bound. The overall shape of the experimentally measured relationship between volume fraction and stiffness is not well described by the bounds.

These observations lead to the requirement for more accurate modeling of the mechanical properties of multiphase materials. Such modeling requires the use of numerical methods. It is important to note the further benefit of numerical modeling: In contrast to analytical techniques, numerical modeling allows the deduction of a complete set of elastic properties for the filled polymer, as described in Section II of this chapter.

The numerical modeling of mechanical properties requires the definition of a unit cell. As described in Section IV, this definition can be based on the assumption of a regular or random distribution of the filler. Termonia [32] used a finite-difference approach to predict the stiffness of glass-filled epoxy resin, assuming a regular distribution of filler. He found good agreement between his predictions and the experimental data [31] used in Fig. 19.1. There are several examples in the literature that

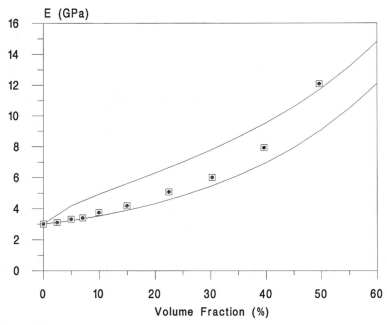

Figure 19.1   Variation of Young's Modulus with the volume fraction of glass spheres reinforcing epoxy resin: Comparison of experimental values from [31] (⊡) to predicted bounds from [11] (—).

use axisymmetric modeling based on a unit cell consisting of a sphere in a cylinder of resin. Results from this model, assuming that the volume fraction can be calculated with the gaps between the cylinders being filled with matrix, have been shown to agree closely with results from three-dimensional modeling that assumes a cubic array [33, 34]. The assumption of a regular cubic array has been found to provide the best description of the experimentally measured relationship between stiffness and included volume fraction for a dispersion of ceramic particles in polyethylene [22]. This model also successfully describes the measured values of shear modulus. Experimental values [35] are compared in Fig. 19.2 to values predicted on the assumption that there is a regular cubic distribution of the filler. Reasonable agreement is found for both Young's modulus and the shear modulus throughout the range of volume fraction. This material is a replacement-bone material, and it is important to define its elastic properties fully, because stress systems within the body are complex.

As an alternative to the aforementioned regular distributions, the distribution of filler particles can be assumed to be random. This materials model should be analyzed using a spherical unit cell; the method of analysis for this type of unit cell has been derived by Guild and Kinloch [36]. The predicted values of Young's modulus and Poisson's ratio for epoxy resin filled with glass beads are compared with measured values [31] in Fig. 19.3. In addition, these comparisons have previously been made using a cylinder as a unit cell [23]; for that analysis, good agreement

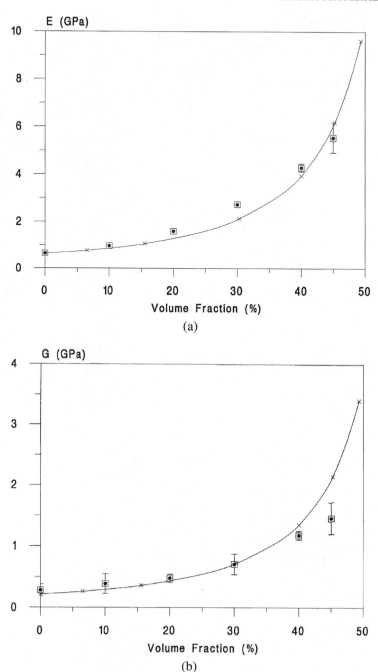

Figure 19.2 Variation of values of the Modulus with the volume fraction of hydroxyapatite particles reinforcing polyethylene: Comparison of experimental values from [35] (⊙) with predicted values from finite-element analysis assuming a cubic material model (×). (a) Young's modulus, (b) the shear modulus.

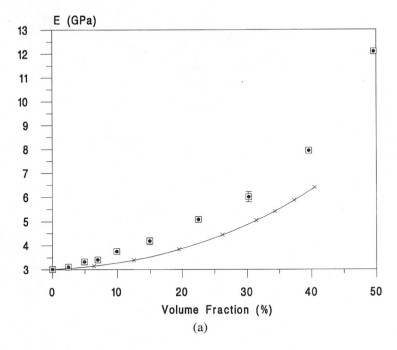

(a)

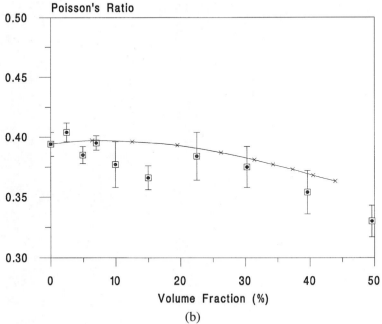

(b)

Figure 19.3 Variation of elastic properties with the volume fraction of glass spheres reinforcing epoxy resin: Comparison of experimental values from [31] (◉) with predicted values from finite-element analysis assuming a cylindrical material model (×). (a) Young's modulus, (b) Poisson's ratio.

was found between measured and predicted values of Young's modulus, but poor agreement was found between measured and predicted values of Poisson's ratio. The results in Fig. 19.3 show that values of Young's modulus predicted using the spherical unit cell underestimate the measured values, particularly at high volume fractions of particles, but the agreement between measured and predicted values of Poisson's ratio is reasonable. The underestimation of Young's modulus must arise from the lack of directionality in models of the interactions; this was the effect deduced for the modeling of the ceramic-filled polyethylene [22]. The improved agreement for the values of Poisson's ratio reflects the strong dependence of this property on the shape of the unit cell.

B. Soft Particles

Modeling of mechanical properties for polymers filled with soft particles has received very great research attention in recent years. This interest arises from the improvement in fracture properties resulting from the addition of soft elastomers to hard, brittle polymers; a well-known example of such materials is rubber-toughened epoxy resin. These materials are used in a wide range of applications, including structural adhesives and high-performance composite materials. For such applications, information about the quasi elastic properties of the filled polymer is needed for analysis and prediction of the behavior of the resulting structures or materials; that is, the properties must be properly defined.

As intuitively expected, the addition of softer particles reduces the overall stiffness of the filled polymer. There are several examples in the literature that compare experimental results with results from predictive modeling for this type of blend. In the presentation of their analytical model, Ishai and Cohen [11] compared experimental values for a porous polymer with the predicted decrease in stiffness. The experimental values were lower than both their predictions and those predictions based on the Hashin [2] and Paul [8] models. Yee and Pearson [37] measured values of Young's modulus of rubber-filled epoxy resin over a range of volume fractions. Their results are compared to the Ishai and Cohen [11] model in Fig. 19.4. The experimental values are close to the predicted upper bound at the lowest volume fractions of rubber and are midway between the bounds at higher concentrations. The measured values show an approximately linear decline with increasing volume fraction, following the same trend as the upper theoretical bound. The change in stiffness with volume fraction of soft particles is reasonably well described by the analytical model, but, as noted in Section II of this chapter, such modeling does not predict the full elastic properties of the filled polymer.

The alternative method of modeling, using numerical methods, has also been applied to this class of materials. Boyce et al. [38] modeled the mechanical properties of the filled polymer, assuming a regular array of axisymmetric cylinders, each containing a sphere at its center. The properties of rubber-toughened epoxy resin have been modeled assuming a random distribution using a unit cell of either cylindrical shape [39] or spherical shape [40]. Comparison of predictions with experimental results [37] reveals a pattern of behavior similar to that observed with hard particles.

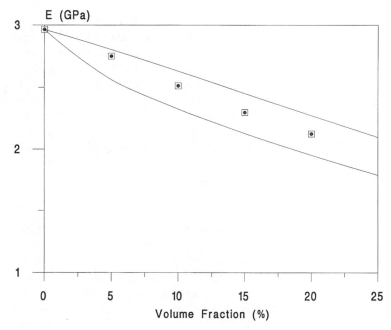

Figure 19.4 Variation of Young's modulus with the volume fraction of rubber spheres reinforcing epoxy resin: Comparison of experimental values from [37] (◉) with predicted the bound from [11] (—).

The cylindrical model predicts a greater change in stiffness with volume fraction; this prediction comes closer to the experimental values. The spherical material model is more logically correct, but it removes all directional interactions. The most accurate prediction probably requires full three-dimensional modeling.

C. More Complex Morphologies

There are several examples of filled polymers with more complex morphologies, which can be divided into two overall categories. In the first category, the complex morphology is found in the filler particles. Examples include the developing wide range of core–shell particles used particularly in the toughening of poly(methylmethacrylate), or PMMA (see Chapter 18), and the naturally occurring "salami" structure of the toughening particles in high-impact polystyrene (HIPS) (see Chapter 22). These filler particles contain some rubbery material as well as some glassy material with properties similar to the matrix. The second category generally arises from the addition of more than one filler to the matrix material. There are several examples of such morphologies in the field of adhesives, particularly those based on epoxy resins, which might, for example, contain comparatively large, hard particles, such as ballotini, to ensure constant bond thickness, as well as smaller, soft particles to improve the toughness of the material. Alternatively, two separate distributions of one type of particle may be present [41].

Presently, there is considerable research on the role of core–shell particles (see Chapter 24) as modifiers for, most particularly, thermoplastics, such as PMMA. These particles can provide improvements in toughness similar to those obtained on addition of simple rubber particles, but their addition is not expected to cause the large deterioration in mechanical stiffness that arises from the addition of pure rubber particles. Further, more accurate matching of optical properties is possible using these complex morphologies (see Chapter 28), leading to levels of transparency close to that obtained from pure PMMA. The appearance of these materials is becoming increasingly important, because health and safety regulations are requiring the use of polymeric materials as replacements of, for example, glass display shelving.

There are many examples in the literature of the measurement of elastic properties for blends containing core–shell particles. As expected, the addition of these particles, which include some rubbery material, reduces stiffness, and for different particle morphologies, the decrease can be correlated with the volume fraction of rubber [42]. However, the measured decrease in stiffness with increasing volume of particles is higher than expected, comparable to that obtained on the addition of pure rubber particles [42–44]. This anomaly requires further investigation, which may best be carried out using dynamic measurements, which lead to insights into detailed particle morphology [45, 46].

As described previously, the "salami" structure of the toughening rubber particles in HIPS, which contain inclusions of polystyrene, is the morphology that occurs in current commercial products [47]. These morphologies are highly variable, so the measured elastic properties are hard to define [43]. More recently, simpler morphologies, similar to the more closely controlled core–shell particles used in toughening PMMA, have been used for toughening polystyrene [48, 49].

The prediction of elastic properties for both categories of more complex materials, using methods based on classical elasticity, can be carried out only in successive stages. For the second type of complex morphology, for which the filler particles are of more than one type or distribution, the more highly dispersed particles must be considered first, in order to define the properties of the "modified matrix." The properties of the modified matrix can then be used to model the properties of the composite, which is treated as containing the second type of particles. Similar steps in the analysis can be used for finite-element modeling of these materials.

The same concepts must be used for classical methods of calculation for the elastic properties of blends containing the more complex particle morphologies. In this case, the properties of the "modified particle" must first be derived, and these properties can then be used to model the properties of the composite. For finite-element modeling of these materials, a similar two-stage process can be used. Alternatively, the complex particle morphology can be included directly into the finite-element model of the material [50]. As described previously, the results of this type of modeling do not agree with experimental measurements of these properties, and this anomaly is the subject of future investigation.

An alternative microstructural transformation approach to the description of microstructure that may be most applicable to these more complex morphologies has been proposed by Fan et al. [51]. The microstructural transformation involves the description of the microstructure in terms of its distinguishing features, such as pure particle, pure matrix, interface regions, etc. These distinguishing features are termed "topological parameters." Thus, this description of the microstructure leads to the definition of the numbers and types of topological parameters that characterize that microstructure. Analysis is carried out for the different structures described by the topological parameters, and the properties of the material can then be derived using a summing process. This method has been used to predict the elastic properties of several particulate-filled materials [52, 53].

VI. CONCLUSIONS

The quasielastic properties of composite materials have been the subject of research for more than a century, yet many unanswered questions remain. The intuitive expectations of increasing stiffness as a greater proportion of stiff particles are added, and decreasing stiffness as a greater proportion of rubbery particles are added, are realized in both experimental measurements and predictive analyses. However, even with the extensive computer facilities available today, it is hard to reproduce experimental results fully in simulations. The anomalies are particularly marked for the new core–shell particles.

It is clear that all simulations must be based on materials models describing the distribution of the particles and the morphology of the particles. Increasing accuracy in results from predictive modeling will be attained only via developments in the field of spatial description of composite materials and via understanding of the morphology of core–shell particles.

VII. ACKNOWLEDGEMENTS

This chapter has arisen from the author's collaboration with Professor W. Bonfield (IRC in Biomedical Materials, Queen Mary & Westfield College, U.K.), Dr. P. J. Davy (University of Wollongong, Australia), Professor A. J. Kinloch (Imperial College, U.K.), Professor B. W. Silverman (University of Bristol, U.K.), Dr. J. Summerscales (University of Plymouth, U.K.) and Professor R. J. Young (Manchester Materials Science Centre, U.K.). The author gratefully acknowledges very many fruitful discussions with these individuals.

VIII. REFERENCES

1. J. N. Goodier, *J. Appl. Mech.*, **55**, 39 (1933).
2. Z. Hashin, *J. Appl. Mech.*, **29**, 143 (1962).

3. J. D. Eshelby, *Proc. Roy. Soc. Series*, **A(241)**, 376 (1957).

4. T. S. Chow, *J. Polymer Sci.*, **16**, 959 (1978).

5. E. H. Kerner, *Proc. Phys. Soc.*, **B(69)**, 808 (1956).

6. A. Reuss, *Zeit. Angew Math. U. Mech.*, **9**, 49 (1929).

7. W. Voigt, *Ann. Phys.*, **33**, 573 (1889).

8. B. Paul, *Trans. AIME*, **218**, 36 (1960).

9. Z. Hashin and S. Shtrikman, *J. Mech. Phys. Solids*, **11**, 127 (1963).

10. J. R. Willis, *J. Mech. Phys. Solids*, **25**, 185 (1977).

11. O. Ishai and L. J. Cohen, *Int. J. Mech. Sci.*, **9**, 539 (1967).

12. A. Margolina and S. Wu, *Polymer*, **29**, 2170 (1988).

13. M. Zheng and X. Zheng, *Metalurgical Trans.*, **A(22)**, 507 (1991).

14. M. J. Fagan, *Finite Element Analysis: Theory and Practice*, Longman, London, 1992.

15. R. D. Cook, *Finite Element Modeling for Stress Analysis*, John Wiley & Sons, Inc., New York, 1995.

16. R. J. Astley, *Finite Elements in Solids and Structure*, Chapman & Hall, London, 1992.

17. G. B. McKenna, *J. Res. NIST*, **99**, 169 (1994).

18. F. J. Guild and J. Summerscales, *Composites*, **24**, 383 (1993).

19. Q. F. Li, R. Smith, and D. G. McCartney, *Mater. Char.*, **28**, 189 (1992).

20. E. E. Underwood, *Quantitative Stereology*, Addison-Wesley, Reading, MA, 1970.

21. J. C. Russ, *Practical Stereology*, Plenum Press, New York, 1986.

22. F. J. Guild and W. Bonfield, *J. Mater. Sci.: Materials in Medicine*, **9**, 497 (1998).

23. P. J. Davy and F. J .Guild, *Proc. Roy. Soc. Series A*, **418**, 95 (1988).

24. A. Getis and B. Boots, *Models of Spatial Processes*, Cambridge University Press, Cambridge, 1978.

25. R. Pyrz, "Stereological Quantification of the Microstructure Morphology for Composite Materials," in *Optimal Design with Advanced Materials*, P. Pederson (ed.), Elsevier, Amsterdam, 1993, p. 81.

26. R. Pyrz, *Composites Sci. Technol.*, **50**, 197 (1994).

27. H. Yang and J. S. Colton, *Polymer Composites*, **15**, 46 (1994).

28. S. Ghosh, Z. Nowak, and K. Lee, *Acta. Mater.*, **45**, 2215 (1997).

29. S. Ghosh, Z. Nowak, and K. Lee, *Composite Sci. Technol.*, **57**, 1187 (1997).

30. R. J. Young, D. L. Maxwell, and A. J. Kinloch, *J. Mater. Sci.*, **21**, 380 (1986).

31. J. C. Smith, *J. Res. Natl. Bur. Stand. (Phys. Chem.)*, **80A**, 1 (1976).

32. Y. Termonia, *J. Mater. Sci.*, **22**, 1733 (1987).

33. B. D. Agarwal and L. J. Broutman, *Fibre Sci. Tehnol.*, **7**, 63 (1974).

34. B. D. Agarwal, G. Panizza, and L. J. Broutman, *J. Amer. Cer. Soc.*, **54**, 620 (1971).

35. M. Wang, S. Deb, K. E. Tanner, and W. Bonfield, *Proc. ECCM-7*, **2**, 455, Woodhead Publishing Ltd., 1996.

36. F. J. Guild and A. J. Kinloch, *J. Mater. Sci. Lett.*, **11**, 484 (1992).

37. A. F. Yee and R. A. Pearson, *J. Mater. Sci.*, **21**, 2562 (1986).

38. M. E. Boyce, A. S. Argon, and D. M. Parks, *Polymer*, **28**, 1680 (1987).

39. F. J. Guild and R. J. Young, *J. Mater. Sci.*, **24**, 2454 (1989).

40. F. J. Guild and A. J. Kinloch, *J. Mater. Sci.*, **30**, 1689 (1995).

41. R. A. Hall and I. Burnstein, "Bimodal-Distribution Models of the Discrete Phase in Toughened Plastics," in: *Toughened Plastics II: Novel Approaches in Science and Engineering*, C. K. Riew and A. J. Kinloch (eds.), American Chemical Society, Washington, DC, 1996, p. 27.

42. P. A. Lovell, M. M. Sherratt, and R. J. Young, "Mechanical Properties and Deformation Micromechanics of Rubber-Toughened Acrylic Polymer," in: *Toughened Plastics II: Novel Approaches in Science and Engineering*, C. K. Riew and A. J. Kinloch (eds.), American Chemical Society, Washington, DC, 1996, p. 211.

43. C. B. Bucknall, I. K. Partridge, and M. V. Ward, *J. Mater. Sci.*, **19**, 2064 (1984).

44. J. Milios, G. C. Papanicolaou, and R. J. Young, *J. Mater. Sci.*, **21**, 4281 (1986).

45. R. Hu, V. L. Dimonie, M. S. AlAsser, R. A. Pearson, A. Hiltner, S. G. Mylonakis, and L. H. Sperling, *J. Polymer Sci.: Part B: Polym. Phys.*, **35**, 1501 (1997).

46. H. J. Ha, Y. J. Park, J. H. An, and J. H. Kim, *Polymer—Korea*, **21**, 438 (1997).

47. R. J. Young and P. A. Lovell, *Introduction to Polymers*, 2nd Ed., Chapman & Hall, London, 1991.

48. M. Schneider, T. Pith, and M. Lambla, *Polym. for Adv. Technol.*, **6**, 326 (1995).

49. M. Schneider, T. Pith, and M. Lambla, *J. Mater. Sci.*, **32**, 6343 (1997).

50. F. J. Guild, R. J. Young, and P. A. Lovell, *J. Mater. Sci. Lett.*, **13**, 10 (1994).

51. Z. Fan, A. P. Midownik, and P. Tsakiropoulos, *Mat. Sci. & Technol.*, **9**, 1094 (1993).

52. Z. Fan, P. Tsakiropoulos, and A. P. Miodownik, *Mat. Sci. & Technol.*, **8**, 922 (1992).

53. Z. Fan, P. Tsakiropoulos, and A. P. Miodownik, *J. Mater. Sci.*, **29**, 141 (1994).

20 Application of Fracture Mechanics for Characterization of Toughness of Polymer Blends

YIU-WING MAI, SHING-CHUNG WONG, and XIAO-HONG CHEN
Centre for Advanced Materials Technology (CAMT)
and
Department of Mechanical and Mechatronic Engineering J07
University of Sydney
Sydney, NSW 2006
Australia

Polymer Blends, Volume 2: Performance. Edited by D. R. Paul and C. B. Bucknall.
ISBN 0-471-35280-2.   © 2000 John Wiley & Sons, Inc.

I. INTRODUCTION

Recent advances in polymer blends have created new frontiers for research into techniques of fracture characterization for many elastic and elastic–plastic multiphase materials. To promote the development of and wider applications for polymer blends, it is necessary to establish the linkage between the macroscopic fracture behavior and the micromorphology. Fracture mechanics provides such a solid foundation for materials selection and a basic understanding upon which to base the microstructural design of novel polymer blends. Unlike some commonly used techniques for toughness assessment, such as Izod impact, drop weight, strain to break, and work to failure, fracture mechanics is established on continuum mechanics and work-rate analysis. It differentiates elastic from plastic fractures and separates the initiation and propagation phases of the total toughness achieved. Although the standardized notched-impact strength measurement is commonly accepted by the plastics industry, Jancar and Dibenedetto [1] have shown that for the majority of particulate-filled and rubber-modified polypropylenes studied, there is no functional dependence of the standardized Charpy or Izod toughness on material variables. (See also Chapter 21.) Only fracture-mechanics-based parameters measured at impact loads can separate the effects due to specimen geometry from those of the intrinsic material properties. As a result, it is emphasized that comparison of the standardized notched-impact toughness of different polymeric materials does not give a reliable indication of their relative fracture toughness and is of little value as a parameter for materials selection [1].

The failure mechanisms that are characteristic of toughened polymers, and blends in general, are best described by fracture mechanics. For example, toughening [2–6] initiated by cavitated second-phase soft particles that subsequently promote shear yielding of the matrix ligament can generate significantly large plastic deformation surrounding the crack tip. (See Chapter 22.) Crack-wake bridging of the PBT phase by the PC phase in PC–PBT blends [7] also gives rise to high levels of toughness. Fracture mechanics parameters, such as the J-integral and the specific essential work of fracture, can provide initiation toughness values under well-defined states of stress. Mechanisms associated with the propagation phase of fracture can be quantified using the crack-resistance curve approach.

In this chapter, we give an overview of the fracture mechanics characterization techniques that have been validated for a range of toughened polymers and polymer blends. Concepts central to the development of these techniques are explained, and areas in which future work is needed are discussed.

II. FRACTURE MECHANICS CONCEPTS

Fracture mechanics is concerned with the quantitative characterization of the conditions under which a load-bearing solid can fail due to enlargement of a sharp crack contained in that solid [8, 9]. Assessments of toughness using fracture mechanics concepts have been widely used for metallic alloys [10, 11]; polymers [12]; fiber composites [13, 14]; and, more recently, concretes and fiber-reinforced cements [15, 16]. Important concepts of fracture mechanics have evolved over the last 50 years, and interested readers may refer to [17, 18].

A. Linear Elastic Fracture Mechanics for Brittle Polymers

1. *The Stress Intensity Factor Approach*   Linear elastic fracture mechanics (LEFM) is now widely used to characterize the fracture behavior of polymers [12]. Relevant fracture mechanics parameters can be derived from either the stress intensity approach or the energy balance approach [9, 17]. However, there are distinct limitations of the fracture mechanics parameters derived from either of these two approaches, and these drawbacks are discussed later. The stress field around a crack tip is associated with a particular mode of crack opening represented by Roman numerical subscripts I, II, and III. These subscripts correspond to crack opening perpendicular to the crack plane, shearing along or parallel to the crack plane, and out-of-plane tearing, respectively. Mode I cracks are most important in engineering design; they are discussed in this chapter. Consider a sharp mode I crack in a linear elastic solid. The stress field close to the crack tip, where higher order terms in θ can be neglected, is sufficiently described by

$$\sigma_{ij} = \frac{K_I}{\sqrt{2\pi r}} f_{ij}(\theta), \tag{20.1}$$

where K_I is the stress intensity factor and f_{ij} depends on the stress state and is a function of the angle θ from the crack plane. At fracture, the critical stress intensity factor K_c is given by

$$K_c = \sigma_c Y \sqrt{a}, \tag{20.2}$$

where σ_c is the critical applied stress, Y is a geometric factor, and a is the length of the crack.

For a single-edge-notch-bend (SENB) specimen loaded in three-point bending with a support span S and width W, Srawley [19] showed that K_I can be expressed by

$$K_I = \frac{PS}{BW^{3/2}} \frac{3\left(\frac{a}{W}\right)^{1/2}\left[1.99 - \frac{a}{W}\left(1 - \frac{a}{W}\right)\left(2.15 - 3.93\frac{a}{W} + 2.7\frac{a^2}{W^2}\right)\right]}{2\left(1 + 2\frac{a}{W}\right)\left(1 - \frac{a}{W}\right)^{3/2}}. \tag{20.3}$$

For the compact tension (CT) specimen K_I is

$$K_I = \frac{PS}{BW^{1/2}} \frac{\left(2 + \frac{a}{W}\right)\left[0.886 + 4.64\frac{a}{W} - 13.32\left(\frac{a}{W}\right)^2 + 14.72\left(\frac{a}{W}\right)^3 - 5.6\left(\frac{a}{W}\right)^4\right]}{\left(1 - \frac{a}{W}\right)^{3/2}},$$

$$\tag{20.4}$$

where P is the applied load and B is the thickness of the specimen. Expressions for K_I that are applicable to specimens of different geometry can be found in [9, 17]. K_c is defined as K_{Ic} if it satisfies specific specimen thickness and size requirements. (See Eq. 20.5.) This variable is called the plane-strain opening-mode fracture toughness, which is a materials property that is independent of specimen geometry and size. For brittle solids, K_{Ic} provides a reliable and reproducible parameter for materials selection and design.

A test method to determine the fracture toughness K_{Ic} of plastics was recently established as ASTM Standard D5045 [20]. Equation (20.3) or (20.4) is first used to determine an apparent K_{Ic}, denoted by K_Q, as set forth in the standard. When measuring K_{Ic}, it is important to distinguish nonlinearity due to plasticity from that due to gradual crack growth preceding fracture. K_Q corresponds to a 2.5% increment of crack growth. The nonlinearity due to excessive plasticity must be excluded. This requirement is expressed clearly in the test standard. If K_Q meets the size requirements for plane-strain fracture toughness, K_Q becomes K_{Ic}. To produce a sharp crack, the standard suggests some specifications for precracks. In lieu of fatigue precracking, as commonly used in metallic materials (ASTM Standard E399 [21]), a sharp precrack is made in plastics by inserting a fresh razor blade into the plastic and tapping it into the machined notch. However, the size requirements for obtaining a valid K_{Ic} remain unchanged for plastics. That is, the specimen dimensions should satisfy the following expression [21]:

$$B, a, (W - a) \geq 2.5 \frac{K_Q^2}{\sigma_y^2}. \tag{20.5}$$

2. The Energy Balance Approach   Griffith's energetic energy balance criterion [17] provides a simple method for evaluating the work of fracture required to extend a crack by a unit area. Energy balance requires that

$$F - U_a = U_\beta + U_k, \tag{20.6}$$

where F is the mechanical work done on the specimen by external forces, U_a is the change in elastic strain energy caused by the introduction of the crack, U_β is the change in elastic surface energy due to the formation of the crack surface, and U_k is the change in kinetic energy of the system. Differentiating Eq. (20.6) with respect to the surface area A of the crack gives

$$\frac{d}{dA}(F - U_a) = \frac{dU_\beta}{dA}. \tag{20.7}$$

Let U represent the total elastic strain energy, or potential energy, stored in the loaded system, that is, $U = U_o + U_a - F$, where U_o is the constant elastic energy of the loaded uncracked plate. The Griffith elastic energy release rate G is defined as follows:

$$G = \frac{1}{B}\frac{d}{da}(F - U_a) = -\frac{1}{B}\frac{dU}{da}, \tag{20.8}$$

where B is the thickness of the specimen. For equilibrium crack spreading, $G = G_c$, which is the critical energy release rate identifiable with the crack resistance R:

$$R = \frac{1}{B}\frac{dU_\beta}{da}. \tag{20.9}$$

From Eqs. (20.7) to (20.9), we find that incipient crack growth occurs when $G = G_c = R$ and that stable crack growth can be maintained only if $dG/dA \leq dR/dA$.

It should be noted that R is not always a constant, and in some materials (such as in short-fiber-reinforced polymer blends), it can vary along the path of the crack. Further, from the principle of virtual crack extension, it can be shown that G_c is related to K_c. And for mode I crack growth,

$$G_{Ic} = \frac{K_{Ic}^2}{E^*},\qquad(20.10)$$

where $E^* = E$ (Young's modulus) for plane stress and $E^* = E/(1 - \nu^2)$ for plane strain, where ν is Poisson's ratio. Equation (20.10) allows G_{Ic} to be calculated from K_{Ic}. More importantly, Eq. (20.8) gives a simple but powerful expression for G_{Ic} that can be related to the total elastic strain energy U [12, 22]. That is,

$$G_{Ic} = \frac{U}{BW\phi},\qquad(20.11)$$

where W is the width of the specimen and ϕ is a correction factor that is dependent on the specimen compliance C given by

$$\phi = \frac{C}{\frac{dC}{d(a/W)}}.\qquad(20.12)$$

ϕ can be obtained from Eq. (20.12) [12] for different specimen geometries with different a/W ratios. In ASTM Standard D5045, numerical values of ϕ for SENB and CT specimens are provided.

Equation (20.11) also allows the evaluation of G_{Ic} directly from the slope of the linear relationship between fracture energies U and $BW\phi$ obtained from a series of specimens with different initial crack lengths. (See Fig. 20.1.) Under impact loading

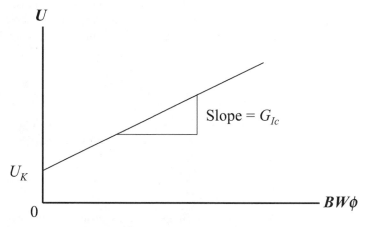

Figure 20.1   Determination of the linear elastic energy release rate G_{Ic}.

conditions, the kinetic energy loss U_k cannot be ignored, and Eq. (20.11) should be modified to

$$U_i = G_{Ic} B W \phi + U_k, \tag{20.13}$$

where U_i is the total impact fracture energy [22–24]. When U_i is plotted against $BW\phi$, a straight line with G_{Ic} as the slope and U_k as the y-intercept can be determined. (See Fig. 20.1.) Note that Eq. (20.13) is valid only for truly brittle fractures that satisfy the requirements of LEFM. For information on impact characterization of semibrittle and semiductile polymers, interested readers may consult references [12, 25–26].

B. Nonlinear Fracture Mechanics for Ductile Polymers

1. J-integral Fracture Toughness   The J-integral is a path-independent contour integral that describes the stresses, strains, and displacements of any path around a singular crack, provided that either linear or nonlinear elastic deformation precedes crack growth. Eshelby [27, 28] defined a number of such contour integrals, but it was Rice [29, 30] who applied the J-integral to elastic–plastic materials and developed the concept as a means of analyzing fracture mechanics problems. For a two-dimensional crack, the J-integral is given by

$$\mathbf{J} = \int_C \left(w \, dy - \mathbf{T} \cdot \frac{\partial \mathbf{u}}{\partial x} ds \right), \tag{20.14}$$

where (x, y) are rectangular coordinates normal to the crack front; y is perpendicular to the surface of the crack; ds is the increment of arc length along a contour C; $\mathbf{T}$ and $\mathbf{u}$ are stress and displacement vectors, respectively, acting on C; and w is the strain energy density and is equal to $\int_0^{\varepsilon_{ij}} \sigma_{ij} d\varepsilon_{ij}$.

The definition of the J-integral given by Eq. (20.14) is useful for mechanics analysis and computational methods. To make the J-integral concept useful and relevant for fracture characterization of materials in laboratories, it is necessary to invoke the energetic interpretation of the contour integral. Mathematically, J can be considered as a nonlinear elastic energy release rate that is equivalent to G in the linear elastic case [17]. By assuming that the plastic deformation under monotonic loading occurs in an identical manner to nonlinear elastic deformation in the presence of a crack, Rice [29] and Landes and Begley [31, 32] showed that Eq. (20.14) is a more general version of Eq. (20.8). Hence, J becomes analogous to G in the nonlinear elastic case. That is, J is related to the potential-energy release rate associated with a moving or extending crack in an elastic material, linear or nonlinear.

For linear elastic fracture, $J_c = G_c$. For elastoplastic fracture, Sumpter and Turner [33] proposed that J can be separated into an elastic part J_e and a plastic part J_p, so that

$$J = J_e + J_p = \frac{\eta_e U_e}{(W - a)B} + \frac{\eta_p U_p}{(W - a)B}, \tag{20.15}$$

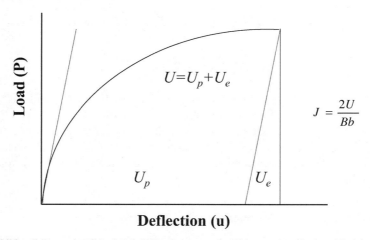

Figure 20.2   Schematic of the load-deflection curve for J-integral analysis with a three-point deeply notched bend specimen.

where U_e and U_p are the elastic- and plastic-energy components of the total energy, respectively, and η_e and η_p are the corresponding geometry correction factors. Equation (20.15) is simply an extension of Eq. (20.11) into the elastic–plastic regime. η_e is, in fact, identical to $(1 - (a/W))/\phi$. Furthermore, it was found that $\eta_e \approx \eta_p \approx 2$ for the deeply cracked ($a/W \geq 0.5$) SENB geometry [17, 33]. Hence,

$$J = \frac{2U}{Bb}, \tag{20.16}$$

where U is the total area under the load-deflection curve (see Fig. 20.2) and $b = (W - a)$. Equation (20.16) has greatly simplified the theoretical analysis required to determine the critical J value J_{Ic} at the onset of crack growth. The significance of Eq. (20.16) is reflected by its widespread use in characterizing the fracture toughness of many ductile materials based on ASTM Standard E813 [34]. For a CT specimen, the expression must be multiplied by a geometrical correction factor $f(a/W)$, which depends on the crack-to-width (a/W) ratio:

$$J = \frac{2U}{Bb} f(a/W). \tag{20.17a}$$

Here, f is defined by

$$f(a/W) = (1 + \alpha)/(1 + \alpha^2), \tag{20.17b}$$

where

$$\alpha = 2\sqrt{(a/b)^2 + a/b + 1/2} - 2(a/b + 1/2). \tag{20.17c}$$

a

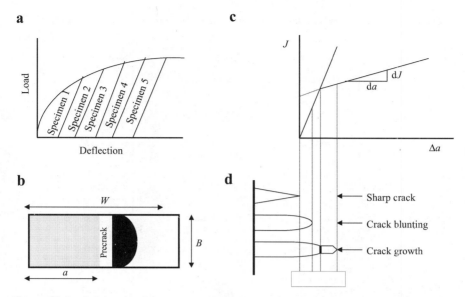

Figure 20.3 Schematic of J_R curve construction and determination of J_{Ic}. (From Landes and Begley [50] and ASTM E813-81 [34].)

a. *J_{Ic} Testing* The energy-based definition of J has permitted its determination directly from the experimental load P against load-point deflection u records using Eq. (20.8). In recent years, there have been many studies on toughness evaluation of rubber-toughened polymers and blends using the J-integral [5, 35–49]. The most widely used technique is based on the multi-specimen crack resistance J_R curve concept developed by Landes and Begley [32, 50]. Guidelines on this technique are given in many textbooks [9, 17] and, most comprehensively, in ASTM Standard E813 [34] for metals and metallic alloys. There are two versions of the multi-specimen method: ASTM E813-81 and ASTM E813-87. The basic differences between the two versions arise from the data analyses used in constructing the J_R curves and the determination of the critical J-initiation toughness. Since a standard procedure for J-integral measurement for polymers is not yet available, most of the research has been focused on the applicability of the ASTM standards originally designed for metals. Huang [39–41] has, however, suggested certain specifications for J_{Ic} testing of toughened plastics.

Figure 20.3 is a schematic representation of the ASTM test methods for experimental measurement of J_{Ic}. The required loading, specimen geometry, and sample preparation procedures are similar to those recommended for K_{Ic} testing. For polymers, a series of specimens with the same initial crack length such that $a/W > 0.5$ are loaded to different points along the P–u curve. After unloading, the region of stable crack growth can be revealed by cryogenic fracture of the remaining liga-

Table 20.1   Major differences in the construction of J–R curves in ASTM standards for plastics

	ASTM E813-81 for metals (commonly adopted for plastics)	ASTM E813-87 for metals (commonly adopted for plastics)	New protocol for plastics (ASTM Task Group, Committee D-20 Group, X-10-155)
Method	Multiple or single specimen	Multiple or single specimen	Multiple or single specimen
R-curve	Linear-regression fit	Power-law regression fit	Power-law regression fit
Blunting line	$2\Delta a\sigma_y$ from 0 crack extension	$2\Delta a\sigma_y$ from 0.2-mm offset of crack extension	A vertical line from 0.2-mm offset of crack extension
Exclusion lines	Parallel to the blunting line at $0.006(W - a)$ and, $0.06(W - a)$	Parallel to the blunting line at 0.15 mm and 1.5 mm	
Others	1. Precrack preparation should be modified in a similar fashion as in ASTM E5045-93	1. Precrack preparation should be modified in a similar fashion as in ASTM E5045-93	1. Precrack preparation should be modified in a similar fashion as in ASTM E5045-93
	2. Side grooves not necessary	2. Side grooves not necessary	2. Side grooves recommended

ment, and the amount of crack extension can be measured with a traveling optical or stereo microscope. The area under the P–u curve for each specimen is used with Eq. (20.16) or (20.17), depending on the specimen geometry, and the results for J are plotted against the measured crack growth Δa. Table 20.1 summarizes the differences in data analysis and J_R curve construction between the two versions of ASTM Standard E813-81 and E813-87 for metals and a proposed ASTM protocol for J_{Ic} testing of plastics.

b. *ASTM E813-81* J_{Ic} is determined from the intersection between the linear regression fit of the J_R curve obtained from Eq. (20.16) or (20.17), and the crack blunting line defined by

$$J = 2\sigma_y\Delta a. \qquad (20.18)$$

Equation (20.18) accounts for the apparent increase in crack length due to crack-tip blunting. It assumes that the crack-tip blunting process takes place such that the initially sharp crack deforms in a semicircular fashion when loaded. (See Fig. 20.3.) Hence, the apparent increase in crack length is equal to the radius of the blunted crack tip, which is half the crack-opening displacement (COD). Relating the COD to J leads to Eq. (20.18).

In the $J-\Delta a$ plot, two exclusion lines are drawn parallel to the blunting line at 0.15 mm and 1.5 mm offsets (or 0.6 and 6% of a 25-mm-thick specimen, respectively) on the abscissa. No data obtained outside these exclusion lines are valid for use in construction of the J_R curve. The upper-bound exclusion line assures that all selected data comply with the J-controlled growth criteria—that is, elastic unloading due to crack extension is sufficiently small or Δa is less than 6% of the ligament length for a geometry-independent J. The lower-bound exclusion line allows precise measurement of crack growth. Nevertheless, Chung and Williams [51] have shown that the maximum permissible crack extension Δa recommended for J-controlled growth is too conservative for polymers. It is clear that the linearity of the J_R curve for tough polymers could extend far beyond a crack extension equal to 6% of the ligament length.

In ASTM E813-81, linear extrapolation of valid J data gives a critical value J_c that intersects the blunting line. J_c represents the rate of potential-energy release at crack initiation, and if it satisfies all of the J-controlled growth conditions and the size requirements of

$$B, b \geq 25\frac{J_c}{\sigma_y}, \tag{20.19}$$

then $J_c = J_{Ic}$, which is the plane-strain fracture toughness.

The blunting-line concept has received many criticisms when applied to polymer blends. Indeed, Narisawa and Takemori [37] proposed abandoning the blunting line for toughened polymers. They found no crack-tip blunting in their experiments. Instead, a large and rather complex damage zone, which included shear yielding, crazing, and void formation near the crack tip, had developed. Huang and Williams [38] contended that crack-tip blunting might not be readily seen in an unloaded specimen. Also, Chung and Williams [51] showed that the blunting-line concept based on a simplified COD model was inappropriate for ductile polymers. Recently, Mai and Powell [52] corroborated the findings of Narisawa and Takemori [37] and confirmed that the blunting line is irrelevant to the fracture initiation process. To conclude, ASTM E 813-81 appears inadequate to provide a valid crack-initiation parameter for fracture characterization of toughened polymers and polymer blends.

c. *ASTM E813-87 and -89*   The revised procedures for measurement of J-integral initiation toughness require a power-law curve fitting of the experimental $J_R-\Delta a$ data. That is,

$$J = C_1 \Delta a^{C_2}, \tag{20.20}$$

where C_1 and C_2 are numerical constant and J_c is defined as the intersection of the $J_R-\Delta a$ curve with the 0.2-mm offset line parallel to the blunting line. In essence, J_c is defined as $J_{0.2}$, at which point 0.2 mm of crack growth has already occurred. Hence, the value of $J_{0.2}$ is consistently higher than the values of J_c obtained from other data analysis schemes [23, 53, 54]. No theoretical justification is associated with the arbitrary value $J_{0.2}$, the advantage of which is its simplicity for materials characterization and comparison. However, the 0.2-mm-offset value of J circumvents the need to identify crack initiation and gives a less conservative estimate for fracture design. In addition, the power-law curve fitting of experimental data has received increasing acceptance in the testing of toughened polymers [39–41], since the crack-growth resistance of different materials can be readily compared.

More recently, Lu and Chang [46] experimented with different functional correlation models, which include linear (ASTM E813-81) and power-law regressions (ASTM E813-87), polynomial functions, and logarithmic and exponential regressions, for fracture characterization of PC–PBT blends. They also compared J_c values as obtained from three different data windows, one of which exceeded the established upper-bound exclusion line. Their results showed that the blunting line at the 0.1-mm offset gave the most consistent values of J_c, regardless of which regression models or data windows were selected. Higher order (third and fourth) polynomial functions provided best fits to all of the data in different windows up to $\Delta a = 3$ mm for 20-mm-wide samples. Both higher order polynomial functions and the power-law regression appeared to be insensitive to the range of the data windows. However, these conclusions are derived from data collected for PC–PBT blends only. Further proof is required for other toughened polymers and blends.

d. *J_R Curve* J_{Ic} provides information only at the onset of crack growth. However, for design purposes, it is also desirable to understand the full potential of toughening in the crack propagation phase. Fracture mechanics methodologies provide a distinct advantage over other commonly used techniques, such as Izod, strain to break, and puncture tests, because the crack initiation and crack propagation phases can be differentiated. Comparison of crack-resistance curves as a means of characterizing crack growth in rubber-toughened plastics has been strongly advocated by Huang [40, 41]. Similar comparisons were applied to short-glass-fiber-reinforced polymers [55–57]. Attempts have been made to characterize materials in terms of the slope of a rising J_R curve at a given amount of crack extension. It is clear that a steeper J_R curve gives rise to a stronger resistance as the crack advances. The slope of the J_R curve is usually quantified by a dimensionless tearing modulus T_R:

$$T_R = \frac{E}{\sigma_y^2} \frac{dJ_R}{da}. \tag{20.21}$$

The subscript R indicates a given value of J on the J_R curve. It is obvious that T_R is meaningful as a material resistance parameter only if the J_R curve is measured under J-controlled growth, which is valid for small crack extensions. As for elastic fracture, crack growth becomes unstable when the applied $T_a = T_R$ and $dT_a/da > dT_R/da$.

2. *Essential-Work-of-Fracture Concept for Brittle and Ductile Polymers*   High impact strength accounts for a significant 38% of the principal properties claimed in patents on newly developed polymer blends and alloys [58]. Toughened blends are an important class of novel materials that require the evaluation of fracture toughness at different testing speeds and sample thicknesses. However, current procedures for J-integral measurement are restricted to quasistatic loading only. It is expensive and difficult, if not impossible, to assess the *impact fracture toughness* of a polymer using the J-integral method. (See, for example, Martinatti and Ricco [59, 60].) Moreover, the ASTM procedure to characterize J-integral fracture toughness usually involves tedious measurements of crack growth. Similarly, the J-integral testing procedure to characterize the fracture of polymer thin films is cumbersome. It is therefore necessary to invoke a more powerful methodology for fracture characterization of ductile polymers, especially in plane stress. The preferred technique is the now widely accepted *essential-work-of-fracture (EWF) method.*

The EWF concept has its origin in Broberg's unified theory of fracture [61, 62]. Stable crack growth occurs because the work input to an autonomous region is filtered through the gradually increasing screening action of the plastic work in the outer region. The total work of fracture for any increment of crack growth includes both the dissipative work in the outer plastic zone (which is geometry dependent) and the essential work in the inner autonomous zone (which is a materials property). The autonomous region is called the fracture process zone (FPZ) hereafter.

Cotterell, Mai, and their colleagues in Sydney [23, 52, 53, 63–66] have developed an elegant, and yet simple, experimental technique to separate these two work components from the total work of fracture. Recently, a European Structural Integrity Society (ESIS) test protocol for essential-work-of-fracture has been established [67].

The EWF method has now been successfully applied to many ductile materials, including metallic alloys [63, 65, 68, 69], papers [70, 71], polymeric films [53, 72–78], toughened polymers and their blends [23, 52, 79–86], and, more recently, glass-reinforced plastics [87, 88], to obtain unambiguous measurements of the specific essential work of fracture, w_e. It has also been extended to determine the impact fracture toughness of PC–PBT–IM blends [23]. In the following discussion, the EWF concept is developed in terms of the physical processes that take place in the FPZ, and its definition is related to the J-integral in Eq. (20.14).

In quasistatic crack growth, the total work of fracture, W_f, can be partitioned into two components: (i) the essential work W_e performed in the FPZ and (ii) the nonessential work W_p performed in the outer plastic zone. When both the FPZ and plastic zone are contained in the ligament, the essential work of fracture is proportional to the ligament length l and the nonessential work of fracture is proportional to the square of the ligament length, l^2. Hence,

$$W_f = w_e l B + \beta w_p l^2 B \qquad (20.22a)$$

and

$$w_f = w_e + \beta w_p l, \qquad (20.22b)$$

where w_f is the specific total work of fracture ($= W_f/lB$), β is the geometry-dependent plastic-zone shape factor, and w_p is the specific nonessential plastic work of fracture. Plotting w_f against l yields a straight line whose y-intercept is w_e and whose slope equals βw_p. Equation (20.22) provides a sound theoretical basis for a simple experimental method to determine w_e from experiments on the total work of fracture using a range of ligament lengths and different specimen geometries [69, 80].

The physical meaning of w_e can be interpreted as follows: Consider a fully developed FPZ of length ρ_c and width d formed at the crack tip. The total work of fracture dissipated in this process zone involves the work to cause necking and the subsequent work necessary to draw out and eventually fracture the necked-down material. Thus, by taking a contour integral around the edge of the fracture process zone according to Eq. (20.14), we have

$$w_e = d \int_0^{\bar{\varepsilon}_n} \overline{\sigma} d\bar{\varepsilon} + \int_{\varepsilon_n d}^{\delta_c} \sigma(\delta) d\delta, \tag{20.23}$$

where $(\overline{\sigma}, \bar{\varepsilon})$ are true stress and true strain, respectively; $\bar{\varepsilon}$ and ε_n are true and engineering necking strains, respectively; δ is the crack-tip opening displacement within the fracture process zone; and δ_c is the critical value of the displacement at tearing.

In a typically brittle glassy polymer, such as polystyrene or polycarbonate, the FPZ is the craze zone at the crack tip. The work required to nucleate, orientate, harden, and fracture the craze fibrils is w_e. In short-glass-fiber-reinforced thermoplastics (SGFRTs), w_e is the work required to debond, slide, and pull out the fibers, to deform and tear the matrix, and, subsequently, to break any material bridging the crack plane inside the FPZ. w_p accounts for all other dissipative processes, which depend on specimen geometry and loading configuration outside the FPZ.

3. *Some Comments on the Method of Essential Fracture Work for Measuring Toughness* Inherent in Eq. (20.22) are four important assumptions that should be satisfied before the EWF method can be applied to determine fracture toughness:

(a) The ligament must be fully yielded prior to crack initiation.
(b) For w_e and βw_p to remain constant independently of ligament length, Eq. (20.22) must apply in theory only to truly plane-stress fracture. The linear relation between w_f and l may not be valid if the fracture is in the mixed-mode region of plane strain and plane stress.
(c) There should be geometric similarity between specimens of different ligament lengths during crack growth, as revealed by curves of load against load-point displacement.
(d) The volume of the outer plastic zone surrounding the fractured ligament must scale with the square of the ligament length. As suggested in the ESIS protocol for measurement of the EWF [67], three different plastic-zone shapes can be obtained—circular, elliptical, and diamond—as shown in Fig. 20.4.

Many investigators [23, 52, 53, 63–66, 68–88] have verified these assumptions and proposed modifications to the testing methodology to suit different materials.

Circular Zone   Elliptical Zone   Diamond-shaped Zone

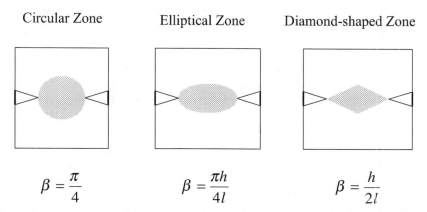

$$\beta = \frac{\pi}{4}$$ $$\beta = \frac{\pi h}{4l}$$ $$\beta = \frac{h}{2l}$$

Figure 20.4 Some examples of the plastic-zone shapes that scale with the ligament length [67].

Studies to date confirm that, given a sheet thickness in plane stress, w_e is a material property. We shall discuss the size requirements for valid measurements of the specific essential work of fracture, followed by other necessary conditions for using the EWF method.

For plane-stress conditions to prevail in the ligament, the ratio between ligament length and sheet thickness should exceed a critical value. It was first suggested that $l \geq 5B$ for cold-rolled low alloy steels [63] and $l \geq 3B$ for ductile polymers [64]. Saleemi and Nairn [79] affirmed this limit ($l \geq 3$–$5B$) when studying rubber-toughened nylons; the net section stress σ_n rises steeply after $l \leq 3B$, indicating an increasing plastic constraint as the ligament length decreases. More recently, Wu and Mai [53] tested 0.285-mm-thick linear low-density polyethylene (LLDPE) films and discovered that σ_n increased rapidly at $l \leq 14B$, which indicates that the plane-stress–plane-strain transition takes place much sooner than expected. Hashemi [75] also reported that the mixed-mode transition took place at a ligament length much greater than $5B$ in polymeric films of 0.125-mm thickness. However, Karger-Kocsis et al. [85] suggested that the lower ligament threshold given by $l \geq 3B$ is too conservative for amorphous copolyester (aCOP) sheets ranging from 0.5 to 3 mm thick; good linearity was still observed with data points taken below the lower threshold. These results have important implications for the size requirements of valid measurements of the essential work of fracture. It has been suggested [53, 67] that there may not be a universal lower-bound limit of the ratio of ligament length to sheet thickness for the determination of plane-stress essential work of fracture. Nevertheless, thicker samples [53, 79] appear to be less sensitive than polymeric thin films to notch-tip plastic constraint [72–78]. The ESIS protocol for measurement of the EWF [67] requires a plot of σ_n versus l to ensure the validity of the experimental data on plane-stress.

The upper limit for valid measurements of plane-stress essential work of fracture stems from the conditions that the ligament should be fully yielded prior to crack initiation [63] and that the plastic zone, with a radius r_p, should be free of interference from specimen edge effects. It is generally accepted that when l is smaller than the minimum of either $W/3$ or $2r_p$, valid data on the work of fracture under plane stress can be obtained [52, 64]. But results shown by Karger-Kocsis et al. [78, 85] indicate that the upper-bound limit is also too restrictive. Good linearity and geometric similarity are still found at larger ligament lengths. Similar findings were reported by Hashemi [75]. It appears that the departure from linearity in the upper region bears no relation to the minimum of $W/3$ or $2r_p$. Judicious selection of test data should be exercised. An arbitrary maximum ligament length of 15 mm is recommended for double-edge-notched-tension (DENT) specimens by the ESIS [67].

Mai and Cotterell [69] verified that EWF is independent of geometry using DENT, deeply center-notched tension (DCNT), deeply single-edge-notched tension (DSEN), and modified double-edge-notched tension (MDENT) specimens on a low work-hardened aluminium alloy. Only the DENT and MDENT specimens satisfied the conditions for valid EWF measurements. However, MDENT geometry was too involved in preparation, and DSEN specimens gave geometrically dissimilar $P–u$ curves, owing to large rotations at small ligaments. Also, DCNT geometry caused buckling under load. Mai and Powell [52] used both DENT and DCNT specimens, the latter being stabilized with an antibuckling device, and obtained identical w_e values for a high density polyethylene (HDPE).

Wu and Mai [53] studied the effect of geometry on w_e in two polymer blends, ABS–PC and PC–PBT–IM, using DENT, CT, and SENB specimens. Figure 20.5 shows the plots of w_f versus l for these two blends in the range $l \leq 6B$. Good linearity is observed in Fig. 20.5a for ABS–PC blends. No clear plane-stress–plane-strain transition can be detected. The plane-stress w_e obtained is independent of specimen geometry. For the PC–PBT–IM blends, however, Fig. 20.5b shows a distinctive plane-stress–plane-strain transition at $l = 3B$, with straight lines fitted to the data above and below this point. Again, the plane-stress w_e is unaffected by specimen geometry. These results indicate that the lower-bound limit for the evaluation of plane-stress w_e is material dependent. Other investigators [72–76, 80] have also verified the independence of w_e on specimen geometry.

a. *Plane-Stress–Plane-Strain Transition and Thickness Dependence of Plane-Stress Essential Work of Fracture*   As the ligament length l continues to shorten relative to the sheet thickness B, the notch-tip plastic constraint increases. Finally, the mixed-mode stress state becomes pure plane strain as l approaches zero. Figure 20.6 is a schematic showing the influence of B on w_e in plane stress and the determination of the plane-strain w_{Ie} in the transition region. Provided that the sheet thickness satisfies the condition specified in the ASTM E813 standard for J_{Ic} measurement, which is

$$B \geq 25 \frac{w_{Ie}}{\sigma_y}, \qquad (20.24)$$

w_{Ie} is the true plane-strain specific essential work of fracture and is identical to the plane-strain J_{Ic}. Otherwise, w_{Ie} is only a "near-plane-strain" toughness that is dependent on B. In the very small-ligament region towards pure plane strain, it is

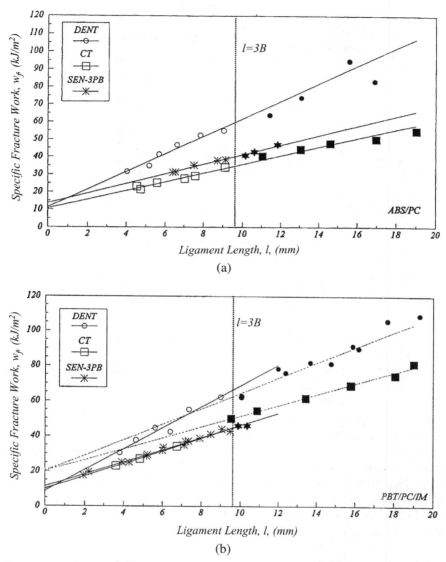

Figure 20.5 Plot of the specific work of fracture, w_f, against the ligament length l for (a) ABS–PC and (b) PBT–PC–IM blends of different specimen geometry. (From Wu and Mai [53]. Reprinted with permission from the Society of Plastics Engineers.)

expected that the fracture data for all values of B under J-controlled crack growth will fall approximately on one single curve and can be extrapolated to a single w_{Ie} ($\equiv J_{Ic}$) value at a ligament length of zero. Note that in the region of plane stress, the slopes that represent βw_p at different values of B can be parallel to each other, provided that the shape of the outer plastic zone surrounding the FPZ is invariant

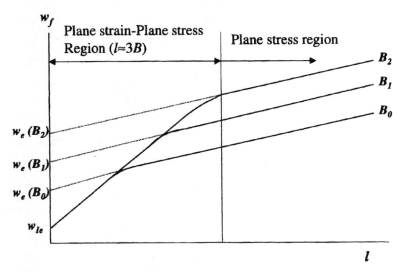

Figure 20.6 Schematic of a plot of w_f versus l for varying specimen thicknesses B in plane stress and plane-stress–plane-strain transition. (From Mai and Powell [52].)

with B. Otherwise, the slopes will decrease with increasing sheet thickness. Extensive results have established that w_{Ie} is thickness independent [76, 79], but w_e for plane stress depends on thickness [72, 73, 76].

The determination of the plane-strain specific essential work of fracture by extrapolation of w_f in the mixed-mode regime to a ligament length of zero was first proposed by Mai and Cotterell [64]. Other investigators [76] have since confirmed this test methodology. Realizing that in the mixed-mode transition region, w_e is a function of both l and B—that is, $w_e = w_e(l, B)$—but that βw_p is invariant with l, regardless of whether failure occurs in plane strain or plane stress, Saleemi and Nairn [79] developed a modified methodology in which they determined that $w_e(l, B) = (w_f - \beta w_p l)$ and plotted it against the ligament length l. This gives a straight line that on extrapolating to $l = 0$ yields the plane-strain specific essential work of fracture, $w_e(0, B) = w_{Ie}$.

Karger-Kocsis et al. [78, 85] recently published results that are somewhat at odds with the established thickness dependence of plane-stress toughness. They contended that most early studies using EWF to characterize the fracture toughness of polymers are incorrect, as there were no distinct load drops in the P–u curves. They believed that the load drop was a necessary condition for incipient ligament necking and that only aCOP displayed this behavior. It is true that necking does occur in thin sheets, but the necking does not usually result in a precipitous load drop. Previous studies by Mai and Cotterell on metals and polymers lend support to this view. What is important is that ligament yielding occur prior to fracture initiation. This factor is best checked by plotting the net section stress against the ligament length at maximum load, as suggested by the ESIS protocol. Karger-Kocsis et al. [78, 85, 89] also proposed that the plane-stress essential work of fracture, w_e, is independent of B and consists of two components: $w_{e,y}$ for yielding and $w_{e,n}$ for deformation from necking to tearing. This proposal follows precisely the same definition for w_e as Eq. (20.23). The first integral is $w_{e,y}$, and the second integral is $w_{e,n}$. Because $w_{e,y}$ is the work to yielding and is proportional to the width d of the process zone, which in turn varies directly with the sheet thickness B, w_e must increase with B. Likewise, $w_{e,n}$ is the work done from the onset of necking to tear initiation. This variable again depends on d (and hence B) and δ_c (which may vary with B). Therefore, in general, w_e will depend on B in the plane-stress region. It is only fortuitous that the plane-stress w_e is found to be invariant with the sheet thickness B in some cases.

b. *Equivalence between J-integral and Essential Work of Fracture*　If J-controlled crack growth prevails, then the resistance to crack growth, J_R, is approximately a linear function of the crack extension Δa and is given by

$$J_R = J_c + \frac{dJ}{da}\Delta a = \frac{dU}{da}. \qquad (20.25)$$

The second equality is based on the energetic definition of J_R. By ignoring the fact that the first equality in Eq. (20.25) applies only to small crack growth, integration of J_R with respect to a (the justification of which is given by Mai and Cotterell [64]) to complete fracture gives the total specific work of fracture, w_f. That is,

$$w_f = J_c + \frac{l}{4}\frac{dJ}{da} \qquad (20.26)$$

for DENT and DCNT specimens. Equation (20.26) is identical to Eq. (20.22b), so that $w_e = J_c$ and $\beta w_p = (1/4)dJ/da$. For DSEN geometry, $\beta w_p = (1/2)dJ/da$. Further details are given in [64]. Note that for small crack growth, dJ/da is approximately constant.

c. *Remarks on Impact Essential Work of Fracture*　Since Wu et al. [23] extended the EWF method to impact testing of ductile polymer blends using SENB specimens, there has been increasing interest and debate on the analysis and testing methodology of impact EWF. While many researchers—for example, Martinatti and Ricco [59, 60]—have shown that Eq. (20.22) is still valid for the evaluation of impact EWF, Vu-Khanh [90] has argued that this method is invalid, because, in several examples he has given, the slope βw_p is negative in plots of w_f versus l. However, the equation for impact energy that he has obtained

$$U = G_i A + T_a A^2/2, \qquad (20.27)$$

where $U = W_f$, $G_i = w_e$, $A = lB$, and $T_a = 2\beta w_p/B$, is identical to Eq. (20.22a) for the EWF method. Hence, if βw_p is negative, so is T_a. There are many physical reasons that βw_p or T_a might be negative during ductile tearing.

Vu-Khanh [90] does not seem to appreciate that Eq. (20.22) applies only when all of the aforementioned conditions are fully satisfied. He quoted the impact results of

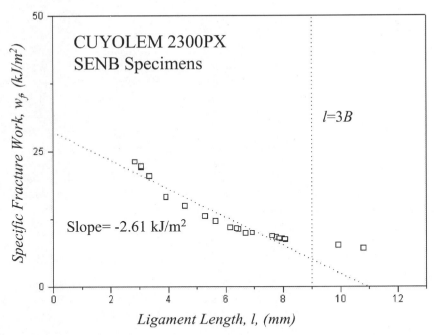

Figure 20.7 Plot of w_f versus l for impact SENB co-PP specimen [93], showing a negative slope, which is caused by the ligament not being fully yielded during fracture.

ABS from Bernal and Frontini [91] to support his proposition that the EWF method could not effectively distinguish the essential work of fracture from the total work of fracture. Private communication with Frontini [92] has clarified that the anomaly arises from violation of the preconditions for valid EWF measurements. First, the data on the ABS samples were collected from specimens of different thicknesses, some with side grooves, and many samples of large ligament length had numerous air bubbles. Hence, it is not surprising that a negative slope was obtained. Recent measurements of impact EWF by Fasce et al. [93] on a 3-mm-thick semicrystalline random PP copolymer have also yielded a negative slope in the plot of w_f versus l. (See Fig. 20.7.) All specimens showed ductile fracture initiation, exemplified by a large plastic zone followed by brittle fracture in the remaining ligament. The size of the plastic zone does not scale with the square of the ligament length, but remains approximately constant. Naturally, the EWF approach, and hence Eq. (20.22), cannot be used under such circumstances.

The real issue with Eq. (20.27) is whether or not T_a is truly a materials property. There has been ample evidence to show that βw_p, and hence T_a, is geometry dependent [52, 53, 69], because the shape of the outer plastic zone changes with specimen type. However, given a fixed geometry and testing condition, T_a and βw_p are both potentially useful in differentiating effects due to differences in microstructure of the blends, such as rubber content, molecular orientation, and so on.

In recent work on PC–ABS blends, Wildes et al. [94] showed that under impact conditions, the specific total impact work of fracture, $w_f (= U/Bl)$, is

$$w_f = U/Bl = u_o + u_d l, \qquad (20.28)$$

where u_o is defined as the limiting specific fracture energy and u_d is the dissipative energy density. Equation (20.28) is identical in form to the Eq. (20.22). But because the test conditions (i.e., dynamic versus quasistatic) and specimen geometry (i.e., SENB versus DENT) are different and do not conform to the yielding conditions required for the EWF analysis, Wildes et al. [94] have cautiously pointed out that the observed intercept u_o and slope u_d may not be identical to w_e and βw_p, respectively. However, it should be noted that Wu et al. [23] successfully used the SENB geometry for impact testing to evaluate the specific essential work of fracture for PC–PBT–IM blends with Eq. (20.22). The justification they gave for this evaluation was that the size of the whitened plastic zone on the fractured specimens evidently scaled with the square of the ligament length. A similar observation of stress-whitened ligaments was also reported by Wildes et al. in their PC–ABS blends [94], which apparently explained why Eq. (20.28) was valid for their study. But what was not conclusively proven in both studies [23, 94] was whether full ligament yielding had occurred prior to crack initiation or not. This question could be answered only if the impact test were stopped immediately after crack initiation. If, on inspection of the specimen, the stress-whitened zone was not fully developed in the unbroken ligament, then Eq. (20.22) would not be valid. Also, if the two halves of the specimen were not completely separated after impact testing, then the EWF analysis of Eq. (20.22) would be difficult to apply. However, it appears that Eq. (20.28) may still be applicable under these two conditions [94].

III. CHARACTERIZATION OF FRACTURE TOUGHNESS OF POLYMER BLENDS USING FRACTURE MECHANICS

In this section, we illustrate the applications of the fracture mechanics parameters described in the previous section to selected polyblend systems under both impact and quasistatic loading conditions.

A. LEFM Characterization of PC–PBT Blends

1. *Quasistatic G_{Ic} of LCP-Reinforced PC–PBT Blends* The reinforcement of single polymers using liquid crystalline polymers (LCPs) has received considerable attention in recent years [95–97]. A comprehensive review of LCP blends is given in Chapter 32. It is an attractive idea to blend LCPs with toughened polymers to form self-reinforced blends (see Chapter 33), which inherit stiffness and toughness from the rigid LCP microfibrils and the toughened matrix, respectively. Most of the results reported on the structure–property relationship of LCP-reinforced blends are primarily concerned with enhancing strength and stiffness [98–105]; a basic understanding of the deformation mechanisms in these self-reinforced blends is still lacking. Nevertheless, a reliable methodology is needed to evaluate the fracture behavior of these new composites. Wong et al. [106] have determined the fracture toughness of LCP-reinforced PC–PBT blends using the LEFM fracture parameters K_{Ic} and G_{Ic}.

a. *Experimental Work*   Polycarbonate (Calibre 300-15) was precompounded with polybutylene terephthalate (Valox 325) at a ratio of 60/40 in a twin-screw extruder. The materials were compounded at a temperature of about 260°C. The LCP (Vectra B950) was procured from Hoechst-Celanese and comprised 58 mol (%) hydroxynaphthoic acid, 21 mol (%) terephthalic acid, and 21 mol (%) para-aminophenol. The LCP was dry blended with precompounded PC–PBT pellets at ratios of 10, 20, 30, 40, and 50 wt% LCP content in a screw injection-moulding machine (JM88MKIII, Chen Hsong Machinery Co. Ltd.) with injection temperatures of 285°C in zone 1 and 270°C in zone 2. The mold temperature was 23°C, the injection time was 5 sec, and the holding time was 44 sec. Prior to each processing operation, all materials were dried in an oven at 120°C for at least 24 h. K_{Ic} was determined using 6-mm-thick SENB specimens at a quasistatic loading rate of 5 mm/min. An initial single-edge notch was introduced midway along each sample, using a diamond saw. Precracks were made by inserting a fresh razor blade into the machined slot, and the a/W ratio was limited to 0.45–0.55 for all tests. The size requirements for valid measurement of K_{Ic} were checked against Eq. (20.5), and K_{Ic} was calculated using Eq. (20.3). G_{Ic} was measured by integration of the load-deflection curve to obtain U, the energy absorbed prior to the onset of fracture for various values of a/W, and then U was plotted as a function of $BW\phi$ to determine G_{Ic} from the slope of the line, using Eqs. (20.11) and (20.12). Ten specimens of different initial crack lengths were prepared and tested for each blend composition.

b. *Results and Discussion*   Figure 20.8 shows typical fracture results for two LCP-reinforced PC–PBT blends. It can be seen that the data for 40% and 50% LCP fall on two separate straight lines radiating from the origin in accordance with Eq. (20.11). The slope of each line gives a G_{Ic} characteristic of that LCP concentration. Independently, values of K_{Ic} were also calculated using Eq. (20.3). Figure 20.9 shows both G_{Ic} and K_{Ic} increasing with the concentration of LCP. But there is a sudden drop at 20% LCP in G_{Ic}, caused by an increase in the elastic modulus at this LCP level. Hence, the fracture toughness trend derived from an energy-based G_{Ic} is not always identical to the trend exhibited by a stress-based K_{Ic}.

2. Comparison of G_c and w_e under Impact Conditions for a PBT–PC–IM Blend
The purpose of this subsection is to show the equivalence between G_c and w_e for brittle fracture and to demonstrate that only the EWF approach can be extended to semiductile and ductile failure of a PBT–PC–IM blend. This section also illustrates the extension of the EWF method to impact testing.

a. *Experimental Work*   A commercial-grade PBT–PC–IM blend supplied by Bayer AG (Australia) in the form of pellets (Makroblend PR52) was studied. The origi-

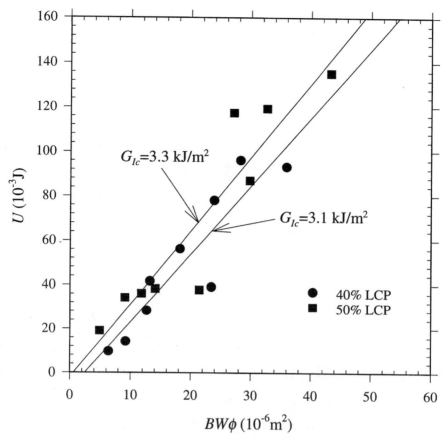

Figure 20.8 U versus $BW\phi$ for LCP-reinforced PC–PBT blends under quasistatic loading.

nal pellets were dried at 120°C for eight hours before being injection molded into plaques at 260°C. The dimensions of the SENB specimens were 8.1 × 60 × 6 mm. All specimens were subsequently notched midway on one side using a saw and then a razor blade driven by a screw. The driving speed was kept slow to avoid inducing excessive plastic deformation at the notch tip. The a/W ratios of the specimens varied from 0.05 to 0.75. Tests were performed on a Zwick 5102 impact tester in the temperature range −196°C to 23°C. After being equilibrated at the required temperature for at least 0.33 h in a chamber controlled by a liquid nitrogen–ethanol bath, the samples were quickly mounted on the specimen holder and fractured. The fracture energy was measured directly from the scale on the machine. The impact speed was 2.96 m s^{-1}.

b. *Results and Discussion*   Figure 20.10 shows a plot of the specific impact fracture energy U/Bl versus the geometry factor ϕ, following Eq. (20.13) for the PC–PBT–IM blends at different temperatures. At −196°C and −100°C, good linearity is ev-

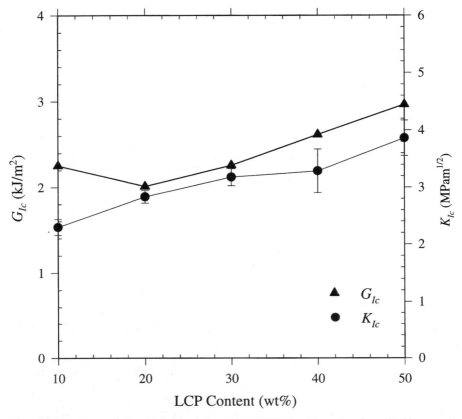

Figure 20.9 G_{Ic} and K_{Ic} for LCP-reinforced PC–PBT blends as a function of LCP content.

ident, validating the LEFM assumptions. G_{Ic} can be determined from the slopes of these curves, which are 2.16 kJm^{-2} and 1.88 kJm^{-2}, respectively. As the temperature rises (see Fig. 20.10c–f), crack-tip plasticity begins to prevail and a straight-line relationship no longer holds.

The same experimental data are also plotted using the EWF method. (See Fig. 20.11.) When values of w_f obtained at low temperatures are plotted against l, an upward curve is observed as the ligament length increases. (See Fig. 20.11a, b.) This upward curvature is attributed to the variation in kinetic energy as a function of the ligament length [23, 24]. To deduct the kinetic energy U_k from the total impact fracture energy U, the former must be measured independently or obtained directly from Ireland's expression [107]. Hence Eq. (20.22) is modified to

$$\frac{(U - U_k)}{B(W - a)} = w_f = w_e + \beta w_p (W - a). \tag{20.29}$$

When the results are analyzed with Eq. (20.29), a horizontal line is obtained, and the

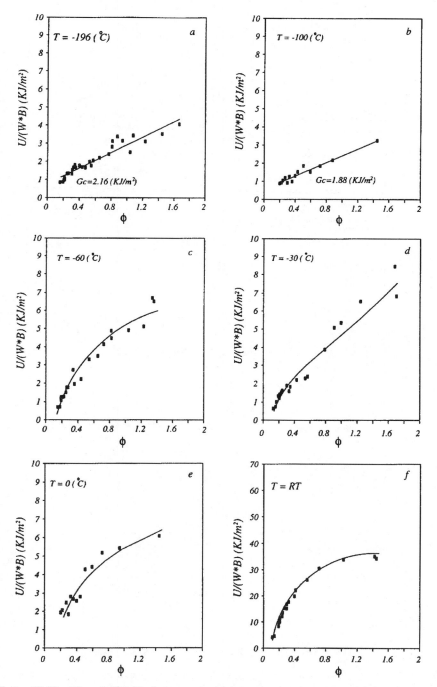

Figure 20.10 The relationship between the impact fracture energy (U/BW) and geometry correction factor ϕ obtained using Williams' model [22] at different temperatures. (From Wu et al. [23]. Reprinted with permission from Kluwer Academic Publishers.)

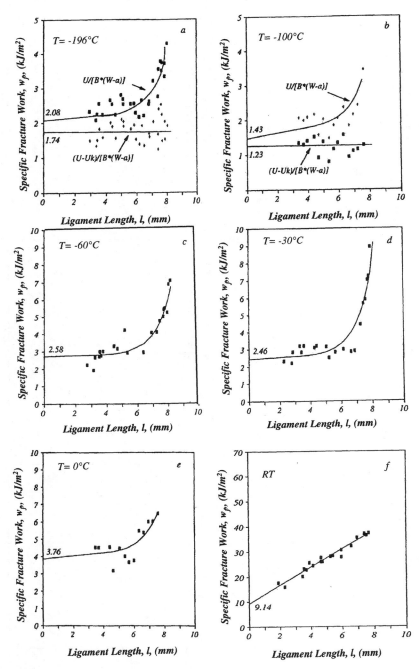

Figure 20.11   Plots of impact fracture energy in terms of the specific work of fracture, w_f, against the ligament length l obtained at different temperatures T of (a) $-196°C$, (b) $-100°C$, (c) $-60°C$, (d) $-30°C$, (e) $0°C$, and (f) $20°C$. (From Wu et al. [23]. Reprinted with permission from Kluwer Academic Publishers.)

intercept gives w_e to be in fairly good agreement with the value of G_{Ic} determined using Williams' method. The zero slope indicates that very little plastic work is done on the ligament during fracture. Indeed, extrapolation of w_f to a ligament length of zero without correction for U_k gives values for w_e of 2.08 and 1.43 kJ/m^2 at $-196°C$ and $-100°C$, respectively, in much closer agreement with G_{Ic}. This method works because at a ligament length of zero, U_k is negligible.

As the temperature continues to rise, both U_k and w_p become important. However, since both terms tend towards zero at a ligament length of zero, it is still possible to obtain w_e from an extrapolation of w_f to zero ligament length. In this case, Williams' technique for the measurement of G_c is inadequate. (See Fig. 20.11c–e.) At 23°C, the effect due to U_k diminishes rapidly, resulting in the straight-line relationship that is expected from Eq. (20.22). Hence, for this PC–PBT–IM blend, w_e is 9.14 kJ m^{-2} and $\beta w_p = 3.64$ MJ m^{-3}. (See Fig. 20.11f.)

The aforementioned results show the distinct advantage of using the EWF technique and SENB geometry to characterize tough polymer blends under impact loading conditions. This method is a useful tool for characterizing the impact toughness of polymer blends with brittle, semiductile, and ductile fracture characteristics. However, much more work is still needed before the impact-EWF methodology can be adequately established and accepted by the technical community. For example, it is necessary to show that the specific EWF (that is, w_e) determined in impact is geometry independent, using tensile impact tests on DENT geometry. Also, it is necessary to explore whether w_e and βw_f are functions of the sheet thickness B, the phase morphology, or the microstructure of the polymer blends.

B. J-integral Toughness Characterization of Rubber-Toughened Blends

LEFM has been successfully applied to describe the fracture toughness of highly cross-linked and toughened epoxies [108–112]. (See Chapter 26.) However, rubber-toughened thermoplastics tend to exhibit gross yielding prior to fracture, in which case postyield fracture parameters, such as J_c and w_e, are required to provide a sensible measurement of the toughness. Assessment of the J-integral fracture toughness has been widely used for rubber-toughened polymer blends [5, 23, 35–49, 51–54] in recent years.

Nair and coworkers [42, 43] found that the J-integral technique is a useful method to elucidate the roles of rubber, rigid polymers, and glass fibers in two rigid–rigid polymer blends; blends of nylon-66 and ABS and of nylon-66 and SAN. Otterson et al. [113, 114] used a modified J-integral method developed by Kim and Joe [115] to study the effects of compatibilizer content [113] and loading rate [114] on the J_R curve for a nylon-6–ABS blend.

Chang and coworkers also investigated the fracture toughness of a series of toughened plastics, including HIPS [116], ABS [44], PC–ABS [48, 49, 117], and PC–PBT [45]. They used a hysteresis method to measure the toughness at crack initiation. The physical basis of the method is that there is a steep rise in the hysteretic energy (HE) at the onset of crack growth, in contrast to the gradual increase of HE caused by viscoelastic and inelastic deformations at the crack-tip region. When the hysteretic energy or the hysteresis ratio ($HR = HE / U$) is plotted against the displacement, where U is the applied work at that displacement, two linear-regression lines can be obtained for the precrack [118, 119] and crack growth, respectively. The intersection point (displacement) of these two lines is used to determine J_{Ic} [44, 45, 48, 49, 116, 117]. It was claimed that this method yielded values of J_{Ic} that are comparable to those based on ASTM E813-81, but that are consistently lower than those based on ASTM E813-87 using a blunting line at 0.2-mm offset. The difficulty with this technique, however, is that it is not always possible to discern the precise point at which the hysteretic line due to precrack changes to the steeply rising hysteretic line due to crack growth. There is always some arbitrariness involved, as it is not clear which data points should be included in the regression analysis of which line. Not all polymer blends will display a distinctive transition in hysteretic energy from precrack to crack growth. Future work is necessary to clarify these ambiguities before this approach can be widely accepted as an alternative method for characterization of toughness.

1. *J-integral Fracture Toughness of PA-66–PP–SEBS-g-MA Blends*

a. *Experimental Work*   The materials investigated in studies on the J-integral fracture thoughness of PA-66–PP–SEB-g-MA blends were novel blends of PA-66 and PP modified with a mixture of maleic anhydride (MA)–grafted and nongrafted styrene-ethylene–butylene-styrene (SEBS). Materials were dry blended simultaneously at weight ratios of 75/25 PA-66/PP, plus 20 wt% SEBS–SEBS-g-MA mixture ranging from 0 to 100 wt% SEBS-g-MA at five different levels. PA-66 (Vydyne 21) was supplied by Monsanto Chemical Company (USA), and PP (Propathene GSE52) was supplied by ICI Australia. The triblock copolymer SEBS (Kraton G1652) and its maleated version (Kraton F1901X) were supplied by Shell Chemical. Compounding of PA-66–PP blends with 20 wt% SEBS–SEBS-g-MA blends was carried out using a high-shear-rate twin-screw extruder (Werner & Pfleiderer ZSK 30) at a temperature range of 260°C to 280°C, depending on the nylon content, followed by injection molding (BOY 22S Dipronic). All of the materials were dried at 60°C in a vacuum oven under reduced pressure prior to compounding and in a regular oven before injection molding into 6-mm-thick rectangular strips.

The J-integral fracture toughness was determined using deeply notched three-point-bend specimens with $0.5 < a/W < 0.65$, $S/W = 4$, and dimensions 82 × 126 × 6 mm. The preparation procedures for the notches and precracks were similar to those for the LCP-reinforced PBT–PC blends. After the specimens were loaded, at 5 mm/min and ambient temperature, to a predetermined value, they were unloaded and submerged in a bath of liquid nitrogen for at least 0.25 h and then fast fractured. The amount of stable crack growth, Δa, was measured using a stereo microscope (WILD Heerbrugg), and J was determined from Eq. (20.16). The data for the J_R curve were fitted to the best power-law relationship according to ASTM E813-87. The initiation toughness J_c was determined from the intersection of the J_R–Δa curve with the blunting line at 0.2-mm offset. For comparison, ASTM E813-81 was also used to analyze the data, and J_c was measured from the intersection between a linear-regression line and the crack-tip-blunting line.

b. *Results and Discussion*   The J_c curve, based on ASTM E813-87 and shown in Fig. 20.12, mirrors the tensile ductility curve [120], shown in Fig. 20.13. A maximum toughness occurs at 0.92% MA-grafted SEBS, and is about 2.5 times that of the nonmaleated blend and six times that of the highest-MA-grafted-SEBS blend. J-integral analysis based on ASTM E813-81 shows a similar pattern of toughening, but yields lower values of toughness. (See Fig. 20.12.) The high ductility of the 0.92%-MA-grafted SEBS blend is caused by the capability of dispersed particles to deform plastically with the matrix prior to interfacial debonding. However, these deformation mechanisms are severely restricted in blends with higher MA-grafted-SEBS content [121], because the different morphology results in quite low values of toughness.

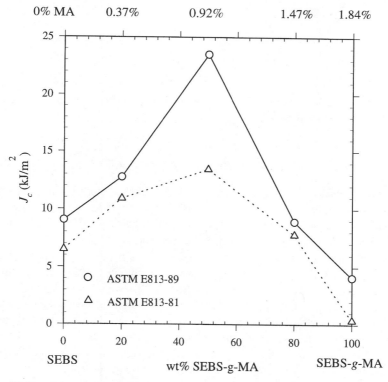

Figure 20.12 J-integral fracture toughness of 75/25 PA-66–PP blends modified with different levels of MA-grafted SEBS copolymers. The total content of SEBS (nongrafted and grafted) is 20 wt%. Values of J-integral initiation toughness were determined using both ASTM E813-87 and ASTM E813-81.

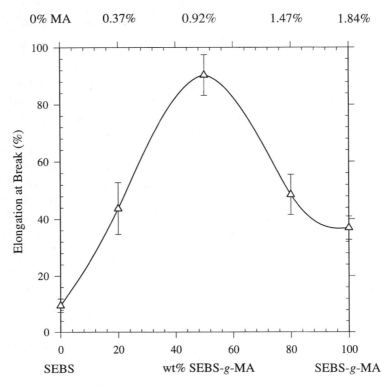

Figure 20.13 Tensile ductility of 75/25 PA-66–PP blends modified with different levels of MA-grafted SEBS copolymers.

C. Specific Essential Work of Fracture of Impact-Modified Rigid–Rigid Blends

1. *PC–PBT Blends* Wu et al. [7] used the EWF method to study the fracture toughness and toughening mechanisms of PC–PBT blends as functions of composition. They found that 60/40 PC–PBT blends showed the highest values of w_e. The toughness reached 16 kJ m^{-2} for 3.2-mm-thick samples, but decreased to a slightly lower value as the sheet thickness increased to 6.4 mm [7]. An extraordinarily effective toughening mechanism is found to occur in this blend composition. It was verified using TEM analysis that the PC domains had stabilized the growing crazes initiated by the crack-tip triaxial stresses. A debonding–cavitation mechanism took place at the PC–PBT interface and promoted shear yielding in the PC and PBT phases. This mechanism absorbed a tremendous amount of energy prior to the onset of crack growth. Most importantly, the PC domains also bridged the crack faces after the crack had initiated. This effect caused a large plastic-damage zone at the crack tip and hence a high crack resistance.

Hashemi [80] also used the EWF methodology to investigate the effects of specimen geometry, size, and strain rate on the fracture behavior of PC–PBT blends. He concluded that w_e is independent of all of these test variables.

2. *PA-6–PP–SEBS-g-MA Blends* Heino et al. [122] employed the EWF method to obtain the essential and nonessential components of the total work of fracture, w_e and w_p, respectively, for injection-molded PA-6–PP blends compatibilized with SEBS-g-MA. It was shown that the EWF technique can successfully be applied to injection-molded toughened polymer blends. Though some extraordinarily high skin deformation may interfere with the EWF measurement [122], this effect, related to the processing conditions, can usually be identified as a "tail" on the curve of load against load-point displacement.

In the rest of this subsection, we demonstrate the equivalence between the EWF method and the J-integral approach for characterization of the fracture behavior of a PA-6–PP–SEBS-g-MA blend.

a. *Experimental Work* PA-6–SEBS-g-MA blends have been studied with and without a dispersed PP phase. Oshinski et al. [123, 124] reported that a mixture of nonfunctionalized (Kraton G1652) and functionalized (Kraton FG 1901X) SEBS at 20/80 wt% is a potent toughener for nylon-6 under impact testing conditions. The enhanced toughness was attributed to the well-balanced rubber particle size when grafted MA is present. This blend composition of SEBS and SEBS-g-MA is collectively denoted SEBS-g-MA hereafter, unless otherwise specified. Materials were dry blended simultaneously for 80/20 nylon 6/PP [123] and nylon 6 alone, in addition to 20 wt% SEBS-g-MA. Nylon-6 (Ultramid B35) was provided by BASF, and PP (Propathene GSE52) was supplied by ICI Australia. Both maleated and nonmaleated SEBS were obtained from Shell Chemical. Compounding and pelletizing were carried out in a similar fashion to the aforementioned procedure for PA-66–PP–SEBS-g-MA blends at 240°C. The pellets were injection molded into three different specimen types: 3-mm-thick dumbbells for tensile tests and SENB ($82 \times 12.6 \times 6$ mm) and DENT ($164 \times 40 \times 3$ mm) geometry for J_c and EWF fracture tests. The EWF analysis for the DENT specimens followed Eq. (20.22), and J_c was obtained using the SENB samples by linearly extrapolating the experimental data to zero crack growth (without intersecting the blunting line).

b. *Results and Discussion* The P–u curves of PA 6–SEBS-g-MA blends with and without the PP dispersed phase show geometric similarity during crack growth for all ligament lengths. As opposed to the results of Heino et al. [122], no skin deformation was observed. Figure 20.14 shows the specific total work of fracture, w_f, plotted against ligament length l for the two blends. There is excellent linearity in both sets of data; however, some data lie in the mixed-mode regime for $l \leq W/3$. Substituting w_e into Eq. (20.24) shows that the thickness requirement for plane-strain toughness is not satisfied. Wu and Mai [53], however, suggested that the plane-strain thickness requirement might be less stringent for these blends. Thus, the ESIS protocol recommendation is used to determine whether a plane-stress–plane-strain transition due to the notch plastic constraint has taken place. However, all values of the net section stress σ_n lie just above $1.15\sigma_y$, where σ_y is the uniaxial tensile stress; no distinct transition is observed. The values of w_e obtained in Fig. 20.14 are therefore plane-

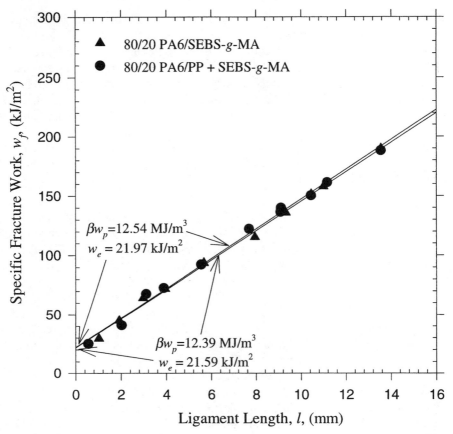

Figure 20.14 The specific work of fracture, w_f, of 80/20 PA-6–SEBS-g-MA and 80/20 PA-6–PP blends modified with 20 wt% SEBS-g-MA, plotted against the ligament length, l.

stress values characteristic of the 3-mm-sheet thickness. Note that the values of w_e for both PA-6–SEBS-g-MA blends are not sensitive to the dispersed PP phase.

Figure 20.15 shows J_R curves for the SEBS-g-MA-modified PA-6 blends with and without dispersed PP. The data were obtained from 6-mm-thick SENB geometry using the multiple-specimen method. The critical values of J_c are much smaller than those obtained under plane-stress conditions, as given by w_e in Fig. 20.14. A check on the thickness requirement for plane-strain toughness shows that both values J_c are very close to true plane-strain values, denoted J_{Ic}. Under plane-strain J-controlled crack-growth conditions, dJ/da may be considered a materials property that is independent of specimen geometry and loading configuration. If we ignore the difference in sheet thickness, we can expect an identical value of dJ/da for the DENT geometry. Hence, we can make use of Eq. (20.26) to estimate $\beta w_p(= 1/4dJ/da)$ for the DENT samples. Therefore, the values of βw_p are 12.85 and 12.58 MJ/m^3 for 80/20 PA-6–SEBS-g-MA and 80/20 PA-6–PP–SEBS-g-MA, respectively, whereas

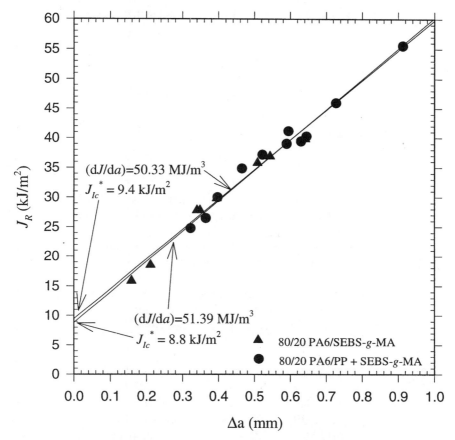

Figure 20.15 J_R curves for 80/20 PA-6–SEBS-g-MA and 80/20 PA-6–PP blends modified with 20 wt% SEBS-g-MA, for the evaluation of J_{Ic} at crack initiation (taken at $\Delta a = 0$).

the measured values from Fig. 20.14 are 12.39 and 12.54 MJ/m^3, respectively. The excellent agreement obtained may be fortuitous, however, because we have assumed that βw_p is independent of sheet thickness [52] and whether the crack tip is in plane-stress or plane strain [79]. A point of contention, however, is whether dJ/da is actually independent of thickness, and this point has to be verified in future work. Table 20.2 summarizes the fracture properties of SEBS-g-MA-modified PA-6 with and without a dispersed PP phase.

D.   Specific Essential Work of Fracture of Short-Fiber and Particulate-Reinforced Blends

Mouzakis et al. [87] recently applied the EWF method to characterize the fracture of PP–glass-bead–SEBS composites. The SEBS used were Kraton G1652 and FG 1901X, as discussed previously. DENT specimens were used to measure w_f from

Table 20.2   Essential work of fracture of SEBS-modified PA-6–PP blends

	80/20 PA-6–SEBS-g-MA	80/20 PA-6–PP + SEBS-g-MA
σ_y (MPa)	35.05	30.27
$1.15\sigma_y$ (MPa)	40.31	34.81
w_e (kJ/m^2)	21.59	21.97
βw_p (MJ/m^3)	12.39	12.54
J_{Ic} (kJ/m^2)	8.8	9.4
$\left.\dfrac{dJ}{da}\right\vert_{\Delta a=1\ mm}$ (MJ/m^3)	51.39	50.33

the load-deflection curves, and the results were analyzed according to Eq. (20.22). Pronounced plastic deformation occurred when glass beads were introduced to the PP matrix. It was found that w_e increased with an increasing amount of rubbery modifier. However, MA-grafted SEBS altered the morphology of the composite and dramatically reduced w_p, which could be calculated from the plastic volume. Because of the difficulties of identifying crack growth in these PP–glass-bead–SEBS composites, the J-integral method is not convenient. The EWF method provides a useful alternative means of fracture characterization. Detailed information on processing, performance characterization and application of other glass-fiber-reinforced polymer blends is given in Chapter 31.

1. *Short-Glass-Fiber-Reinforced PA-66–PP–SEBS-g-MA Blends*

a. *Experimental Work*   The matrix materials used in studies of short-glass-fiber-reinforced (SGFR) PA-66–PP–SEBS-g-MA blends were the same as those described for the unreinforced PA-66–PP–SEBS-g-MA blends. Short glass fibers (6 mm in length and 13 μm in diameter) were precompounded with the dry-blended matrices at a ratio of 20/80 glass–polymer blend. (All composition variations are denoted by wt%.) The fiber-reinforced PA-66–PP–SEBS blends were dried at 80°C in a vacuum oven at reduced pressure prior to compounding at 280° to 285°C. The pellets were then injection molded at 260°C to 6-mm-thick rectangular bars.

SENB geometry was chosen to examine the equivalence between w_e and J_c. For EWF tests, specimens of different ligament lengths were machine notched using a diamond saw, and precracks in the range $0.10 < a/W < 0.88$ were made by tapping fresh razor blades into the notch tip. The J-integral toughness was also determined using the same SENB geometry, but a/W was limited to 0.5 and 0.65 [30], and S/W was set to 4. ($L = 82$ mm, $W = 12.6$ mm, and $B = 5.8$ mm.)

b. *Results and Discussion*   Table 20.3 summarizes w_{Ie}, J_{Ic}, βw_p and dJ/da for the SGFR PA-66–PP blends modified with different levels of maleated SEBS. Recall that different MA levels were achieved by blending nonfunctionalized SEBS (0% MA content) with functionalized SEBS-g-MA (1.84% MA content). (See Section 3.2 of this chapter.) Good linear regression lines between w_f and l are obtained for all SGFR blends (see Figure 20.16), indicating that W_p scales reasonably well with l^2.

Table 20.3   Fracture toughness values of SGFR 75/25 PA-66–PP blends with SEBS-*g*-MA

MA Content (%)	ASTM E813-87 $J_{0.2}$ (kJ/m^2)	Linear extrapolation for J_{Ic} (kJ/m^2)	dJ/da (MJ/m^3)	Specific essential work of fracture w_{Ie} (kJ/m^2)	βw_p (MJ/m^3)
0	not available	not available	not available	4.4	0.47
0.37	12.1	10.8	5.70	9.6	1.02
0.92	8.0	7.6	4.40	6.2	0.90
1.47	4.8	4.4	4.10	3.6	0.86
1.84	7.1	7.1	2.00	6.9	0.47

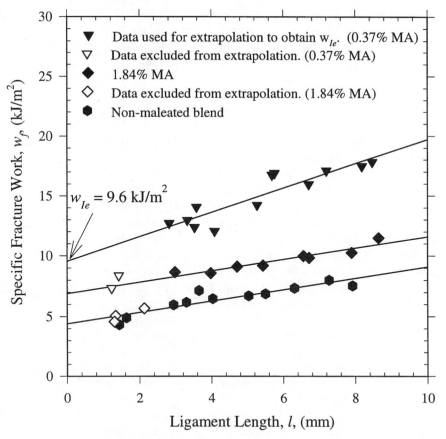

Figure 20.16   The specific work of fracture, w_f, of SGFR 75/25 PA-66–PP blends modified with different levels of MA-grafted SEBS copolymers, plotted against the ligament length l.

Extrapolating w_f to zero ligament length for the blend with an MA level of 0.37 wt% gives $w_{Ie} = 9.6$ kJ/m^2. This number is a plane-strain value, as the required criterion for specimen thickness according to Eq. (20.24) is satisfied. That is, at an MA level of 0.37%, $\sigma_y = 45$ MPa; hence the minimum value of B is 5.3 mm, which is less than the 6-mm thickness of the SENB samples. (Similar calculations for other blend compositions show that plane-strain w_{Ie} values are indeed obtained with the EWF method.) Note that for valuse of l less than 2 mm in the 0.37% maleated blend, two data points lie below the straight line. This is a direct result of the skin–core effect in injection-molded SFRT, for which the specimen skins have an insufficient fiber concentration [88]. A more uniform fiber concentration is, however, observed in the central layer at $0.2 < l/W < 0.8$. The aforementioned observations apply to the 1.84%-MA-grafted SEBS blends.

Hence, it is necessary to exercise care in applying the EWF method to SGFR blends that exhibit skin–core morphologies. The EWF data for specimens in which the skin effect dominates should not be included in the analysis based on Eq. (20.22). Otherwise, incorrect values of w_e will be obtained, because the value of w_f for very small ligaments near the free surface of the sample is not representative of the fracture process at the center of the plaque.

Figure 20.17 shows the J–R curves of the SGFR maleated PA-66–PP blends. Clearly, the value of J_{Ic} obtained using the multiple-specimen method and straight-line extrapolation to zero ligament length (i.e., 10.8 kJ/m^2) agrees very well with the value of w_{Ie} obtained by the EWF method (i.e., 9.6 kJ/m^2). (See Table 20.3.) The ASTM E813-87 procedure defined by the 0.2-mm-offset line gives $J_{Ic} = 12.1$ kJ/m^2, which is somewhat higher than the aforementioned values and is not a true crack-initiation value, as discussed previously [52, 53]. Nevertheless, the difference is not dramatic for these SGFR blends.

For the SGFR PA-66–PP blend with nonmaleated SEBS, it was impossible to use the multiple-specimen J-integral method to assess the plane-strain fracture toughness; neither stress-whitened zones nor stable crack growth could be found. Thus, the EWF technique is a more powerful toughness-characterization method for short-fiber reinforced thermoplastics (SFRT), as shown in Fig. 20.16.

Table 20.3 also compares data on βw_p for all SGFR blends, and it can be seen that these values decrease with increasing MA level. Interestingly, values of βw_p for 0%- and 1.84%-maleated blends are equal, indicating poor plastic energy absorption outside the FPZ. On comparing values of dJ/da with values of βw_p for the blends, a constant ratio of approximately 4 is obtained. From Eq. (20.25) we have shown that this ratio is 4 for DENT and DCNT geometry, but 2 for DSEN geometry. A similar calculation for the SENB specimens used in these tests also gives a ratio of 2. However, experimental results show that the plastic-deformation-zone shape factor β in SENB geometry (elliptical to circular) is unlike that in DSEN geometry (triangular), but instead resembles that in DENT geometry (circular). Hence, for the same ligament length, βw_p for SENB geometry is less than that for DSEN geometry and is close to the value for DENT geometry. Consequently, the ratio should approach 4 rather than 2, as these results demonstrate. A theoretical basis for these observations, however, has yet to be developed.

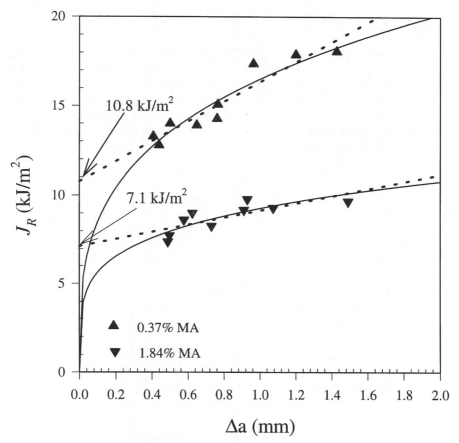

Figure 20.17 *J–R* curves for the determination of J_{Ic} for SGFR 75/25 PA-66–PP blends modified with different levels of MA-grafted SEBS copolymers. J_{Ic} at crack initiation is taken at $\Delta a = 0$.

IV. CONCLUDING REMARKS

In this chapter, we have presented a review of the application of LEFM and non-linear fracture mechanics (NLFM) parameters to the toughness characterization of polymer blends. In particular, it has been shown that the EWF method can be applied to both brittle and tough blends, including short-glass-fiber-reinforced blends, and to both quasistatic and impact fractures. The specific essential work of fracture, w_e, and the specific plastic work absorption w_p have the capability to distinguish between the effects of phase morphology and concentration, molecular microstructure and orientation, and the fiber content of these toughened polymer blends. While extension of the EWF concept to impact testing is very promising, further work is needed to establish the methodology and theoretical foundation for the evaluation of specific impact EWF.

V. ACKNOLWEDGMENTS

We would like to thank the Australian Research Council (ARC) for the continuing support of this polymer blends project. X.-H. Chen acknowledges the tenure of a postdoctoral research fellowship and S.-C. Wong a postgraduate research scholarship, both funded by the ARC. We also thank the various suppliers for providing the polymer blends, either in the form of raw materials or as commercial blends, for the experimental program reported in this chapter. Finally, the Sydney University Electron Microscope Unit kindly made available its facilities for our research, and this help is much appreciated.

VI. REFERENCES

1. J. Jancar and A. T. Dibenedetto, *Polym. Eng. Sci.*, **34**, 1799 (1994).
2. A. M. Donald and E. J. Kramer, *J. Mater. Sci.*, **17**, 1765 (1982).
3. R. A. Pearson and A. F. Yee, *J. Mater. Sci.*, **21**, 2475 (1986).
4. H.-J. Sue and A. F. Yee, *J. Mater. Sci.*, **24**, 1447 (1989).
5. D. S. Parker, H.-J. Sue, J. Huang, and A. F. Yee, *Polymer*, **31**, 2267 (1990).
6. B. Majumdar, H. Keskkula, and D. R. Paul, *J. Polym. Sci.: Part B: Polym. Phys.*, **32**, 2127 (1994).
7. J. Wu, Y.-W. Mai, and A. F. Yee, *J. Mater. Sci.*, **29**, 4510 (1994).
8. M. F. Kanninen and C. H. Popelar, *Advanced Fracture Mechanics*, Oxford University Press, New York, 1985.
9. T. L. Anderson, *Fracture Mechanics: Fundamentals and Applications*, 2nd Ed., CRC Press, Boca Raton, FL, 1995.
10. D. Broek, *Elementary Engineering Fracture Mechanics*, Martinus Nijhoff Publishers, Dordrecht, The Netherlands, 1986.
11. J. F. Knott, *Fundamentals of Fracture Mechanics*, Butterworths, London, 1973.
12. J. G. Williams, *Fracture Mechanics of Polymer*, Ellis Horwood, Chichester, United Kingdon, 1987.
13. K. Friedrich (ed.), *Applications of Fracture Mechanics to Composite Materials*, Vol. 6 of Composite Materials Series, R. B. Pipes (series ed.), Elsevier Science Publisher, Amsterdam, 1989.
14. J. K. Kim and Y.-W. Mai, *Engineered Interfaces in Fiber-reinforced Composites*, Elsevier Science, Oxford, 1998.
15. S. P. Shah, S. E. Swartz, and C. Ouyang, *Fracture Mechanics of Concrete*, John Wiley, New York, 1995.
16. B. Cotterell and Y.-W. Mai, *Fracture Mechanics of Cementitious Materials*, Chapman & Hall, Glasgow, United Kingdon, 1996.
17. A. G. Atkins and Y.-W. Mai., *Elastic and Plastic Fracture*, Ellis Horwood, Chichester, United Kingdon, 1985.
18. H. P. Rossamanith (ed.), *Fracture Research in Retrospect*, AA Balkema, Rotterdam, The Netherlands, 1997.
19. J. E. Srawley, *Int. J. Fract.*, **12**, 475 (1976).

20. D 5045-93, *Standard Test Methods for Plane Strain Fracture Toughness and Strain Energy Release Rate of Plastic Materials*, American Society for Testing and Materials, Philadelphia, 1993.

21. E 399-90, *Standard Test Method for Fracture Toughness of Metallic Materials*, American Society for Testing and Materials, Philadelphia, 1990.

22. G. P. Marshall, J. G. Williams, and C. E. Turner, *J. Mater. Sci.*, **8**, 949 (1973).

23. J.-S. Wu, Y.-W. Mai, and B. Cotterell, *J. Mater. Sci.*, **28**, 3373 (1993).

24. J.-S Wu, "Fracture Toughness and Toughening Mechanisms of Poly(butylene terephthalate)/Polycarbonate (PBT/PC) Blends," Ph.D. Dissertation, University of Sydney, 1994, pp. 46–101.

25. E. Plati and J. G. Williams, *Polym. Eng. Sci.*, **15**, 470 (1975).

26. M. W. Birch and J. G. William, *Int. J. Fract.*, **14**, 69 (1978).

27. J. D. Eshelby, "The Energy Momentum Tensor in Continuum Mechanics," in *Inelastic Behavior of Solids*, M. F. Kanninen, W. F. Alder, A. R. Rosenfield, and R. I. Jaffee (eds.), McGraw-Hill, New York, 1970.

28. J. D. Eshelby, "The Continuum Theory of Lattice Defects," in *Solid State Physics*, Vol. III, F. Seitz and D. Turnbull (eds.) Academic Press, New York, 1956.

29. J. R. Rice, *J. Appl. Mech.*, **35**, 379 (1968).

30. J. R. Rice, P. C. Paris, and J. G. Merkle, *ASTM STP* **536**, 231, American Society for Testing and Materials, Philadelphia, 1973.

31. J. A. Begley and J. D. Landes, *ASTM STP* **514**, 1, American Society for Testing and Materials, Philadelphia, 1972.

32. J. D. Landes and J. A. Begley, *ASTM STP* **514**, 24, American Society for Testing and Materials, Philadelphia, 1972.

33. J. D. G. Sumpter and C. E. Turner, *ASTM STP* **601**, 3, American Society for Testing and Materials, Philadelphia, 1976.

34. E 813, *Standard Test Method for J_{Ic}, A Measure of Fracture Toughness*, American Society for Testing and Materials, Philadelphia, 1987.

35. S. Hashemi and J. G. Williams, *J. Mater. Sci.*, **26**, 621 (1991).

36. E. J. Moskala, *J. Mater. Sci.*, **27**, 4883 (1992).

37. I. Narisawa and M. T. Takemori, *Polym. Eng. Sci.*, **29**, 671 (1989).

38. D. D. Huang and J. G. Williams, *Polym. Eng. Sci.*, **30**, 1341 (1990).

39. D. D. Huang, *ASTM STP* **1114**, 290, American Society for Testing and Materials, Philadelphia, 1991.

40. D. D. Huang, "Fracture-Toughness Testing of Toughened Polymers," in *Toughened Plastics I: Science and Engineering*, Advances in Chemistry Series **233**, C. K. Riew and A. J. Kinloch (eds.), American Chemical Society, Washington, DC, 1993, p. 39.

41. D. D. Huang, *Polym. Eng. Sci.*, **36**, 2270 (1996).

42. S. V. Nair, S.-C. Wong, and L. A. Goettler, *J. Mater. Sci.*, **32**, 5335 (1997).

43. S. V. Nair, A. Subramaniam, and L. A. Goettler, *J. Mater. Sci.*, **32**, 5347 (1997).

44. M.-L. Lu, C.-B. Lee, and F.-C. Chang, *Polym. Eng. Sci.*, **35**, 1433 (1995).

45. M.-L. Lu and F.-C. Chang, *J. Appl. Polym. Sci.*, **56**, 1065 (1995).

46. M.-L. Lu and F.-C. Chang, *Polymer*, **36**, 4639 (1995).

47. M.-L. Lu, K.-C. Chiou, and F.-C. Chang, *Polymer*, **37**, 4289 (1996).

48. M.-L. Lu, K.-C. Chiou, and F.-C. Chang, *Polym. Eng. Sci.*, **36**, 2289 (1996).

49. M.-L. Lu, K.-C. Chiou, and F.-C. Chang, *J. Appl. Polym. Sci.*, **62**, 863 (1996).

50. J. D. Landes and J. A. Begley, *ASTM STP* **560**, 170, American Society for Testing and Materials, Philadelphia, 1974.

51. W. N. Chung and J. G. Williams, *ASTM STP* **1114**, 320, American Society for Testing and Materials, Philadelphia, 1991.

52. Y.-W. Mai and P. Powell, *J. Polym. Sci.: Part B: Polym. Phys.*, **29**, 785 (1991).

53. J.-S. Wu and Y.-W. Mai, *Polym. Eng. Sci.*, **36**, 2275 (1996).

54. S.-W. Kim, "Fracture Toughness Studies of Rubber-Toughened Polycarabonate," Ph.D. Dissertation, Massachusetts Institute of Technology, 1993, p. 60.

55. M. L. Shiao and S. V. Nair, "The Role of Matrix on the Mechanical Behaviour of Glass-Fiber Reinforced Thermoplastic Composites," in *Frontiers of Polymer Research*, P. N. Prasad and J. K. Nigam (eds.), Plenum Press, New York, 1991, pp. 505–517.

56. S. V. Nair, M. L. Shiao, and P. D. Garrett, *J. Mater. Sci.*, **27**, 1085 (1992).

57. S.-C. Wong, S. V. Nair, L. H. Vestergaard, L. A. Goettler, and L. A. Gustafson, *Plast. Eng.*, January 1995, pp. 23–25.

58. L. A. Utracki, *Polymer Alloys and Blends*, Hanser, Munich, 1989, p. 10.

59. F. Martinatti and T. Ricco, *Impact and Dynamic Fracture of Polymers and Composites*, ESIS 19, J. G. Williams and A. Pavan (eds.), Mechanical Engineering Publications, London, 1995, pp. 83–91.

60. F. Martinatti and T. Ricco, *J. Mater. Sci.*, **29**, 442 (1994).

61. K. B. Broberg, *J. Mech. Phys. Solids*, **19**, 407 (1971).

62. K. B. Broberg, *J. Mech. Phys. Solids*, **23**, 215 (1975).

63. B. Cotterell and J. K. Reddel, *Int. J. Fract.*, **12**, 267 (1977).

64. Y.-W. Mai and B. Cotterell, *Int. J. Fract.*, **32**, 105 (1986).

65. Y.-W. Mai, *Int. J. Mech. Sci.*, **35**, 995 (1993).

66. Y.-W. Mai, "Fracture Mechanics and Fracture Toughness Measurements of Polymer Blends," in *Proceedings of the International Symposium on Polymer Alloys and Composites*, Hong Kong, December 1992, C. L. Choy and F. G. Shin (eds.), Hong Kong Polytechnic, Hong Kong, 1994, pp. 104–112.

67. Test Protocol for Essential Work of Fracture (Version 5), European Structural Integrity Society, 5 October 1997.

68. Y.-W. Mai and B. Cotterell, *Int. J. Fract.*, **24**, 229 (1984).

69. Y.-W. Mai and B. Cotterell, *Eng. Fract. Mech.*, **21**, 123 (1985).

70. Y.-W. Mai, *Appita*, **36**, 461 (1983).

71. Y.-W. Mai, H. He, R. Leung, and R. S. Seth, "In-Plane Fracture Toughness Measurement of Paper," *Fracture Mechanics*, Vol. 26, *ASTM STP* **1256**, 587, American Society for Testing and Materials, Philadelphia, 1995.

72. S. Hashemi, *Plast. Rub. Comp. Proc. Appl.*, **20**, 229 (1993).

73. S. Hashemi, *J. Mater. Sci.*, **28**, 6178 (1993).

74. W. Y. F. Chan and J. G. Williams, *Polymer*, **35**, 1666 (1994).

75. S. Hashemi, *J. Mater. Sci.*, **32**, 1563 (1997).

76. G. Levita, L. Parisi, A. Marchetti and L. Bartolommei, *Polym. Eng. Sci.*, **36**, 2534 (1996).

77. G. Levita, L. Parisi, and S. Mcloughlin, *J. Mater. Sci.*, **31**, 1545 (1996).

78. J. Karger-Kocsis and T. Czigány, *Polymer*, **37**, 2433 (1996).

79. A. S. Saleemi and J. A. Nairn, *Polym. Eng. Sci.*, **30**, 211 (1990).

80. S. Hashemi, *Polym. Eng. Sci.*, **37**, 912 (1997).

81. G. Levita, L. Parisi, and A. Marchetti, *J. Mater. Sci.*, **29**, 4545 (1994).

82. J. Karger-Kocsis and E. J. Moskala, *Polym. Bull.*, **39**, 503 (1997).

83. J. Karger-Kocsis, *Polym. Eng. Sci.*, **36**, 203 (1996).

84. J. Karger-Kocsis and J. Varga, *J. Appl. Polym. Sci.*, **62**, 291 (1996).

85. J. Karger-Kocsis, T. Czigány, and E. J. Moskala, *Polymer*, **38**, 4587 (1997).

86. J. Karger-Kocsis, T. Czigány, and E. J. Moskala, *Polymer*, **39**, 3939 (1998).

87. D. E. Mouzakis, F. Stricker, R. Mülhaupt, and J. Karger-Kocsis, *J. Mater. Sci.*, **33**, 2551 (1998).

88. S.-C. Wong and Y.-W. Mai, "Essential Fracture Work of Short Glass Fiber-Reinforced Polymer Blends," *Polym. Eng. Sci.*, **39**, 356 (1999).

89. J. Karger-Kocsis, *Polym. Bull.*, **37**, 119 (1996).

90. T. Vu-Khahn, *Trends in Polymer Science*, **5**, 356 (1997).

91. C. R. Bernal and P. M. Frontini, *Polym. Eng. Sci.*, **35**, 1705 (1995).

92. P. M. Frontini, private communication, March 1998.

93. L. Fasce, P. M. Frontini, C. Bernal, and Y.-W. Mai, "On the Impact Essential Work of Fracture of Ductile Polymers," in *Proc. Structural Integrity and Fracture: Australian Fracture Group Symposium*, 21 September 1998, Melbourne, Australia.

94. G. Wildes, H. Keskkula, and D. R. Paul, *Polymer*, 1998, submitted for publication.

95. C. S. Brown and A. T. Alder, "Blends Containing Liquid Crystalline Polymers," in *Polymer Blends and Alloys*, M. J. Folkes and P. S. Hope (eds.), Blackie Academic & Professional, Chapman & Hall, Glasgow, United Kingdom, 1993, p. 195.

96. D. Dutta, H. Fruitwala, A Kohli, and R. A. Weiss, *Polym. Eng. Sci.*, **30**, 1005 (1990).

97. G. Kiss, *Polym. Eng. Sci.*, **27**, 410 (1987).

98. P. R. Subramanian and A. I. Isayev, *Polymer*, **32**, 1961 (1991).

99. G. Crevecoeur and G. Groeninckx, *Polym. Eng. Sci.*, **30**, 532 (1990).

100. K. G. Blizard, C. Federici, O. Federico, and L. L. Chapoy, *Polym. Eng. Sci.*, **30**, 1442 (1990).

101. J. X. Li, S. Silverstein, A. Hiltner, and E. Baer, *J. Appl. Polym. Sci.*, **44**, 1531 (1992).

102. R. A. Weiss, W. Huh, and L. Nicolais, *Polym. Eng. Sci.*, **27**, 684 (1987).

103. A. I. Isayev and M. J. Modic, *Polym. Compos.*, **8**, 158 (1987).

104. R. A. Weiss, W. Huh, and L. Nicolais, "Rheology and Properties of Self-Reinforcing Polymer–Liquid Crystalline Polymer Blends," in *High Modulus Polymers*, A. E. Zachariades and R. S. Porter (eds.), Marcel Dekker, New York, 1988, p. 145.

105. K. G. Blizard and D. G. Baird, *Polym. Eng. Sci.*, **27**, 673 (1987).

106. S.-C. Wong, Y.-W. Mai, and Y. Leng, *Polym. Eng. Sci.*, **38**, 156 (1998).

107. D. R. Ireland, "Procedures and Problems Associated with Reliable Control of the Instrumented Impact Test," in *Instrumented Impact Testing*, ASTM STP **563**, 3, American Society for Testing and Materials, Philadelphia, 1974.

108. R. A. Pearson and A. F. Yee, *J. Mater. Sci.*, **26**, 3828 (1991).

109. H.-J. Sue, E. I. Garcia-Meitin, D. M. Pickelman, and P. C. Yang, "Optimisation of Model-I Fracture Toughness of High Performance Epoxies by Using Designed Core-Shell Rubber Particles," in *Toughened Plastics I: Science and Engineering*, Advances in Chemistry Series **233**, C. K. Riew and A. J. Kinloch (eds.), American Chemical Society, Washington, DC, 1993, p. 259.

110. H.-J. Sue, *Polym. Eng. Sci.*, **31**, 275 (1991).

111. A. J. Kinloch, M. L. Yuen, and S. D. Jenkins, *J. Mater. Sci.*, **29**, 3781 (1994).

112. B. Geisler and F. N. Kelley, *J. Appl. Polym. Sci.*, **54**, 177 (1994).

113. D. M. Otterson, B. H. Kim, and R. E. Lavengood, *J. Mater. Sci.*, **26**, 1478 (1991).

114. D. M. Otterson, B. H. Kim, and R. E. Lavengood, *J. Mater. Sci.*, **26**, 4855 (1991).

115. B. H. Kim and C. R. Joe, *Eng. Fract. Mech.*, **34**, 221 (1989).

116. C.-B. Lee, M.-L. Lu, and F. C. Chang, *J. Appl. Polym. Sci.*, **47**, 1867 (1993).

117. M.-L. Lu and F. C. Chang, *Polymer*, **36**, 2541 (1995).

118. F.-C. Chang and H.-C. Hsu, *J. Appl. Polym. Sci.*, **43**, 1025 (1991).

119. F.-C. Chang and H.-C. Hsu, *J. Appl. Polym. Sci.*, **47**, 2195 (1993).

120. S.-C. Wong and Y.-W. Mai, *Key Eng. Mater.*, **137**, 55 (1998).

121. S.-C. Wong and Y.-W. Mai, *Polymer*, **40**, 1553 (1999).

122. M. Heino, P. Hietaoja, J. Seppälä, T. Harmia, and K. Friedrich, *J. Appl. Polym. Sci.*, **66**, 2209 (1997).

123. A. J. Oshinski, H. Keskkula, and D. R. Paul, *Polymer*, **33**, 268 (1992).

124. A. J. Oshinski, H. Keskkula, and D. R. Paul, *Polymer*, **33**, 284 (1992).

21 Characterizing Toughness using Standard Empirical Tests

C. B. BUCKNALL

Cranfield University
Bedford MK43 0AL
United Kingdom

I. WHY USE EMPIRICAL TESTS?

Fracture resistance is an important requirement in most applications of plastics, one that has largely driven the development of new polymer blends. However, fracture is

Polymer Blends, Volume 2: Performance. Edited by D. R. Paul and C. B. Bucknall.
ISBN 0-471-35280-2. © 2000 John Wiley & Sons, Inc.

a complex process, involving a continuous and evolving sequence of mechanical and thermal interactions in the various zones surrounding the crack tip and between the cracked body and its environment. In view of this complexity, there are obvious difficulties in defining a single scalar quantity that summarizes the fracture behavior of the material. Rather than being tests of the material (as, for example, in density measurements), all fracture tests are essentially tests on a structure. The results depend, among other things, upon the dimensions and geometrical shape of the specimen.

One solution to this problem is provided by linear elastic fracture mechanics, which is reviewed in Chapter 20. A particular type of geometry is chosen, in which a deep, sharp notch is cut into a thick specimen to obtain the maximum degree of constraint upon yielding, and attention is focused on the conditions for initiation of a crack from the notch tip. This approach has obvious advantages, in that it assesses all materials under similar conditions, where they exhibit the lowest possible toughness. Because it allows engineers to calculate the minimum stresses under which cracks can propagate in a given component, fracture mechanics is an essential design tool in critical applications such as pressure vessels and class I aircraft structures, where the consequences of any failure would be extremely serious.

The problem facing the plastics industry is a different one. In the majority of applications for rigid polymers, the consequences of failure are far from catastrophic, and there is a need to balance fracture resistance against other factors, especially weight and cost. For example, a suitcase should be able to withstand a reasonable level of rough handling, but should also be lightweight and not too expensive. In this application, product testing provides the best possible information about performance. However, it would be impossible to conduct trials on all candidate materials, and product testing must therefore be preceded by materials selection based on standardized test data. Linear elastic fracture mechanics measurements, which reveal serious deficiencies even in the toughness of polycarbonate, the most robust of rigid polymers, are not an ideal source of such data.

In view of the need to characterize all aspects of fracture behavior, and not simply the worst case, it is not surprising that the plastics industry continues to use a range of different tests for measuring toughness. Of these tests, the most widely used are undoubtedly the notched Charpy and Izod impact tests, which are simple to perform, but very difficult to translate into engineering design. In their most basic form, as laid down by the various standards organizations, they provide at best a semiquantitative basis for materials selection. They may be used either for ranking existing materials or for assessing improvements in experimental formulations, including blends. However, their limitations must always be borne in mind, a comment that applies equally to the other standard empirical methods for assessing toughness.

Standard empirical tests offer a number of advantages. First, they are simple to perform and require little in the way of calculation or analysis. With the aid of instrumentation and high-speed video recorders, they can be used to perform detailed studies on the fracture behavior of test specimens. Furthermore, in comparison to the sharp notching of bars and plates required for fracture mechanics, the preparation of reproducible specimens containing a radiused notch is relatively straightforward. Apart from all of these advantages, however, empirical methods suffer from one major disadvantage: The toughness data that they provide are specific to the specimen geometry chosen, so that they cannot be used in engineering design or evaluation. As measured in standard tests, tensile elongation to break, Izod and Charpy impact strength, dart-drop impact behavior, and the other measurements used so widely to define toughness in polymer blends find no place in design textbooks. However, this is not to say that they cannot provide valuable information; the present chapter examines the problem of making full use of standard test data with the aid of supplementary experimental techniques and theoretical analysis. Toughness tests are reviewed under four headings: tension, flexure of unnotched bars, flexure of notched bars, and fracture of discs.

II. TENSILE TESTS

A. Measures of Toughness

Of all the standard methods for measuring fracture resistance, tensile testing is the simplest and most easily analyzed. Three ASTM standards define conditions for this type of test, at moderate, high, and impact speeds, respectively [1–3]. The specimens have a parallel gauge section of rectangular cross section, broadening at each end to provide attachment points for the grips, which are pulled apart at constant velocity. In most major laboratories, contacting or remote extensometers are used to measure the separation L between two points on a gauge length of initial cross-sectional area A_o, while a load cell records the applied force F. Most modern equipment incorporates a computer to record and analyze the resulting data, which are usually shown as plots of engineering stress σ $(= F/A_o)$ against nominal strain ε $(= \Delta L/L_o)$. The true stress σ^*, which is defined as F/A, where A is the current cross-sectional area of the bar, is always higher than the engineering stress, but cannot be calculated unless separate measurements are made to determine A. No additional information is needed to calculate Young's modulus, the engineering yield stress, the engineering stress and strain at fracture, and the work to break.

Of these quantities, w_v, the work to break per unit volume of material, provides the most effective measure of toughness in a ductile material, provided that proper safeguards are applied. In determining w_v, the underlying assumptions are the extension to break ε_b is a reproducible intrinsic property of the material under the chosen test conditions and that a valid work to break per unit volume can be obtained by integrating the area under the force–extension curve. This method is applicable only if the whole of the section monitored by the extensometer yields and draws to approximately the same extent before the specimen finally fails. If yielding is confined to a localized neck or to other regions of high strain, more detailed analysis of the data is required in order to determine w_v.

B. Strain Localization

It is widely recognized that the calculation of a strain from the percentage extension, as measured in the tensile test, is meaningful only at small strains. Beyond the load maximum (the engineering yield point σ_{TY}, where the subscript TY denotes tensile yield), many polymers form a neck, so that the (local) axial strain varies markedly along the gauge length. The cross-sectional area in the neck is typically less than half that in the remainder of the gauge section, which means that the material forming the neck has more than doubled in length as a result of yielding. Extensions of this magnitude are better expressed in terms of the extension ratio λ, which, for a material deforming with negligible change in density, can be written as

$$\lambda = \frac{L}{L_o} = \frac{A_o}{A},\tag{21.1}$$

where L and L_o are the current and original length of the element under consideration, respectively.

In some ductile materials, notably the semicrystalline thermoplastics, such as nylons and polypropylene, the neck extends in both directions, essentially at constant applied load, until it occupies the whole of the gauge length and meets the shoulders of the specimen. At this point, the load rises; the strain-hardened material in the neck then extends further and finally breaks, often at an elongation of over 200%. Under these conditions, the percentage elongation at break, ε_b, and the work to break per unit volume, w_v, are meaningful quantities. However, in other cases, wherein the specimen breaks in the neck at a much earlier stage, before it has propagated along the gauge length, the nominal value of ε_b is low and considerably underestimates the energy-absorbing capacity of the polymer.

If the material in the neck is a continuum, and therefore has not suffered a significant decrease in density, the (local) extension at break can still be determined from the measured minimum cross-sectional area $A_{\min}$, using Eq. 21.1. In this case, Considère's construction can be used to calculate the engineering yield stress and natural draw ratio in the neck, from a plot of true stress σ^* against λ (see Fig. 22.1) [4, 5]. However, yielding of polymer blends is often accompanied by internal cohesive failure, through cavitation of rubber particles, debonding of rigid inclusions, or crazing of the matrix, so that the current area A cannot be used to determine λ. A good example of this type of behavior is provided by high-impact polystyrene (HIPS), which yields and draws through rubber particle cavitation and matrix crazing with little or no change in cross-sectional area. (See Chapter 22.) Even more complex behavior is shown by ABS (acrylonitrile-butadiene-styrene polymer), which at room temperature yields through a combination of rubber particle cavitation, crazing, and shear yielding, with the result that it forms a neck in tension, but fractures in the neck at a relatively low total extension (relative to the whole gauge portion), typically about 20%.

Two recent reviews discuss in depth the effects of necking and other yield phenomena on local strain rates, internal heating, and other responses of ductile polymers during a tensile test, with especial reference to polymeric glasses [6, 7]. One of the most interesting developments has been the introduction of computer-aided video extensometry to monitor true stresses and strains in a cylindrical tensile specimen with a radiused "hourglass" central section [8]. Necking is localized at the

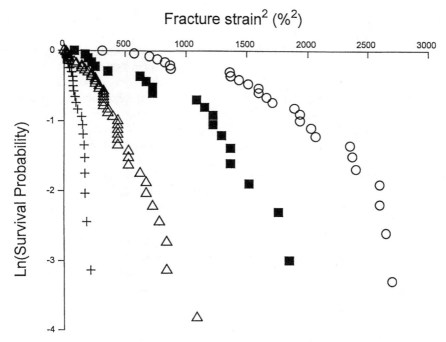

Figure 21.1   Probability of survival P_{SV} as a function of creep strain ε for four sets of HIPS specimens subjected to fixed tensile load at 23°C. Blends contain (+) 2%, ($\triangle$) 4%, ($\blacksquare$) 6%, and (o) 8 wt% of rubber in the form of salami particles. (See Chapter 22.) All of these HIPS blends were made from the same batch of HIPS. Applied stresses, matched to the yield stresses of the blends, are 26.5, 22.5, 18.5, and 14 MPa, respectively. From [9], reprinted with permission.

midpoint, and local strain rates at this point can be held constant with the aid of suitable software. This technique offers a possible method for the determination of w_v at high tensile strain rates, provided that adequate checks are made to ensure that internal void formation has not taken place in the neck.

A different kind of strain localization is shown by HIPS. During the early stages of a tensile test on this material, whitened bands extend across the specimen on planes roughly normal to the tensile stress, marking regions in which extensive crazing and rubber particle cavitation have taken place. Under ideal conditions, these zones are stabilized mechanically by fibrillated rubber particles, which act as strain-hardening elements (see Chapter 22), the whole gauge portion becomes uniformly whitened, and the material reaches relatively high elongations (up to about 60%). However, because stretched rubber particles break when the local strain reaches a critical level [9, 10], it is possible to observe fracture at quite low tensile strains.

Sjoerdsma and Boyens have shown that there is a correlation between the probability of fracture in a HIPS tensile specimen and its tensile strain, which is essentially unaffected by stress and temperature [11]. Their experiments involved subjecting large batches of tensile bars to the same constant stress and observing the extension at break, ε_b, for each bar. The probability of survival, P_{SV}, at any given strain in these creep rupture tests was defined as the fraction of specimens remaining unbroken at that strain. They found a linear dependence of $\log(P_{SV})$ on ε_b^2. In a later study, O'Connor and Bucknall also found correlations between $\log(P_{SV})$ and ε_b^x for a range of HIPS and ABS materials, but with different values for the exponent x. They concluded that the correlations result from failures of overloaded rubber particles, which give rise to a Weibull distribution of breaking strains in ABS and HIPS. Figure 21.1 shows the relationship between P_{SV} and ε_b^2 for four HIPS materials containing different concentrations of rubber particles [9]. The rubber content was varied by blending the original HIPS with plain polystyrene in a compounding extruder. The data show that there is a maximum extension to break in HIPS that increases with rubber content, but also that the work to break in tension is often well below the maximum attainable value.

In summary, it is clear that w_v, the work to break per unit volume, is a useful measure of toughness in ductile polymer blends. However, it is valid only if failure takes place at high strains, and it follows deformation involving the whole of the measured gauge length, with no evidence of strain localization in the form of necking or nonuniform internal cavitation.

III. FLEXURAL TESTING OF UNNOTCHED BARS

A. Elastic Deflection

Flexural testing is widely used to assess the mechanical properties of rigid polymers, because of the simplicity of the test procedure, as described in ASTM method D 790 and other standards [12]. A plain rectangular bar is simply supported near each end and subjected to three-point bending through application of load at the midpoint, as illustrated in Fig. 21.2, or to four-point bending using two loading "noses" placed closely and symmetrically on either side of the midpoint. Both supports and loading noses are radiused to minimize indentation. The data are used to calculate flexural strength or flexural yield stress, depending on the mode of failure, and also to determine a "flexural modulus."

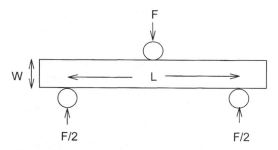

Figure 21.2   Schematic diagram of the flexural test.

Sources of error in flexural tests have been reviewed in some detail by Heap and Norman [13]. One such source is large elastic deflection, which causes both a horizontal component of end thrust and a shortening of the span as the two halves of the specimen rotate in opposite directions and the points of contact between test bar and radiused supports come closer [14]. Another possible problem arises in thick beams, especially those made of fiber-reinforced composites, where significant additional deformation may take place through shear [15, 16]. A third source of error is anticlastic curvature, which occurs when there is lateral contraction across the tensile face of the flexed bar and corresponding lateral expansion on the compressive face. In wide sheets subjected to bending, anticlastic curvature is prevented, thereby setting up a state of plane strain in the sheet [16, 17]. Finally, elastic beam theory is based on linear elasticity, whereas polymers show significant deviations from linearity.

Strikingly, the values quoted in commercial data sheets for the flexural strengths and yield stresses of polymers are usually much higher, sometimes by as much as 50%, than corresponding values obtained from tensile tests. Some of the aforementioned errors might contribute to this discrepancy, but there are other possibilities as well.

Flexure produces tensile stresses in the convex, lower surface of the bar, which is known as the "outer fiber" region of the bar. These stresses are balanced by compressive stresses on the opposite face. The maximum outer-fiber stress occurs at the midspan and can be calculated using beam theory. The ASTM standard gives expressions for three-point and four-point loading, with provision for beams undergoing large deflections ($>10\%$ of the support span) [12]. For the most frequently encountered case, a bar in three-point bending with (relatively) small deflections, the maximum outer-fiber stress σ_{max} is given by

$$\sigma_{max} = \frac{3FL}{2BW^2},\qquad(21.2)$$

where F is the applied force at the midpoint, L is the distance between supports, B is the width of the bar, and W is the depth of the bar (between tension and compression faces). (See Fig. 21.2.) Fracture strengths and yield stresses are obtained from the calculated outer-fiber stress, as given in Eq. 21.2. Strictly, flexural measurements should be terminated when the maximum outer-fiber strain $\varepsilon = 6dW/L^2$ reaches 5%. (d is the load-point deflection, corrected for indentation.) The ASTM standard considers data obtained beyond this limit to be invalid and recommends an alternative test—for example, the tensile test—if strains exceed 5% before yield or fracture [12].

Brittle fracture is associated with low failure strains, and therefore, flexural tests ending in low-energy fracture usually meet the ASTM strain criterion. The observed differences between flexural and tensile strengths must originate elsewhere. The most obvious source of variation is the degree of anisotropy in the regions where crazes and cracks originate. Because injection molding is the preferred method for preparing test bars, levels of molecular orientation are high in the surfaces of both flexural and tensile test bars. However, in flexural tests, cracks necessarily initiate in the outer fibers, where the imposed stresses are at a maximum. By contrast, in tensile specimens, the first evidence of crazing and fracture is usually seen in the core region of the molding, where molecular orientation is low and thermal contraction effects generate substantial internal tensile stresses parallel to the length of the bar. In flexure, the enhanced strength of the highly oriented surface layers dominates brittle fracture behavior in a way that does not apply to tensile bars, thereby raising the stress required for brittle fracture. To a lesser extent, the same considerations apply to modulus measurements, which are also affected by the marked nonlinearity in the stress–strain behavior of polymers. In uniaxial tension, the resulting reduction in stiffness applies to the whole specimen, whereas in flexure only one surface is subjected to a high tensile strain, and the remainder of the bar is either at a lower tensile strain or in compression.

B. Plastic Collapse in Bending

The comparison between tensile and flexural tests becomes more interesting when we consider the measurement of yield stress. In tension, the sequence of events is fairly clear, as discussed in Section II. The stress increases until local plastic deformation becomes sufficiently developed to cause yielding through a complete cross section of the specimen, so that the load-deflection curve reaches a maximum, which defines the engineering yield stress. In most ductile polymers, tensile yielding is associated with the formation of a neck. The sequence of events is quite different in flexure. Yielding in the outer fibers does not produce a load maximum, because the less highly stressed subsurface layers continue to support the applied load. Instead, the load-deflection curve continues to rise, showing an increasing departure from linearity as yield zones spread inward from the tensile and compressive surfaces. Only when the two wedge-shaped yield zones meet at the neutral surface in the central region of the bar and the specimen is fully plastic throughout its complete central cross section does the load reach a maximum. (See Fig. 21.3.) From measurements

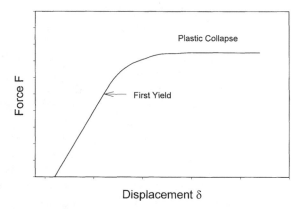

Figure 21.3 Force-deflection curve for an unnotched bar undergoing plastic collapse in bending.

of the maximum applied force, it is, in principle, possible to calculate the yield stress of the material.

Most analyzes of plastic collapse in bending are based on the behavior of metals, which have almost the same yield stress in tension and compression, so that the neutral surface remains at the center of the specimen throughout elastic loading and subsequent yielding. The load maximum in this case, obtained using this simplified approach [16], is

$$F_{max} = \frac{BW^2 \sigma_{TY}}{L}. \tag{21.3}$$

For polymers, yield stresses are always substantially higher in compression (σ_{CY}) than in tension (σ_{TY}), because of the pressure dependence of yielding. (See Chapter 22.) Compression increases the density of the polymer, thereby hindering molecular motion. However, in the case of polymer blends, the differences between tension and compression are much more pronounced. Whereas $(\sigma_{CY})/(\sigma_{TY}) \approx 1.3$ for a typical shear-yielding polymer, such as polycarbonate, the ratio might be as high as 3 for multiphase polymers, such as HIPS, which yield in tension by a combination of matrix crazing and rubber particle cavitation.

Using the standard approach to plastic collapse, but allowing for the differences between tension and compression, gives the following expression for the maximum load F_{max} [18]:

$$F_{max} = \frac{2BW^2}{L} \left(\sigma_{CY} - \frac{\sigma_{CY}^2}{(\sigma_{CY} + \sigma_{TY})} \right). \tag{21.4}$$

This expression reduces to Eq. 21.3 when $\sigma_{CY} = \sigma_{TY}$. The potential errors in assuming that flexural yielding follows Eq. 21.3 can be appreciated by substituting values of $\sigma_{TY} = 40$ MPa and $\sigma_{CY} = 60$ MPa into Eq. 21.4 and $\sigma_{CY} = \sigma_{TY} = 48$ MPa into Eq. 21.3. The same value of applied load F is obtained in the two cases. Application of a corrected yield equation, taken together with the greater dominance of surface effects in flexural yielding, goes some way towards explaining the discrepancies between flexural and tensile yield results.

The flexural test is sometimes used to assess toughness, notably in so-called unnotched Charpy impact measurements [19]. It can be effective as a product simulation test, but has a number of limitations. If the material is really ductile, the bar might simply undergo plastic collapse, slip between the supports, and remain unbroken at the end. Fracture occurring before the bar reaches the point of plastic collapse is an indication of a lower degree of ductility. In both cases, the area under the load-deflection curve up to the load maximum can be regarded as a measure of toughness, although it is rarely used as such. In the context of toughness measurements, the main importance of flexural testing is in providing reference data with which the fracture energies of round-notched Charpy and Izod specimens can be compared.

IV. FLEXURAL TESTING OF NOTCHED BARS

A. Advantages of Standard Charpy and Izod Tests

The notched Izod impact test is probably the most widely used method for assessing the fracture resistance of thermoplastics. The notched Charpy test is less popular, but as it is very similar to the notched Izod test in its main features, it is convenient to discuss the two tests together. In both cases, a radiused notch is located centrally on the narrow face of a rectangular bar, which is broken in bending, with the notched surface in tension [20]. The Charpy bar is subjected to three-point bending, in a similar manner to the flexural specimen illustrated in Fig. 21.2. By contrast, the Izod specimen is tested as a cantilever, held with its long axis vertical and the lower half clamped. The stresses introduced through clamping near the notch make the Izod test less attractive than the Charpy test as a basis for sharp-notch fracture mechanics studies [21, 22].

In the notched Izod test, a striker head mounted on a pendulum impacts the free end of the bar, and the energy absorbed in fracture is recorded by means of a pointer and scale. The Charpy test is similar, except that the specimen is laid horizontally across a gap bounded by two uprights and is struck centrally at the point opposite the notch. Many laboratories now use more sophisticated, instrumented equipment to measure energy absorption. Instrumented impact testing provides force-time curves that can be analyzed further with the aid of computer packages and allows the pendulum to be replaced with a falling dart or other type of impact head as preferred.

Standard notched Izod and Charpy tests have several advantages over fracture mechanics tests. They are easy to carry out, and they use the same specimen geometry and dimensions regardless of the yield stress or other properties of the material. Thus, they can be used for studies of fracture behavior over a range of temperatures and blend compositions. Very little effort is required to calculate the "impact strength" (which is actually an impact energy), and the various standards raise no questions concerning the validity of the data, as in LEFM. Most importantly, the presence of a rounded notch tip of defined radius (usually 250 μm) generally ensures that the results are very reproducible, so that impact strengths can be quoted to a precision of about 5%, unless the measurements are made close to a transition.

This reproducibility of notched Izod and Charpy data is important in many practical situations, including the evaluation of new polymer blends, the maintenance of quality standards, and legal disputes. The large scatter in results that often arises when LEFM trials are conducted between different laboratories is rarely encountered in standard notched Izod tests, because the problem of producing a sharp notch tip is not an issue. However, this is not to say that the energy absorbed in standard impact tests cannot be influenced by the notching procedure. Some laboratories injection mold a plain rectangular bar and machine the notch into the bar at a later stage, whereas others mold the bar with the notch in place as a feature of the molding. The resulting orientation in the region surrounding the notch tip inevitably raises the fracture resistance of the bar above that found with a machined notch, and the level of orientation can be varied by changing the molding conditions. Arguments between suppliers and customers about whether a particular batch of material reaches an agreed level of toughness can obviously be affected by this dependence on processing conditions.

B. Brittle and Ductile Fractures

Instrumented tests on notched Izod and Charpy specimens show three distinct types of response, which may be labeled brittle, semibrittle, and ductile. Typical force-deflection curves for these three types of fracture are shown in Fig. 21.4. Brittle fracture is characterized by an approximately linear loading curve, ending in a sharp drop in load within about 0.1 ms. In this case, plastic deformation at the notch tip is very limited and insufficient to cause a detectable deviation from linearity. Inspection of the fracture surface reveals a broken pattern with little or no evidence of ductility. Fractures of this kind, involving the formation of a single craze at the notch tip, are typical of glassy polymers, such as polystyrene (PS) and poly(methyl methacrylate) (PMMA). Assuming that the fracture energy $G_{Ic} = 300$ J·m^{-2} for these polymers, a total of about 0.02 J is required to form and break a ligament measuring 6×10 mm. This value is about one fifth of the energy absorbed in Izod impact tests on 6-mm-thick specimens of PS and PMMA.

With a striker velocity at impact of 3.46 m·s^{-1}, as prescribed in the ASTM Izod test, the time to peak load is on the order of 1 ms. At the point of crack initiation, the stored energy in the bar is substantially greater than is needed to break the bar in half, and the crack accelerates to its terminal velocity, which is about 500 m·s^{-1}. This limit is imposed by inertia: A crack cannot exceed the velocity of the transverse wave in a solid, and as it approaches this velocity, it tends to split into two, an effect that gives rise to shattering in glasses, including polymeric glasses. Relatively low dynamic fracture toughness G_D is observed even in normally tough polymers at speeds above about 100 m·s^{-1} [23, 24].

When there is some yielding around the notch tip, but final fracture occurs in a brittle manner through rapid crack propagation, the response of the impact specimen may be described as semibrittle. As shown in Fig. 21.4b, yielding near the notch causes a gradual decrease in the slope of the force-deflection curve, but absorption of energy from the striker is terminated when a crack initiates and the load falls abruptly (within 0.1 ms). The behavior of the crack is governed by three factors: (a) the increase in the energy release rate G with increasing crack length in three-point bend specimens, which is a result of the test geometry (see Chapter 20); (b) a limited increase in crack growth resistance as the crack extends, which can be expressed in fracture mechanics terms as a low value of dJ/da; and (c) the low dynamic fracture energy G_D in polymers at crack speeds above about 50 m·s^{-1}.

Leevers has developed a thermal decohesion model to account for the low values of G_D observed even in very ductile semicrystalline polymers, including, for example, toughened polyoxymethylene [23, 24]. In fast fracture of rigid polymers, the crack front is preceded by a cavitated yield zone, which, in the case of glassy polymers, can be described as a craze, but in ductile semicrystalline polymers should perhaps be given another name (e.g., quasicellular line zone), because of its greater thickness and rather coarser morphology. Whatever these fibrillated planar zones are

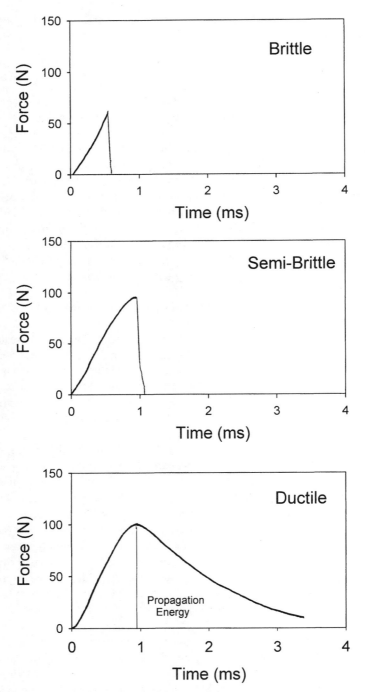

Figure 21.4   Force–time curves for notched Izod bars, showing (a) brittle fracture, (b) semi-brittle fracture, and (c) ductile fracture.

called, they have similar characteristics: Deformation occurs principally in an active interfacial zone where material is being pulled from the bulk polymer under adiabatic conditions. Thus, while both the bulk polymer and the fibrillated, strain-hardened material within the cavitated zone are cold, a large amount of heat is generated at the interface between the fibrillar zone and the bulk. At high crack speeds, this interfacial layer melts, setting an upper limit on the fracture surface energy, typically in the range of 3 to 5 $kJ\cdot m^{-2}$. This is not a particularly low value, but it is substantially lower than G_D for ductile tearing in polyethylene and similar semicystalline polymers at lower crack speeds. In the case of crazes in glassy thermoplastics, G_D values are closer to 0.5 $kJ\cdot m^{-2}$ [25]. Leevers pointed out that heating within the active zone at lower crack speeds should result in crack blunting, as heat generated at the craze–bulk boundary diffuses into the bulk of the specimen, melting it to a depth of a few microns [24]. Blunting effects of this kind have been observed by Gaymans in toughened nylons and polypropylene. (See Chapter 25.) A further implication of the analysis is that a crack cannot initiate until the strain energy release rate G exceeds G_D.

In ductile fracture, the crack never reaches high velocities. Instead, failure occurs through a process of stable crack propagation over a period of a few milliseconds, as recorded in Fig. 21.4c. The ligament length in standard Izod specimens is about 10 mm, which means that tearing over periods between 1 and 4 ms represents average crack speeds in the range from 10 to 2.5 $m\cdot s^{-1}$. At these speeds, heating to depths of a few microns below the crack surface can lead to crack-tip blunting.

When bending of the impact bar is followed by stable crack propagation, as in Fig. 21.4c, it becomes possible to divide energy absorption into two separate stages. The first stage, up to the load maximum, has generally been regarded as defining a crack initiation energy (from which a blunt-notch fracture resistance G_B could be calculated [22]), while the second stage of energy absorption has been assigned to crack propagation. (See Fig. 21.4c.) This is a very approximate basis upon which to discuss the data, but it is at least better than regarding the energy absorbed in impact as a single, indivisible quantity. The limitations of dividing the energy in this rather arbitrary manner are fairly obvious. In ductile fracture, the point of crack initiation usually occurs before the load maximum (see Chapter 20), so that it is difficult to define an initiation energy with any precision. Furthermore, the energy stored elastically in the specimen at the load maximum is released during subsequent unloading, when it is absorbed in crack propagation.

A paper by Jancar and Dibenedetto emphasizes the importance of recognizing that energy is absorbed in two stages during Charpy and Izod tests [26]. These authors determined the Charpy notched impact strength (CNIS) and the (sharp-notched) fracture energy G_C for a range of polypropylene blends and plotted the ratio CNIS $/G_C$ against mineral filler content, rubber content, and level of matrix-inclusion grafting. Not surprisingly, they found that in some series of blends, the ratio was dependent upon composition. Their procedure allows for variations in the contributions from the plane-strain and plane-stress regions of the specimen, which have similar effects on the measured G_C and on the initiation stage of the Charpy test. For brittle and semibrittle fracture, the ratio CNIS $/G_C$ therefore remains approximately constant.

However, the ratio varies when there are major changes in toughness with increasing rubber content, because the CNIS for tougher blends includes a substantial amount of energy absorbed in crack propagation, while G_C does not. Criticisms of standard impact tests that rest simply on the grounds that the tests do not measure exactly the same properties as those measured by fracture mechanics tests are unjustified. The validity of these criticisms is even more questionable when it is recognized that widely available instrumented equipment is capable of resolving the difficulties.

C. Effect of Notch Radius

The main difference between standard notched Charpy and Izod bars and their fracture mechanics equivalents lies in their notch-tip radii. Standard ASTM specimens have notch radii of 250 μm (i.e., 0.01 inch ≡ 10 mil), whereas razor-notched specimens have tip radii on the order of 10 μm or less. The effects of the notch-root radius ρ were analyzed by Williams, who obtained a blunt-notch value of the strain energy release rate G_B in terms of the distance l_o ahead of the notch tip at which the stress reaches a critical level [21, 22]. The relationship between G_B and the corresponding value for a sharp notch, G_C, is

$$\frac{G_B}{G_C} = \frac{\left(1 + \dfrac{\rho}{2l_o}\right)^3}{\left(1 + \dfrac{\rho}{l_o}\right)^2}, \tag{21.5}$$

which for $\rho \gg l_o$ can be approximated to

$$\frac{G_B}{G_C} = \frac{1}{2} + \frac{\rho}{8l_o}. \tag{21.6}$$

This analysis predicts a linear relationship between G_B and the notch-root radius, which has been demonstrated in rigid PVC [21, 22]. The preferred method for determining G_B is from the total fracture energy U (which equals $BW\phi G_c$, as described in Chapter 20 and Eq. 20.11). This method is effective only when crack initiation is followed by unstable brittle fracture, during which the load drops rapidly, and no further energy is absorbed from the striker. However, blunting inevitably reduces the constraint ahead of the notch tip and often results in stable crack propagation, so that additional energy is absorbed after crack initiation. A good example of notch blunting has been reported by Bramuzzo, who carried out three-point-bending experiments using impact loading on sharply notched specimens of polypropylene–rubber blends [27]. Despite the introduction of razor-sharp notches, video recordings clearly showed very ductile behavior in the toughest of these blends, with extensive yielding and blunting at the notch tip, and the corresponding force-deflection curves emphasize the importance of the blunting response in increasing the energy absorption. Another, comparable polypropylene blend, which was not tough enough to produce extensive notch-tip blunting, was semibrittle.

Crack initiation ahead of blunt notches in glassy polymers has been studied by Mills [28] and by Narisawa et al. [29]. The latter group showed that yielding in thick specimens of polycarbonate occurred through the formation of intersecting curved shear bands, to form a "slip-line field," as described by Hill [30]. When the local stress reached a critical level, a disc-shaped craze nucleated at the tip of this plastic zone, from which a crack was initiated. Similar mechanisms of brittle fracture were observed in PVC and PMMA.

The effects of notch-tip radius on fracture behavior have been studied by Havriliak et al., who carried out notched Izod impact tests on specimens of polycarbonate and toughened grades of rigid PVC and nylon 6 [31]. The notch radii chosen were 25,250 and 2500 μm (1, 10 and 100 mil). Force–time curves and photographic records show that the transition from semibrittle to ductile fracture occurs with increasing notch radius.

D. Effects of Ageing on Impact Strength

Under certain conditions, the blunt notches of standard notched Charpy and Izod specimens become sharpened, so that the bars respond to stress in a manner very similar to that of the sharp, single-edge-notched bend (SENB) specimens of linear elastic fracture mechanics. All that is necessary is for a single craze to form at low strain and initiate a brittle crack. This is effectively what happens when the material at the notch tip becomes degraded as a result of outdoor ageing. Polymer blends containing diene rubbers are particularly prone to this problem. Ultraviolet light from the sun (or from a sun lamp) initiates an oxidative chain reaction in the diene rubber, causing it to cross-link and become resistant to cavitation. (See Chapter 22.) The depth of the degraded surface layer can be measured by examining osmium-stained sections cut normal to the surface [32, 33]. In the bulk of the material, OsO_4 reacts with the double bonds of the diene groups to form an osmate ester, and the rubber particles appear dark in the TEM. (See Chapter 9.) However, rubber particles close to the exposed surface are not stained in this way, because the diene groups are eliminated through oxidation [32]. The depth of polymer affected increases with exposure time, but after two years of ageing, it might reach 50 to 100 μm. Both the absorption of UV light by the polymer and the limitations imposed by oxygen diffusion limit the depth of the oxidized layer. Nevertheless, the results can be dramatic in toughened plastics, such as ABS. Instead of reaching an extension of 50% at the tip of the notch during the initial stages of the impact test, the oxidized layer behaves in a more brittle manner than the unmodified glassy PSAN matrix, forming a sharp, propagating crack at a strain on the order of 1% [33, 34].

The response of the specimen then resembles that of a fracture mechanics specimen with a sharp notch of length $a = 2.54$ mm (0.1 in.) and a width $W = 12.7$ mm (0.5 in.). At low temperatures, where the yield stress is high, the fracture is essentially brittle, whereas at higher temperatures, the yield zone can expand as the crack extends (with dJ/da increasing steeply), and a substantial amount of energy absorption takes place after the load maximum is reached. At and above room temperature, ABS specimens containing 20 wt% rubber usually retain a satisfactory level of impact resistance after ageing, because in 6-mm-thick Izod bars the yield stress is low enough to allow extensive yielding around the tip of the propagating crack. The data for these higher temperatures suggest that fracture initiates under what are effectively plane-stress conditions. However, between -60 and $0°C$, most of the toughening observed is due to restricted yielding around the notch tip during the initial stages of the test, and there is no energy input from the pendulum once a crack has been initiated. At these lower temperatures, therefore, the Charpy or Izod bar approximates to a LEFM specimen tested under plane-strain conditions.

E. Plastic Collapse in Bending of Notched Specimens

At the opposite extreme to these notch sharpening effects in aged specimens, there are cases in which extensive blunting occurs at the notch tip and Izod or Charpy bars behave much more like unnotched flexural specimens. The presence of the notch does, of course, reduce the stiffness of the bar, but this effect can be allowed for by treating the bar as having a smaller effective width $(W - a)$, where W is the total width and a is the notch depth. Thus, a standard 12.7-mm-wide ASTM Izod bar, with a notch depth of 2.54 mm, can be compared to a rectangular bar of width $W = 10.16$ mm (0.4 in.). This approach has been used by Carpinteri to develop a criterion for notch sensitivity in polymers [35]. The maximum force applied to Charpy specimens during impact is measured for a range of notch depths and compared to the force necessary to cause either first yield or plastic collapse in an unnotched flexural specimen of the same overall dimensions, under similar conditions of loading rate and temperature. If the material shows no notch sensitivity, then the maximum applied load on the notched bar, $F(a)_{max}$, corresponds to plastic collapse on the net section of length $(W - a)$. $F(a)_{max}$ is obtained by substituting the effective width $(W - a)$ for W in Eq. 21.3, giving

$$F(a)_{max} = \frac{BW^2 \left(1 - \dfrac{a}{W}\right)^2 \sigma_{TY}}{L}. \tag{21.7}$$

Thus, $F(a)_{max}$ should be proportional to $(1 - a/W)^2$. The data are analyzed as shown in Fig. 21.5, by plotting the ratio $F^*(a) = F(a)_{max}/F_Y(0)$ against a/W, where $F_Y(0)$ is the force required to produce first yield in the unnotched bar (i.e., $a = 0$). The construction line $F^*(a) = (1 - a/W)^2$ defines the case for which the notched bar meets conditions equivalent to first yield at the notch tip, while $F^*(a) = 1.5(1 - a/W)^2$ corresponds to the more important case in which plastic collapse occurs across the net section of the notched bar. The schematic diagram is based on data for a toughened PC blend, which comes close to exhibiting zero notch sensitivity at $23°C$.

This analyzis is relevant to the ongoing debate about the interparticle spacing effect in supertough nylons and similar materials. (See Chapter 26.) The brittle–tough transition reported by Wu [36] and later studied by Gaymans et al. (see Chapter 25) is a transition from semibrittle fracture to ductile failure, above which Izod speci-

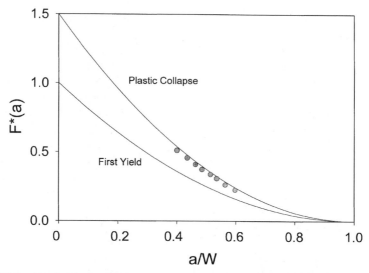

Figure 21.5   Carpintiri's method for assessing the notch sensitivity of materials subjected to impact tests [35]. The ratio of notch depth a to specimen width W is varied, and the maximum force sustained by the specimen is compared with the stresses to produce first yield and plastic collapse in unnotched bars. (See Eq. 21.7.)

mens fail by plastic collapse on the net section, absorbing large amounts of energy in the process. Below the transition, the specimen is not completely brittle, as it absorbs substantial amounts of energy before reaching the load maximum. However, this early stage of the test is succeeded by rapid crack propagation at low energy, so that the total amount of energy absorbed in impact is less than optimum.

F. Limitations of Standard Charpy and Izod Test Procedures

The foregoing discussion emphasizes the strengths and weaknesses of the Izod and Charpy impact tests. Because there is no requirement in the standard empirical test methods for specimen dimensions to be matched to materials properties, nor are there strict rules for validity to be applied, as in fracture mechanics, users are often unaware of what exactly is being measured and how it might relate to product performance. One well-known weakness of empirical impact tests on notched specimens concerns the units in which the data are expressed. The ASTM standard provides for impact energies to be divided by specimen thickness, to give results in J/m (or ft-lbf/in) of notch, on the assumption that impact energy is proportional to thickness [20]. However, it is well known that many polymers show strong departures from proportionality. The best known example is polycarbonate, which has a notched Izod impact "strength" of 8.5 $J \cdot cm^{-1}$ (16-ft-lbf/in notch) when the specimen is 3 mm (0.125 inch) thick, but only 1.6 $J \cdot cm^{-1}$ (3-ft-lbf/in notch) when the thickess is 6 mm (0.25 in). In other words, the thinner specimens absorb nearly three times as much total energy as the thicker ones. Clearly, the failure processes taking place are completely different in the two cases, and the tests are measuring different aspects of the polymer's fracture resistance. The 3-mm-thick specimens show a high level of ductility and reach the point of plastic collapse on the net section, whereas the 6-mm-thick specimens are thick enough to develop plane-strain conditions at the center of the bar, initiating a craze in the region ahead of the notch tip, which then turns into a crack.

Similar problems arise when the impact energy is expressed as a specific total work of fracture w_f, in kJ·m^{-2}, on the assumption that the total energy absorbed by a specimen is proportional to the area of the fracture surface and that w_f is a materials constant. There is no generally applicable justification for this form of data presentation. Where fracture is brittle or semibrittle (with yielding confined to a small region near the notch tip), the energy to fracture is controlled by initiation, as in linear elastic fracture mechanics, and is certainly not proportional to fracture surface area. In cases of ductile fracture, w_f usually increases with the square of the ligament length, as discussed in Chapter 20.

Notched impact testing is discussed at some length by Williams [22], who shows that stable crack growth can result in a linear relationship between U, the total energy absorbed in an impact test, and $B(W - a)$, the ligament area. Examples of this relationship are given for ABS. However, the line does not pass through the origin, so that U is not proportional to ligament area in the case of stable crack growth, and the relationship deviates strongly from linearity when crack growth is unstable.

In comparing notched Izod or Charpy data given in different units, the best policy is probably to remember that $(W - a)$ is, in most cases, of standard length (usually 10.16 mm, or 0.4 in) [20]. On this basis, it is possible to convert energy per unit notch length into energy per unit area, provided that the dimensions of the test bar are fully specified by the authors. However, it is important to stress once again that both quantities are characteristics of the fractured structure (i.e., the test bar) and fall some way short of providing a geometry-independent measurement of fracture resistance. Where that degree of rigor is required, it is necessary to use fracture mechanics.

V. FALLING-DART IMPACT TESTS

Falling-dart impact tests on disc or plate specimens are widely used by the plastics industry to characterize the "practical" fracture resistance of rigid thermoplastic blends. They have the advantage of being simple to perform experimentally, but are quite difficult to analyze mathematically. Essentially, they are formalized product tests on representative components, rather than measurements of basic materials properties. In this respect, they are in marked contrast to fracture mechanics measurements. They are intended to reproduce the real behavior of molded parts under impact loading.

Test specimens are typically flat discs about 3 mm thick and 100 mm in diameter, which are either simply supported by an annular, flat-topped ring or clamped between two matching rings. Impact loading is directed at the center of the disc. The falling dart is nowadays almost invariably instrumented in order to record the force-deflection curve for each specimen and transfer the data to a computer for further processing. The head of the dart is a cylindrical rod with a hemispherical tip about 20 mm in diameter. The dart is mounted on a carriage that runs between vertical guide rods. The kinetic energy of the striker can be varied by adding weights to the carriage or by changing the drop height. Measurements can be made over a range of temperatures by preconditioning the specimens, transferring them to the impact machine, and testing within 5 seconds.

A complex pattern of stresses is set up in the specimen during the impact test. At the center of the disc, immediately below the point of impact, the lower surface of the disc is in equibiaxial tension, and the stresses are at their highest. This small dome-shaped zone, which is convex on the lower side of the disc, is surrounded by two other distinct regions: a local, highly deformed ring of material, in which the curvature is convex on the upper side, and the remainder of the disc, in which the material is subject to lower levels of both flexural and tensile stress. The effects of this complex stress pattern are seen most clearly when the disc fractures in a brittle manner. On reassembling the shattered specimen, it can be seen that some cracks run radially from the central point of impact, while others follow a circular path around the same point. (See Fig. 21.6.) Circular cracks occur where the depression caused by the striker produces a bending moment and a high level of out-of-plane curvature.

Truly brittle behavior is observed in polymers such a polystyrene (PS) and poly(styrene-*r*-acrylonitrile), which are glassy thermoplastics with very high yield stresses. For these polymers, the force-deflection curve in dart impact tests is basically linear, and there is no visible sign of ductility on the shattered specimens. Blends designed to yield before fracture are generally found to be semibrittle rather than completely brittle, even at very low temperatures. In a study on the falling-dart impact behavior of ABS, Casiraghi et al. observed stress whitening on the surface below the impact point at temperatures down to $-80°C$ [37]. At this temperature, force-deflection curves were basically linear, ending in a sharp drop when the disc shattered. On raising of the temperature, the size of the whitened yield zone increases, and the maximum load on the specimen increases accordingly. The load-deflection curve then turns upwards as the main section of the clamped disc begins to deform like a membrane, and its stiffness therefore increases. The fracture energy rises with

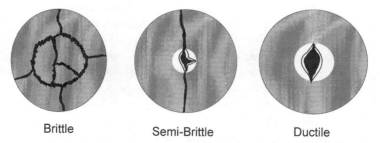

Brittle   Semi-Brittle   Ductile

Figure 21.6 Schematic diagram showing the appearance of discs subjected to falling-dart impact tests and undergoing brittle, semibrittle, and ductile failure.

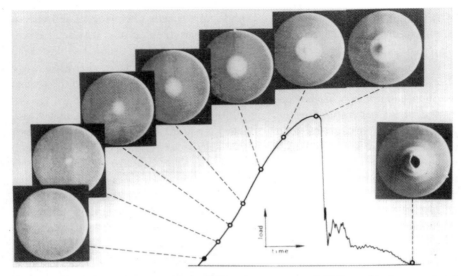

Figure 21.7   A force–time curve from a falling-dart test on an ABS disc, with photographs of the disc at various stages of loading. Stress whitening is first observed below the impact head at a very early stage of loading, marked by (•) on the force-time curve. From [37], reprinted with permission.

temperature as the maximum stress and deflection become larger, but it is still possible for the final fracture to be brittle (see Fig. 21.6), because yield stresses in biaxial tension are high and strains to break are relatively low. Under these conditions, small cracks can accelerate to their terminal velocities (on the order of 500 m·s^{-1}), and energies to fracture are therefore low.

At yet higher temperatures, tough blends, such as ABS, become fully ductile. Yielding and extensive ductile drawing occur both in the central zone, where the polymer is in contact with the striker, and in the surrounding region of high stress. (See Fig. 21.7.) The impact head acts very much like a forming tool, producing a whitened, hemispherical depression in the middle of the disc. A large amount of energy is absorbed before the most highly strained section of the yield zone, at the tip of the striker, fails by tearing. At this point, the impact head penetrates the disc, but tearing is localized, and there is no rapid crack propagation, as at lower temperatures (see Fig. 21.6).

The force-deflection response of a ductile thermoplastic subjected to falling-dart impact loading has been modeled by Nimmer, using a finite-element approach [38]. The deformation behavior of the polymer was represented by a tensile stress–strain curve divided into three stages: linear elastic extension, defined by a Young's modulus E_1; drawing at constant stress to a draw ratio λ; and strain hardening, represented by a linear process with an effective modulus E_3. Using this model and specific values for E_1, λ, and E_3, Nimmer was able to obtain approximate agreement with experimental force-deflection curves.

A complication that was not included in Nimmer's analyzis was the effect of internal heating in the highly strained yield zone around the hemispherical tip of the dart. Rapid deep drawing in this region dissipates enough energy to make the central region of the disc quite hot to the touch, so that the test is far from isothermal when the material responds in a ductile manner. A major factor in this heating is the rapid release of stored elastic energy from the outer part of the disc once the central region has begun to yield and strain soften. Other contributory factors are frictional heating as the polymer stretches in contact with the striker, conduction of heat into the metal impact head, and changes in the rate of energy dissipation with increasing temperature and strain in the yield zone. The development of a complete model for the response of a ductile polymer in the dart impact test will clearly require a major effort, both in obtaining the constitutive relationships necessary to describe the behavior of the polymer over a range of temperatures and patterns of straining and in applying those data in a finite-element program.

Most users of dart impact tests are not interested in obtaining such deep insights into the behavior of the specimen. Their aim is to rank materials in terms of their expected performance in standard applications—automobiles, luggage, packaging, etc.—where good, all-round impact resistance is required, but the consequences of fracture are not so serious that the very conservative criteria of fracture mechanics have to be applied. The test does not provide quantitative information that can be used in engineering calculations, but it does identify possible problems of brittle or low-energy fracture in the absence of a sharp notch of significant dimensions.

In order to exhibit a high dart impact strength, a material must be able to reach high strains in biaxial deformation and then strain harden to prevent excessive thinning near the point of impact, so that the yield zone spreads. Rubber-toughened polypropylenes and nylons, which are able to reach high strains in tensile tests, provide good examples of this type of behavior. All else being equal, toughened nylons will give higher impact energies than toughened polypropylenes, because they have higher yield stresses. However, if both materials undergo extensive deep drawing around the impact head, then the difference in energy absorption might be unimportant in practice. Users will probably be more interested in the transition temperature between semibrittle and ductile behavior.

In all dart impact tests, the level of energy absorption is strongly affected by the degree of anisotropy in the test specimen. Because the area below the striker is in equal biaxial tension, splitting will always occur in the weakest direction. This effect is particularly apparent when the test discs are injection molded. High levels of chain orientation in the surface layers of the molding can result in a marked decrease in impact energy.

Another factor that can cause embrittlement of an otherwise ductile sheet is the presence of a brittle surface layer, where sharp cracks can initiate in bending. Embrittlement of this kind can result from outdoor ageing or from the application of a brittle surface coating, in the form of a paint layer, laminated surface film, or metallic coating. As discussed in Section III, the extent to which these surface modifications affect fracture behavior depends upon the resistance of the subsurface material to a short, sharp, propagating crack. One advantage of the falling-dart impact test, an advantage that it shares with the standard Charpy and Izod impact tests but not with sharply notched fracture mechanics specimens, is that it provides a convenient method for studying surface embrittlement and methods of countering it through the use of additives.

VI. CONCLUSIONS

Standard tests should be regarded as complementary to fracture mechanics as methods for measuring the toughness of polymers. Fracture mechanics tends to concentrate on the deficiencies of materials, characterizing their response to applied stress under the most extreme conditions, with the maximum level of geometric constraint. Standard empirical tests, on the other hand, allow materials to exhibit their ability to absorb energy under less extreme conditions. By subjecting materials to tests of varying severity, from slow uniaxial tension to impact loading of notched bars, it is thus possible to form a picture of both their strengths and weaknesses.

In the area of polymers, one of the most valuable roles of fracture mechanics has been in providing a theoretical framework for the interpretation of standard test data. The test specimens discussed in this chapter can be regarded as practical structures subjected to well-defined loading conditions under which they eventually fail. Engineering techniques, including fracture mechanics, finite-element analyzis, and plasticity theory, can all be used to develop an insight into the failure behavior of these structures. Thus, the effects of specimen thickness on notched Izod impact strength can be explained in terms of a plane-stress–plane-strain transition. Similarly, reductions in the Izod and falling-dart impact strengths of rubber-toughened plastics as a result of ageing can be related directly to a fall in G_{Ic}.

This approach is perhaps the best way to move forward in the development of test methods for plastics. Instead of regarding standard empirical tests as poor substitutes for fracture mechanics measurements, we should regard them as legitimate methods for characterizing the response of materials in the form of structures of defined geometry. Valuable information can be gained if enough effort is devoted to interpreting this response.

This chapter discusses some of the ways in which this extra effort should be directed, and further examples are given in Chapter 25. Instrumentation of impact test equipment, the use of video recorders, and studies of the morphology of fracture surfaces all aid in the interpretation of standard test data. Variations in specimen geometry, especially notch radius and specimen thickness, are also helpful in developing an understanding of the way in which the material behaves. Finally, the use of these test procedures in combination with systematic changes in composition is usually effective in identifying the directions in which improvements in blend formulations should be sought.

Before concluding this chapter, it is important to place the subject of toughness in context. Problems arise when the fracture test data used for materials selection do not relate directly to the specific engineering requirement. It is often assumed that levels of energy absorption in standard empirical tests should be as high as possible and that high values necessarily ensure high fracture resistance in service. However, this assumption is often not true. For example, plain, unblended nylon is notch sensitive, absorbing only modest amounts of energy in the Izod test, whereas polycarbonate has a very high notched-impact strength. However, in applications involving repeated load cycling, nylons are generally preferred, because they show better resistance to fatigue.

Similar problems are likely to arise whenever fracture data for two dissimilar materials are compared. In evaluating the performance of new polymer blends, the emphasis is usually upon achieving the highest possible levels of impact energy absorption, which is regarded as a goal in itself. However, the most common cause of failure in polymeric components is not impact but slow, subcritical crack growth, assisted by ageing, thermal cycling, and, in some cases, absorption of aggressive liquids. In developing new polymer blend formulations, it is essential that research teams should evaluate all aspects of their fracture resistance, rather than concentrating on just one aspect of toughness.

VII. REFERENCES

1. ASTM Standard D638, American Society for Testing and Materials, Philadelphia, issued annually.

2. ASTM Standard D2289, American Society for Testing and Materials, Philadelphia, issued annually.

3. ASTM Standard D1822, American Society for Testing and Materials, Philadelphia, issued annually.

4. I. M. Ward, *Mechanical Properties of Solid Polymers*, 2nd Ed., John Wiley, New York, 1983.

5. N. G. McCrum, C. B. Bucknall, and C. P. Buckley, *Principles of Polymer Engineering*, 2nd Ed., Oxford University Press, 1997, p. 186.

6. B. Crist, "Yield Processes in Glassy Polymers," in *The Physics of Glassy Polymers*, 2nd Ed., R. N. Haward and R. J. Young (eds.), Chapman and Hall, London, 1997, Ch. 4.

7. M. C. Boyce and R. N. Haward, "The post-yield deformation of glassy polymers," in *The Physics of Glassy Polymers*, 2nd Ed., R. N. Haward and R. J.Young (eds.), Chapman and Hall, London, 1997, Ch. 5.

8. C. G'Sell, J. M. Hiver, A. Dahouin, and A. Souahi, *J. Mater. Sci.*, **27**, 5039 (1992).

9. B. O'Connor, "High Strain Deformation and Ultimate Failure of HIPS and ABS Polymers," Ph.D. Thesis, Cranfield University, 1997.

10. B. O'Connor, C. B. Bucknall, and J. Hahnfeld, *Plast. Rubber Compos. Process. Appl.*, **26**, 360 (1997).

11. S. D. Sjoerdsma and J. P. H. Boyens, *Polym. Eng. Sci.*, **34**, 86 (1994).

12. ASTM Standard D790, American Society for Testing and Materials, Philadelphia, issued annually.

13. R. D. Heap and R. H. Norman, *Flexural Testing of Plastics*, The Plastics Institute, London, 1969.

14. J. W. Westwater, *Proc. ASTM*, **49**, 1092 (1949).

15. R. E. Chambers and F. J. McGarry, *ASTM Bull*, May 1959, 18.

16. J. G. Williams, *Stress Analysis of Polymers*, Longman Group, London, 1973.

17. D. G. Ashwell, *J. Royal Aeronaut. Soc.*, **54**, 708 (1950).

18. D. S. Ayre and C. B. Bucknall, *Polymer*, **39**, 4785 (1998).

19. H. S. Kim, B. Cotterell, and Y. W. Mai, *Polym. Eng. Sci.*, **27**, 277 (1987).

20. ASTM Standard D256 and ASTM Standard D790, American Society for Testing and Materials, Philadelphia, issued annually.

21. E. Plati and J. G. Williams, *Polym. Eng. Sci.*, **15**, 470 (1975).

22. J.G. Williams, *Fracture Mechanics of Polymers*, Ellis Horwood, Chichester, United Kingdom, 1983.

23. P. S. Leevers, *Int. J. Fracture*, **73**, 109 (1995).

24. P. S. Leevers, *Polym. Eng. Sci.*, **36**, 2296 (1996).

25. O. Julien, P. Béguelin, L. Monnerie, and H. H. Kausch, "Loading Rate Dependence of the Fracture Behavior of Rubber-Modified Poly(Methyl Methacrylate)," in *Toughened Plastics II*, Advances in Chemistry Series 252, C. K. Riew and A. J. Kinloch (eds.), American Chemical Society, Washington, DC, 1997.

26. J. Jancar and A. T. Dibenedetto, *Polym. Eng. Sci.*, **34**, 1799 (1994).

27. M. Bramuzzo, *Polym. Eng. Sci.*, **29**, 1077 (1989).

28. N. J. Mills, *J. Mater. Sci.*, **11**, 363 (1976).

29. I. Narisawa, M. Ishikawa, and H. Ogawa, *J. Mater. Sci.*, **15**, 2059 (1980).

30. R. Hill, *Mathematical Theory of Plasticity*, Oxford University Press, Oxford, 1950.

31. S. Havriliak, C. A. Cruz, and S. E Slavin, *Polym. Eng. Sci.*, **36**, 2327 (1996).

32. E. Priebe and J. Stabenow, *Kunststoffe*, **64**, 497 (1974).

33. C. B. Bucknall, *Toughened Plastics*, Applied Science Publishers, London, 1977.

34. C. B. Bucknall and D. G. Street, *J. Appl. Polym. Sci.*, **12**, 1311 (1968).

35. A. Carpinteri, *Proceedings of the International Conference on Analytical and Experimental Fracture Mechanics*, Rome, Italy, 1980, Sijthoff and Nordhoff, 1981, p. 785.

36. S. Wu, *Polymer*, **26**, 1855 (1985).

37. T. Casiraghi, G. Castiglioni, and G. Ajroldi, *Plast. Rubber Process. Appl.*, **2**, 353 (1982).

38. R. P. Nimmer, *Polym. Eng. Sci.*, **27**, 263 (1987).

22 Deformation Mechanisms in Rubber-Toughened Polymers

C. B. BUCKNALL

Cranfield University
Bedford, MK43 0AL
United Kingdom

Polymer Blends, Volume 2: Performance. Edited by D. R. Paul and C. B. Bucknall.
ISBN 0-471-35280-2. © 2000 John Wiley & Sons, Inc.

I. INTRODUCTION

The technology of rubber toughening, which involves blending small amounts of rubber with rigid polymers in order to increase their fracture resistance, has been used commercially since the late 1940s and has played a major part in the growth of the polymer industry. Manufacturers of both thermoplastics and thermosets now offer rubber-toughened grades of almost every type of rigid polymer, including glassy thermoplastics, semicrystalline thermoplastics, and thermosetting resins.

The fracture behavior of these rubber-toughened polymers is complex and varies enormously with composition, morphology, and testing conditions. Consequently, progress in understanding the mechanisms of toughening has been relatively slow. Even today, several important aspects of structure–property relationships are not properly understood and remain the subject of controversy. However, novel experimental methods combined with new approaches to modeling have resulted in significant advances over the past decade, and there is now every prospect of placing the subject on a proper quantitative basis. This chapter reviews the current state of knowledge and examines some of the remaining difficulties.

II. BASIC PRINCIPLES OF TOUGHENING

In a materials science context, toughness may be defined as the ability to resist fracture by absorbing energy. It is usually expressed in terms of the work done in forming a unit area of fracture surface. For rigid polymers, published values range from 50 $J \cdot m^{-2}$ in highly cross-linked epoxy resins (see Chapter 26) to 80 $kJ \cdot m^{-2}$ for toughened nylon blends (see Chapter 25). To put these figures into perspective, it is instructive to calculate the approximate total thicknesses of material involved in energy absorption. As yield and fracture stresses for rigid polymers are generally between 25 and 100 MPa, calculations will be based, for purposes of illustration, on a yield stress of 50 MPa. With this figure, it follows that the brittle epoxy resins mentioned earlier fail at a crack-opening displacement on the order of 1 μm, while for very tough rubber-modified nylons, the figure may be over 1 mm (i.e., 1000 μm). These displacements may be the result of large local strains in a small volume of material near the crack tip, as in craze formation, or an accumulation of smaller strains over a much larger volume.

To estimate the total thickness of material contributing to the crack-opening displacement, it is necessary to assign an average extension ratio λ ($= L/L_o$) to the deformation bands or zones involved. Maximum extension ratios are determined by the concentration of chain entanglements or, in the case of thermosets, the cross-link density, both of which can be obtained from the "plateau modulus" of the polymer above its T_g [1, 2], provided that the material is noncrystalline. The best documented evidence for this relationship comes from studies of crazes in glassy thermoplastics, which typically have fracture surface energies G_{Ic} between 300 and 1000 $J \cdot m^{-2}$ and maximum extension ratios of 2 to 4 [3–5]. These figures indicate that L_o, the initial thickness of material undergoing large strains, is on the order of 3 μm on each side of the crack. Locally, within the craze or deformation zone, energy absorption per unit volume of polymer is high (on the order of 100 MJ·m^{-3}), but the amount of material involved in that absorption is very limited, and fracture is therefore essentially brittle. In a fracture mechanics test, the force-deflection curve is linear. (See Chapter 20.)

The need to involve larger volumes of material in energy absorption is the basic challenge to anyone wishing to toughen rigid polymers. Polystyrene (PS) and poly(methyl methacrylate) (PMMA) are, in principle, capable of absorbing large amounts of energy, but only on a local scale. The amount of polymeric material involved in energy absorption under standard test conditions, or in everyday use, is too small to give satisfactory fracture resistance. The problem is most acute when the specimen or structure contains a crack or sharp notch.

A basic requirement for toughness is that the most highly stressed regions of the material, on reaching a critical elastic strain, should not only strain soften, but also eventually strain harden. In other words, when the true stress is plotted against the extension ratio, the slope should first decrease and then increase, as shown in Fig. 22.1a. If the material at the tip of the notch does not soften, yield, and draw, little energy will be dissipated: Very high resistance to yielding is characteristic of strong, covalently, or ionically bonded materials, like diamond and glass, which come close to being ideally brittle. If, on the other hand, the material yields and draws, but does not eventually strain harden, the stress concentration at the notch tip will cause fracture at a relatively early stage of deformation. Both types of behavior are illustrated in Fig. 22.1.

Thermoplastics, and especially semicrystalline thermoplastics, are well able to meet these requirements for high toughness, provided that they have a reasonably high molecular weight. The weak interactions between polymer chains offer limited resistance to segmental motion, while orientation of the chain segments as the material yields provides an effective mechanism of strain hardening. The difficulty is in realizing this potential, especially under the constrained conditions at a crack tip.

Both strain softening and strain hardening contribute to crazing. Dilatational stresses, which are present both at the craze tip and at the boundary between craze fibrils and undeformed polymer, promote softening of the polymer, allowing it to reach high draw ratios ($\lambda = 4.2$ for polystyrene [3–5]). Molecular orientation in the network of entangled polymer chains then generates high resistance to further extension. Instead of becoming even thinner and then breaking, the strain-hardened craze fibrils draw fresh polymer from the craze walls, thereby increasing the volume of material involved in energy absorption. This principle of stabilizing the region of highest stress and strain, so that the yield zone expands outwards, is fundamental to toughening in any material. Unfortunately, in the case of crazing, the extent of this expansion is limited by failure at the craze–bulk interface, probably as a result of adiabatic heating. (See Chapter 21.) Much higher levels of toughness are achieved when the principal mechanism of energy absorption is shear yielding.

Shear yielding and subsequent drawing involve major changes in the shape of each element in the yield zone, with little change in volume: In a tensile test, material becomes longer and thinner as it is drawn into the neck. However, at a crack tip,

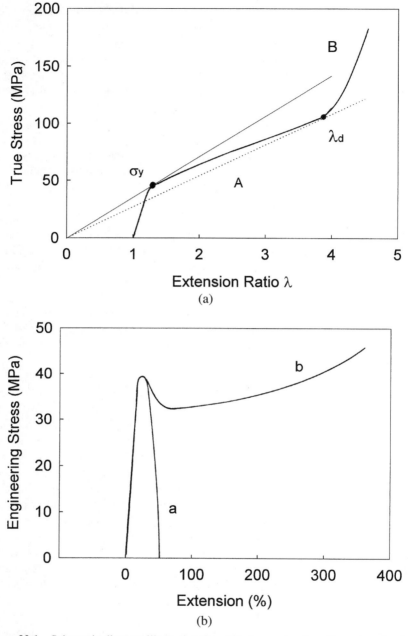

Figure 22.1   Schematic diagram illustrating Considère's construction. True stress is plotted against extension ratio, and tangents are drawn through the origin. The first tangent defines the strain at which the tensile load reaches a maximum (the engineering yield stress σ_y). The second tangent marks the point at which strain hardening stabilizes the neck (the natural draw ratio λ_d). Failure at point A, between the two tangents, gives a low extension to break (curve a). Failure beyond the second tangent gives curve b.

significant changes in shape are resisted by the surrounding, less highly stressed material, especially when the specimen is thick. Consequently, the stress field generated in the middle of the specimen has a large degree of triaxiality; that is, the components of stress in the x, y, and z directions are all of similar magnitude. This means that they have a much stronger tendency to increase the volume of each element, a process absorbs little energy and ultimately leads to crack growth, than to promote shear yielding, which dissipates large amounts of energy.

In recent years, it has become clear that optimization of rubber toughening requires the rubber particles to undergo both strain softening and strain hardening in turn. First, the formation of holes in the rubber phase, called rubber particle cavitation, serves to weaken the particles' resistance to deformation, initiating yielding in the matrix at reduced stresses and allowing it to cold draw despite the constraints imposed at a crack tip. Then, at a later stage of deformation, stretching of rubber fibrils within the cavitated particles introduces a significant degree of strain hardening. This combination of responses to applied stresses makes rubber particle cavitation an effective response to the triaxial tensile stresses generated at a crack tip, because it allows the surrounding matrix to deform by all available mechanisms. The matrix may respond by shear yielding, multiple crazing, or both, depending upon its properties, and the strained rubber particles then stabilize the most highly strained regions of the polymer. Current understanding of these three key processes—cavitation, shear yielding, and multiple crazing—is reviewed next.

III. RUBBER PARTICLE CAVITATION

The formation of holes in the rubber phase is one of the most important ways in which toughened plastics can respond to tensile stress. The phenomenon of rubber particle cavitation was first noted in the 1970s [6], but its significance was not generally recognized until some 10 years later, for several reasons. Firstly, most researchers could see little distinction between cohesive failure, which means that voids are formed within the rubber phase, and adhesive failure, which means that voids are formed at the particle–matrix boundary. Poor adhesion between particles and matrix had long been recognized as resulting in low fracture resistance, and the load-bearing capacity of the cavitated rubber particles was considered to be negligible. Secondly, particle cavitation, where it occurred, was thought to be a secondary process, caused by extensive yielding or crazing in the surrounding matrix, rather than a primary response to stress. Thirdly, many electron microscopists regarded evidence of particle cavitation in microtomed thin sections as inconclusive, on the grounds that it could result from artifacts of the sectioning procedure—a problem that still affects the interpretation of electron micrographs today.

Opinions gradually changed as the experimental evidence accumulated. Reports of cavitation became so numerous that the phenomenon had to be taken seriously, and several commentators pointed out that the formation of holes in the rubber phase freed the material from the constraints imposed upon shear yielding at crack tips in thick sections. The process was increasingly recognized as important, but some reservations remained: Cavitation appeared to be a secondary deformation mechanism, initiated by yielding in the matrix, and the effects of cohesive failure in the rubber particles were believed by many to be indistinguishable from those of adhesive failure, i.e., particle–matrix debonding. Before we discuss these issues, which are central to any discussion of rubber particle cavitation, it is necessary first to review the experimental evidence.

A. Microscopy

The clearest evidence for rubber particle cavitation has come from studies of toughened epoxy resins, which typically contain simple, spherical particles of carboxyl-terminated butadiene-acrylonitrile (CTBN) rubber. (See Chapter 26.) When the fracture surfaces of these resins are examined in the scanning electron microscope, epoxy and CTBN phases are easily distinguished, and obvious cavities can be seen in the rubber particles, as illustrated in Fig. 22.2 [7]. Rubber-toughened thermoplastics generally have more complex fracture surfaces, which offer less conclusive evidence of cavitation.

When it is compared with fracture surfaces formed at low temperatures, Fig. 22.2a also provides evidence that the radii of the CTBN particles increase as the voids expand, a process that can occur only as a result of yielding in the resin matrix. In creating space for the growing void, the surrounding rubber is stretched biaxially and is therefore subjected to substantial hoop stresses, especially in the region close to the void. These stresses are released when the crack cleaves through the particle, so that the stretched rubber shell retracts, leaving a craterlike depression in the fracture surface, as shown in Fig. 22.2b. This observation marks a difference between cavitated and debonded rubber particles: The presence of biaxial tensile stresses in the cavitated rubber phase provides a degree of strain hardening to the system. The key question is whether this strain hardening is sufficient to affect fracture behavior.

In early studies of cavitation, the most convincing evidence for void formation in crack-tip yield zones came from the work of Pearson and Yee, who used optical microscopy to examine relatively thick slices of toughened epoxy resin cut on planes lying normal to fracture surfaces [8]. (See also Chapter 26.) In plain light, cavitated particles show up clearly as dark circles, while in polarized light both the stretched rubber shells surrounding each void and the shear bands connecting them shine brightly. To avoid cutting damage, Pearson and Yee treated their specimens like rock samples, polishing them down to the required thickness using the methods used by geologists. Shear bands of the type observed in this study have sometimes been called "croids," but are better described as "dilatation bands"—the term used by metallurgists to describe shear bands full of holes. The importance of these bands will be discussed in more detail later.

A different type of cavitation, which requires different methods of observation, is seen in high-impact polystyrene (HIPS). As illustrated in Fig. 22.3, the rubber particles in these materials are much more complex than the simple spheres formed in toughened epoxy resins. In HIPS, about 80% of each "salami" rubber particle consists of hard polystyrene subinclusions, which are embedded in a continuous matrix

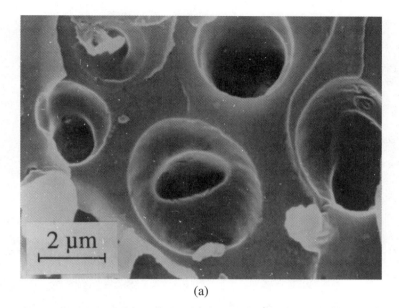

(a)

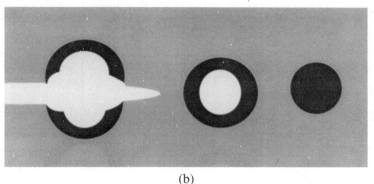

(b)

Figure 22.2   Scanning electron micrograph of toughened epoxy resin containing CTBN rubber, showing cavitation of the particles on the fracture surface, with a sketch explaining the processes involved. Micrograph reproduced from [7], with permission.

of cross-linked rubber. The rubber forms membranes about 30 nm thick, which are cross-linked and grafted to the polystyrene phase during manufacture, so that the resulting interfaces are very strong. Cavitation in the membranes results in a crazelike morphology, with rubber fibrils forming bridges between neighboring subinclusions and between subinclusions and the matrix, as in Fig. 22.3. Since true stresses of over 300 MPa can be achieved even in macroscopic samples of highly stretched rubber, these fibrillated membranes are well able to carry large nominal stresses on the order of 30 MPa or more.

When a fibrillated "salami" particle breaks, it leaves little evidence of cavitation on the fracture surface. It is therefore necessary to use other methods of observation.

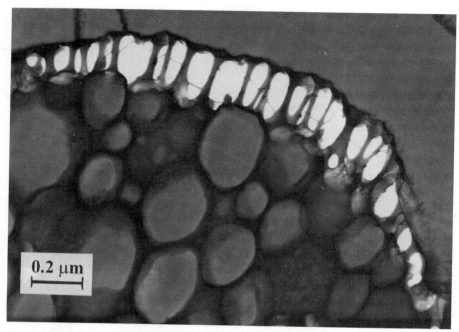

Figure 22.3 Section of HIPS showing fibrillation of polybutadiene membrane between the matrix and adjacent subinclusions. The material was strained and treated with OsO$_4$ before sectioning. *Micrograph courtesy of Peter Logan.*

One very effective method is to cut thin sections and stretch them on the stage of an electron microscope, as shown in Fig. 22.4. This technique has been employed by a number of authors. However, it is open to the criticism that processes taking place in thin sections do not necessarily represent the behavior of the bulk polymer. One alternative is to deform the HIPS sample first, treat it with osmium tetraoxide to harden and stain the rubber, and then make the sections, which was the method used to prepare the section in Fig. 22.3. There are always some concerns about introducing damage during sectioning, but these concerns can be countered by making comparisons to sections from fresh specimens that have not been deformed.

In HIPS, and in some other toughened plastics, cavitation of the rubber particles is usually accompanied by crazing of the matrix, which is discussed in Section V. As illustrated in Fig. 22.4, when yielding is well developed it can be difficult to distinguish between crazes and fibrillated rubber, which are very similar in appearance.

B. Light Scattering

Light scattering in multiphase polymers is discussed in detail in Chapter 28. In the case of toughened plastics, the key parameters are the refractive index mismatch between matrix and particles and the sizes of the particles. Through careful choice of compositions and particle structure, it is possible to match refractive indices and

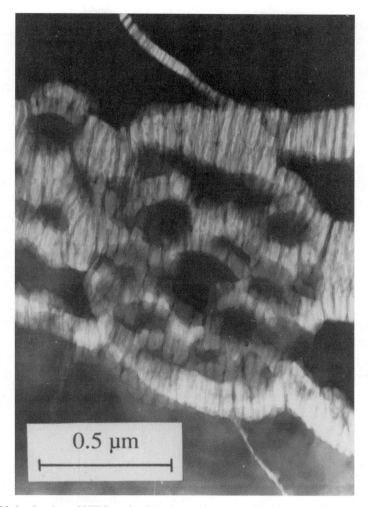

0.5 µm

Figure 22.4   Section of HIPS strained *in situ* on the stage of an electron microscope. *Micrograph courtesy of G. H. Michler and J. Laatsch.*

thus produce rubber-toughened plastics that are transparent over a certain range of temperatures. (See Chapter 24.) The formation in the rubber phase of extremely small voids (diameters on the order of 10 nm) will cause a reduction in its refractive index, which will fall further as the voids expand. Subsequent increases in particle diameter will further increase the intensity of light scattering, leading eventually to intense whitening that renders transparent or translucent materials completely opaque.

Some of the earliest evidence for cavitation of rubber particles came from light-scattering experiments on toughened PVC [9]. More recently, reductions in the intensity of transmitted light have been used by several groups to detect cavitation in toughened plastics, especially toughened nylons and PMMA. However, except in the very early stages, it is difficult to obtain specific information about the cavitation process in this way, because multiple scattering rapidly reduces the mean free path of light, even in the most transparent of starting materials.

Recent work on transparent RTPMMA by Schirrer et al. has shown that the sizes, shapes, and concentrations of cavitated particles can be determined using light transmission in combination with coherent back scattering [10, 11]. This technique takes advantage of coherent interference, which enhances the intensity of light scattered at angles of about 180° to the incident beam. The average void contents of particles increase to about 20% with increasing strain in the specimen [11]. Yielding in the matrix allows the rubber fibrils to stretch and the rubber particles to expand, until strain hardening finally stabilizes the void content. This response is typical of three-stage particles with a rigid polymeric core, a rubbery inner shell, and a thin outer shell to provide adhesion to the matrix. Each rubber fibril is anchored at one end in the rigid core and at the other end in the matrix, in a similar manner to that illustrated in Figs. 22.3 and 22.4.

C. Thermal Contraction and Expansion

Recent studies have shown that measurements of thermal contraction and expansion in toughened plastics can also provide valuable information about rubber particle cavitation. The basis of the technique is the large difference in expansion coefficients between rubbers and rigid polymers. The volumetric expansion coefficient β_{tp} of a rubber-toughened plastic containing a volume fraction ϕ of rubber particles is given by [12]

$$\beta_{tp} = \beta_m + \phi(\beta_p - \beta_m)\left(\frac{4G_m K_p + 3K_m K_p}{4G_m K_m(1 - \phi) + K_p(4G_m \phi + 3K_m)}\right), \quad (22.1)$$

where G_m and K_m are the shear and bulk moduli of the matrix and K_p is the bulk modulus of the rubber particle. Typical values of β, the volumetric coefficient of thermal expansion, are 1.4×10^{-4} K^{-1} for PS or PSAN (poly(styrene-co-acrylonitrile)) and 7.6×10^{-4} K^{-1} for polybutadiene, so that the measured value of β for a typical ABS polymer, consisting of 20 vol% of polybutadiene rubber particles in a PSAN matrix, is 50% higher than that of the parent PSAN.

If the rubber particle debonds from the matrix or cavitates in such a way that the stresses at the particle–matrix boundary disappear, K_p effectively falls to zero and β_{tp} becomes equal to β_m. Thus, the relative volume V_{rel} of an ABS specimen, defined as the ratio of its current volume $V(T)$ to its reference volume $V(T_g)$ at 100°C, depends not only on the temperature T but also on the nature and extent of rubber particle cavitation. As illustrated in Fig. 22.5, the lower bound on V_{rel} is represented by the contraction curve for void-free ABS, while the upper bound is the contraction curve for PSAN, which coincides with the contraction curve of a fully cavitated ABS with no remaining stress in the rubber phase. When cooled from room temperature, some ABS polymers exhibit a transition between the two curves, as illustrated in Fig. 22.6 [12]. This transition is a clear indication that voids are

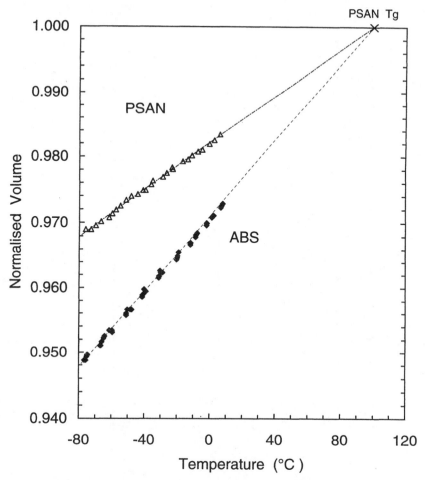

Figure 22.5 Thermal contraction curves for PSAN and ABS, normalized with respect to their volumes at 100°C. For further details, see [12].

forming in the rubber phase as a result of thermal contraction stresses. On reheating, the volume-vs.-temperature curve is approximately linear, with no sign of the anomalous behavior observed during cooling, thus demonstrating that void formation is not thermodynamically reversible.

In the aforementioned experiments, the ABS specimens were compression molded, so that both the rubber particles and the matrix were fully relaxed at the T_g of PSAN, which is about 100°C. The interface between the phases should therefore be stress free at this temperature. However, on cooling below the of the T_g matrix, differential thermal contraction generates tensile stresses in the rubber particle; these stresses act on the particle–matrix boundary and pull the surrounding shell of rigid matrix inward. One result of this interaction is that the volume of the ABS becomes

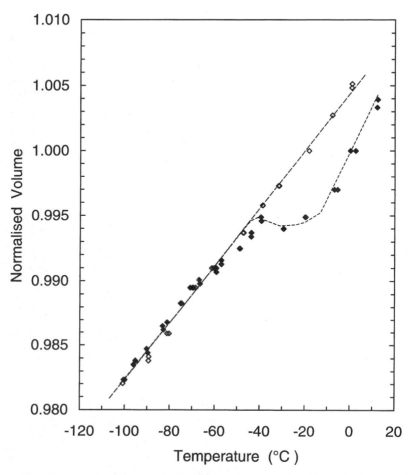

Figure 22.6   Volume changes in an ABS specimen due to cooling (◆) and subsequent heating (◇), showing evidence of cavitation during the cooling stage. Data normalized to the initial volume at 0°C [12].

progressively reduced in comparison with PSAN, hence the increase in the thermal expansion coefficient. Another consequence is that the rubber phase has a lower density and shorter relaxation times than it has in the unstressed state. This change in relaxation behavior is the basis of another technique for detecting rubber particle cavitation, described next.

D. Dynamic Mechanical Thermal Analysis

The dynamic mechanical thermal analysis (DMTA) method is routinely used to determine the glass transition temperature T_g (or, more correctly, the alpha transition temperature T_α) in the rubber phase of a toughened plastic, which produces a promi-

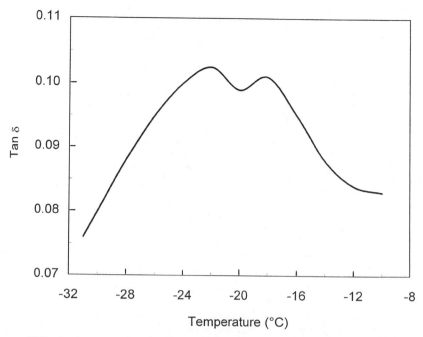

Figure 22.7 Dynamic loss curve for rubber-toughened PMMA, showing splitting of the low-temperature tan δ peak, indicating cavitation of a proportion of the rubber particles. *Courtesy of Rosi Rizzieri.*

nent secondary peak in the dynamic loss curve. However, in some cases, the α transition of a single rubber modifier can give rise to two loss peaks some 5–10 K apart, as illustrated in Fig. 22.7. This peak splitting is a consequence of cavitation. The lower peak is due to intact rubber particles, which are under large triaxial tensile stresses as a result of differential thermal contraction; the resulting decrease in density reduces relaxation times and shifts T_α to lower temperatures. The upper peak, on the other hand, is due to cavitated particles, in which the rubber has returned, at least approximately, to its natural unstressed density. The associated decrease in free volume increases relaxation times, so that the loss peak shifts back to a higher temperature. When the voids are closed up by applying an axial compressive stress to the DMTA specimen, only one peak is observed.

Shifts in the low-temperature loss peak of ABS were first reported by Morbitzer et al., who attributed them to debonding rather than internal cavitation of the rubber particles [14, 15]. They also observed splitting of the loss peak and showed that the separation between the maxima of the split peak increased as the concentration of rubber particles decreased, because the matrix then imposed greater constraints on thermal contraction of the particles. Liquid additives further assisted cavitation, presumably by lowering the surface energy of the voids [15].

Recent studies have shown that significant shifts in T_α can be obtained by applying large axial tensile or compressive stresses to DMTA specimens. Tension shifts

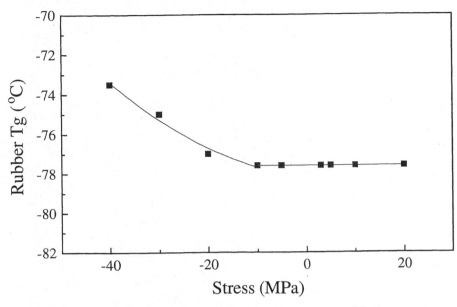

Figure 22.8   Effects of superimposed axial stress on the temperature of the low-temperature loss peak in HIPS. Rubber particle cavitation makes the T_g of the polybutadiene phase independent of stress when tested in tension. From [16]; reproduced with permission.

the peak to lower temperatures, while compression shifts it upwards. This principle has been applied by Lin et al. to detect cavitation in HIPS specimens that exhibit only one secondary loss peak [16]. In some toughened plastics, the T_α of the rubber is found to be independent of the magnitude of the tensile stress, indicating that the particles either contain voids or have become debonded from the matrix. Under compression, the voids close up, and the peak shifts linearly with applied stress, as illustrated in Fig. 22.8.

E.   Stresses in Cavitated Particles

It is sometimes suggested that rubber particles completely lose their capacity to sustain a stress once they have cavitated, but this is true in only a limited number of cases. The fibrils shown in Figs. 22.3 and 22.4 are clearly load-bearing elements of the structure, which make a major contribution to strain hardening. The magnitude of stresses reached in fibrillated rubber membranes has been demonstrated by Starke et al., using sections cut from a PSAN blend containing core–shell rubber particles that were stretched on the stage of an electron microscope [17]. The particles in question consisted of a spherical PMMA core surrounded by an inner shell of rubber and a thin outer shell of PMMA. (See Chapter 24.) In this experiment, the inner shell of rubber not only fibrillated, but also extended those fibrils by pulling PMMA from the core, leaving a lens-shaped remnant acting as an anchor for the fibrils. This evidence indicates that the stresses in the fibrils reached 30 MPa or more. It is clear from the micrographs that every element in each rubber particle is subjected to very high stresses and strains.

Stress distributions within a simple homogeneous rubber particle are much less uniform after cavitation, when each void is surrounded by a continuous, thick, spherical shell of rubber. This shell can be treated as being made up of a series of concentric thin layers, all in equibiaxial tension. After the formation of a hole and subsequent modest expansion of the particle, the stretch ratio λ ($= L/L_o$) near the particle–matrix boundary might, for example, be 1.2, while at the surface of the hole, λ could in principle reach infinity, because the void is created from a continuum. In practice, of course, the chains of cross-linked rubber will break on reaching a critical strain λ_f, which is probably on the order of 4. Thus, the particle is supporting quite high stresses within a very small volume of rubber close to the void, but much lower stresses elsewhere in the particle.

This nonuniform distribution of stresses within the rubber phase contrasts strongly with that found in a composite particle containing rigid, polymeric subinclusions. The fibrillation mechanism illustrated in Figs. 22.3 and 22.4 allows every element of the rubber phase to reach approximately the same level of stress at the same time, thereby ensuring effective strain hardening. When dilatational stresses are applied to a homogeneous rubber particle containing a single void, there is much less resistance to expansion. Instead, the most highly stressed elements of the rubber phase close to the void fail through progressive tearing, a process that has been modeled by Fond et al. [18].

F. Modeling

The experiments described in Section III.C show that the elastic stresses generated in a toughened plastic during cooling are sometimes sufficient to initiate voids in the rubber phase. In order to develop a quantitative theory of toughening, it is necessary to understand the criteria governing this process. A simplified energy balance approach to this problem, based on thermodynamic principles, has been developed by Lazzeri and Bucknall [19].

The basic principle of the model is that the expansion of a void within a rubber particle is driven by the release of energy stored both in the particle itself and in the rigid material immediately surrounding it. For an isotropic material, the energy stored in any volume element can be divided into two parts: a shear component, which changes the shape of the element but not its volume, and a dilatational component, which changes its volume but not its shape. In analyzing the energy available to the void, the energy balance model considers only the dilatational component of strain energy.

The volumetric strain energy U_V in a rubber element of volume dV is given by

$$U_V = \frac{\sigma_m \Delta_V}{2} dV = \frac{K_R \Delta_V^2}{2} dV, \tag{22.2}$$

where K_R is the bulk modulus of the rubber phase. As shown in Eq. 22.2, the volume strain Δ_V $(= \Delta V / V)$ is proportional to the mean stress σ_m, which is the average of the three principal stresses acting in the 1, 2, and 3 directions.

When a spherical void of radius r is formed in a spherical particle of radius R, it occupies a volume fraction r^3 / R^3 relative to the particle. If the radius of the rubber particle is first increased from R_o to R and then held constant during cavitation, the final volume strain in the rubber phase is $(\Delta_V - r^3 / R^3)$, and the volumetric strain energy density of the strained particle falls accordingly. In the simple model, as first introduced, the energy stored locally within the particle itself is the sole driving force for cavitation. Later papers built upon the basic physical principle by including both the energy released from the matrix and the volumetric strain energy generated by differential thermal contraction, as discussed in Section III.C [20, 21].

The energy released through relaxation of volume strain in the rubber is absorbed in three ways: (a) the creation of a new void surface of area $4\pi r^2$, (b) biaxial stretching of the rubber surrounding the void, and (c) tearing of spherical shell elements close to the void that have exceeded the critical extension ratio λ_f. The stretching energy required is $2\pi r^3 G_R F(\lambda_f)$, where the function $F(\lambda_f)$ arises from the integration of energy terms over all extension ratios up to a limit of λ_f. The value generally lies between 0.7 and 1.3 and for practical purposes may be taken as 1.0.

This thermodynamic, energy balance approach follows the principles applied by Gent and Wang to void growth in bulk rubbers [22]. One very significant difference between the two models is that Gent and Wang based their analysis on a large block of rubber containing voids about 1 μm in diameter. This dimension is clearly unrealistic in the case of rubber-toughened plastics, wherein the rubber particles themselves often have diameters well below 1 μm. The Lazzeri–Bucknall model therefore assumes that the rubber phase contains no defects larger than those associated with its free volume and that the voids form within an elastomeric continuum.

For the simplified case in which a rubber particle of radius R is held at a fixed volume strain Δ_V and forms a single void of radius r, the model gives the total energy $U(r)$ as:

$$U(r) = \frac{2\pi}{3} R^3 K_R \left(\Delta_V - \frac{r^3}{R^3} \right)^2 + 4\pi r^2 \Gamma + 2\pi r^3 G_R F(\lambda_f). \qquad (22.3)$$

Equation 22.3 has been used to produce the curves shown in Fig. 22.9, in which the normalized energy of the particle, U/U_o, is plotted against void size for a range of different particle sizes, taking typical values for the other parameters of the equation. Assuming that cavitation occurs whenever $U < U_o$ (i.e., $U(r) < U(0)$), Eq. 22.3 predicts that voids will form at a critical particle size that is a function of the initial applied volume strain $\Delta_V(0)$; the shear modulus of the rubber phase, G_R; and the surface energy of the rubber, Γ. Conversely, for each particle size, there is a critical volume strain. Figure 22.10 shows the relationship predicted by Eq. 22.3 between critical volume strain and rubber particle size, for a range of different shear moduli, G_R. Although the critical volume strain increases with decreasing particle size,

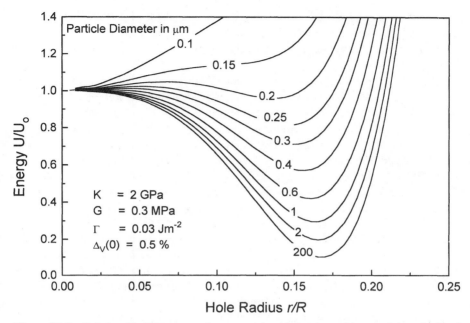

Figure 22.9   Relationship between total energy and void size, in a rubber particle, calculated using Eq. 22.3, for a range of particle sizes. From [38]; reproduced with permission.

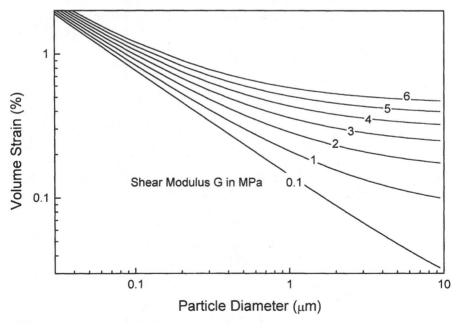

Figure 22.10   Relationship between particle size and critical volume strain at cavitation, calculated using Eq. 22.3 for a range of shear moduli. From [38]; reproduced with permission.

it is clear that there is no theoretical lower limit on the size of particle that can cavitate. Of course, there is a practical limit, because if the particles are extremely small, the sample will fail by yield or fracture before it can reach the very high volume strain required for cavitation.

The inclusion of a term in G_R is important, because it explains why cross-linking of the rubber particles—for example, through exposure to sunlight—has such a dramatic effect on toughness and why rubbers of widely different moduli have such different effects on the temperature of the ductile–brittle transition in toughened polyamides. (See Chapter 25.) The same term in G_R also accounts for the brittle–ductile transition observed in many toughened plastics near the T_g of the rubber phase. A simplified version of Eq. 22.3, suggested by Dompas et al., also accounts for the increasing difficulty of generating voids in rubber particles as their diameters are reduced [23–26]. However, it is of limited validity, because it omits the modulus term and thus ignores effects related to the stretching behavior of the rubber phase.

As noted previously, several simplifying assumptions were used in deriving Eq. 22.3, and in order to develop a complete quantitative theory of cavitation, it is necessary to take several other energy terms into account. The most obvious of these terms derives from elastic interactions with the surrounding matrix as the void expands and the particle becomes more compliant; some of the volumetric strain energy stored in the matrix is released to the particle during this process. Calculations show that these interactions greatly increase the energy available to the void [21]. Another substantial contribution comes from the volumetric strain energy generated on cooling a toughened thermoplastic from the glass transition temperature of the matrix. Thus cavitation is the result of both mechanically applied stresses and internal stresses due to differential thermal contraction of the two phases. When specimens have been cooled rapidly from the melt, as in injection molding, it is also necessary to take account of the additional internal stresses resulting from differences in rates of cooling between skin and core.

The aforementioned energy balance model correctly predicts observed trends in the deformation and fracture behavior of toughened plastics with variations in structure and test conditions, and in this respect, it represents a major advance in understanding. However, in both thermal contraction studies and flexural yield tests, the particles have proved to be much more resistant to cavitation than predicted by the (updated) model [12, 21]. There are at least two possible explanations for the discrepancy. One relates to the assumption that cavitation proceeds through the initiation and expansion of a single void in each rubber particle. This assumption makes calculations simpler, but is probably not valid in the majority of cases. There is some evidence, notably from electron microscopy, that cavitation is initiated at many separate sites in multilayer particles, which have rigid cores enclosed in rubber shells. When the void content of each particle is divided in this way into a number of smaller cavities, the effects of surface forces become increasingly important, causing the particle to resist expansion much more strongly than is predicted by the simple energy balance model involving a single void. This discrepancy can be demonstrated by replacing the terms (r^3/R^3), $r^2\Gamma$, and r^3G_R in Eq. 22.3 with (nr^3/R^3), $nr^2\Gamma$, and nr^3G_R, respectively, and plotting the equation for $n = 1$ and, say, $n = 10$.

The second explanation is a more fundamental one, namely that the energy balance model defines necessary, but not sufficient, conditions for cavitation. The basic problem is that the model is based entirely upon thermodynamic considerations and makes no allowance for kinetic effects. In other words, it defines minimum values for the volume strain required to cause cavitation, but does not consider the rate at which voids might form under these conditions. The model shows that there is a small energy barrier to be overcome, as with many nucleation phenomena (e.g., crystallization), before the newly formed feature (in this case, a void) becomes large enough to expand freely. The existence of this energy barrier might account for the high resistance to cavitation exhibited by rubber-toughened polymers.

At present, there is no conclusive evidence concerning the relative importance of these two explanations. However, the thermal contraction and expansion data presented in Fig. 22.6 support the suggestion that kinetic effects control cavitation, giving rise to a supercooling effect. Expansion of the rubber phase on heating from $-40°C$ causes the voids to become smaller and would be expected to result in their closure at $0°C$ if the volume changes were determined purely by thermodynamic factors. The observation that ABS does not follow the same volume–temperature curve on heating as on cooling suggests that there is a real energy barrier to be overcome in both opening and closing the voids.

IV. SHEAR YIELDING

A. Factors Affecting Yielding

Shear yielding is the process by which most ductile materials extend to high strains in standard tests. Atoms or chain segments slip past each other in response to shear stresses, with the result that small elements of material in the yield zone change shape while remaining close to constant volume. These plastic deformation processes generate heat, which can produce significant rises in temperature, especially in polymers. Minor changes in density usually accompany shear yielding, because molecular packing is disturbed, but there is no loss of cohesion, as there is in crazing.

The most widely used criterion for shear yielding was proposed by von Mises, who stated that yielding occurs on reaching a critical value of effective stress σ_e, defined as

$$\sigma_e \equiv \left(\frac{(\sigma_1 - \sigma_2)^2 + (\sigma_2 - \sigma_3)^2 + (\sigma_3 - \sigma_1)^2}{2} \right)^{1/2} \geq \sigma_o, \qquad (22.4)$$

where σ_1, σ_2, and σ_3 are principal stresses—that is, stresses acting normal to the faces of a small cube of material, in the 1, 2 and 3 directions. If σ_1 is taken as the tensile yield stress σ_{TY} and $\sigma_2 = \sigma_3 = 0$, it can be seen that the critical effective stress σ_o in this inequality is constant and equal to σ_{TY}.

Equation 22.4 is applicable to many metals, but requires some modification to fit the behavior of polymers, for which the critical effective stress is not a constant. To a first approximation, it varies linearly with the mean stress σ_m, sometimes called negative pressure. The von Mises critical strain energy density criterion can therefore be written as follows:

$$\sigma_e > \sigma_o - \mu\sigma_m \equiv \sigma_o - \frac{\mu(\sigma_1 + \sigma_2 + \sigma_3)}{3}. \qquad (22.5)$$

Another way in which the yield behavior of polymers differs from that of metals is in the extent to which it is affected by the strain rate $\dot{\varepsilon}$ and the temperature T. In both classes of materials, shear yielding is an activated rate process, which follows the Eyring equation

$$\sigma_{TY} = \frac{2}{\gamma V}\left[\Delta H + 2.303 RT \log\left(\frac{\dot{\varepsilon}}{\dot{\varepsilon}_o}\right)\right], \qquad (22.6)$$

where ΔH is the activation energy; γ is a stress concentration factor, which relates the local stress acting on the flow element to the nominal applied stress; V is the activation volume; and R is the gas constant.

Equations 22.4–22.6 provide a useful basis for discussing the yield behavior of rigid polymers, but it is important to note that other factors, notably strain softening and strain hardening, also influence the development of shear deformation. One result of strain softening is that yielding tends to become localized into planar zones of high shear strain, known as "shear bands." As the test proceeds, elastically strained material around the edges of each quasiplanar band becomes strain softened, thereby allowing the band to extend in area. Shear strains of 1.0 or more are commonly observed in these bands; as in crazes, hardening due to chain orientation eventually imposes a limit on the strain achieved. The intensity of strain localization increases with the degree of strain softening exhibited by the polymer.

One inevitable effect of adding soft rubber particles to a rigid polymer is that its shear yield stress (σ_o in Eq. 22.5) is lowered. At room temperature, typical unfilled rubbers have shear moduli on the order of 1 MPa and therefore set up significant stress concentrations in the matrix, which reach a maximum at the particle–matrix boundary. In simple cases—for example, an isolated particle in an elastic matrix—stress distributions can be calculated using analytical expressions, but most studies now rely on finite-element analysis (see Chapter 19). Both methods show that stress concentration factors are relatively insensitive to G_p, the shear modulus of the particle, provided that it remains less that one tenth of the shear modulus of the matrix, G_m. The shear moduli of rigid polymers are typically on the order of 1 GPa, so that the ratio G_p/G_m is unlikely to exceed 0.1 unless the rubber phase is close to or below its glass transition temperature.

This transition from rubbery to glassy behavior in the modifier particles is of great practical importance, because it defines the lowest temperature at which the particles are able to reduce the yield stress enough to produce significant toughening. In impact tests, the first clear improvements in fracture resistance are usually seen some 10 K above the true T_g of the rubber phase, because it is necessary for the rubber to relax and for subsequent yielding to take place in the matrix, on time scales comparable with those over which load is applied to the specimen. However, even when the modifier is below its T_g, the notched Izod impact energies of blends are often a little higher than those of the neat matrix (see Chapter 25), which suggests that the (now rigid) dispersed particles are sufficiently ductile to reinforce the brittle matrix, like thermoplastic particles in epoxy resin (see Chapter 26).

When G_p/G_m is small, the high stress concentrations generated in the matrix due to the presence of rubber particles can cause dramatic increases in the rates of plastic deformation, following Eq. 22.6. When shear yielding takes place in the absence of cavitated particles, the effects on yield behavior can be modeled simply by increasing the stress concentration factor γ in this equation. In creep experiments on PMMA, polypropylene (PP), and PA-6.6, the blends all follow the same creep curves as the neat matrix, but at lower applied stresses σ_{appl} [26, 27]. Eyring plots of the logarithm of creep rate coefficient against stress are linear and can be superimposed by replacing σ_{appl} with $\gamma\sigma_{\text{appl}}$. To a first approximation, the stress concentration factors obtained from these creep experiments can be fitted to the Ishai–Cohen equation, which is based on the assumption that cracks and deformation bands propagate on planes of least resistance, running diametrically through the centers of soft particles lying in their path. The Ishai–Cohen model considers a unit cube with a spherical particle of radius R at its center. The minimum path through the matrix then occupies area $(1 - \pi R^2)$, and the stress concentration factor is given by

$$\frac{1}{\gamma} = 1 - \pi R^2 = 1 - \pi\left(\frac{3\phi}{4\pi}\right)^{2/3} = 1 - 1.21\phi^{2/3}. \tag{22.7}$$

An alternative expression for the relationship between yield stress and composition was obtained by Gloagen et al., who carried out compression tests on rubber-toughened PMMA, thus ensuring that the material deformed by shear yielding, uncomplicated by rubber particle cavitation or matrix crazing [29]. They found that the compressive yield stress σ_{cy} decreased linearly with the volume fraction of rubber particles, ϕ_p, as follows:

$$\sigma_{\text{cy}}(\phi_p) = (1 - 1.375\phi_p)\sigma_{\text{cy}}(0). \tag{22.8}$$

In the absence of other deformation mechanisms, a similar dependence of yield stress on rubber content is to be expected in both pure shear and uniaxial tension. Comparison of Eqs. 22.7 and 22.8 shows that they predict very different dependences of yield stress upon rubber content, reflecting the difficulties involved in correlating data on the large strain behavior of polymer blends when they are tested under different loading conditions.

Equations 22.7 and 22.8 apply to blends in which the rubber particles are uniformly dispersed in the rigid matrix. Significantly higher local stress concentrations, leading to higher impact energies, can be obtained with quite modest amounts of added rubber by forming a "pseudonetwork" morphology, as illustrated in Fig. 22.11. In this morphology, domains of material A, containing no rubber, are dispersed in a "matrix" of material B, which is itself a blend, consisting of a rigid polymer matrix and a high volume content of rubber particles. For example, domains of polycarbon-

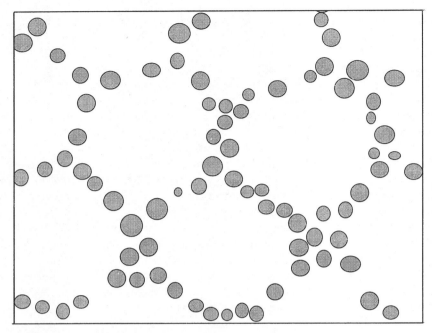

Figure 22.11 Schematic diagram showing "pseudonetwork" morphology in a rubber-toughened polymer. From [30]; reproduced with permission.

ate (material A) might be dispersed in ABS (material B). A more striking example of this morphology is described by Liu et al., who made blends in which material A was PVC and material B was rubber-toughened PVC [30]. These blends exhibited higher levels of toughness than did blends of the same overall composition, but with well-dispersed particles.

In the absence of voids, the shear yield behavior of rubber-toughened plastics appears to be insensitive to the sizes of the particles, provided that they are large enough to constitute a completely separate phase. Even the presence of rigid thermoplastic subinclusions within the rubber particles appears to have only a minor effect upon the initial yield stress, although it might influence the subsequent processes of cold drawing and strain hardening.

Thus, the main factors controlling the shear yield stress of void-free toughened plastics are the yield behavior of the rigid matrix, the dispersion and volume fraction of the rubber particles, and, near its T_g, the shear modulus of the rubber. Under certain conditions, the simple reduction in shear yield stress brought about by adding rubber particles to a rigid polymer is in itself sufficient to increase toughness. In particular, reducing the shear yield stress is important when test conditions are close to the plane-strain–plane-stress transition. In that sense, adding rubber has a similar effect to increasing the temperature or reducing the strain rate. However, the overall improvements in toughness achieved in this way are incremental rather than radical and are much less dramatic than those observed when the particles are able to cavitate.

B. Dilatation Bands

Once the rubber particles have cavitated, the surrounding matrix is free to yield and stretch in a way that was previously impossible. The shell of rigid polymer enclosing the rubber phase expands through biaxial extension, thereby increasing the dimensions of the cavitated particle. Obviously, there are limits on the extent of yielding that can occur around a single isolated particle. Cavitation is most effective when it occurs in an array of closely spaced rubber particles, so that the ligaments between particles become fully yielded. The comparable case of yielding in a porous ductile solid was analyzed by Gurson, who obtained a relationship between σ_e and f, the volume fraction of voids [31, 32]. Lazzeri and Bucknall later introduced a slight modification of Gurson's equation, which includes a parameter μ to account for the pressure dependence of yielding in polymers [19, 33]. Their expression is

$$\frac{\sigma_e}{\sigma_o} + \frac{\mu\sigma_m}{\sigma_o}\left(2 - \frac{\mu\sigma_m}{\sigma_o}\right) + 2f\cosh\left(\frac{3\sigma_m}{2\sigma_o}\right) - f^2 = 1, \tag{22.9}$$

which reduces to Eq. 22.5 when $f = 0$ and to Eq. 22.4 when both f and μ are zero. In later papers, Steenbrink et al. [34] and Lazzeri and Giulini [35] consider additional modifications of the Gurson equation, which account for the interactions between neighboring voids and for the internal tensile stresses σ_R acting across the rubber–matrix boundary. Both groups include the parameters q_1 and q_2 suggested by Tvergaard [36, 37]. One modified version of Eq. 22.9 [35] is as follows:

$$\frac{\sigma_e}{\sigma_o} + \frac{\mu\sigma_m}{\sigma_o}\left(2 - \frac{\mu\sigma_m}{\sigma_o}\right) + 2fq_1\cosh\left(\frac{3q_2(\sigma_m - \sigma_R)}{2\sigma_o}\right) - q_1^2 f^2 = 1. \tag{22.10}$$

The assumption that cavitated particles have exactly the same effect upon yielding as does the same volume fraction of voids is valid only when the stresses acting across the particle–matrix boundary are close to zero. As discussed earlier, cavitated particles are able to sustain quite high tensile stresses under certain conditions, notably when the rubber forms fibrils that are securely anchored at both ends, either in rigid subinclusions or in the surrounding matrix, as shown in Figs. 22.3 and 22.4. The other relevant case is when void sizes are small enough for surface-energy effects to generate significant closure forces. Calculations based on treating cavitated rubber particles as equivalent to voids will therefore give the lower bound on yield stress.

There are several other important differences between cavitated rubber particles and real voids. Most obviously, the actual void content in the cavitated particle is initially very small, and in the early stages of the process the resulting density changes are also relatively small, as demonstrated in the thermal contraction data presented in Fig. 22.6. Secondly, introduced voids are present before the specimen is loaded, whereas rubber particle cavitation usually occurs in response to an applied tensile stress, beginning with the largest particles and progressively affecting the smaller ones. Thirdly, the effects of cavitation are seen only under dilatational conditions

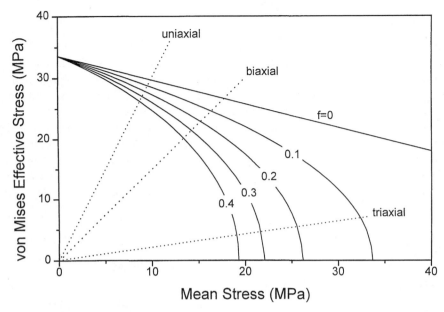

Figure 22.12 Yield envelopes for a toughened plastic containing various volume fractions f of cavitated rubber particles. Dotted lines define stress states of interest in mechanics. From [38]; reproduced with permission.

(i.e., positive values of σ_m). In compression, the voids close up and have no effect on the yield stress.

The relationships between rubber particle cavitation and yielding behavior can be analyzed using "cavitation diagrams," as illustrated in Figs. 22.12–22.14. This type of diagram is based on two principles: (a) cavitation occurs at a critical volume strain, which can be translated into a critical mean stress; and (b) yielding in polymers is governed by the combination of effective stress σ_e and mean stress σ_m, as shown in Eq. 22.5. Both processes can be represented in terms of these two stress invariants, which, in an isotropic solid, are responsible for shape changes and volume changes, respectively.

Several stress states of practical interest are marked in Fig. 22.12. The ordinate axis corresponds to a state of pure shear ($\sigma_m = 0$), and construction lines are also drawn to represent uniaxial tensile stress, equibiaxial tension, and the triaxial stress state set up immediately ahead of the crack tip in an elastic solid under plane strain conditions. (See Chapter 20.) It is obvious from the diagram that the depression in yield stress due to cavitation intensifies as the ratio σ_m/σ_e increases. In pure shear ($\sigma_m = 0$), particle cavitation leaves the yield stress unchanged, whereas at a crack tip, it has a profound effect on yield behavior. A comparison between the yield envelopes represented in Fig. 22.12 and the mean stresses required for cavitation, which are shown schematically in Fig. 22.13, provides considerable insight into the deformation behavior of toughened plastics over the whole range of stress states.

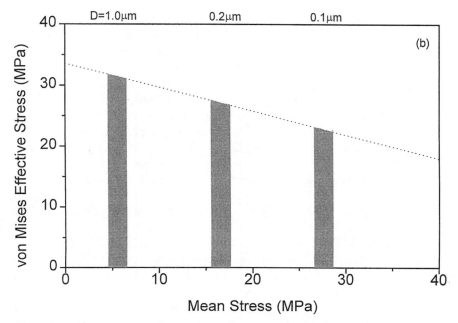

Figure 22.13   Schematic diagram showing shifts in the mean stress at cavitation upon reduction of the rubber particle size. From [38]; reproduced with permission.

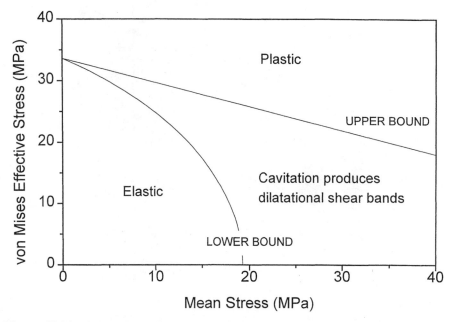

Figure 22.14   Schematic cavitation diagram combining information from Figs. 22.12 and 22.13 to define upper- and lower-bound yield envelopes. From [38]; reproduced with permission.

The basic issues relating to yielding are represented more simply in Fig. 22.14, which defines three regions of deformation behavior. Below the lower bound, which follows Eq. 22.7 with f set equal to the volume fraction of particles, ϕ_p, the deformation of the matrix is elastic. Above the upper bound, which follows Eq. 22.5, the matrix deforms plastically. Between the two bounds, the behavior of the matrix may be elastic or plastic, depending on the extent of rubber particle cavitation. The locations of the two bounds are, of course, dependent upon the rubber content, the properties of the rigid matrix, and the test conditions, especially temperature and strain rate. Cavitation diagrams provide a very convenient format for discussing the effects of these variables on blends containing different concentrations and types of rubber particles. A more complete discussion of this approach is given in a previous review [38].

Reference is made in Section III.A to shear bands connecting cavitated rubber particles in toughened epoxy resins. A stylistic representation of such a band is presented in Fig. 22.15. The angle between such a "dilatation band" and the principal stress axes depends upon its void content. For an isotropic material with no pressure sensitivity ($\mu = f = 0$ in Eq. 22.7), the angle θ between the principal tensile axis and the normal to the band is 45°. For polymers, in which yielding is pressure dependent, θ is about 38° in uniaxial tension. The introduction of voids into the shear bands greatly increases the pressure sensitivity of yielding and causes a further reduction

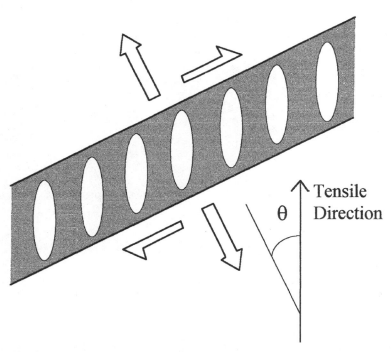

Tensile Direction

θ

Figure 22.15   Schematic representation of a dilatational shear band. From [38]; reproduced with permission. See also the corresponding micrograph in Fig. 25.28.

in θ, so that dilatation bands respond to stress by both increasing in thickness and undergoing shear in plane, as illustrated in the schematic diagram in Fig. 22.15. In the triaxial stress field ahead of a sharp crack, a void content as low as 0.04 is sufficient to bring θ down to zero—hence, the appearance of "croids" in fracture mechanics tests on toughened epoxy resins.

As with ordinary shear bands, dilatation bands propagate by generating peripheral zones of high elastic strain, where the matrix becomes strain softened and yields more easily. Dilatation bands form when the enhanced strains are also sufficient to cause rubber particle cavitation. An example of a dilatation band in rubber-toughened PA-6 is shown in Fig. 25.28 [33]. It has some similarities to a craze, but there are important differences. In particular, the voids are not interconnected, and their free surfaces are formed in the rubber phase, not in the matrix. Other differences between dilatation bands and crazes are discussed in Section V.

V. MULTIPLE CRAZING

A. Evidence for Multiple Crazing

Multiple crazing is the dominant mechanism of toughening in high-impact polystyrene (HIPS) [39]. It is also important in ABS and rubber-toughened poly(methyl methacrylate) (RTPMMA), especially at low temperatures. In all three classes of polymer, the matrix is a brittle thermoplastic, which forms crazes at strains between 0.3 and 1.0%, and fractures shortly afterwards. The adjective "glassy" is very appropriate for these transparent polymers, because they behave essentially like linear elastic solids up to the point of fracture. However, as noted in Section I, they are all ductile on the scale of 1 μm, within a single craze, and could absorb large amounts of energy if this ductility could be extended over a larger volume of material, a result that can be achieved by greatly increasing the numbers of crazes formed.

Multiple crazing was first observed in thin sections of high-impact polystyrene (HIPS) that were stretched on the stage of an optical microscope [39]. The highly strained crazes and associated rubber particles were strongly birefringent and therefore clearly visible in polarized light. Since these early observations, most studies of crazing in toughened plastics have been based on transmission electron microscopy of thin sections. Sections are either stretched on the stage of the microscope (*in situ* deformation) or taken from specimens that have already been strained. In the latter case, the specimen is usually treated with osmium tetroxide (OsO_4), which hardens the rubber phase (see Chapter 9) and deposits osmium in the crazes, providing good electron contrast in both regions of the specimen.

Microscopists are always conscious of the possibility of introducing artifacts into their specimens and the need to avoid misinterpretation of their data. This is especially true when the specimen contains holes, as in crazes and cavitated rubber particles. Sectioning can introduce damage, and the structures seen in very thin sections are not necessarily representative of the bulk material. However, numerous TEM observations, which are supported by other experimental evidence, confirm that multiple crazing does take place in HIPS, ABS, and some other toughened plastics. Large strains (over 50%) can be reached if the applied stress is shared between the craze fibrils in the matrix and load-bearing fibrils in the rubber, which often look very similar. (See Fig. 22.4.)

Independent evidence for multiple crazing in HIPS has been obtained by Buckley et al., who used realtime small-angle X-ray scattering (SAXS) to monitor deformation mechanisms in tensile impact specimens, a technique that requires a high-intensity X-ray source [40, 41]. Similar experiments were later conducted at lower strain rates by Magalhaes and Borggreve [42]. A characteristic scattering pattern due to the craze fibrils was obtained in both studies and used to determine the fractional contribution of crazing to both the volume strain and the extension of the test bar. Both groups concluded that crazing of the matrix accounted for no more than 50% of the volumetric expansion in HIPS specimens and played an even more minor role in the deformation of ABS. These results are in approximate agreement with the evidence obtained from studies of thin sections subjected to *in situ* straining in the electron microscope. As illustrated in Fig. 22.4, these studies showed extensive crazing of the matrix, especially in HIPS, and well-developed fibrillation of the rubber phase, which should also produce a characteristic SAXS pattern. Both techniques confirm the importance of rubber particle cavitation and matrix crazing in the tensile yielding of HIPS and ABS. It is clear from TEM studies that no other kinds of void formation contribute significantly to yielding in these materials.

B. Kinetics of Multiple Crazing

The kinetics of tensile deformation due to multiple crazing are quite distinctive. They are most clearly seen in creep tests, as illustrated in Fig. 22.16. On application of stress, the material extends instantly, in a (quasi)elastic response. This is followed by a period of comparatively slow but accelerating extension, which has been termed the "induction period." On reaching a strain of 1.5 to 2.0%, there is a sharp upturn in the creep curve, which becomes almost linear. Beyond a strain of about 20%, the creep rate begins to decrease, and eventually the specimen breaks.

This pattern of behavior can be understood by considering the steps involved in craze initiation and growth. Crazes are known to propagate in a similar manner to cracks, through the continuous extension of a free surface. The highly stressed material at the craze tip strain softens, and the craze front then extends fingers into the fluid, dense polymer ahead of the tip, as a result of "meniscus instability" [43]. One implication of this mechanism is that craze initiation should occur only at a free surface.

On the basis of this hypothesis, cavitated rubber particles can be seen as the only effective sites for initiation of crazes within the bulk of HIPS or ABS specimens. According to this interpretation, application of stress causes the largest rubber particles to cavitate and initiate the first crazes, which grow outwards, increasing in both area and thickness. The resulting deformation increases the elastic strain on other, smaller rubber particles close to the craze plane, thereby causing further (secondary) cavitation followed by craze initiation. During this early stage of creep (the induc-

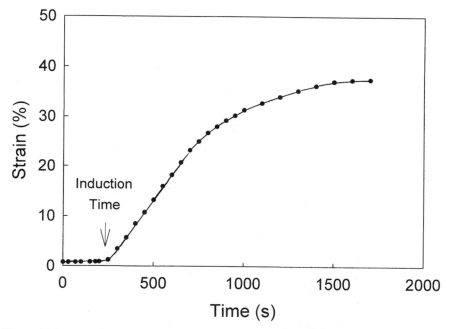

Figure 22.16   Creep curve for HIPS containing a relatively high volume fraction of "salami" rubber particles, showing the various stages of deformation.

tion period), the crazes are separate and subject to closure forces from the elastically strained continuous polystyrene phase. Consequently, creep rates are low.

More rapid craze thickening occurs during the second stage of creep, after crazes lying close to a single plane have begun to connect together. The beginning of this stage can be regarded as a percolation threshold, at which each system of interconnected crazes has some similarities to a moderately large microcrack. Once this threshold has been reached, the principal mechanism of deformation is craze thickening, which involves drawing fresh polymer from the walls of existing crazes. While the stresses acting at the craze–matrix interface remain constant, the strain rate is also constant. The strains in the fibrillated rubber phase and associated crazes increase with the strain in the matrix, but the relationship between stress and strain is different in the two cases. The stress in the rubber fibrils increases with strain until cavitation is initiated in a parallel rubber membrane within the same particle, at which point the stress in both cavitated layers falls. The cycle of straining followed by fresh cavitation is then repeated until the particle is fully fibrillated or until fracture intervenes. One consequence of this mechanism is that the average strain in the rubber phase increases steadily with the total strain in the HIPS.

A third stage of creep is observed when the volume fraction of rubber particles is relatively high (over about 0.3). The creep rate begins to fall, clearly indicating that the stresses acting on the crazes are beginning to decrease. It follows that the average stress on the rubber phase must be rising, a stage that will begin when there are no longer enough intact rubber membranes to provide a "safety valve" against overloading of existing fibrillated material. This change in behavior is associated with an increase in the probability of fracture, as discussed in Chapter 21.

Thus, the rubber particles interact with crazes in three different ways. In the early stages, the larger particles cavitate and form craze nuclei. In the middle stages, fibrillation occurs in an increasing fraction of the rubber particle, allowing the rubber particles to stretch in parallel with the crazes, without sustaining exceptionally high stresses. Finally, as fibrillation of the rubber phase approaches saturation point, the stresses acting on the fibrillated rubber rise rapidly, thereby (in a creep test) unloading the neighboring crazes. In a standard tensile test, the average stress on the crazes remains approximately constant, thereby ensuring that the extension rate is kept constant, and the average stresses on the rubber particles, and therefore on the specimen, rise steadily with strain.

This interpretation of typical creep curves for HIPS and ABS provides a suitable basis for discussing the effects of stress and rubber content on the kinetics of deformation in these blends. Crazing, like shear yielding, is a thermally activated rate process, which follows the Eyring equation. Plots of the logarithm of the rate against applied stress give straight lines from which the apparent activation volume γV can, at least in principle, be calculated. By comparing Eyring plots of $\log(dV/dt)_{max}$ against stress to corresponding plots of $\log \tau^{-1}$, Bucknall and Clayton showed that the same kinetic parameters apply to the induction period as to the subsequent rapid stage of creep in HIPS [44].

Later studies showed that the kinetics of crazing in HIPS are more complicated than the kinetics of shear yielding in blends based on PA-6, PMMA, and PP; see Section IV.A. In contrast to these materials, it was not possible to superimpose creep data from a coordinated series of HIPS blends, each of different rubber content, simply by plotting the logarithm of the creep rate against $\gamma \sigma_{appl}$ [45]. All of the blends gave a linear plot of the logarithm of the creep rate against the applied stress, but some additional factor affected the kinetics of deformation.

The most likely explanation is that the overall creep rate is controlled by a combination of two processes: (a) craze drawing, which is a stress-driven flow process following the Eyring equation, and (b) craze initiation from cavitated particles, which is a quite different mechanism, controlled by volume strain rather than applied stress. Although both mechanisms show a strong dependence on applied stress, in the case of rubber cavitation the rate-controlling factor in a creep test is not σ but σ/E, where E is Young's modulus. If rubber particle cavitation determines the rate of craze initiation, a rapid increase in creep rate is to be expected over a narrow range of applied stress, simply because an increasing fraction of particles is undergoing cavitation. However, because rates of craze initiation are dependent upon strain rather than stress, simple correlations are unlikely to be found between creep rate and applied stress for a series of blends of differing composition.

As discussed in Chapter 21, total strain rather than applied stress controls final failure in blends that deform by multiple crazing, notably ABS and HIPS [46, 47]. This dependence appears to be associated with failure in the fibrillated rubber particles when the local tensile strain exceeds their capacity to stretch further.

VI. STRUCTURE–PROPERTY RELATIONSHIPS

The basic reason for studying deformation mechanisms in toughened plastics is to understand the relationships between structure and mechanical properties, with particular reference to fracture resistance. Many structural variables have been identified as being of significance under critical conditions, including the volume fraction of rubber particles, the size and size distribution of the rubber particles, the average interparticle spacing, the internal morphology of the rubber particles, and the strength of the interface between particles and matrix. Unfortunately, the effects of these variables are often confusing and contradictory. For example, increasing the average particle size from 0.2 to 1.0 μm raises the Izod impact strength of HIPS, but causes a dramatic fall in the Izod impact strength of rubber-toughened polyamide. Most of the difficulties can be traced to the complex interactions between the rubber particles and the matrix, and the response of the matrix is therefore of primary concern.

A. Matrix Properties

The key to understanding structure–property relationships in toughened polymers lies in the deformation and failure behavior of the rigid matrix, which is where most of the energy is absorbed during deformation and fracture. This point can best be illustrated by comparing two familiar families of thermoplastics, the polystyrenes and the polyamides.

Because of its molecular structure, polystyrene has a relatively high shear yield stress and a low resistance to craze propagation and failure. Its low chain-entanglement density and low surface free energy make it susceptible to crazing, while the large cross-sectional area of its molecules increases the probability of chain rupture under a given applied stress. Plain polystyrene has a K_{Ic} of about 0.9 MPa $m^{1/2}$ [48]. In contrast, dry polyamides have lower shear yield stresses, higher resistance to crazing, and a smaller chain cross-sectional area; packing densities for PA-6.6 can reach 5.22 bonds per nm^2, as compared to 1.35 bonds per nm^2 for PS [49]. Typical K_{Ic} values for polyamides are between 2.5 and 3.0 MPa $m^{1/2}$ [48].

B. Minimum Particle Size

Because of these differences in the deformation behavior and fracture toughness of the matrix polymer, the optimum sizes for toughening rubber particles are different in the two cases. Toughening of polystyrene works best with quite large (1–2 μm in diameter) rubber particles, which can cavitate in relatively weak crack-tip stress fields, whereas toughening of polyamides works most effectively with 0.2 to 0.3 μm particles, which require stronger crack-tip stress fields for cavitation. In both PS and PA, making the particle size smaller than the previously quoted range results in a sharp drop in impact strength, which can be attributed to an excessively high resistance to rubber particle cavitation in relation to the critical crack-tip stress field. The same effect is seen when cavitation resistance is increased in some other way, for example, by choosing a rubber with an intrinsically high shear modulus or by increasing the cross-link density of the rubber phase through thermal or photochemical oxidation (e.g., ageing in sunlight). In a sharply notched specimen, the stresses responsible for cavitation decrease with radial distance from the notch tip and also become smaller in the plane-stress regions at the sides of the specimen. Making it more difficult to form voids by reducing particle size or hardening the rubber phase therefore reduces the size of cavitated zone around the crack tip and, in thick specimens, leads to a marked drop in fracture resistance.

The importance of cavitation in toughening is well illustrated in experiments by Purcell, who measured notched Izod impact strengths for toughened PVC using preformed rubber particles between 50 and 280 nm in diameter [50]. In 3-mm bars, plane-stress conditions prevailed, and impact energies remained above 100 J·m^{-1} for all of the blends. By contrast, in 6-mm bars, where cavitation is needed to alleviate plane-strain conditions, the Izod values fell from 12 to 2 J·m^{-1} with decreasing particle size.

C. Interparticle Spacing

Work by Wu [51] has focused attention upon interparticle spacing rather than particle size as the factor controlling notched Izod impact behavior, especially in toughened PA-6. An explanation of this relationship has recently been advanced by Argon et al., who demonstrated that PA-6.6 forms a layer some 150 nm thick at PA–rubber interfaces, with its crystallographic axes parallel to the surface [52, 53]. Similar layers of different thicknesses are found in PA-6 and HDPE. On the basis of these observations, they concluded that the critical interparticle spacing is a permeation threshold, marking the point at which material of enhanced mobility, crystallized in thin layers around the rubber particles, is able to form continuous pathways through the polymer matrix.

The problem with this explanation is that measurements by Gaymans, Borggreve, and coworkers have shown that the critical interparticle spacing is not constant for one matrix, but varies systematically with the test temperature and the modulus of the rubber particles [54, 55]. The presence of oriented crystalline layers next to the matrix–rubber interface necessarily affects toughness, but the balance of evidence suggests that correlations between impact transitions and interparticle spacings probably have a more fundamental, mechanically based origin, deriving from the interaction between particle cavitation and shear yield stress. If toughening is optimized when rubber particle cavitation coincides with the lower-bound yield envelope (see Fig. 22.14), then the critical particle size should not be a constant, but should vary with rubber content (as observed by Wu [51]) and temperature (as found by Gaymans et al. [54, 55]). Further work is needed to settle this controversial point.

D. Large Rubber Particles and Internal Morphology

Large homogeneous rubber particles are ineffective in toughening PA, PP, and a number of other ductile matrixes. In this context, the term "large" refers to particles over 1 μm in diameter. However, there is ample experience to show that complex "salami" particles as large as 10 μm provide high levels of toughness in HIPS. The essential difference between PA and PS lies in the way in which the rubber is deployed after cavitation. Through comprehensive formation of fibrils that are anchored to either the PS subinclusions or the matrix, it is possible to ensure that the maximum amount of strain-hardened rubber contributes to stabilizing the associated crazes against failure. In order to achieve this result, it is, of course, necessary to ensure that the interfaces between rubber and polystyrene are strong by introducing graft or block copolymers..

It is possible that the same considerations also apply to large particles in other matrixes. If these particles cavitate at stresses well below the lower yield envelope, they might then initiate crazes in the matrix. If they are also homogeneous in structure, therefore lacking the stabilizing features provided in complex "salami" or core–shell particles, they will be unable to reinforce the crazes through controlled fibrillation and thus might act as Griffith flaws, initiating microcracks in the matrix.

E. Cross-Linking of the Rubber Phase

The energy balance model has clarified another aspect of rubber toughening, relating to cross-linking of the rubber phase. It has long been known that some cross-linking is necessary to preserve the structure of the particles during melt processing, but that excessive cross-linking impairs fracture resistance, notably by reducing the notched impact strength. Very high levels of cross-linking shift the T_g of the rubber upwards towards room temperature, and it has been assumed that this shift is responsible for the brittleness observed in toughened plastics exposed to prolonged sunshine, an effect known as "ageing." However, it is now clear that quite modest amounts of cross-linking can cause a dramatic change in the resistance of the rubber phase to cavitation, while raising the T_g by no more than 5–10°C. Efforts to combat ultraviolet degradation of rubber-toughened plastics should therefore be directed at ensuring that the modulus of the rubber phase remains low enough to allow cavitation of the particles at a critical stage in the deformation process.

VII. CONCLUSIONS

This chapter has shown that rubber toughening involves a variety of responses within both the rubber particles and the surrounding matrix. Modest increases in toughness may be achieved simply by reducing the yield stress, without loss of continuity in either phase or at the interface between them. These improvements are similar to those observed on raising the temperature by 10 or 20°C or on reducing the strain rate by a few decades and are due to minor shifts in the location of the transition from plane-strain to plane-stress conditions at break. However, really dramatic gains in toughness are achieved only when cohesive failure within the rubber particles induces greatly accelerated shear yielding in the matrix, followed by strain hardening of the yield zone due to stretching of both rubber and rigid polymer. Cohesive failure in the matrix, through the formation of crazes, also provides toughness, but is much more dependent upon strain hardening in the rubber phase and, therefore, upon the strength of the interface between rubber particles and matrix. Whatever the precise mechanisms of deformation involved, the basic requirement is to modify the response of the material so that it is able to reach large strains when subjected to dilatational stress fields and to alter the kinetics of deformation to match the strain rates encountered during loading. Strain softening, strain hardening, and fracture resistance in the strain-hardened zone are all of critical importance in achieving optimum toughening.

VIII. REFERENCES

1. J. D. Ferry, *Viscoelastic Properties of Polymers*, 3d Ed., Wiley, New York, 1980.
2. R. N. Haward and R. J. Young, "Introduction," in *The Physics of Glassy Polymers*, 2nd Ed., R. N. Haward and R. J. Young (eds.), Chapman & Hall, London, 1997.
3. A. M. Donald and E. J. Kramer, *J. Polym. Sci.—Polym. Phys.*, **20**, 899 (1982).
4. A. M. Donald and E. J. Kramer, *Polymer*, **23**, 461 (1982).
5. A. M. Donald, "Crazing," in *The Physics of Glassy Polymers*, 2nd Ed., R. N. Haward and R. J. Young (eds.), Chapman & Hall, London, 1997.
6. W. D. Bascom, R.L. Cottington, R .L. Jones, and P. Peyser, *J. Appl. Polym. Sci.*, **19**, 2545 (1975).
7. A. J. Kinloch, *Proc. Instn. Mech. Engrs.*, **211**, 307 (1997).
8. R. A. Pearson and A. F. Yee, *J. Mater. Sci.*, **24**, 2571 (1989).
9. H. Breuer, F. Haaf, and J. Stabenow, *J. Macromol. Sci.—Phys. B*, **14**, 387 (1977).
10. R. Schirrer, R. Lenke, and J. Boudouaz, *Polym. Eng. Sci.*, **37**, 1748 (1997).
11. S. Géhant and R. Schirrer, *J. Polym. Sci.—Phys.*, **37**, 113 (1999).
12. C. B. Bucknall, D. S. Ayre, and D. J. Dijkstra, *Polymer*, submitted for publication.
13. C. B. Bucknall, R. Rizzieri, and D. R. Moore, *Polymer*, submitted for publication.
14. G. Humme, D. Kranz, L. Morbitzer, and K. H. Ott, *J. Appl. Polym. Sci.*, **20**, 2691 (1976).
15. L. Morbitzer, G. Humme, K. H. Ott, and K. Zabrocki, *Angew. Makromol. Chem.*, **108**, 123 (1982).
16. C. S. Lin, D. S. Ayre, and C. B. Bucknall, *J. Mater. Sci. Lett.*, **17**, 667 (1998).
17. J. U. Starke, R. Godehardt, G. H. Michler, and C. B. Bucknall, *J. Mater. Sci.*, **32**, 1855 (1997).
18. C. Fond, A. Lobbrecht, and R. Schirrer, *Int. J. Fract.*, **77**, 141 (1996).
19. A. Lazzeri and C. B. Bucknall, *J. Mater. Sci.*, **28**, 6799 (1993).
20. C. B. Bucknall, A. Karpodinis, and X. C. Zhang, *J. Mater. Sci.*, **29**, 3377 (1994).
21. D. S. Ayre and C. B. Bucknall, *Polymer*, **39**, 4785 (1998).
22. A. N. Gent and C. W. Wang, *J. Mater Sci.*, **26**, 3392 (1991).
23. D. Dompas and G. Groeninckx, *Polymer*, **35**, 4743 (1994).
24. D. Dompas, G. Groeninckx, M. Isogawa, T. Hasegawa, and M. Kadokura, *Polymer*, **35**, 4750 (1994).
25. D. Dompas, G. Groeninckx, M. Isogawa, T. Hasegawa, and M. Kadokura, *Polymer*, **35**, 4760 (1994).
26. C. B. Bucknall and C. J. Page, *J. Mater. Sci.*, **17**, 808 (1982).
27. C. B. Bucknall, P. S. Heather, and A. Lazzeri, *J. Mater. Sci.*, **24**, 1489 (1989).

28. O. Ishai and L. J. Cohen, *J. Compos. Mater.*, **2**, 302 (1968).

29. J. M. Gloagen, P. Steer, P. Galliard, C. Wrotecki, and J. M. Lefebvre, *Polym. Eng. Sci.*, **33**, 748 (1993).

30. Z. H. Liu, X. D. Zhang, X. G. Zhu, Z. K. Y. Li, Z. N. Wang, and C. L. Choy, *Polymer*, **39**, 5035 (1998).

31. A. L. Gurson, *J. Eng. Mater. Technol., Trans. ASME*, **99**, 2 (1977).

32. A. L. Gurson, "Porous Rigid-Plastic Materials Containing Rigid Indusions–Yield Function, Plastic Potential and Void Nucleation," in *ICF4 Fracture 1977*, Vol. 2A, D. M. R. Taplin (ed.), Pergamon, Oxford, 1977, p. 357.

33. A. Lazzeri and C. B. Bucknall, *Polymer*, **36**, 2895 (1995).

34. A. C. Steenbrink, E. van der Giessen, and P. D. Wu, *J. Mech. Phys. Solids*, **45**, 405 (1997).

35. A. Lazzeri and D. Giulini, "Yielding Kinetics in Rubber Toughened Polymers," in *10th International Conference on Deformation, Yield & Fracture*, Institute of Materials, London, 1997, p. 446.

36. V. Tvergaard, *Int. J. Fracture*, **17**, 389 (1981).

37. V. Tvergaard, *Int. J. Fracture*, **18**, 237 (1982).

38. C. B. Bucknall, "Rubber Toughening," in *The Physics of Glassy Polymers*, 2nd Ed., R. N. Haward and R. J. Young (eds.), Chapman & Hall, London, 1997.

39. C. B. Bucknall, *Toughened Plastics*, Applied Science Publishers, London 1977.

40. R. A Bubeck, D. J. Buckley, E. J. Kramer, and H. R. Brown, *J. Mater. Sci.*, **26**, 249 (1991).

41. D. J. Buckley, "Toughening Mechanisms in the High Strain Rate Deformation of Rubber-Modified Polymer Glasses," Ph.D. Thesis, Cornell University, 1993.

42. A. M. L. Magalhaes and R. J. M. Borggreve, *Macromolecules*, **28**, 5841 (1995).

43. A. S. Argon and M. M. Salama, *Phil. Mag.*, **36**, 1217 (1977).

44. C. B. Bucknall and D. Clayton, *J. Mater. Sci.*, **7**, 202 (1972).

45. C. B. Bucknall, P. Davies, and I. K. Partridge, *J. Mater. Sci.*, **21**, 307 (1986).

46. S. D. Sjoerdsma and J. P. H. Boyens, *Polym. Eng. Sci.*, **34**, 86 (1994).

47. B. O'Connor, C. B. Bucknall, and J. L. Hahnfeld, *Plast. Rubber Compos. Process. Appl.*, **26**, 360 (1997).

48. J. G. Williams, *Fracture Mechanics of Polymers*, Wiley, New York, 1984.

49. P. I. Vincent, *Polymer*, **13**, 558 (1972).

50. T. O. Purcell, *Amer. Chem. Soc. Polym. Prepr.*, **13(1)**, 699 (1972).

51. S. Wu, *J. Appl. Polym. Sci.*, **35**, 549 (1988).

52. O. K. Muratoglu, A. S. Argon, and R. E. Cohen, *Polymer*, **36**, 2143 (1995).

53. A. S. Argon, Z. Bartczak, R. E. Cohen, and O. K. Muratoglu, "Novel Methods of Toughening Semicrystalline Polymers," in *18th Risø International Symposium on Materials Science*, Risø National Laboratory, Roskilde, Denmark, 1997.

54. R. J. M. Borggreve, R. J. Gaymans, J. Schuijer, and J. F. Ingen-Housz, *Polymer*, **28**, 1489 (1987).

55. R. J. M. Borggreve, R. J. Gaymans, and J. Schuijer, *Polymer*, **30**, 71 (1989).

23 Strengthening Polymer–Polymer Interfaces

HUGH R. BROWN

BHP Steel Institute
University of Wollongong
Wollongong, NSW 2522
Australia

I. INTRODUCTION

The mechanical properties of polymer blends are strongly influenced by the strength of the interfaces between the different phases. The purpose of this chapter is to summarize current understanding of available methods for increasing the toughness of polymer–polymer interfaces. Most information in this area has come from experiments using polymer bilayer or sandwich specimens that have a planar interface between the phases. The well-defined mechanical experiments required to obtain reliable data on interfacial toughness and failure are much easier to perform on such

Polymer Blends, Volume 2: Performance. Edited by D. R. Paul and C. B. Bucknall.
ISBN 0-471-35280-2.   © 2000 John Wiley & Sons, Inc.

bilayer samples than on actual blends, in which the interfaces are on a fine scale and highly curved.

Polymeric materials gain their strength and toughness from entanglement or cross-linking between the chains. The main resistance to crack propagation originates in the energy dissipated in the inelastic deformation around the crack. This deformation is triggered by the high stresses borne by the chains right at the crack tip. In un-cross-linked polymers, crack-tip chains that are not entangled on both sides of the interface often experience facile pullout, initiating little inelastic deformation and contributing almost nothing to toughness. Interfaces between polymer phases (in immiscible polymers) are normally narrow compared to the average distance between entanglements along a chain, so there is little cross-interface entanglement. Hence, the strengths or toughnesses of interfaces are normally low. Exceptions are to be found where the polymer–polymer χ parameter is less than about 0.02, so that the materials are close to miscibility and form broad interfaces.

The most common way to strengthen a polymer–polymer interface is to place or synthesize a block or graft copolymer at the interface, so that one of the two types of block or graft chains mixes with each of the two substrate polymers. As long as the block or graft chains are long enough to entangle with the substrate polymers, or at least to pull out at a high force, the copolymer will increase the mechanical coupling and normally strengthen the interface. The other main reason for introducing block or graft copolymers into polymer blends is to refine the phase size by decreasing the interfacial tension and suppressing phase coalescence. Effective phase size control does not require the block or graft copolymer chains to be long enough to entangle with those of the substrate polymers, so selection of optimum blend compatibilizers requires consideration of both phase size and interfacial strength issues.

II. MEASUREMENT OF INTERFACIAL TOUGHNESS

As the interface between the two materials normally has significantly lower toughness than either of the bulk materials, one might expect the crack to propagate along the interface rather than through one of the bulk materials. One might also expect, as long as the crack does propagate along the interface, that the measured toughness will be independent of the geometry of the sample. However, neither of these expectations is invariably borne out in practice, and exceptions are observed, particularly in materials that fail by crazing.

Interfacial toughness is most easily measured in systems for which the interface is planar. For situations in which both polymers are glassy, most published work has used sandwich-type samples that are cut up to form double-cantilever-beam specimens [1]. Sample loading by opening with a wedge, often a razor blade, has been found to be convenient. Compact tension specimens have also been used [2]. In crazing materials, it is often necessary to choose the thicknesses of the two beams carefully (see Fig. 23.1) in order to ensure that the crack is driven along the interface, with no excursions of either the primary crack or subsidiary crazes into either bulk

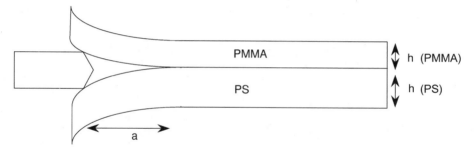

Figure 23.1   Typical double-cantilever-beam geometry.

material, and thereby to obtain a minimum value for the toughness. Much of the published work in the last few years has used such asymmetric double-cantilever-beam
(aDCB) samples. The assumption that the most relevant quantity is the minimum
value of toughness is a conservative concept originating in the fracture mechanics of
structures. It is not really obvious that it is applicable to the design of blends, where
interfaces are curved, and localized interfacial failure is not necessarily a problem.

The elastic stress pattern at a crack tip can be described in terms of the two relevant components of the stress intensity, K_I and K_{II}. As discussed in Chapter 20,
K_I refers to opening mode loading, wherein the stresses acting on the crack plane in
front of the crack are tensile, and K_{II} describes shear loading, wherein the stresses
in front of the crack are mainly in-plane shear. The ratio between the two modes is
described by the mixity ψ, where $\tan \psi = K_{II}/K_I$. The author of this chapter, who
originally proposed the use of aDCB specimens for polymer interface studies, initially believed that in order to obtain the minimum toughness, the amplitude of the
mixity simply needed to be above a minimum value, as long as the sign was chosen so that the crack was being pushed towards the material with the higher crazing
resistance (which is, by convention, negative). However, later work showed that the
toughness tends to show a minimum as a function of mixity [3]. In many systems,
this minimum tends to be very broad [4, 5], so that a range of sample geometries
can give similar results. When one of the materials has a low crazing resistance,
such as polystyrene, the minimum can be quite sharp, as shown for the example
in Fig. 23.2 [6].

For crazing materials, the increase in toughness from the minimum is caused by
the growth of subsidiary crazes away from the interfacial plane into the material of
low crazing resistance. When the mixity is more positive than the value at which
the toughness reaches a minimum, the subsidiary crazes tend to grow at a small
angle to the crack propagation direction along the interface, with the crack sometimes
propagating away from the interface along such a craze, before jumping back to
the interface [5, 7, 8]. Considerable energy can be dissipated in these processes. As
the mixity continues to increase, the crack tends to propagate into the material of
low crazing resistance and thereby break off an arm of the sample. In this case, the
crack clearly does not choose the path of minimum energy; instead, its direction is
controlled by the direction of the craze propagation.

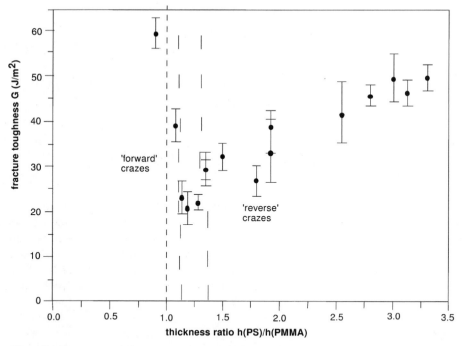

Figure 23.2 Variation of measured toughness with mixity of PS–PMMA interface reinforced with PS-r-PMMA.

As the mixity is made more negative than the value required to give minimum toughness [5], crazes tend to initiate on the interface in front of the crack and its primary craze. (See Section III.) These crazes initiate in the region where the stress is mainly hydrostatic tension, but at low mixity there is also a small shear component. The crazes grow into the easily crazed material normal to the maximum principal stress—thus, at about 45° to the interface—and back towards the crack. These backward-sloping crazes are inevitably limited in size and thus do not cause the toughness to increase as rapidly with mixity as is the case with forward-sloping crazes.

For materials that fail by yielding or by a combination of yielding and crazing, the interfacial toughness is not as sensitive to mixity as in the case of crazing materials, but mixity can nevertheless be important [4].

III. MODELS OF INTERFACIAL FAILURE AND TOUGHNESS

When a crack propagates in a polymeric material, the polymer chains that actually span the crack plane must either break or pull out from one side. For systems of very low toughness, the energy involved in this breakage or pullout can be a major fraction of the total energy dissipated during crack propagation. For tougher systems, most of the energy is dissipated in the formation and growth of either a primary crack-tip craze or a yield zone at the crack tip. An understanding of resistance to crack propagation hence requires both a knowledge of the chain failure mechanism and an understanding of the relationship between that process and the energy dissipated. In this section, we shall consider models for three energy dissipation processes: (a) chain pullout, (b) chain scission in an elastomer, and (c) crazing, wherein the chain process can be either pullout or scission.

The energy involved in chain pullout from a glassy or an elastomeric surface can be estimated if it is assumed that each chain pulls out independently and that the resistance to pullout depends on both the length of chain still embedded and the pullout rate. In a glass, it seems reasonable that there should be a yieldlike static monomer friction coefficient f_{mono} (force per monomer at zero pullout rate). This assumption gives the toughness at a low crack propagation rate as

$$G_c \approx \Sigma f_{mono} N^2 l_o / 2, \tag{23.1}$$

where N is the degree of polymerization of the pulled-out chain and l_o is the monomer repeat length [9]. In an elastomer, chain pullout is expected to be viscous, so f_{mono} (defined at zero pullout rate) should be zero. However, a finite G_c at zero pullout rate has been predicted [10] based on an assumption that in the pullout process, single-chain fibrils must be formed.

At finite crack propagation rates, the friction increases, and sophisticated mechanical analysis is required to find the relationship between crack velocity and chain pullout rate [9, 10]. In general, G_c is predicted to increase linearly with crack speed.

Crack propagation in a cross-linked system inevitably involves chain scission. The classic Lake and Thomas model [11] for the threshold (zero rate) toughness of a cross-linked elastomer should apply just as well to an interface as it does to bulk material. The model is concerned only with the chains that must be broken because they have adjacent cross-links on either side of the crack path. It is assumed that all of the energy required to stretch these chains to the point at which a bond breaks will be lost. Hence, G_c increases linearly with Σ and N.

Many glassy and semicrystalline polymers fail by crazing. Most of the energy is dissipated in the growth of a single primary craze in front of the crack tip. The author of this chapter has proposed a simple model that relates the energy dissipated in the growth of the craze to the chain failure processes that must occur at the crack tip [12]. This model is based on the realization that craze fibrils are interconnected and thus form a network structure. Hence, a stress concentration must exist at the tip of the crack within the craze. (See Fig. 23.3.) The crack tip stresses therefore depend both on the stress at the craze–matrix interface (a material parameter) and on the craze thickness. The energy dissipated in growth of the craze increases linearly with the craze thickness. The maximum stress that can be borne by the fibril closest to the crack tip is assumed to be controlled by the number of coupling chains in the fibril and by f_b, the force to break or pull out a chain. For mature crazes, this model, which assumes a simple continuum model for the craze mechanics, predicts that the

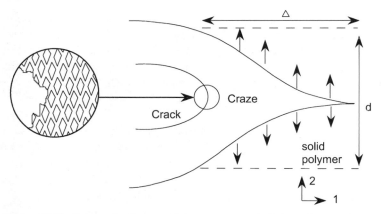

Figure 23.3   Schematic diagram of a craze at a crack tip in a glassy polymer.

interface toughness is given by

$$G_c = C(\Sigma f_b)^2, \tag{23.2}$$

where C is a constant for a particular material that depends on the structure and mechanical properties of the craze. Experimental evidence for the range of validity of this model will be discussed later.

The aforementioned model for crazing failure has been extended and improved by a number of workers [13–16]. A more accurate mechanical solution to the continuum model has been proposed and extended to weak and narrow (low energy) crazes [13, 14]. Crazes actually consist of discrete fibrils, and so the continuum model may be in error when it predicts considerable stress variation over a single interfibrillar distance. This problem has been approached by modeling the craze as a network of springs [15, 16]. The simple model given by Eq. 23.2 is expected to be applicable to large crazes and tough interfaces, but to overestimate the toughness of weak interfaces.

IV. DIBLOCK COPOLYMERS AT GLASSY POLYMER INTERFACES

The main picture of the effects of copolymers at interfaces has been obtained from experiments using premade diblock copolymers at interfaces between glassy polymers. The two main series of experiments have used diblocks of polystyrene (PS) and poly(methyl methacrylate) (PMMA) [12, 17–22] and diblocks of PS and poly(vinylpyridine) (PVP) [3, 23–27]. The PS–PVP diblocks were used between PS and PVP homopolymers, while the most informative of the experiments with PS–PMMA diblocks used them to couple between poly(phenylene oxide) (PPO) and PMMA. The PS block of the diblock has a favorable mixing enthalpy with the PPO homopolymer, thereby driving the system to organization, in spite of the fact that the

PPO is never heated above its T_g. The important variables are the degree of polymerization of the blocks in the copolymer and the areal density of the copolymer, Σ. Most information has been obtained by using a combination of the mechanical techniques described previously with techniques to examine the chain failure processes. The diblock molecules were synthesized with one of the fully deuterated blocks. Secondary ion mass spectrometry (SIMS) or forward recoil ion scattering (FRES) were then used to examine the locations of the blocks on the fracture surfaces.

Consider first the situation in which each block is long enough to entangle with chains of the relevant homopolymer. FRES and SIMS experiments have shown clearly that the copolymers fail by scission [17]. If Σ is small, the copolymer molecules break near their junction points with little dissipated energy. As Σ is increased, the stress at the interface increases until, at $\Sigma = \Sigma^*$ (where Σ^* is defined so that the crazing stress $\sigma_c = \Sigma^* f_b$), a craze or yield zone can form at the interface. The toughness then increases rapidly [3], even discontinuously, as shown in Fig. 23.4. This transition can give a reliable value of f_b, the force to break a single chain, of about 2 nN. The measured values of G_c obtained for $\Sigma > \Sigma^*$ are in good agreement with the predictions of the crazing model discussed earlier and vary linearly with Σ^2 at high values of Σ [20], as shown in Fig. 23.5. As the copolymer fails by scission, the toughness is independent of the molecular weight of the copolymer [20].

In the crazing regime, the location of the failure does depend on the system and, in particular, on the chain entanglement density between the crazing homopolymer and the relevant block of the copolymer. In the PS–PVP system, for which the craze is in the PS, the copolymer fails within its PS block. In the system with PMMA and PPO homopolymers as substrates, for which the craze is in the PMMA, the copolymer fails close to the junction point. This difference in failure point probably has its origin in the difference in entanglement density, defined by the entanglement degree of polymerization, N_e, between the polymers. The entanglement density of PMMA is about twice that of polystyrene. When a craze is formed, a certain amount of chain breakage is inevitable. This chain breakage reduces the entanglement density more in PS craze fibrils than in PMMA craze fibrils, moving the most likely failure point in PS slightly away from the interface.

As the degree of polymerization is reduced, at a block length of about $2N_e$, the local failure mode changes to chain pullout. This transition is easily observed using SIMS or FRES. When the stress on the interface, $\Sigma N f_{mono}$, is less than the crazing stress, the failure energy is fairly low and is given approximately by the predictions of the pullout model, Eq. 23.1 [26, 28]. However, this model does rather underestimate G_c, suggesting that some energy is lost in local plastic processes.

There is a limit to the areal density Σ_{sat} of chains that can form a single layer at the interface. This limit is similar to the density of chains within one lamella of the neat diblock. It is often possible to see a transition to crazing failure while the local process is chain pullout [24]. For chains shorter than N_e, this transition normally does not exist, because $\Sigma_{sat} N f_{mono}$ is less than the crazing stress, so crazing failure cannot occur with local chain pullout. Sometimes, as Σ approaches Σ_{sat}, the mixing between the block polymer and homopolymer chains decreases, and so the

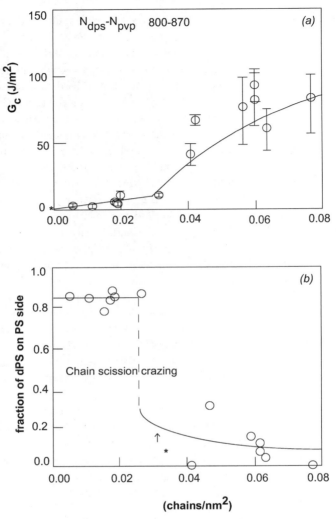

Figure 23.4 Fracture toughness at the interface between PS and PVP reinforced with PS-b-PVP copolymers. The transition to crazing is clearly seen at a Σ of 0.03 chains/nm^2. Redrawn with permission from E. J. Kramer, *Plastics and Rubber and Composites Processing and Applications*, **26**, 241 (1997).

simple approximation of independent pullout ceases to be valid. The pullout stress will decrease, and again crazing may not be observed.

When the amount of copolymer at the interface is increased above Σ_{sat}, the interfacial toughness tends to decrease. The origin of this decrease has been investigated in both of the experimental systems discussed previously and has been found to be system specific. When $\Sigma > \Sigma_{sat}$, the copolymer forms multilayers at the interface. In the system for which the interface between PMMA and PPO is toughened with

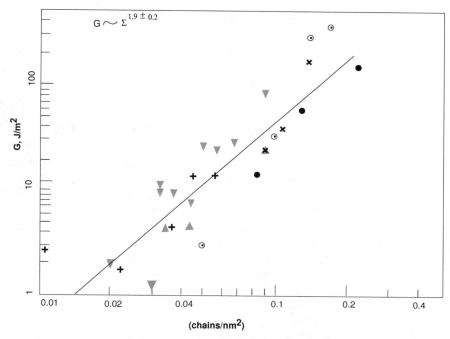

Figure 23.5   Variation of G_c with Σ showing the Σ^2 relation at high coverages.

PS–PMMA copolymer, the multilayer contains internal PS lamellae [22]. Owing to the low entanglement density of PS, these lamellae tend to fail at a low stress, weakening the interface. In the PS–PVP system the internal PS layer tends to be swollen with low-molecular-weight PS from the polydisperse homopolymer, so again it is a weaker layer [25].

It is possible to estimate the molecular weight of the copolymer that will give maximum interfacial toughness. In the chain breakage regime, the maximum toughness varies as Σ_{sat}^2. It seems likely that, at saturation, the diblock chains pack at the interface similarly to the way in which they pack in the neat diblock for which Σ varies as $N^{-0.4}$. (The long period L varies as $N^{0.6}$, and $\Sigma \approx L/N$.) Therefore, in this regime the maximum toughness is expected to vary as $N^{-0.8}$. In the chain pullout (and crazing) regime, the maximum toughness is expected to vary as $N^2\Sigma_{\text{sat}}^2$ or as $N^{1.2}$, and so the maximum toughness is at about the transition from pullout to scission. The position of this transition is not entirely clear, but seems to occur when N is between $2N_e$ to $5N_e$.

So far, in our discussion of the interface toughened with a diblock copolymer, it has been assumed that the copolymer is organized at the interface so that one block is mixed with one homopolymer and the other block with the other homopolymer. Such organization is not easy to obtain in systems wherein χ is small, such as a PS–PMMA diblock blended with PS and PMMA homopolymers. Here, high-molecular-weight diblocks were found to organize only slowly at 150°C, particularly if Σ was large [21]. Interestingly, the interfacial toughness was found to decrease as the diblock became organized, and disorganized high-molecular-weight diblock has given stronger interfaces in this system than have been observed with any organized diblock layer. It is these results on disorganized layers that inspired the experiments with random copolymers discussed in the section.

V. TRIBLOCK, RANDOM, AND OTHER POLYMER GEOMETRIES

Diblock copolymers can clearly be very effective in toughening an interface, but it is not obvious that copolymers of a different architecture might not be more effective. For optimum interfacial toughening, as discussed previously, diblock copolymers must have a fairly low molecular weight and thus contribute a significant density of chain ends close to the interface. As chain ends are sources of weakness, it seems possible that multiblock systems might be more effective than diblocks.

The mechanical effects of *ABA* triblock copolymers have been examined in the PS–PVP system [29]. At coverages lower than saturation, the PVP-b-PS-b-PVP copolymer was found to form a staple structure with both PVP blocks in the PVP. The copolymer behaved in a very similar manner to that expected if each PS block were cut in the middle to form two molecules of diblock. However, there was evidence that the PS center block, when forming a loop, entangled more strongly with the homopolymer than would be expected for a PS end block of half the molecular weight. Very different results were obtained when an *ABC* triblock was placed between PPO and PMMA, where the *A* and *C* blocks were PS and PMMA and the *B* block was polyisoprene (PI). The copolymer weakened rather than toughened the interface, because the presence of the elastomeric center block suppressed crazing [30].

Random copolymers have been found to be quite effective in toughening polymer interfaces in favorable situations [6, 20, 21, 31–38]. In the initial system studied, PS-r-PMMA between PS and PMMA, the adhesion is caused by the ability of a random copolymer to form relatively broad interfaces with each of the homopolymers. The repulsive interaction between the two homopolymers is not strong ($\chi \approx 0.035$). Adhesion and interfacial thickness have been examined as a function of monomer ratio in the copolymer. Maximum adhesion is found with a copolymer containing 70% PS, as shown in Fig. 23.6. This copolymer is thermodynamically balanced, as it has similar (low) solubility in both homopolymers. If the copolymer forms a discrete layer having a separate interface with each homopolymer, then the homopolymer-to-homopolymer toughness will be controlled by the thinner of these two interfaces. Neutron reflectivity has shown that, owing to the thermodynamic balance, the 70% PS copolymer maximizes the width of the narrower of the two interfaces [33, 34]. The situation in the PS–PVP system is not as clear. Although the homopolymers are quite strongly immiscible ($\chi \approx 0.1$), high-molecular-weight random copolymers are still very effective coupling agents, with the chains failing by scission [31, 36]. The toughness is found to vary rapidly with copolymer concentration and show a pronounced maximum close to 50% PS. This strong coupling might have its origin

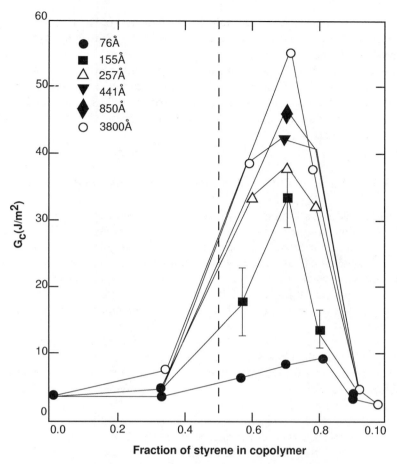

Figure 23.6   Variation of G_c with volume fraction of PS in a PS-r-PMMA copolymer at the PS–PMMA interface.

in the composition drift that occurred when the copolymer chains were synthesized. The chains might have organized so that there is a composition gradient across the interface, giving an overall interface that is highly entangled.

A related strategy for interfacial toughening is to use, instead of a random copolymer, a third polymer that is miscible, or close to miscible, with the two substrate polymers. This technique has been shown to be very effective when PMMA is used to improve the adhesion between polycarbonate and poly(vinylidiene fluoride) [39] and to be useful in the polycarbonate–styrene acrylonitrile system [40].

In none of the cases considered so far have there been specific interactions between repeat units in the copolymer at the interface and repeat units in one of the substrate homopolymers. The effect of such interactions has been studied, again in the PS–PVP system, when the copolymer contains p-hydroxystyrene (PPHS) units that interact strongly with PVP. Random PPHS–PS copolymers were found to be very effective coupling agents, more effective than random or block PS–PVP copolymers, with an optimum PPHS fraction of just 3% [37]. Failure at a toughness close to that of bulk PS occurred by crazing followed by chain pullout from the PS craze. The low optimum PPHS level is explicable in terms of the hydrogen bonding interactions between the PPHS units and between PPHS and PVP units. Measurements have also been made on the toughening effects of multiblock and graft PPHS–PS copolymers at the PS–PVP interface, but none of the polymers tried proved effective in interfacial toughening [38].

VI. REACTIVE SYSTEMS AND SEMICRYSTALLINE POLYMERS

In many blends, and also in multilayer coextrusions, the interfaces between the different phases are toughened using block copolymers that are formed *in situ* at the interface by reaction between components that exist in both phases. This type of interfacial toughening has been examined in a number of systems [4, 41–54], but the picture is not as clear as that for premade block copolymers, as the situation is more complicated and difficult to study. In many of the experimental systems, one or both polymers was semicrystalline, but the simpler situation occurs when both bulk phases are glassy.

The broad picture that provides an understanding of interfacial toughening by the addition of premade block copolymers can form the basis for understanding toughening in reactive systems. However, there are important extra issues connected with the kinetics of formation and the topology of the copolymer at the interface. Many of the relevant copolymers are formed by the end reaction of one polymer, often a polyamide, with grafting sites on a second polymer. For such a polymer to be mechanically effective, it is necessary that both the backbone and the grafts entangle with the relevant bulk polymers. For the grafts, the molecular-weight criterion for effectiveness is presumably the same as that for diblocks. However, if the reaction enthalpy is very large, it might be possible to generate a grafting density so large that entanglement between the grafts and the homopolymer becomes poor. For the backbone, the situation is not clear. It would seem likely from the experiments on triblocks described previously that entanglement with homopolymer chains will occur only when the lengths of the polymer loops between grafting points are at least $2N_e$, and perhaps longer. The end blocks will presumably behave in a manner similar to a diblock. Hence, good entanglement of the backbone with the relevant homopolymer will require either a low grafting density, perhaps just one graft per backbone, or a backbone of very high molecular weight.

The issue of chain entanglement between a block copolymer and a homopolymer is not well understood for semicrystalline polymers. However, it would seem likely that chains that are much too short to form a tough homopolymer will also fail to entangle (or cocrystallize causing strong coupling [55]) with the homopolymer. A second important issue with these materials is that the crystallizing conditions, and hence crystal structure, can have a profound effect on the deformation properties and toughness of the homopolymer.

The rate of end reaction to form a copolymer varies rapidly with temperature. Clearly, sufficient chain mobility is required, and so neither material can be glassy. If one of the materials is noncrystalline and at a temperature above its T_g and the other material is semicrystalline, then some reaction would be expected in the amorphous regions. Reaction presumably occurs most rapidly when both materials are molten. In many of the published experiments, the kinetics are complicated by the fact that a thin layer of one of the reactive materials is placed at the interface and then covered with a thicker layer of a similar but nonreactive material. In this situation, a competition exists between interfacial reaction and diffusion of the reactive material away from the interface. Hence, toughness can either decrease or increase with increasing temperature, depending on the relative temperature dependencies of the reaction and diffusion rates.

Perhaps the clearest experimental situation in a reactive system was obtained when a small amount of carboxy-terminated d-PS was reacted with a partially cured epoxy resin [45, 46]. Excess ungrafted chains were then washed off, and the epoxy slab with a grafted PS layer at the surface was fusion bonded to either a PS or HIPS substrate. Crack-tip deformation and failure occurred within the PS or HIPS. The areal density Σ of grafted chains and the location of failure were measured using ion scattering. The toughness and failure mechanisms were found to be consistent with those observed using diblock copolymers. As the maximum obtainable value of Σ was found to decrease linearly with grafted chain length, the toughness showed a maximum at intermediate chain lengths.

Experiments on the use of a layer of styrene maleic anhydride (SMA) to cause coupling between an amorphous polyamide and PS have shown the existence of an optimum amount of interfacial reaction [42, 43]. The SMA is thought to diffuse into the PS as well as react at the interface. Too much interfacial reaction causes multiple grafts per chain and poor entanglement. Similar conclusions were obtained in experiments on adhesion between PA-6 and polypropylene (PP) containing a small amount of maleic-anhydride-grafted PP (PP-g-MA) [4]. Less adhesion was obtained with PP-g-MA, with three grafts per chain, than with a material with just one graft per chain. These experiments all used relative long joining times. It is interesting to speculate what might be the practical situation for the formation of blends and coextrusions. Here, the temperature is high, but only for a short time. If the interfacial reaction is far from completion, then multiply grafted material may be optimum.

The PP–PA-6 system was also used to study the relationship between Σ and G in a reactive system for which the crack-tip deformation is in the semicrystalline PP [4]. Perhaps surprisingly, in most cases the toughness was found to vary as Σ^2 (see Fig. 23.7), as it would in a glassy material. However, an exception was observed when the joint was made at a high temperature (220°C) and a long-chain PP-g-MA material was used. In this case, the toughness increased by a factor of 4 from the expected result, perhaps because a different PP crystal structure was formed [41] but recent unpublished work from the same group has cast doubt on the presence of the different crystal structure. Effects of the amount of copolymer, as measured by interfacial thickness λ, have also been examined in a polysulfone (PSF)–amorphous PA (a-PA) system, where λ was measured by ellipsometry [52]. Toughness was found

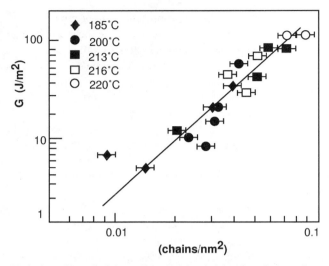

Figure 23.7   Variation of interfacial toughness with areal density of *in situ* formed copolymer at an interface between PP and PA-6. The PP contained a small amount of PP chains grafted with maleic anhydride. Reprinted from [4], with permission.

to vary approximately as λ^2, but at a given λ, the toughness was significantly higher for an end-reactive PSF than for a grafted PSF.

The effects of the joining temperature have been examined in a number of semi-crystalline systems. In the PP-g-MA–PA-6 system, no interfacial toughness is observed until the joining temperature is above the melting point of PP. Toughness, as shown in Fig. 23.8, then tends to increase as more reaction occurs and then jumps to very high values when the PA is melted [4, 49]. This latter jump has been ascribed to interfacial roughening. If a-PA is used instead of PA-6, the reaction starts to occur as the temperature is increased above the T_g of the a-PA [51].

The effect of copolymer chain topology has been examined in a number of experiments in which the copolymers formed were neither simple grafts nor diblocks. The mechanical effect of a dicarboxy-terminated PS, containing carboxyl groups at each end of the PS chain, was compared with the effect of a number of monocarboxy-terminated PS's of different chain lengths at an interface between PS and poly (glycidyl methacrylate). The difunctional material, which was believed to form loops, provided a stronger interface than did any of the monofunctional materials [44]. The effect of having both chains multiply reactive was studied in a PVP–PS system in which the PS was lightly sulfonated [47]. The sulfonated groups are thought to react with the PVP repeat unit forming multigrafted chains. Optimum interfacial toughness was observed at about 7 mol% sulfonation. It is thought that excess reaction at high sulfonation levels causes lack of entanglement, as is seen in other grafting systems.

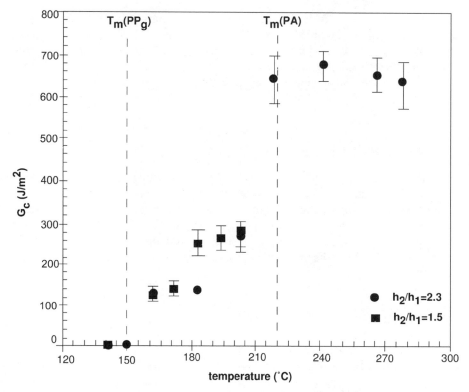

Figure 23.8   Toughness of an interface between reactive PP and PA-6 as a function of joining temperature, showing the rapid increase in toughness at the melting point of the PA. Reprinted from [49], with permission.

VII. ELASTOMERIC POLYMERS

In many fully elastomeric systems, interfacial toughness is obtained simply by cross-reaction. If the two elastomers cross-link using the same chemistry, as, for example, in sulfur vulcanization, then cross-reaction can occur within the interfaces. No extra interfacial toughening is required.

In systems for which one or both of the polymers are elastomeric, the deformation processes normally occur within the elastomer. Interfacial coupling can occur through the presence or formation of block copolymers, so the interfacial toughness is normally controlled by the coupling between the elastomeric block and the relevant homopolymer. If the copolymer block is reacted into the homopolymer, then interfacial failure requires chain scission; if not, chain pullout can probably occur for all reasonable chain lengths.

Effects of PS–PI diblock copolymers at interfaces between PS and PI and also of PS–PDMS (polydimethyl siloxane) copolymers at interfaces between PS and PDMS have been examined both with and without reaction between the elastomeric block and bulk elastomer [56–59]. The chain pullout in the elastomer produces weak but measurable interface toughening in the PS–PI system, but very little effect in the PS–PDMS system. When the block and homopolymer are reacted together, considerable adhesion could result, consistent with the model of elastomer failure discussed previously. Toughening by chain pullout has also been observed in a reactive system [60].

VIII. CONCLUSIONS

Polymer interfaces can be toughened using copolymers with a range of chain topologies, which are either premade or formed *in situ*. The simplest system, when a diblock copolymer is used to toughen the interface between glassy polymers, is fairly well understood. As the systems become more complicated, with different copolymer topologies, semicrystalline polymers, and reactive systems, the level of understanding decreases. However the broad rules derived from the simplest systems sometimes still apply.

IX. REFERENCES

1. H. R. Brown, *J. Mater. Sci.*, **25**, 2791 (1990).
2. F. Ramsteiner, *Polym. Test.*, **15**, 573 (1996).
3. C. Creton, E. J. Kramer, C.-Y. Hui, and H. R. Brown, *Macromolecules*, **25**, 3075 (1992).
4. E. Boucher, J. P. Folkers, H. Hervet, L. Léger, and C. Creton, *Macromolecules*, **29**, 774 (1996).
5. F. Xiao, C.-Y. Hui, J. Washiyama, and E. J. Kramer, *Macromolecules*, **27**, 4382 (1994).
6. B. Bernard, H. R. Brown, C. J. Hawker, A. J. Kellock, and T. P. Russell, *Macromolecules*, 1999, in press.
7. K. Cho, H. R. Brown, and D. C. Miller, *J. Polym. Sci., Polym. Phys. Ed.*, **28**, 1699 (1990).
8. Q. Wang, F. P. Chiang, L. Guo, M. Rafailovich, and J. Sokolov, *Mat. Res. Soc. Symp. Proc.*, **409**, 275 (1996).
9. D.-B. Xu, C.-Y. Hui, E. J. Kramer, and C. Creton, *Mech. Mater.*, **11**, 257 (1991).
10. E. Raphaël and P. G. de Gennes, *J. Phys. Chem.*, **96**, 4002 (1992).
11. G. J. Lake and A. G. Thomas, *Proc. R. Soc. London, A*, **300**, 108 (1967).
12. H. R. Brown, *Macromolecules*, **24**, 2752 (1991).
13. C. Y. Hui, A. Ruina, C. Creton, and E. J. Kramer, *Macromolecules*, **25**, 3949 (1992).
14. Y. Sha, C. Y. Hui, A. Ruina, and E. J. Kramer, *Macromolecules*, **28**, 2450 (1995).
15. Y. Sha, C. Y. Hui, A. Ruina, and E. J. Kramer, *Acta Mater.*, **45**, 3555 (1997).
16. F. Xiao and W. A. Curtin, *Macromolecules*, **28**, 1654 (1995).
17. H. R. Brown, V. R. Deline, and P. F. Green, *Nature*, **341**, 221 (1989).
18. H. R. Brown, *Macromolecules*, **22**, 2859 (1989).
19. H. R. Brown, W. F. Reichert, and K. Char, *A.C.S. Polym. Pre.*, **33**, 72 (1992).
20. K. Char, H. R. Brown, and V. R. Deline, *Macromolecules*, **26**, 4164 (1993).

21. H. R. Brown, K. Char, V. R. Deline, and P. F. Green, *Macromolecules*, **26**, 4155 (1993).

22. C. Creton, H. R. Brown, and V. R. Deline, *Macromolecules*, **27**, 1774 (1994).

23. C. Creton, E. J. Kramer, and G. Hadziioannou, *Macromolecules*, **24**, 1846 (1991).

24. J. Washiyama, E. J. Kramer, and C. Y. Hui, *Macromolecules*, **26**, 2928 (1993).

25. J. Washiyama, C. Creton, E. J. Kramer, F. Xiao, and C. Y. Hui, *Macromolecules*, **26**, 6011 (1993).

26. J. Washiyama, E. J. Kramer, C. Creton, and C. Y. Hui, *Macromolecules*, **27**, 2019 (1994).

27. C.-A. Dai, E. J. Kramer, J. Washijama, and C.-Y. Hui, *Macromolecules*, **29**, 7536 (1996).

28. J. C. Cho, K. H. Kim, and W. H. Jo, *Polym. J.*, **28**, 1060 (1996).

29. C.-A. Dai, K. D. Jandt, D. R. Iyengar, N. L. Slack, K. H. Dai, W. B. Davidson, E. J. Kramer, and C.-Y. Hui, *Macromolecules*, **20**, 549 (1997).

30. H. R. Brown, U. Krappe, and R. Stadler, *Macromolecules*, **29**, 6582 (1996).

31. C.-A. Dai, B. J. Dair, K. H. Dai, C. K. Ober, E. J. Kramer, C.-Y. Hui, and L. W. Jelinski, *Phys. Rev. Lett.*, **73**, 2472 (1994).

32. M. Sikka, N. N. Pellegrini, E. A. Schmitt, and K. I. Winey, *Macromolecules*, **30**, 445 (1997).

33. R. Kulasekere, H. Kaiser, J. F. Anker, T. P. Russell, H. R. Brown, C. J. Hawker, and A. M. Mayes, *Macromolecules*, **29**, 5493 (1996).

34. R. Kulasekere, H. Kaiser, J. F. Ankner, T. P. Russell, H. R. Brown, C. J. Hawker, and A. M. Mayes, *Physica A*, **221**, 306 (1996).

35. K. Cho, T. O. Ahn, H. S. Ryu, and K. H. Seo, *Polymer*, **37**, 4849 (1996).

36. C.-A. Dai, C. O. Osuji, K. D. Jandt, B. J. Dair, C. K. Ober, and E. J. Kramer, *Macromolecules*, **30**, 6727 (1997).

37. Z. Xu, E. J. Kramer, B. D. Edgecombe, and J. M. J. Frechet, *Macromolecules*, **30**, 7958 (1997).

38. B. D. Edgecombe, J. A. Stein, J. M. J. Frechet, Z. Xu, and E. J. Kramer, *Macromolecules*, **31**, 1292 (1998).

39. N. Moussaif, P. Marechal, and R. Jerome, *Macromolecules*, **30**, 658 (1997).

40. M. Schaffer, V. Janarthanan, Y. Deng, J. La Scala, M. Rafilovich, and J. Sokolov, *Macromolecules*, **30**, 1225 (1997).

41. E. Boucher, J. P. Folkers, C. Creton, H. Hervet, and L. Léger, *Macromolecules*, **30**, 2102 (1997).

42. Y. Lee and K. Char, *Macromolecules*, **31**, 7091 (1998).

43. Y. Lee and K. Char, *Macromolecules*, **27**, 2603 (1994).

44. Y. Lee, C. Sim, and K. Char, *Macromolecules*, 1999, in press.

45. L. J. Norton, V. Smigolova, M. U. Pralle, A. Hubenko, K. H. Dai, E. J. Kramer, S. Hahn, C. Bergland, and B. DeKoven, *Macromolecules*, **28**, 1999 (1995).

46. Y. Sha, C.-Y. Hui, E. J. Kramer, S. F. Hahn, and C. A. Bergland, *Macromolecules*, **29**, 4728 (1996).

47. N. C. Beck Tan, D. G. Peiffer, and R. M. Briber, *Macromolecules*, **29**, 4969 (1996).

48. N. C. Beck Tan, S.-K. Tai, and R. M. Briber, *Polymer*, **37**, 3509 (1996).

49. J.-E. Bidaux, G. D. Smith, N. Bernet, J.-A. E. Månson, and J. Hilborn, *Polymer*, **37**, 1129 (1996).

50. K. Cho, K. H. Seo, T. O. Ahn, J. Kim, and K. U. Kim, *Polymer*, **19**, 4825 (1997).

51. K. Cho and F. Li, *Macromolecules*, **31**, 7495 (1998).

52. H. Koriyama, H. T. Oyama, T. Ougizawa, T. Inoue, M. Weber, and E. Koch, *Macromolecules*, 1999, in press.

53. D. Mäder, J. Kressler, and M. Weber, *Macromol. Symp.*, **112**, 123 (1996).

54. D. Mäder, M. C. Coen, J. Kressler, R. Mülhaupt, and M. Weber, *J. Appl. Polym. Sci.*, **65**, 567 (1997).

55. H. E. Hermes, D. G. Bucknall, J. S. Higgins, and R. L. Scherrenberg, *Polymer*, **39**, 3099 (1998).

56. W. F. Reichert and H. R. Brown, *Polymer*, **34**, 2289 (1993).

57. H. R. Brown, W. Hu, J. T. Koberstein, and Y. Gallot, *Macromolecules*, 1998, submitted for publication.

58. H. R. Brown, *Macromolecules*, **23**, 1666 (1993).

59. C. Creton, H. R. Brown, and K. R. Shull, *Macromolecules*, **27**, 3174 (1994).

60. M. H. Chung and G. R. Hamed, *Rubber Chem. Technol.*, **62**, 367 (1989).

24 Core–Shell Impact Modifiers

CARLOS A. CRUZ-RAMOS

Plastics Additives Research Department
Rohm and Haas Company
Route 13 and Old State Rd.
Bristol, PA 19007

Polymer Blends, Volume 2: Performance. Edited by D. R. Paul and C. B. Bucknall.
ISBN 0-471-35280-2.   © 2000 John Wiley & Sons, Inc.

I. INTRODUCTION

Core–shell polymers were commercially introduced as PVC impact modifiers in 1958 [1]. Since that time, their use has continuously expanded into new toughening applications, which now include a wide variety of engineering polymers. In general, the demand for core–shell polymers is expected to increase steadily as the overall plastics market keeps growing [2].

A major distinction between core–shell particles and other types of impact modifiers is that their size is set during the synthesis process and remains the same after they are dispersed in a host matrix, whereas the final particle size of linear polymers and bulk rubbers after blending depends largely on processing conditions. (See Chapter 16.)

The purpose of this chapter is to offer general ideas about the architecture and synthesis of core–shell polymers and their use as impact modifiers. Current notions on how they disperse and interact with polymeric matrices are briefly addressed, and examples of their use in various types of polymers are included as well. This survey does not pretend to be exhaustive, but rather is a starting point from which one may then consult other works, including several chapters in this book and the material already reviewed in excellent fashion in previous publications [1, 3–7].

II. GENERAL CONCEPTS

The typical architecture of core–shell polymers is described schematically in Fig. 24.1, wherein a soft core, made up of a rubbery polymer, is surrounded by a "shell" of rigid polymer that is grafted to the core. The chemical makeup of the core, or inner phase, in commercial materials is usually a cross-linked rubber based on poly(butyl acrylate) (PBA), or poly(butadiene), generally copolymerized with styrene. This central nucleus of the particle provides the soft second phase that induces toughening. The shell of the particles, occasionally referred to as the outer or hard stage, consists of a polymer that is chemically grafted onto the core and generally has a much higher glass transition temperature T_g than that of the core. Typical commercial examples of polymers used in the outer stage

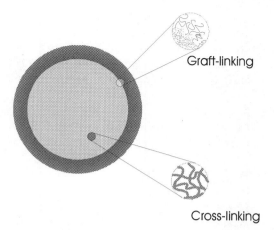

Graft-linking

Cross-linking

Figure 24.1   Diagram of a typical core–shell modifier particle. Notice the distinction between cross-linking used in the rubber and graft-linking, to chemically bind the shell polymer.

are poly(methyl methacrylate) (PMMA) homopolymers and copolymers, as well as styrene–acrylonitrile copolymers, SAN [1].

The shell polymer has two primary functions. First, it must allow isolation of the particles from the emulsion by providing a hard coating that keeps the rubbery cores from adhering to one another during the drying process. Second, as the particles are dispersed in the host matrix, the shell functions as the layer that physically binds the matrix to the rubber core. PBA–poly(methyl methacrylate)-based core–shell tougheners are known as all-acrylic impact modifiers (AIMS), while poly(butadiene-styrene)–poly(methyl methacrylate) combinations are referred to as MBS copolymers. AIMS are generally used as toughening agents when weatherability is important, whereas MBS modifiers are preferred for low-temperature applications.

The well-defined particle size and narrow particle-size distribution of core–shell particles are a direct result of the emulsion polymerization process, wherein each colloidal particle is individually polymerized and cross-linked during the synthesis process. Once the particles have been dispersed in the melt, they retain their individual dimensions set in the reactor.

III. PREPARATION

Core–shell modifiers are produced by emulsion polymerization of free-radical-initiated systems. The versatility of the polymers prepared using emulsion technology is reflected in the wide variety and relevance of products in which they are applied, including films, coatings, adhesives, paints, and various kinds of additives. Next, a brief outline of how this preparation process takes place is presented. Numerous references that provide detail about the polymerization process in emulsion can be found in the literature [8–12].

A. Chemistry

Free-radical polymerization is well known to consist of the following steps, where I–I is an initiator, M is a monomer molecule, P is the polymer being formed, C is a chain transfer agent, and all of the moieties with a ($\bullet$) next to them are free radicals:

Initiation: $I–I \rightarrow 2I\bullet$
                $I\bullet + M \rightarrow I–M\bullet$
Propagation: $I–M\bullet + M \rightarrow P\bullet$

                . . . .

                $P_x\bullet + M \rightarrow P_{x+1}\bullet$
Termination: $P_y\bullet + P_z\bullet \rightarrow P_{y+z}$
                $P_y\bullet + P_z\bullet \rightarrow P_y + P_z$
Chain transfer: $P_x\bullet + C \rightarrow P_x + C\bullet$

Figure 24.2 provides a traditional schematic view of how a polymerization reaction takes place in emulsion [13].

Emulsion processes rely on the monomer, together with the polymer being produced to form stable emulsified colloids in an aqueous medium with the aid of a surface active ingredient (emulsifying agent or surfactant). Maintaining a proper balance of temperature, agitation, and other reaction conditions, such as the pH of the aqueous medium, is the key to running the polymerization successfully. The water phase provides the heat transfer that helps keep the exothermic polymerization process under control.

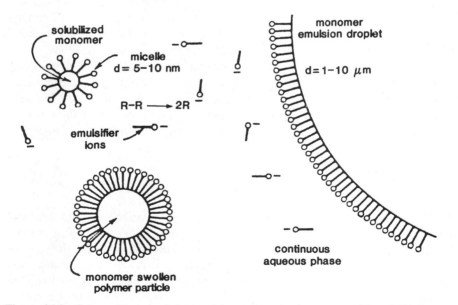

Figure 24.2   Schematic view of the emulsion polymerization process. From [12]. Reproduced with permission from the publisher.

Some of the typical monomers used in emulsion processes, with particular emphasis on those used to produce commercial core–shell tougheners, are listed in Table 24.1, along with some of their characteristic properties [14–17].

Surfactant is first added to the water phase, where it forms micelles if it is above its "critical micelle concentration." Upon this addition, the bulk of the monomer becomes dispersed as large droplets well in excess of 1 μm in diameter. Part of the monomer swells the micelles, and a small portion is dissolved in the aqueous phase. A water-soluble initiator generates free radicals that remain mainly in the water phase and form oligomers with the small amount of monomer present. These oligomers precipitate in the micelles, which far exceed the monomer droplets in number. It is currently believed that oligomers containing free radicals penetrate into the micelles, where the propagation step of the free-radical process takes place [10]. The solubility of the monomer in the water medium and its diffusion are fundamental in driving this process. Once the monomer has reached the polymerizing droplet, free-radical polymerization continues as described previously.

Three types of surfactants are commonly used [10]: electrostatic surfactants, either anionic or cationic, which stabilize the latex particles by electrostatic repulsion; polymeric surfactants (steric stabilizers), which maintain stable latex particles via entropic repulsion; and electrosteric stabilizers, which combine the properties of the other two types of surfactants. Metal salts of arylalkyl carboxylic, sulfates, sulfonates, and sulfocarboxylates (such as, for instance, sodium laurate, potassium oleate, sodium n-hexadecyl sulfate, sodium dodecyl sulfate, sodium dodecylbenzene sulfonate, and sodium dihexyl sulfosuccinate) are common anionic surfactants used in industry for "seeded" emulsion polymerization—that is, emulsions that contain prepolymerized particles that serve as the loci for polymerization. The initiator is usually soluble in water. In so-called thermal initiation, an often-used initiator is potassium persulfate, which liberates $SO_4^-\bullet$ free radicals at temperatures in excess of 50°C. Also used in practice is redox initiation, wherein an Fe^{2+} salt is dissolved in the aqueous phase, along with a peroxide, like cumene hydroperoxide, to provide a source of free radicals at temperatures lower than 50°C.

Other important ingredients for properly carrying out the process are buffering agents and chain transfer agents. The former help control changes in pH that the medium experiences during the generation of free radicals by the redox initiator system. The latter allow the control of molecular weight. Long-chain mercaptans are typically used for chain transfer. Finally, "chasers" are added to reduce the amount of residual monomer at the end of the polymerization and may consist of a shot of the redox system used for initiation.

Particles with sizes on the order of 0.1 and up to 0.5 μm are commonly produced via the emulsion route. Particle size is controlled by the amount of surfactant present in relation to the polymerizing monomer added. For a given amount of monomer, a larger amount of surfactant produces a larger number of micelles, and the monomer ends up being distributed amongst all micelles. Consequently, smaller particles are formed. Reducing the concentration of surfactant causes larger particles to form. The micelle size is determined by the packing of the surfactant molecules, and the

Table 24.1 Relevant Properties of Typical Monomers and Their Homopolymers used to Produce Core–Shell Particles[a]

Monomer	Monomer Properties			Polymer Properties			
	Molecular Weight, g/mole	Water Solubility g_monomer/100 g H_2O	Heat of Polymerization, kJ/mole	Density, g/cm^3	Refractive Index (23°C)	Glass Transition Temperature, °C	Solubility Parameter (cal/cm^3)
Ethyl Acrylate (EA)	100.1	1.51	77.8	1.12	1.464	−8	9.2
Butyl Acrylate (BA)	128.2	0.2	77.4	1.06	1.474	−45	8.9
Methyl Methacrylate (MMA)	100.1	1.59^{20}	57.7	1.19	1.489	105	9.2
Butyl Methacrylate (BMA)	142.2	0.1	59.3	1.06	1.483	32	8.7
Styrene (St)	104.2	0.032	60.8	1.07	1.589	107	9.2
Acrylonitrile (AN)	53.1	7.9^{20}	72.4	1.18	1.519	105	12.7
Butadiene (Bd)[b,c]	54.1	0.081^d	73^e	0.892	1.518	−85 to −58	8.3

[a]Reference [14]. [b]Reference [15]. [c]For the polymer, mixture of cis- and trans-poly(1,4-butadiene). [d]Reference [16]. [e]Reference [17].

monomer is distributed in a uniform manner throughout the micelles. Consequently, the particle-size distribution in these systems tends to be monodisperse.

The process becomes more complex when it includes a second monomer, since variations in composition or chain microstructures of the final polymer occur, because the monomers have different transport properties and different thermodynamic interactions with the polymers being formed in the micelles.

B. Formation of the Core–Shell Structure

To maintain the integrity of the rubber particles during processing and produce an elastomeric behavior, cross-linkers are added to the first stage of polymerization. This factor is particularly valid in the case of butyl acrylate. Butylene glycol dimethacrylate, ethylene glycol dimethacrylate, trimethylolpropane trimethacrylate, pentaerythritol tetra-acrylate, and divinylbenzene are common cross-linking agents. Polybutadiene cores do not usually require a cross-linker, as the conjugated double bonds in the monomer induce the formation of a network.

Monomers that contain two double bonds, each polymerizing at a different rate, are added in the latter part of the first stage to induce the grafting of the shell. Examples of these monomers are allyl, methallyl, and crotyl esters of acrylic acid, methacrylic acid, and maleic acid. Heatley et al. have studied the very special reaction kinetics of allyl methacrylate and other graft-linking molecules [18].

The seemingly straightforward process of adding a second monomer or monomer mix to form a shell around the rubber core might lead to different morphologies, as illustrated in Fig. 24.3 [19]. This process has a thermodynamic basis: Hydrophilic monomers tend to remain on the outside of the particle, more in contact with the aqueous phase, whereas hydrophobic monomers tend to penetrate the core. A core–shell structure is favored by immiscibility between the polymers present in the core and those forming the shell and by a low solubility of the shell monomer in the core polymer [20]. When polymerizing a hydrophobic monomer, like styrene, on top of an already-formed hydrophilic core, such as poly(butyl acrylate), the more hydrophobic monomer tends to penetrate the core and form inclusions similar to those shown in Fig. 24.3, because of the immiscibility of the two polymers.

If the interfacial tension between the incoming monomer and the prepolymerized core is large, as might be the case for a hydrophilic monomer polymerizing on a polybutadiene core, the monomer might form domains, instead of spreading evenly about the core surface, as shown in Fig. 24.3. Okubo [21] showed how different morphologies can be obtained by taking advantage of the hydrophobic–hydrophilic interactions in systems wherein two monomers are polymerized sequentially, and other investigators have proposed thermodynamic models to predict and control desired morphologies in such situations [22, 23].

The polymerization kinetics of the second-stage monomers and the availability of reactive sites on the prepolymerized core can be used to anchor morphologies that are not thermodynamically favored. Rudin [24] has shown examples of this type of "kinetic" control of the morphology. In Rudin's view, the formation of a core–shell morphology is a strong function of the following factors:

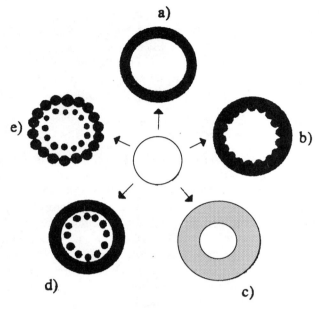

Figure 24.3   Departures from ideal core–shell morphology upon grafting of the shell. (a) Ideal core–shell morphology, (b) wavy internal interface, (c) interface with composition gradients, (d) microdomains or occlusions, (e) island shell structures and occlusions. From [19]. Reproduced with permission from the publisher.

(a) miscibility and hydrophilicity of the different monomer–polymer combinations present,
(b) type of surfactant,
(c) volume ratio of first and second stages,
(d) type of process (batch or semibatch),
(e) process temperature and individual glass transition temperatures of the polymers involved, and
(f) cross-linking agents.

C. Manufacturing

The production of core–shell modifiers for commercial use as tougheners generally involves at least an isolation step beyond the reaction phase. Figure 24.4 is a simple flow diagram of the operations that are normally carried out to produce these types of polymers. In most industrial applications to produce core–shell polymers, an agitated reactor is used in "semibatch" mode. The reactor is equipped with a cooling system and inlets for inert gas and monomer streams. The semibatch, or semicontinuous, mode consists of charging a portion of the ingredients first and adding the remaining portions at key points during the polymerization. Running the operation

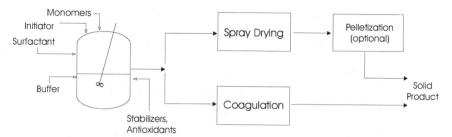

Figure 24.4 A simplified flow diagram of the commercial production of core–shell modifiers. Either spray drying or coagulation is carried out to obtain dry powders. Pelletization of the powders is sometimes carried out.

in this manner adds versatility to the synthetic process. Models designed to facilitate the control of the reaction process have been published [25, 26].

Isolation of the polymeric substrate from the emulsion is an important part of the process, as the form of solid substrate produced (powders, granules, or pellets) can have a significant effect on its ease of incorporation into the matrix and on particle shape. A general method of modifier isolation is spray drying, which involves injecting the emulsion with hot air and forcing it rapidly through a rotating nozzle, to evaporate the water quickly [27]. Although highly efficient, this method has the disadvantage of carrying along with the desired product most of the nonvolatile elements added during the reaction, such as emulsifier, inorganic salts added to control pH, and any ions present in the water, which might affect the resin to be toughened. To this end, a full range of technology has been developed around controlled coagulation of the emulsion, followed by filtration of the aqueous phase and final drying of the resulting wetcake. A much cleaner product can be produced in this manner. More detail on these methodologies is available elsewhere [28–31].

IV. CHARACTERIZATION

A review of the characterization methods applicable to emulsion polymers in general was recently published [32]. Particle size and particle-size distribution are two of the most important parameters that can be measured directly in the latex. Collins [33] recently reviewed the field, with particular emphasis on practical aspects, and provided an excellent guide for instrument selection. Collins makes note that the determination of particle size by scattering measurements would be relatively simple if all particles in the emulsion were spherical. If the shapes of the particles are not known, the routine application of these techniques can be limited.

For the normal range of particle sizes and shapes observed for core–shell particles in the emulsion, normally below 0.6 μm, certain techniques can be used effectively. Quasielastic light scattering (QELS) is a convenient tool to assess particle diameters with precision, although it relies on the presence of spherical particles and assumes a predefined particle-size distribution. Hydrodynamic chromatography is another fre-

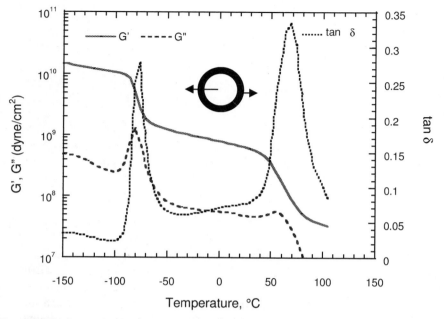

Figure 24.5 Dynamic mechanical analysis (DMA) temperature scan of an MBS modifier. The clearly defined two-phase structure of the core–shell particle is manifested in the sharp temperature transitions in the loss angle (tan δ). The arrows in the figure point out the assignment of each transition to the core and shell portions of the particle.

quently used technique, based on the different transport rates exhibited by colloidal particles of different sizes when passing through a packed bed. Capillary hydrodynamic fractionation (CHDF) provides relatively fast and reliable measurements when the emulsion particles are below 1.5 μm.

Characterization in the solid state involves the determination of thermal and viscoelastic properties. Differential scanning calorimetry (DSC) uses the dry powder isolated from the emulsion and gives a first indication that two phases with different T_g's are present in the polymer. A preliminary assessment of the ratio of the two phases can be gained by comparing the magnitudes of the respective T_g's (see Chapter 10) [34]. Further use of DSC allows one to rate the thermal stability of modifiers on a relative basis and also with respect to the polymers with which they will be blended.

Viscoelastic characterization methods, such as dynamic mechanical analysis (DMA) provide a more accurate method of determining this type of information. Melt pressing the solid powder to form a plaque or film is required to carry out this test. Figure 24.5 shows a DMA temperature scan of an MBS modifier. The core and shell of the modifier are clearly differentiated by the drop in storage modulus as each phase softens and by the maxima in loss modulus and in loss tangent tan δ. Both the DSC and DMA traces of core–shell polymers are similar to those of block or graft copolymers that are phase separated and to immiscible polymer blends. Core–shell polymers are a special kind of graft copolymer, with the difference being that one of the two polymers is cross-linked and the phases remain separated at all temperatures, in contrast to other architectures. (See Chapters 13–15.)

Verification of the core–shell structure might require a combination of latex-based and solid-state techniques, as described previously [35]. The process of polymerizing MMA monomer to form a PMMA shell on top of polybutadiene latex particles shows that a minimum amount of the latter polymer is needed to form a full core–shell structure. The swell index, or degree of swelling of the core–shell structure, is determined by measuring the solvent uptake of the particle and is related to the cross-link density of the rubber. If a suitable solvent is used, a determination of the amount of shell polymer actually grafted to the structure can be made, as shown elsewhere [36].

Microscopic characterization to determine the structure of core–shell particles has involved mainly transmission electron microscopy (TEM) (see Chapter 9) [21, 37] and atomic force microscopy (AFM) [38]. X-ray scattering [39] and small-angle neutron scattering (SANS) (see Chapter 12) [40] are also useful characterization methods. The aim of these techniques is to verify the existence of a core–shell structure and obtain further detail about the interaction of the rubber core with the shell polymer. Because each method presents its own limitations, combinations of the methods are frequently used. TEM, for example, may require staining of the samples and its standard resolution of 10 nm might not detect smaller features. (See Chapter 9.) SANS makes use of partially deuterated samples for contrast. (See Chapter 12.) AFM might offer a distinct advantage in samples that are difficult to stain, as long as the phases have different-enough elastic moduli, as Fig. 24.6 shows. Fluorescence nonradiative energy transfer (NRET) provides evidence of the core–shell structure and the sharpness of the boundary between the core and the shell, but makes use of core–shell particles based on fluorinated polymers to monitor these subtle differences [41].

More advanced nuclear magnetic resonance (NMR) techniques, such as a two-dimensional combination of the proton line shape (^{1}H–^{1}H dipolar-dipolar coupling) and the ^{13}C chemical shift [19], have been applied to core–shell particles. Substructures in the formation of poly(butyl acrylate)–PMMA core–shell particles have been detected that would otherwise be difficult to pinpoint by other methods. Although these and similar findings have not yet been correlated with toughening performance, these tools have the potential to explain the role that subtle changes in the core–shell architecture have on the macroscopic properties of their blends with other polymers.

V. MATRIX–MODIFIER INTERACTION

A key factor in the performance of core–shell particles as impact modifiers is their adhesion to the matrices in which they are dispersed, which is determined by the composition of the shell polymer. Inspection of Flory–Huggins interaction parameters or interaction energy density values (see Chapters 2–5) provides a simple guide to assessing how effective a core–shell modifier can be to toughen a given matrix. A low value of the interaction parameter (or a negative interaction energy density) for

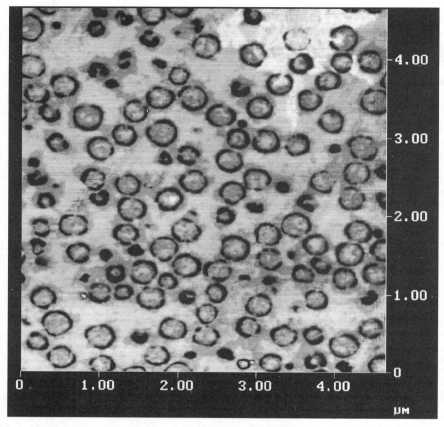

Figure 24.6   Atomic force micrograph of rubber-toughened PMMA. The dark ring domains are made up of the softer phase, butyl acrylate rubber, whereas the core of each particle is glassy, with a modulus close to that of the surrounding matrix. Micrograph courtesy of Dr. R. Antrim, Rohm and Haas Company.

the shell–matrix polymer combination, indicative of miscibility, generally translates into effective toughening. For instance, when the shell of the toughener is made of PMMA, the toughener will tend to perform well in PVC, which is fully miscible with PMMA, and in PC, which is partially miscible with PMMA. A less effective performance of the same core–shell toughener would be expected in the case of PBT, and even no toughening whatsoever would be expected if the toughener were used in PET or nylon, which are both highly immiscible with PMMA. Introducing an intermediate "compatibilizing" polymer layer has a dramatic enhancement on the efficiency of the core–shell impact modifier in these situations. (See Chapter 17.) This is only an elementary description of the role of the interface at the boundary between the core–shell and the host matrix. Increasingly more sophisticated models and experimental approaches are being proposed to understand this interface in more detail, as discussed next.

A. Thermodynamics

1. *Grafting Degree* Aoki and collaborators [42, 43] studied how the equilibrium dispersion of core–shell modifiers varies as a function of the degree and density of grafting onto the shell, defined as follows:

$$\kappa = \text{Graft Density} = \frac{\text{weight of shell polymer grafted to core}}{\text{core surface area}}$$

$$\theta = \text{Grafting Degree} = \frac{\text{weight of shell polymer grafted to core}}{\text{weight of core}}.$$

Aoki et al.'s experiments were carried out in an ABS system, wherein the shell grafted on a butadiene rubber consisted of a styrene–acrylonitrile (SAN) copolymer with the same composition as that of the matrix. It is important to stress that in ABS systems, a portion of the shell chains may not be grafted; that is, there might be free SAN copolymer. In all-acrylic or MBS core–shell impact modifiers, this situation is less likely to occur. In Aoki et al.'s work, rheological and microscopic measurements show that an optimum grafting degree and an optimum graft density of shell polymer are required to achieve the best possible dispersion of core–shell particles in the matrix. If the graft density increases beyond this optimum value, the particles tend to agglomerate, whereas lowering the graft density below the optimum value also leads to poor dispersion. As shown in Fig. 24.7 [42], the shear storage modulus G' at low frequencies was used to monitor the degree of dispersion of core–shell particles in the matrix as a function of the graft density of the shell polymer. A minimum in G' was observed when the core–shell particles showed the highest degree of dispersion by microscopy. The optimum grafting density does not appear to depend on particle size nor on the amount of modifier present, at least within the ranges tested [43].

Bertin et al. [44] confirmed the existence of an optimum degree of grafting in an ABS system and found that the value of this parameter is higher for smaller particles. The origin of the optimum degree of grafting was attributed to the lack of proper coverage of the rubber core when the graft density is low and to the expulsion of chains that are not chemically bound when the grafting density increases to very high values. These ideas, described schematically in Fig. 24.8, are in agreement with a concept derived for polymer-grafted particles in a polymer solution [45].

Recently, Hasegawa, Aoki, and Doi [43] looked at the problem using colloid stabilization concepts. The authors proposed that repulsion between the grafted chains keeps the particles apart, while a van der Waals–type interaction causes an attraction between the particles. When the graft density is low, attraction between the cores prevails and leads to agglomeration. Higher graft densities decrease the attractive potential, and the particles tend to disperse in the matrix by steric stabilization. As the graft density increases further, however, the chains become more stretched and form a dense brush. Attraction between the shell surfaces predominates, and the particles tend to agglomerate. As a first approximation, the authors proposed that the optimum graft density κ_c depends on the degree of polymerization N_g and a, the

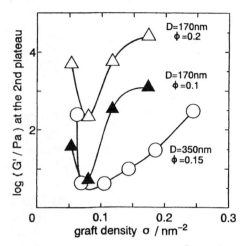

Figure 24.7   Behavior of the storage modulus G' at 10^{-3} sec^{-1} in ABS systems wherein graft density of the core–shell particles is varied. The highest degree of dispersion, after shearing of the samples, was observed when G' showed a minimum. See [43]. Reproduced with permission from the publisher.

segment length of the graft polymer, as follows:

$$\kappa_c = N_g^{-1/2} a^{-2}. \tag{24.1}$$

Therefore, as the molecular weight of the graft increases, the graft density that induces optimum dispersion becomes smaller. If smaller chains are grafted to the core, a higher graft density is required to achieve optimum dispersion. The graft density and the degree of grafting are related to one another via the particle diameter and the molecular weight of the graft [43]. One would expect that the critical degree of grafting θ_c varies according to the following equation:

$$\theta_c \approx \frac{1}{R} N_g^{-1/2}. \tag{24.2}$$

This relationship qualitatively follows the results observed by Aoki [42] and Bertin et al. [44].

2. Matrix–Shell Interaction   The situation more often encountered in the use of core–shell particles as impact modifiers is their dispersion in matrices that are dissimilar to the grafted shell, which includes most relevant commercial cases. Commercially available core–shell additives, which usually contain poly(methyl methacrylate) and styrene–acrylonitrile copolymers in the shell, are used to toughen poly(vinyl chloride), poly(butylene terephthalate), and other matrices with a chemical makeup quite different from that of the shell. Solubilization of the matrix polymer into the grafted layer must occur for the proper interfacial adhesion to develop. This solubi-

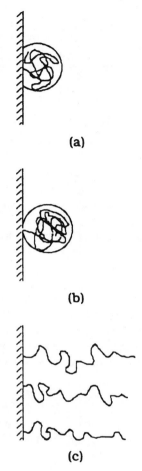

(a)

(b)

(c)

Figure 24.8 Interpretation of Bertin et al. [44] of the behavior of the grafted shell polymer as the degree of grafting varies: (a) low degree of grafting ("mushroom"), (b) full coverage, (c) overgrafting (fully stretched brush). The structure in the middle tends to produce optimum dispersion of the particles in a matrix of the same material as that of the shell. Reproduced with permission from the publisher.

lization process is far more complex than the mere miscibility between the same two polymers when none of them is grafted, because the shell polymer is immobilized on one end by being grafted onto the core.

Mechanistic treatments that analyze the specific parameters involved in the solubilization process and their role had not been developed until recently, when Lu et al. [46] proposed a thermodynamic model in which core–shell particles, with a shell made up of polymer A, are dispersed in matrix polymer B, as described in Fig. 24.9. The degree of solubilization of polymer B into polymer A was determined by computing the total free energy for the process, ΔG_{total}. The calculation of the total free

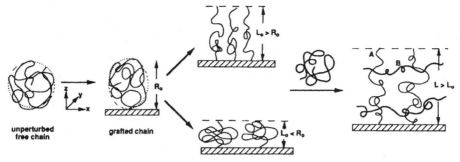

Figure 24.9 Representation of the conformational change (stretching) suffered by a polymer chain A upon grafting. The chains can become highly stretched ("brush" regime) if the grafting density is high enough (see the upper central figure) or adopt a flatter ("mushroom") regime) conformation if the grafting density is low (see the lower central figure). Swelling (solubilization) of polymer B into the grafted chains leads to stretching of the chains. Taken from the work of Lu et al. [46]. Reproduced with permission from the publisher.

energy includes the sum of the free energy of mixing of the two polymers and the entropy differences that account for changes in the conformation of both A and B to allow the penetration of free polymer B into the tethered chains of polymer A. The tethered condition keeps the A chains from adopting conformations otherwise available to them in the free state. Consideration of these factors leads to the expression

$$\Delta G_{\text{total}} = L_o S B \phi_B + \frac{L_o S \phi_B \rho_B}{\phi_A M_B} RT \ln \phi_B$$
$$+ \frac{L_o S \rho_A RT}{2 M_A} \frac{L_o^2}{R_o^2} (\phi_A^{-2} - 1) - \frac{L_0 S \phi_B \rho_B}{\phi_A M_B} RT \ln K, \quad (24.3)$$

where L_o is the shell thickness before polymer B is incorporated; S is the surface area of the core; B is the interaction energy density between polymers A and B (see Chapter 3); ρ_A and ρ_B are the respective bulk densities of polymers A and B; M_A and M_B are the molecular weights of polymers A and B, respectively; ϕ_A and ϕ_B are the corresponding volume fractions of polymers A and B in the swollen shell; K is the partition coefficient for B chains as they undergo a conformational rearrangement to penetrate the grafted A layer; R_o is the unperturbed end-to-end distance of a single grafted molecule; R is the gas constant; and T is the temperature. The authors of this model have looked at the degree of solubilization of polymer B in the shell, which is given by ϕ_B, by deriving the corresponding expression for the chemical potential of polymer B.

The combinations of parameter values that can lead to a favorable free energy of mixing are numerous. Still, high degrees of solubilization of the matrix polymer into the shell are favored by negative values of the interaction energy density, high molecular weight of the shell relative to the matrix, and low initial shell thickness. To best illustrate the interplay of these factors, the authors of this model followed

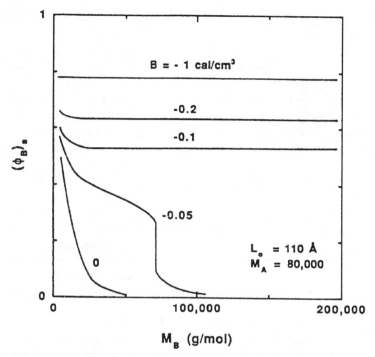

Figure 24.10   Predicted maximum degree of solubilization $(\phi_B)_s$ of a polymer B of molecular weight M_B on the shell, for different values of the interaction energy density B (in cal/cm^3). The initial shell thickness L_0 and the molecular weight of the shell M_A are fixed. Note that favorable enthalpic interactions between the two polymers (the more negative values of B) lead to a higher amount of polymer B being dissolved in the shell, independent of molecular weight. From [46]. Reproduced with permission from the publisher.

the maximum extent of solubilization of the matrix, or "free" polymer, into the shell. Figure 24.10 shows that negative values of the interaction energy density B—that is, high miscibility of the shell polymer and the matrix—can override the other factors. As this parameter becomes closer to zero the role of the relative molecular weights becomes more important. The need for a strong driving force to obtain a high degree of solubilization is illustrated for the case in which B becomes zero. This would be the equivalent of making the matrix polymer chemically identical to the shell polymer. The model indicates that the maximum degree of solubilization is relatively low, compared to systems in which a high enthalpic driving force exists.

This model provides a series of excellent guidelines for optimizing the molecular architecture of core–shell polymers to enhance their dispersion and helps us understand the shell–matrix interaction process. Nevertheless, the authors of this model have cautioned that their calculations still require further refinements [46]. As they experimentally verified their predictions, the authors found that it is difficult to detect the maximum degree of solubilization when a high degree of miscibility exists between the matrix and the shell polymers [47]. Detection of the upper limit, especially, might require sophisticated experimental techniques.

For core–shell impact modifiers, the model offers a far more advanced tool for product design than to solubility parameters alone. Nevertheless, it is important to bear in mind which factors in core–shell synthesis can be controlled within practical limits so that they significantly influence dispersion and adhesion. Furthermore, even though a large degree of solubilization is a guarantee that the matrix and the shell will adhere well to one another, the generation of a strong interface might not necessarily require that large proportions of the matrix are physically incorporated into the shell.

B. Dynamics of Dispersion and Related Issues

Even though the particle size of core–shell polymers is fixed during the emulsion process, proper dispersion in the melt must be achieved to obtain optimum toughening performance. Kayano et al. [48] showed that blends of a core–shell modifier in polycarbonate consistently showed better dispersion when melt blended in a twin-screw extruder than in a single-screw extruder. Better dispersion translated into higher notched Izod impact strength and a lower ductile–brittle temperature transition.

Gaymans and van der Werff reported that even nonreactive core–shell particles can be efficiently dispersed in a nylon matrix when the right extrusion conditions are used [49]. The same authors showed that a similar degree of toughening can be achieved by melt processing the blend in a twin-screw extruder and recognizing that the molecular weight of the matrix can play a significant role in the process. Lu et al. [50] presented experimental evidence that higher molecular-weight nylon matrices are indeed more effective in dispersing core–shell modifiers. According to Lu et al., the higher shear stress imposed by a matrix of higher viscosity can be more effective in breaking up particle clusters to disperse the individual particles in the matrix.

The problem of how the dispersion of linear and rubbery polymers into matrix polymers occurs in the melt has been studied extensively (see Chapter 16) [51]. Core–shell polymer dispersion has not received the same kind of attention nor systematic study. Because of their special physical nature, it is difficult to fit core–shell polymers into schemes that are more appropriate for linear polymers, for which the concept of balancing viscous and surface forces has been successfully applied. (See Chapter 16.) Nonetheless, some efforts along those lines have shown some degree of success in rationalizing the core–shell dispersion process [52]. Yee and collaborators [53] have made efforts to quantify the degree of dispersion of core–shell modifiers and correlate it with fracture toughness in PC and epoxy resins.

Core–shell polymers might be more akin to solid-particle agglomerates, such as carbon black, and breakup and subsequent dispersive mixing might be more appropriately treated as such [54]. Interestingly, the size of typical powdery core–shell agglomerates is on the order of 100 μm, which is similar to that of carbon-black agglomerates [55]. Carbon-black agglomerates are made up of aggregates about 0.1 μm in size, the order of magnitude of core–shell emulsion particles. The polymeric nature of core–shell particles must be taken into account, however, both in terms of their tendency to deform under stress and in terms of the special nature of the grafted polymeric shell.

VI. TOUGHENING POLYMERIC MATRICES WITH CORE–SHELL MODIFIERS

Particle size, glass transition temperature, and proper shell composition to achieve adequate adhesion to the matrix are determinant for modifier performance. Some of the specific issues are discussed throughout this section. Other minor components can also affect performance in an indirect fashion. Stabilizer and antioxidant packages, the vast majority of them of proprietary compositions, are often added in the reactor to core–shell modifiers, particularly when butadiene-based cores are used. Toughening of engineering resins with core–shell polymers would not be feasible without stabilization of this nature, because the modifier must survive the high-temperature melt processes associated with those materials. The stabilization of core–shell and other polymers falls beyond the scope of this work, but relevant detail and references can be found in [56] and [57].

A. Major Factors in Core–Shell Toughening

1. *Particle Size*   It is well known that the particle dimensions necessary to toughen a rigid polymer depend on the inherent fracture mechanism of the matrix. As a general rule, brittle glassy matrices that tend to craze benefit more from large rubber particle size, i.e., in excess of 1 μm. (See Chapter 22.) On the other hand, matrices that can absorb fracture energy via shear yielding are effectively toughened with relatively small-particle-size modifiers, on the order of 0.5 μm or less. Emulsion polymerization lends itself to the preparation of small particles, and therefore, core–shell polymers are more often used in shear yielding matrices, such as PVC and engineering thermoplastics. Nevertheless, the preparation of large particles by emulsion polymerization techniques is receiving a large amount of attention, as large particles prepared in this manner could broaden core–shell applications in styrenics and similar brittle glassy matrices.

PBA rubber particles well above 1 μm have been obtained in emulsion [58, 59]. On the MBS side, carefully adding coagulant to a stable rubber latex can induce the agglomeration of several small particles to produce large rubber particles [60]. Relatively large and monodisperse polybutadiene particles can be generated *in situ* if suitable coagulants are added during the polymerization of the rubber [61]. Alternative postrubber polymerization process schemes have also been reported [36, 62].

A significant body of work has convincingly established that the particle size of the modifier must lie within an optimum range to toughen a given matrix (see Chapter 25) [63–67]. Wu has proposed that the optimum particle size for toughening can be related to inherent molecular parameters of the matrix [64]. For nylon 6, the optimum particle size is from about 0.2 and up to 0.5 μm. (See Chapter 25.)

However, most of the published studies to determine optimum particle-size ranges have been carried out with non-core–shell polymers, with two exceptions involving PMMA, and PVC. For PMMA, an optimum size range is found between 0.2 and about 0.3 μm [64, 65]. In PVC, an optimum size of 0.2 μm has been observed [66].

The special attribute of core–shell modifiers that allows manufacturers to set their particle size and make it independent of processing conditions when melt blended can permit in principle, to devise modifiers with the optimum particle size for a given matrix. However, even though one would expect the behavior of core–shell particles in general to parallel the results for other additives, few experimental facts are available in the open literature that objectively establish these ranges for core–shell polymers. Perhaps the unavailability of this information for core–shell particles underscores how important knowing these ranges, is beyond purely scientific reasons. One of the few reports available indicates, without specifying them, that optimum particle-size ranges do exist for toughening PVC, PBT, and PC–PET blends with core–shell modifiers, all of them well below 1 μm, in general agreement with results for other types of modifiers [67]. It is important to recognize that certain properties of the core–shell structure are coupled to their particle size and complicate this analysis to some extent. For example, shell thickness increases as particle diameter increases, at a fixed weight ratio of core and shell.

2. Glass Transition Temperature   Maintaining a consistent T_g in the polymer core as well as in the shell is one of the key factors that determine the quality of an impact modifier. As a general rule for core–shell particles, rubber cores with lower glass transition temperatures induce better toughening at lower temperatures One of the most interesting alternative rubbers for higher-performing modifiers is poly(dimethyl siloxane) (PDMS), given its low glass transition temperature (about $-100°C$), weatherability, and chemical resistance. Obtaining core–shell polymers in emulsion with PDMS is not straightforward, because of the low surface tension of the polymer and its high interfacial tension with the monomers and polymers generally used to form shells [68]. Nonetheless, these difficulties appear to have been overcome in commercial materials, such as Mitsubishi Rayon's Metablen [TM]S-2001 [69], that contain silicone–based rubber cores. These polymers are prepared by polymerizing n-butyl acrylate in the presence of a dimethyl siloxane–based latex, prepared by emulsion polymerization. A hard shell is grafted onto the latex particles to help isolation and provide compatibility with the host matrix. Figure 24.11 shows that the composite rubber particles that contain 50 wt% or less of silicone provide the most efficient degree of toughening in PVC. The higher cost of the silicone moiety in these polymers is an important factor to consider, if their usage is to increase. However, in principle, their efficiency over traditional acrylic impact modifiers (AIMS) and methacrylate-butadiene-styrene (MBS) modifiers, as well as their superior weatherability and performance, can offset their steeper price [69–71].

3. Adhesion   The lack of miscibility of the most widely available PMMA or styrenic shells with engineering resins has prompted the need to use alternative strategies to enhance the dispersion and adhesion of core–shell particles in these

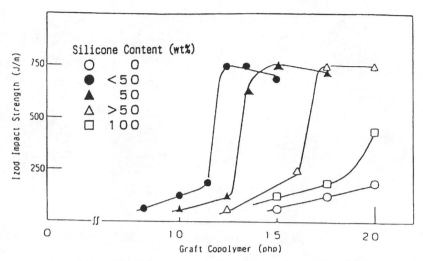

Figure 24.11   Toughening of PVC with a silicone-containing butyl acrylate core–shell particle. About 50% of silicone rubber in the core produces optimum results. See [69]. Reproduced with permission from the publisher.

types of polymers. The most well-known method is to take advantage of the end groups present in engineering polymers. Amine and carboxyl groups in nylons and carboxyl and hydroxyl groups in polyesters offer potentially reactive sites for suitable chemical moieties placed on the shell. Functional monomers, such as acrylic and methacrylic acids, glycidyl methacrylate (GMA), hydroxyethyl acrylate (HEA) or methacrylate (HEMA), and maleic anhydride (MAH), have been used to induce chemical adhesion with the matrix in this fashion [72, 73]. The incorporation of these types of monomers into the shell via emulsion polymerization is not trivial, because their high water solubility hinders chemical bonding to the shell during polymerization. Another option for enhancing the adhesion between the matrix and the shell is to add a compatibilizing agent, which is a well-documented method. (See Chapter 17.)

4. *Microstructural Variants*   Significant examples of these variants, which have been available commercially for some time, are the multilayer particles used to toughen PMMA [7, 74, 75]. These particles provide an optimum balance of stiffness and impact resistance. Attempts at using more sophisticated types of morphology inside the rubber cores, such as interpenetrating polymer networks (IPN) of two soft polymers with different T_g's, are also interesting, although their performance in early experiments is similar to that of typical core–shell polymers [76].

B. Toughening of Thermoplastic Matrices

1. *Poly(Methyl Methacrylate)*   Rubber-toughened PMMA (RT-PMMA) has been the focus of commercial and scientific interest for many years [7, 75, 77–80]. Besides

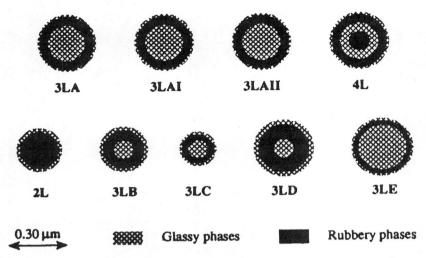

0.30 µm ⟷   ▨▨▨ Glassy phases   ■■■ Rubbery phases

Figure 24.12 Multilayer structures that combine glassy and rubbery phases in different sequences and distributions. Particles 3LE, with one glassy phase at the center and one on the outside, provide an optimum balance between toughness and stiffness. From [79], with permission from the publisher.

improving the fracture resistance of the polymer, core–shell modifiers for PMMA must also maintain the high stiffness of the matrix as well as its clarity. For that reason, suitable combinations of butyl acrylate and styrene monomers are used in the rubbery phase, while a PMMA copolymer is used in the glassy phase, to match the refractive index of the matrix. Particles on the order of 0.2 to 0.3 µm are normally used. The level of modifier particles present can vary from 10 to 40% by weight of the matrix. In commercial materials, optimum toughening is obtained with a multilayer particle morphology that consists either of a rubbery core and a glassy shell or a glassy core, an intermediate rubbery layer, and an outer glassy shell [7, 79]. In general, RT-PMMA is sold commercially as a final blend, as with HIPS and ABS resins. Figure 24.6 is an atomic force micrograph that clearly shows the typical microstructure of RT-PMMA.

Lovell, Young, and collaborators carried out a detailed study of how different sequences of multilayer structures offer different degrees of ductility in PMMA while maintaining stiffness and clarity [81, 82]. Figure 24.12 shows schematic diagrams of cross sections of the types of particles used by these authors to toughen PMMA. The decrease in elastic modulus depends on the amount of rubber present in the modifier and not on the sequence of hard and soft layers. An optimum balance between stiffness and impact was obtained for the particles with a hard core and hard shell, with an intermediate rubber layer, as shown in Fig. 24.13.

2. Poly(Vinyl Chloride) The miscibility of poly(vinyl chloride) (PVC) with PMMA, a common type of shell in commercial modifiers, is a driving force that facilitates dispersion and promotes adhesion between the particles and the matrix.

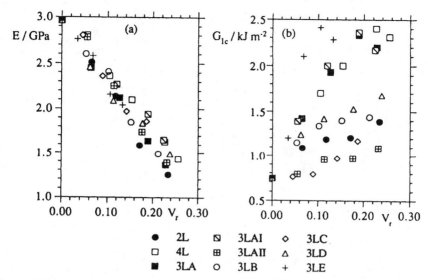

Figure 24.13   Young's modulus E and the critical energy release rate G_{Ic} in rubber-toughened PMMA for the structures shown in Fig. 24.12 [79], as a function of the volume fraction of the modifier, V_f. The 3LE particles tend to induce higher toughness at a lower concentration. Reproduced with permission from the publisher.

Lutz and Dunkelberger have published an extensive review of the impact modification of PVC [1].

The degree of impact modification achievable in PVC is intimately linked to processing and formulation parameters. The well-known structure of PVC (see Chapter 18) and the extent to which it can be fused during processing can make a difference in the intrinsic toughening of the matrix. (See, for example, Chapter 1 in [1].)

Typical levels of core–shell polymers added to PVC are on the order of 5 to 10 parts per hundred (phr) of total formulation. The notch sensitivity of PVC to impact, typical of various plastics, can be more than adequately handled by toughening with core–shell polymers. Figure 24.14 is illustrative of impact modification of PVC by an MBS modifier [83]: As the notch in the sample becomes sharper (i.e., its radius decreases), neat PVC is less capable to absorb impact energy. In contrast, the sample modified with the core–shell additive can clearly sustain its toughness even when very sharp notches are present.

All-acrylic impact modifiers (AIMS) are normally used in opaque applications for which weathering of the part is an issue, as in building and construction. Methacrylate-butadiene-styrene (MBS) polymers are used when transparency is required, as, for example, in packaging applications. The refractive index of PVC (1.564) and its essentially amorphous character after processing allows suitable combinations of monomers in the core–shell structure (see Table 24.1), coupled with a small particle size of the modifier, to render transparent blends [1].

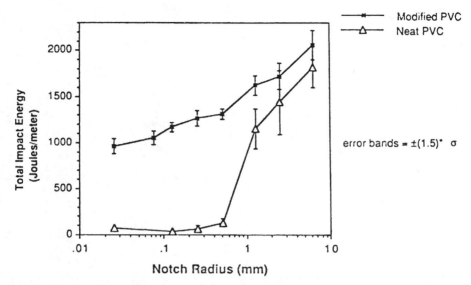

Figure 24.14 Effect of a core–shell MBS modifier on the ductile-to-brittle transition of PVC, as a function of the notch radius in the sample. The modifier, loaded at 6 phr, makes the matrix less sensitive to failure by sharp notches. From [83], with permission from the publisher.

3. *Bisphenol A Polycarbonate* Given the inherent impact resistance of polycarbonate, its toughening is usually directed at maintaining toughness when Bisphenol A polycarbonate (PC) is used in thick sections. Typical use levels fall in the range of 5 to 10 weight percent for commercial modifiers [5]. PC blends with even small amounts of core–shell modifiers are opaque or translucent. This factor occurs because of the relatively high refractive index of PC (1.585), compared to that of commercially available core–shell polymers.

Physical aging of polycarbonate at temperatures close to and below T_g causes significant embrittlement. The beneficial effect of adding commercial core–shell modifiers to delay embrittlement was demonstrated in a systematic study by Cheng et al. [84]. Figure 24.15 illustrates how the presence of an MBS impact modifier extends the tough behavior of the resin well beyond the time when failure occurs in neat polycarbonate. As seen in Fig. 24.15, however, the modifier also depresses matrix toughness. Chemical changes in the modifier and chemical interactions between the acrylic shell of the modifier and polycarbonate that reduce the molecular weight of the matrix tend to counterbalance the beneficial effect of the modifier [84].

Maintaining depth of color in pigmented PC while increasing its toughness is another important situation. Certain types of core–shell polymers appear to be more suitable to provide this quality [85]. Finally, attempts at increasing the effectiveness and efficiency of core–shell polymers in PC by varying particle size, the cross-link density of the core, and the composition of the shell have been reported in the literature [7].

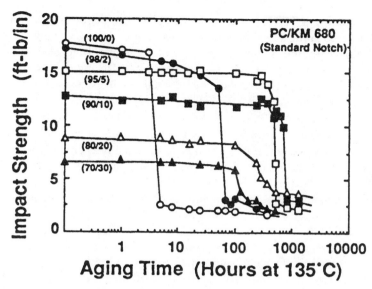

Figure 24.15 Behavior of the notched Izod impact strength of PC with and without modification by an MBS core–shell polymer (Paraloid^{TM}KM 680), upon ageing in an air oven at 135°C. From [84], with permission from the publisher.

4. *Poly(Butylene Terephthalate)* An increasing commercial interest in poly(butylene terephthalate) (PBT), especially in the automotive, electric, and electronic markets, is also driving the use of its core–shell-modified versions. Traditionally, PBT is impact modified with ABS resins. However, specific processing conditions are required to obtain maximum toughness [86]. Core–shell polymers offer an interesting alternative to ABS. Normally, 10 to 25 wt% MBS toughener or 15 to 30 wt% AIMS, both with PMMA-based shells, can bring about adequate levels of toughening that are at least comparable to those obtained with the addition of ABS resins.

The use of functional monomers on the shell, such as glycidyl methacrylate (GMA), can significantly improve the performance of the modifier [87]. Alternatively, compatibilization of the nonfunctionalized core–shell polymer can be carried out with a third polymer. 10 wt% or less of PC greatly facilitates the dispersal of core–shell impact modifiers in PBT and leads to tough blends, even at low temperature [88, 89].

5. *Poly(Ethylene Terephthalate) and Copolymers* Poly(ethylene terephthalate) (PET) is becoming an increasingly important commodity, especially in packaging applications. Various types of PET have appeared on the market to satisfy specific needs. Crystalline PET (CPET), which may contain nucleating agents to induce as high a degree of crystallinity as possible, is used in thermoformed ovenable trays. Amorphous PET (APET) is produced by including small amounts of isophthalic acid during the polymerization, to reduce crystallinity. It finds uses in packaging. Copolymers of ethylene glycol and terephthalic acid with cyclohexane-dimethanol, generically known as PETG, are generally tougher than the other two varieties of PET and are used in packaging films and bottles. A brief overview of this subject can be found elsewhere [90]. In addition, an abundant supply of recycled PET, sometimes known in the industry as RPET, is also available. Upgrading this material via judicious processing and the addition of core–shell polymers, as well as other types of additives, is also important [91].

The impact resistance of PET can be improved significantly by biaxial drawing, as occurs in bottles made by stretch blow molding. Extrusion blow molding and sheet extrusion, however, are less capable of inducing biaxial orientation, and the use of impact modifiers is an attractive option to enhance toughness. Furthermore, PET and especially APET are prone to become embrittled upon thermal ageing below the glass transition temperature. Impact modifiers can decrease this inherent sensitivity to physical ageing, as for PC [91].

Toughening PET with conventional core–shell modifiers is difficult because its interaction with the shells used in commercial materials is poor and the low viscosity of the matrix is not conducive to appropriate modifier dispersion. Efforts to resolve these problems have involved both the use of compatibilizing agents [88, 92] and the copolymerization of functional monomers on the shell, such as hydroxy-containing acrylates or methacrylates [93].

One further challenge to toughening APET and PETG is to maintain their high degree of clarity. Suitable refractive index matching and the use of small core–shell particles, as done with PVC, can help achieve this goal [91, 93].

6. Polyamides   Toughened nylon 6 and nylon 66 are used in the electrical and electronic markets, as well as for transportation, appliances, and several consumer goods [94]. Glass-fiber-reinforced grades, containing up to 30% glass fiber, are often used. Toughening of these different types and grades is needed, because of the notch sensitivity of the resin. Due to the low degree of compatibility between nylon, a highly hydrogen-bonded polymer, and the typical acrylic- or styrenic-based shell present in commercial modifiers, it is necessary to use special strategies to achieve good dispersion and adhesion of core–shell tougheners and other tougheners in the matrix [3, 95, 96]. Terpolymers of styrene, methyl methacrylate, and maleic anhydride (MAH) have been successfully used as compatibilizing agents for blends of nylon with ABS-type core–shell polymers, as they are miscible with the SAN shell of the core–shell particles and the MAH reacts with the end groups of the matrix to form *in situ* graft copolymers (see Chapters 17 and 25) [96].

Core–shell products that do not require the use of a compatibilizer to toughen nylon have been made by using acid-functionalized shells [72, 73, 97]. Their recommended use level is about 20 to 25 wt%. Core–shell polymers with GMA bound to the shell have also been disclosed as effective nylon tougheners [98]. A recent paper claims that the degree of dispersion of these types of polymers in nylon 6 is as good as that found in compatibilized systems [99].

Yet another way to disperse nonreactive core–shell polymers in nylon 6 is to add the core–shell polymer directly to the reactor, before effecting the polymerization [49]. Other methods to disperse core–shell particles in nylon 6 were mentioned in Section V.B of this chapter.

7. Thermoplastic Blends   Blends of polycarbonate with poly(butylene terephthalate), known as PC–PBT blends, are widely used in the automotive industry. They combine the stiffness and chemical resistance of PBT and the ductility of PC [100]. About 10 to 15 wt% of a core–shell polymer is generally used to increase the toughness of the blend significantly.

The influence of adding core–shell modifiers to PC–SAN blends has been systematically studied by at least two different groups [101, 102]. A use level of 10 to 15% of the modifier, normally an MBS type, is beneficial for toughness. At a 75/25 concentration ratio, PC–SAN blends have poor impact resistance as compared to PC–ABS blends. In addition, Debier et al. detected a size reduction of the SAN phase and proposed that the particles acted as compatibilizers [102]. Significant increases in toughness were observed by both groups.

The distribution of core–shell particles in two-phase polymer blends has been studied extensively, especially in polycarbonate-based blends [101, 103–105]. When compounded into these materials, core–shell modifiers with PMMA shells reside almost exclusively in the polycarbonate phase instead of in the PBT, PET, polystyrene, or nylon phases. The high degree of compatibility, although not miscibility, of PMMA with polycarbonate is at the root of these results. By contrast, MBS particles dispersed in a PC–SAN blend locate at the interface between PC and SAN. PMMA is highly compatible with PC and fully miscible with SAN.

The location of the modifier can be controlled by including a small amount of functional monomer in the shell. In PC–nylon 6 blends, the core–shell modifier with a PMMA shell stays in the PC domains, while, if a similar modifier with maleic anhydride copolymerized in the shell is used, the modifier is located in the nylon phase [104], because of a reaction between maleic anhydride and the nylon amine end groups. Figure 24.16 shows micrographs from that work. The blends in which the modifier resides in the polycarbonate phase show a greater degree of toughness.

In PET blends with polycarbonate, the modifier with a PMMA shell and no functionalization remains in the PC phase, while the inclusion of GMA in the shell drives the modifier into the PET domains [105]. The blends with modifier in the PC phase were again found to be tougher. A similar inversion of the modifier location was observed when a GMA-based polymer, miscible with the shell, was used to compatibilize the modifier with the PET shell.

Hobbs et al. [106] proposed a model based on the interplay of interfacial tensions and showed that in blends of PC, PMMA, and PBT, polycarbonate tends to coat the PMMA domains, while the PBT covers the PC layer. Cheng et al. [101] extended this thermodynamic model to successfully explain the distribution of modifier in a given phase or in the interfacial region.

8. Other Thermoplastic Resins   Polyoxymethylene (polyacetal), a polymer with a high degree of crystallinity, is commonly toughened by blending with thermoplastic urethanes [87]. Core–shell polymers have been demonstrated to be suitable for this

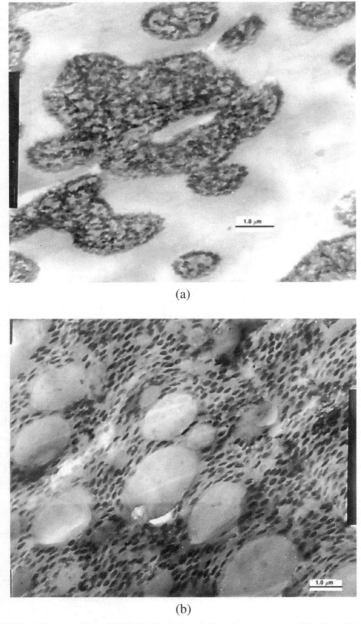

(a)

(b)

Figure 24.16 Micrographs of 30/70 PC–nylon 6 blends, impact modified by 30 additional parts of (a) a nonreactive MBS and (b) a similar MBS with maleic anhydride on the shell. The PC disperses into relatively large domains (> 1 μm) in the nylon matrix. The modifier (darkest domains, < 1 μm in diameter) resides almost exclusively in the phase with which it interacts to a higher extent: (a) physically with PC and (b) chemically binding to nylon. From [104]. Reproduced with permission from the publisher.

application. A use level of 20 wt% is typical for toughening polyacetal [107, 108]. Careful processing of the resin–modifier blend is of particular importance, because polyacetal tends to degrade around 200°C. Special stabilizer packages can help overcome many of these problems [109].

High-impact polystyrene (HIPS) and ABS, although already considered to be "tough" materials, can benefit from the addition of core–shell polymers. Keskkula et al. [110] observed that the addition of MMA-grafted latex rubber particles to ABS copolymer and HIPS significantly increased their impact strength. The toughness enhancement in HIPS was somewhat unexpected, because of the well-known immiscibility of PMMA and polystyrene. A recent patent that teaches the synthesis of butadiene rubbers grafted with essentially pure polystyrene [111] shows that the addition of these particles to HIPS can also enhance the toughness of the polymer.

High-service-temperature polymers (HTPs) offer a special challenge to core–shell modification, owing to their high processing temperatures. Polysulfone, polyarylate, polyethereimide, poly ether-ether ketone (PEEK), polyaromatic amides, and polyethersulfone fall into this category. With proper stabilization, AIMS can increase the toughness of some of these resins. Sulfone polymers appear to be more amenable to impact modification with commercially available core–shell modifiers [5].

Polyketone polymers, made by copolymerizing ethylene and carbon monoxide, are a new type of engineering resin. Toughening of these polymers with 20% of an MBS-type core–shell modifier has been claimed [112]. Another new matrix on the market that shows growth potential is syndiotactic polystyrene (SPS). Here, combinations of a styrene-hydrogenated butadiene-styrene triblock copolymer with a butadiene-based-core–styrene-based-shell modifier have been shown to produce a synergistic effect on toughness [113].

C. Toughening of Thermoset Resins

The toughening of epoxy resins is becoming an important area of application for core–shell polymers (see Chapter 26) [114–122]. Using core–shell particles in these systems is attractive because they increase the fracture toughness of the material while preserving its stiffness. In commercial epoxy systems, the levels of core–shell modifier addition are normally about 5 to 10 wt%. Table 24.2 illustrates how the addition of a core–shell polymer enhances the toughness of a typical epoxy resin, in comparison to a conventional carboxy-terminated butadiene rubber (CTBN). The glass transition temperature of the resin is hardly affected by the presence of the core–shell

Table 24.2   Effect of Toughener Type on Epoxy Modification

Material	K_{Ic}, MPa·m$^{0.5}$	T_g, °C
Neat Epoxy	0.75	80
Epoxy + 10% All-Acrylic Core–Shell	2.35	80
Epoxy + 10% CTBN rubber	1.75	66

Formulations cured for 15 hours at 100°C. Epoxy: Ciba Geigy's Araldite™ GY 250 and HY 2963. CTBN: B. F. Goodrich's Hycar™ X13. Data taken from [118].

particles. The CTBN telechelic rubber is dissolved in the liquid resin mixture and, after curing, forms separate domains. However, a small portion of this material remains dissolved and plasticizes the resin. Similar results have also been found when core–shell particles were used at a 15 wt% loading [110]. Particles with sizes on the order of 0.2 to 0.5 μm produce a similar degree of toughening in terms of the critical stress intensity factor [118]. On the other hand, recent claims suggest that for a given rubber type, using particles on the order of 0.050 μm can significantly increase the toughness of an epoxy resin, as measured by a falling-dart flat-plate method [119].

Core–shell particles that consist of a glassy core, an intermediate rubbery layer, and a hard shell, produce a greater degree of toughening in the system over conventional soft-core–hard-shell modifiers [74]. Functionalization of the shell with suitable monomers has also been used to optimize adhesion to the matrix. Glycidyl methacrylate (GMA) [120] and methacrylic acid (MAA) [121] have been used for this purpose.

Maximizing the dispersion of core–shell polymers in an epoxy resin does not necessarily translate into a tougher system. Qian et al. [120] observed that core–shell polymers that form clusters of particles instead of a uniform morphology induce a higher fracture toughness in the resin.

Unsaturated polyester resins, which are commonly used as glass-reinforced materials, are toughened with liquid nitrile rubbers or polyurethane elastomers. The use of conventional core–shell rubbers does not appear to have extended in this type of material. Still, one patent has shown that three-layer core–shell compositions can be useful toughening agents in unsaturated polyesters [123].

VII. EFFECTS ON OTHER PHYSICAL PROPERTIES

The presence of core–shell tougheners brings about other property changes, which are not always desirable. This tradeoff can sometimes be resolved, but various situations still defy solution. This section addresses some of these issues.

Toughening with any type of rubber implies placement of a soft phase in a rigid matrix and a consequent decrease in stiffness. An examination of available data, as shown in Fig. 24.17, indicates that, as a general rule, acrylic-type and MBS-type core–shell polymers reduce the flexural moduli in various engineering resins by a little over 1% for every weight percentage of modifier added. Similarly, a decrease in the tensile modulus of about 10 to 15% and a 15 to 20% loss in compressive yield stress have been observed in epoxy systems when toughened with about 15 wt% modifier [117].

The optical properties of the matrix polymer are usually affected by the addition of core–shell particles, with a consequent loss in transparency. (See Chapter 28.) The matching of refractive indices of the overall modifier to that of the matrix has been used to create toughened transparent blends of PVC and polyesters with core–shell particles. However, this strategy might not necessarily offer both minimum haze and maximum transmission at the same time. The particle size of the modifier, which should preferably lie below the wavelength of visible light, plays an important role

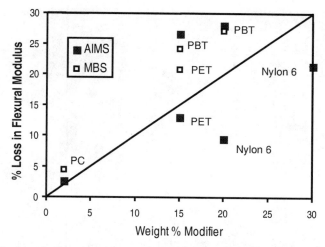

Figure 24.17 Loss in flexural modulus with the use of cross–shell impact modifiers, in various types of polymers. Both the all-acrylic (AIMS) and butadiene-based particles (MBS) toughen the matrices significantly, while reducing their stiffness. The reference line in the figure represents a 1% loss in modulus per every percent of modifier added. Data taken from [5].

in meeting these objectives [124]. The minimum particle size that can be used could be limited to the minimum size requirement for toughening or for effecting practical dispersion.

An important byproduct of core–shell toughening might be an undesirable rise in the viscosity of the system, which can be detrimental to processing. Memon [125] has shown that the severity of viscosity increase in core–shell–PC–PBT systems is higher at shear rates below 100 sec^{-1} and decreases substantially at higher shear rates. In nylon systems, core–shell polymers with reactive moieties tend to induce particularly high viscosity. Formulation and processing adjustments can help alleviate this problem [4].

Environmental stress crack resistance (ESCR) is an important practical feature for several polymers, especially those in contact with aggressive media that tend to interact and induce failure of the polymer while under strain. As an example, food oils in contact with HIPS tend to crack the polymer, especially when it is subjected to a tensile strain. Significant improvements in ESCR have been reported when a core–shell polymer made essentially of a 60/40 butadiene–styrene ratio is blended with HIPS [111], even though the addition of core–shell polymers normally produces only a modest increase in resistance in most other matrices.

VIII. CAVITATION IN CORE–SHELL PARTICLES

The relevance of voiding, or "cavitation," in impact modifiers is amply covered in Chapter 22. Essentially, during impact, cavitation of the particles triggers shear de-

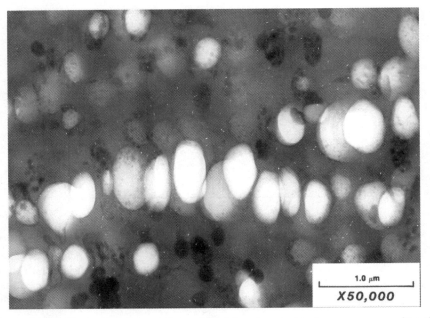

Figure 24.18 Cavitation of an acrylic modifier (clear domains) in the process zone of toughened nylon 66. The sample showed ductile fracture behavior.

formation in shear yielding matrices. In polymers prone to crazing, the rubbery phase is also auspicious in craze formation. Cavitation of the rubber phase of MBS particles has been well studied in blends of PC with core–shell particles [126]. Cavitation of core–shell polymers in other matrices has been studied less frequently. Nonetheless, toughening associated with particle cavitation has been observed. Figure 24.18 shows an acrylic core–shell modifier for which particles are cavitating near the fracture surface in a nylon matrix. The particles toughen the matrix effectively.

Lazzeri and Bucknall [127] have proposed that the smaller the size of the toughener, the more difficult it is to induce its cavitation. These authors related voiding to the local stress state around the modifier in the matrix and illustrated how cavitation of the modifier, coupled with the sensitivity of polymers to yield under hydrostatic tension provide the right conditions for shear yielding. (See Chapter 22.) Dompas and Groeninckx [63] also proposed a model for cavitation and related the existence of a minimum particle size for toughening in nylon 6 and in PVC to the minimum particle size required for cavitation. A minimum particle size required for toughening, between 0.2 and 0.25 μm, has been determined or proposed for various matrices, including PVC, PMMA, nylon, and polycarbonate.

Extensive cavitation of core–shell particles might not necessarily translate into higher performance, nor is extensive cavitation always an indispensable condition for effective toughening. For example, when comparing the toughening effectiveness of two core–shell modifiers in polycarbonate, Cheng et al. observed that the modifier that produced the lower ductile-to-brittle transition temperature had greater

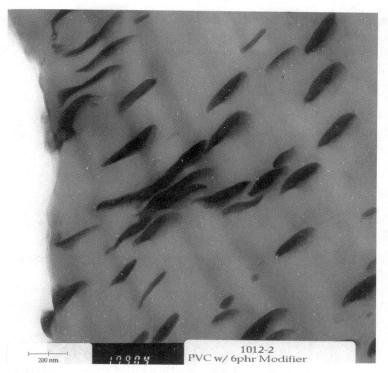

Figure 24.19 Behavior of butadiene-based (MBS) modifier particles (dark domains) at the fracture zone in toughened PVC. No cavitation was detected. Only high particle elongation was observed to occur. The sample failed in a ductile fashion.

resistance to cavitation [128]. Lazzeri and Giuliani showed that matrix yielding can occur without previous particle cavitation under certain stress conditions [129]. Such is perhaps the situation illustrated by Fig. 24.19, which shows highly deformed MBS particles at the fracture zone in PVC. The core–shell particles effectively toughen PVC, although cavitation is not observed. A similar case was shown for nylon 6 [130].

As a consequence of the above observations, the role of core–shell particles during the toughening process might be different for different situations. Cavitation of the particles to release energy-absorbing mechanisms in the matrix appears to be generally desirable. However, under certain conditions, yielding of the matrix may occur even in the absence of cavitation.

IX. ACKNOWLEDGEMENTS

The author is grateful to the editors of this book, Profs. Paul and Bucknall, for their comments on the manuscript and many discussions on the topic over the years.

The author acknowledges the support and many helpful comments from his colleagues at the Plastics Additives and Analytical Research Departments of Rohm and Haas. Thanks are especially due to Mr. R. H. Weese, Dr. R. Antrim, and Dr. E. P. Dougherty for discussions and material they kindly provided. Thanks are also due to Prof. C. Scott of the Massachusetts Institute of Technology, for his comments on dispersion of core–shell particles, and to Prof. F. C. Chang of National Chiao Tung University, Taiwan, for providing original photographic material. Finally special thanks are due to the Rohm and Haas Company for supporting the publication of this chapter.

X. REFERENCES

1. J. T. Lutz, Jr. and D. L. Dunkelberger (eds.), *Impact Modifiers for PVC. The History and Practice*, John Wiley & Sons, Inc., New York, 1992.

2. J. Hoffman, "Impact Modifiers are Solid," in *Chemical Marketing Reporter*, Schnell Publishing Company, Inc., June 12, 1995, p. SR11.

3. H. Keskkula and D. R. Paul, "Toughening Agents for Engineering Polymers," in *Rubber Toughened Engineering Plastics*, A. A. Collyer (ed.), Chapman & Hall, London, 1994, Ch. 5.

4. A. P. Berzinis, "Impact Modifiers," in *Thermoplastic Polymer Additives. Theory and Practice*, J. Lutz (ed.), Marcel Dekker, Inc., New York, 1989, Ch. 5.

5. J. T. Lutz, Jr., and C. B. Hemenway, "Acrylic Derivatives," in *Plastics Additives and Modifiers Handbook*, J. Edenbaum (ed.), Van Nostrand Reinhold, New York, 1992, Ch. 43.

6. Y. Nakamura, M. Ohta, and M. Okubo, *Nippon Setchaku Gakkaishi*, **32**, 104 (1996).

7. P. A. Lovell and D. Pierre, "Rubber-toughened Plastics," in *Emulsion Polymerization and Emulsion Polymers*, P. A. Lovell and M. S. El-Aasser (eds.), John Wiley & Sons, Ltd., Chichester, England, 1997, Ch. 19.

8. G. W. Poehlein, "Emulsion Polymerization," in *Encyclopedia of Polymer Science and Technology*, Vol. 6, H. F. Mark, N. M. Bikales, C. G. Overberger, and G. Menges (eds.), John Wiley & Sons, New York, 1989, p. 1.

9. E. S. Daniels, E. D. Sudol, and M. S. El-Aasser (eds.), *Polymer Latexes*, ACS Symposium Series, Vol. 492, American Chemical Society, Washington, DC, 1992.

10. R. G. Gilbert, *Emulsion Polymerization. A Mechanistic Approach*, Academic Press Limited, San Diego, 1995.

11. D. C. Blackley, *Polymer Latices. Science and Technology*, 2nd Ed., Chapman and Hall, London, 1997.

12. P. A. Lovell and M. S. El-Aasser (eds.), *Emulsion Polymerization and Emulsion Polymers*, John Wiley & Sons, Ltd., Chichester, England, 1997.

13. J. M. G. Cowie, *Polymers: Chemistry & Physics of Modern Materials*, 2nd Ed., Blackie Academic & Professional, London, 1991.

14. P. M. Lesko and P. R. Sperry, "Acrylic and Styrene–Acrylic Polymers," in *Emulsion Polymerization and Emulsion Polymers*, P. A. Lovell and M. S. El-Aasser (eds.), John Wiley & Sons, Ltd., Chichester, England, 1997, Ch. 18.

15. D. W. van Krevelen, *Properties of Polymers*, 3rd Ed., Elsevier Science B. V., Amsterdam, 1990.

16. E. Penzel, "Polyacrylates," in *Ullman's Encyclopedia of Industrial Chemistry*, Vol. A21, B. Evers, S. Hawkins, and G. Schulz (eds.), VCH Publishers, Inc., Weinheim, Germany, 1992, p. 157.

17. G. Odian, *Principles of Polymerization*, 3rd Ed., John Wiley & Sons, Inc., New York, 1991.

18. F. A. Heatley, P. A. Lovell, and J. McDonald, *Eur. Polym. J.*, **29**, 255 (1993).

19. K. Landfester, C. Boeffel, M. Lambla, and H. W. Spiess, *Macromolecules*, **29**, 5972 (1996).

20. M. S. Silverstein, Y. Talmon, and M. Narkis, *Polymer*, **30**, 416 (1980).

21. M. Okubo, *Makromol. Chem., Macromol. Symp.*, **35/36**, 307 (1990).

22. Y.-C. Chen, V. Dimonie, and M. S. El-Aasser, *J. Appl. Polym. Sci.*, **42**, 1063 (1991).

23. Y. G. Durant and D. C. Sundberg, *J. Appl. Polym. Sci.*, **58**, 1607 (1995).

24. A. Rudin, *Macromol. Symp.*, **92**, 53 (1995).

25. E. P. Dougherty, *J. Appl. Polym. Sci.*, **32**, 3079 (1986).

26. E. P. Dougherty, *J. Appl. Polym. Sci.*, **32**, 3051 (1986).

27. K. Masters, *Spray Drying Handbook*, 5th Ed., Longman Scientific and Technical, New York, 1991.

28. R. J. Grandzol, A. J. McFaull, H. Wanger, and I. S. Rabinovic, U.S. Patent 4,463,131 (1984).

29. H. Morikawa, S. Kato, H. Yasui, M. Hasegawa, and T. Shimizu, "Innovation of MBS Powder," in *SPE Annual Technical Conference*, 1987, p. 669.

30. H. Yasui and K. Higashitani, *J. Coll. Interf. Sci.*, **125**, 472 (1988).

31. H. Yasui, W. Okada, Y. Miki, and T. Mitani, U.S. Patent 4,997,911 (1991).

32. A. L. German, A. L. van Herk, H. A. S. Schoonbrood, and A. M. Aerdts, "Latex Polymer Characterization," in *Emulsion Polymerization and Emulsion Polymers*, P. A. Lovell and M. S. El-Aasser (eds.), John Wiley & Sons, Ltd., Chichester, England, 1997, Ch. 11.

33. E. A. Collins, "Measurement of Particle Size and Particle Size Distribution," in *Emulsion Polymerization and Emulsion Polymers*, P. A. Lovell and M. S. El-Aasser (eds.), John Wiley & Sons, Ltd., Chichester, England, 1997, Ch. 12.

34. X. Z. Kong, C. Pichot, J. Guillot, and J. Y. Cavaille, "Correlation of the Extent of Chain Transfer to Polymer with Reaction Conditions for Emulsion Polymerization of n-Butyl Acrylate," in *Polymer Latexes*, E. S. Daniels, E. D. Sudol, and M. S. El-Aasser (eds.), ACS Symposium Series, Vol. 492, American Chemical Society, Washington, DC, 1992, p. 163.

35. D. T. Nzudie, L. Delmotte, and G. Riess, *Macromol. Chem. Phys.*, **195**, 2723 (1994).

36. K. Kishida, N. Yamamoto, K. Nishida, T. Narita, and Y. Sato, U.S. Patent 4,513,111 (1985).

37. G. A. Vandezande and A. Rudin, *J. Coat. Technol.*, **66**, 99 (1994).

38. F. Sommer, T. M. Duc, R. Pirri, G. Meunier, and C. Quet, *Langmuir*, **11**, 440 (1995).

39. M. Ballauf, *Makromol. Chem., Macromol. Symp.*, **87**, 93 (1994).

40. G. D. Wignall, V. R. Ramakrishnan, M. A. Linne, A. Klein, L. H. Sperling, M. P. Wai, R. A. Gelman, M. G. Fatica, R. H. Hoerl, L. W. Fisher, S. M. Melpolder, and J. M. O'Reilly, *Mol. Cryst. Liq. Cryst.*, **180A**, 25 (1990).

41. P. Marion, G. Beinert, D. Juhue, and J. Lang, *Macromolecules*, **30**, 123 (1997).

42. Y. Aoki, *Macromolecules*, **20**, 2208 (1987).

43. R. Hasegawa, Y. Aoki, and M. Doi, *Macromolecules*, **29**, 6656 (1996).

44. M. P. Bertin, G. Marin, and J. P. Monfort, *Polym. Eng. Sci.*, **35**, 1394 (1995).

45. A. P. Gast and L. Leibler, *Macromolecules*, **25**, 2557 (1986).

46. M. Lu and D. R. Paul, *Polymer*, **37**, 115 (1996).

47. M. Lu, H. Keskkula, and D. R. Paul, *Polymer*, **37**, 125 (1996).

48. Y. Kayano, H. Keskkula, and D. R. Paul, *Polymer*, **37**, 4505 (1996).

49. R. J. Gaymans and J. W. van der Werff, *Polymer*, **35**, 3658 (1994).

50. M. Lu, H. Keskkula, and D. R. Paul, *J. Appl. Polym. Sci.*, **59**, 1467 (1996).

51. I. Manas-Zloczower and Z. Tadmor (eds.), *Mixing and Compounding of Polymers: Theory and Practice*, Hanser/Gardner Publications, Inc., Cincinnati, 1994.

52. C. H. Lai, W. G. Carson, and L. Schmidt, "Morphology of Crosslinked Polymer Particles in Thermoplastic Matrix," in *SPE ANTEC Technical Papers*, Vol. 37, 1991, p. 994.

53. J. Huang, Y. Shi, and A. F. Yee, "Correlating Dispersion Characteristics with Toughening in Polymer Alloys: A Unit-Cell-Strain-Energy Approach," in *PMSE Spring Meeting, San Diego, CA*, Vol. 70, American Chemical Society, 1994, p. 258.

54. C. S. Scott, personal communication, 1998.

55. I. Manas-Zloczower, "Dispersive Mixing of Solid Additives," in *Mixing and Compounding of Polymers: Theory and Practice*, I. Manas-Zloczower and Z. Tadmor (eds.), Hanser/Gardner Publications, Inc., Cincinnati, 1994, p. 55.

56. S. Al-Malaika, "Antioxidants and Stabilizers for Hydrocarbon Polymers: Past, Present, and Future," in *Handbook of Polymer Science and Technology*, Vol. 2, N. P. Cheremisinoff (ed.), Marcel Dekker, Inc., New York, 1989, p. 261.

57. P. R. Paolino, "Antioxidants," in *Thermoplastic Polymer Additives. Theory and Practice*, J. Lutz (ed.), Marcel Dekker, Inc., New York, 1989, Ch. 1.

58. D. G. Cook, A. Rudin, and A. Plumtree, *J. Appl. Polym. Sci.*, **46**, 1387 (1992).

59. K. J. O'Callaghan, A. J. Paine, and A. Rudin, *J. Polym. Sci.: Part A: Polym. Chem.*, **33**, 1849 (1995).

60. D. L. Dunkelberger and E. P. Dougherty, *J. Vinyl Tech.*, **12**, 212 (1990).

61. K. R. Kidder, U.S. Patent 5,294,659 (1994).

62. S. Richards, W. L. Wills, D. M. Wetzel, and A. Rosado, U.S. Patent 5,336,720 (1994).

63. D. Dompas and G. Groeninckx, *Polymer*, **35**, 4743 (1994).

64. S. Wu, *Polym. Eng. Sci.*, **30**, 753 (1990).

65. C. Wrotecki, P. Heim, and P. Gaillard, *Polym. Eng. Sci.*, **31**, 213 (1991).

66. A. Takaki, H. Yasui, and I. Narisawa, *Polym. Eng. Sci.*, **37**, 105 (1997).

67. C. A. Bertelo and M. Mori, "Particle Size Control: The Key to Impact Modification of Engineering Resins and Blends," in *Third International Conference: Advances in Additives and Modifiers for Polymers and Blends*, C. L. Beatty, C. A. Bertelo, A. Golovoy, and C. E. Scott (eds.), Executive Conference Management, Clearwater Beach, FL, 1994.

68. W. He, C. Cao, and C. Pan, *J. Appl. Polym. Sci.*, **61**, 383 (1996).

69. M. Ito, A. Yanagase, N. Yamamoto, and M. Mori, "A New Silicone Based Impact Modifier for PVC, Engineering Resins, and Blends," in *Third International Conference: Advances in Additives and Modifiers for Polymers and Blends*, C. L. Beatty, C. A. Bertelo, A. Golovoy, and C. E. Scott (eds.), Executive Conference Management, Clearwater Beach, FL, 1994.

70. H. Hosoi, N. Miyatake, and H. Yoshino, International Patent 9,710,283 (1997).

71. J. L. DeRudder and I. W. Wang, European Patent Application 537,014 (1992).

72. J. S. Clovis and F. H. Owens, U.S. Patent 4,148,846 (1979).

73. J. S. Clovis and F. H. Owens, U.S. Patent 3,984,497 (1976).

74. C. K. Riew, A. R. Siebert, R. W. Smith, M. Fernando, and A. J. Kinloch, "Toughened Epoxy Resins: Preformed Particles as Toughener," in *PMSE Spring Meeting*, Vol. 70, American Chemical Society, San Diego, 1994, p. 5.

75. F. Vazquez, M. Schneider, T. Pith, and M. Lambla, *Polym. Int.*, **41**, 1 (1996).

76. V. Tanrattanakul, E. Baer, A. Hiltner, R. Hu, V. L. Dimonie, M. S. El-Aasser, L. H. Sperling, and S. G. Mylonakis, *J. Appl. Polym. Sci.*, **62**, 2005 (1996).

77. F. H. Owens, U.S. Patent 3,808,180 (1974).

78. F. H. Owens, U.S. Patent 3,843,753 (1974).

79. N. Shah, *J. Mater. Sci.*, **23**, 3623 (1988).

80. C. He, A. M. Donald, and M. F. Butler, *Macromolecules*, **31**, 158 (1998).

81. P. A. Lovell, M. N. Sherratt, and R. J. Young, "Mechanical Properties and Deformation Micromechanics of Rubber-Toughened Acrylic Polymers," in *Toughened Plastics II: Novel Approaches in Science and Engineering*, C. K. Riew and A. J. Kinloch (eds.), ACS Advances in Chemistry Series, American Chemical Society, Washington, DC, 1996, Ch. 15.

82. P. A. Lovell, J. McDonald, D. E. J. Saunders, and R. J. Young, *Polymer*, **34**, 61 (1993).

83. S. Havriliak, Jr., C. P. Hemenway, and G. T. Beswick, *J. Vinyl Tech.*, **12**, 174 (1990).

84. T. W. Cheng, H. Keskkula, and D. R. Paul, *J. Appl. Polym. Sci.*, **45**, 531 (1992).

85. *Paraloid*™ *Additives Technical Literature*, Rohm and Haas Co., Philadelphia, 1995.

86. E. Hage, W. Hale, H. Keskkula, and D. R. Paul, *Polymer*, **38**, 3237 (1997).

87. J. Oshima and I. Sasaki, *Polymer News*, **16**, 198 (1991).

88. H. C. Fromuth and K. M. Shell, U.S. Patent 4,264,487 (1981).

89. A. J.Brady, H. Keskkula, and D. R. Paul, *Polymer*, **35**, 3665 (1994).

90. Anonymous, "What's cooking in PET?," in *Plastics World*, Melville, NY, P. T. N. Publishing Corporation, July 1995, p. 44.

91. J.-P. Meyer, D. Leblanc, and K. Nees-Brand, *Kunststoffe*, **85**, 452 (1995).

92. H. C. Fromuth and K. M. Shell, U.S. Patent 4,180,494 (1979).

93. W. G. Carson, C. D. Lai, and E. J. Troy, U.S. Patent 5,409,967 (1995).

94. P. Mapleston, "Resins '96," in *Modern Plastics International*, Lausanne, Switzerland, McGraw-Hill, January, 1996, p. 59.

95. M. I. Kohan, M. Hewel, H. D. Torre, H. A. Scheetz, H. Keskkula, D. R. Paul, F. J. Rietmeijer, G. Stoppelmann, and P. N. Richardson, "Modification," in *Nylon Plastics Handbook*, M. I. Kohan (ed.), Carl Hanser, Munich, 1995, p. 361.

96. R. E. Lavengood, A. R. Padwa, and A. F. Harris, U.S. Patent 4,713,415 (1987).

97. M. Baer, U.S. Patent 4,584,344 (1986).

98. B. S. Y. Chen and D. E. Henton, International Application Patent WO 88/01635 (1988).

99. A. M. Aerdts, G. Groeninckx, H. F. Zirkzee, H. A. M. van Aert, and J. M. Geurts, *Polymer*, **38**, 4247 (1997).

100. Y. Nakamura, R. Hasegawa, and H. Kubota, U.S. Patent 3,864,428 (1975).

101. T. W. Cheng, H. Keskkula, and D. R. Paul, *Polymer*, **33**, 1606 (1992).

102. D. Debier, J. Devaux, R. Legras, and D. Leblanc, *Polym. Eng. Sci.*, **34**, 613 (1994).

103. S. Y. Hobbs, M. E. J. Dekkers, and V. H. Watkins, *J. Mater. Sci.*, **23**, 1219 (1988).

104. F.-C. Chang and D.-C. Chou, "Rubber Toughening of Polycarbonate–Nylon Blends," in *Toughened Plastics II: Novel Approaches in Science and Engineering*, C. K. Riew and A. J. Kinloch (eds.), ACS Advances in Chemistry Series, American Chemical Society, Washington, DC, 1996, Ch. 18.

105. K. P. Lin and F. C. Chang, *Polym. Networks Blends*, **4**, 51 (1994).

106. S. Y. Hobbs, M. E. J. Dekkers, and V. H. Watkins, *Polymer*, **29**, 1598 (1988).

107. M. Fuji and H. Hirata, Japan Kokai Patent 61023640 (1986).

108. Y. Fukute and T. Yamada, Japan Kokai Patent 08325432 (1996).

109. N. A. Memon, R. H. Weese, and U. E. Ziegler, U.S. Patent 5,599,860 (1997).

110. H. Keskkula, D. R. Paul, K. M. McCreedy, and D. E. Henton, *Polymer*, **28**, 2063 (1987).

111. W. L. Wills and A. P. Berzinis, U.S. Patent 5,686,528 (1997).

112. J. P. Machado and W. P. Gergen, European Patent Application 459,587 (1991).

113. S. Havriliak, Jr., C. D. Lai, and N. A. Memon, U.S. Patent 5,654,365 (1997).

114. Y. Nakamura, H. Tabata, H. Suzuki, K. Ito, M. Okubo, and T. Matsumoto, *J. Appl. Polym. Sci.*, **32**, 4865 (1986).

115. D. E. Henton, D. M. Pickelman, C. B. Arends, and V. E. Meyer, U.S. Patent 4,778,851 (1988).

116. H.-J. Sue, *J. Mater. Sci.*, **27**, 3098 (1992).

117. L. Becu, H. Sautereau, A. Maazouz, J. F. Gerard, M. Pabon, and C. Pichot, *Ann. Chim. Fr.*, **19**, 363 (1994).

118. J. Wiersma, J. Meyer, D. Leblanc, and R. Berardino, "Toughening of Epoxy Resins Using Core–Shell Modifiers," in *Fourth International Conference Additives '96*, S. M. Andrews, C. A. Cruz, A. Golovoy, and C. A. Wilkie (eds.), Executive Conference Management, Houston, 1996.

119. J.-W. Kim, J.-Y. Kim, and K.-D. Suh, *J. Appl. Polym. Sci.*, **63**, 1589 (1997).

120. J. Y. Qian, R. A. Pearson, V. L. Dimonie, and M. S. El-Aasser, *J. Appl. Polym. Sci.*, **58**, 439 (1995).

121. A. Maazouz, H. Sutereau, and J. F. Gerard, *Polym. Bull. (Berlin)*, **33**, 67 (1994).

122. G. L. Shaffer, R. Bagheri, J. Y. Qian, V. Dimonie, R. A. Pearson, and M. S. El-Aasser, *J. Appl. Polym. Sci.*, **58**, 465 (1995).

123. R. H. Backderf and C. D. Guiley, Jr., U.S. Patent 4,082,895 (1978).

124. R. Alexander-Katz, *J. Appl. Polym. Sci.*, **31**, 663 (1993).

125. N. A. Memon, *J. Appl. Polym. Sci.*, **54**, 1059 (1994).

126. D. S. Parker, H.-J. Sue, and A. F. Yee, *Polymer*, **31**, 2267 (1990).

127. A. Lazzeri and C. B. Bucknall, *J. Mater. Sci.*, **28**, 6799 (1993).

128. C. Cheng, A. Hiltner, E. Baer, P. R. Soskey, and S. G. Mylonakis, *J. Appl. Polym. Sci.*, **52**, 177 (1994).

129. A. Lazzeri and D. Giuliani, "Yielding Kinetics in Rubber Toughened Polymers," in *10th International Conference on Deformation, Yield and Fracture of Polymers, Cambridge, UK*, Vol. 10, The Institute of Materials (Printed by The Chameleon Press Ltd), Cambridge, UK, 1997, p. 446.

130. S. J. Havriliak, Jr., C. A. Cruz, Jr., and S. E. Slavin, *Polym. Eng. Sci.*, **36**, 2327 (1996).

25 Toughening of Semicrystalline Thermoplastics

R. J. GAYMANS

Faculty of Chemical Technology
University of Twente
7500 AE Enschede
The Netherlands

Polymer Blends, Volume 2: Performance. Edited by D. R. Paul and C. B. Bucknall.
ISBN 0-471-35280-2.   © 2000 John Wiley & Sons, Inc.

I. INTRODUCTION

Semicrystalline polymers have a high craze stress and deform easily by shear yielding; however, with a notch, they nearly all fracture in a brittle manner. An efficient way to toughen these materials is to incorporate a dispersed rubbery phase, which increases the fracture energy by severalfold, at the cost of some loss of the modulus and the yield strength. Such blends, alloys, and copolymers are used in engineering applications (e.g., car bumpers) for which super toughness is required. The fracture energy of these toughened materials as measured by the notched Izod method is on the order of 50–100 kJ/m^2. The materials deform in a ductile manner under the conditions for which these materials are employed. Important parameters that affect the ductile fracture behavior are the following:

(a) matrix material

    (i) molecular weight

    (ii) crystallinity

    (iii) entanglement density

(b) dispersed-phase material

    (i) cavitational behavior

    (ii) concentration

    (iii) particle size

(c) test conditions

    (i) sample dimensions

    (ii) notch dimensions

    (iii) test speed

    (iv) test temperature.

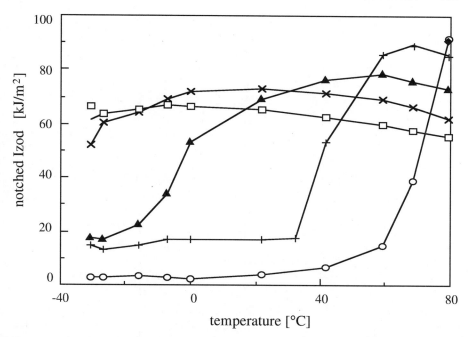

Figure 25.1 Notched Izod measurements as a function of the temperature of PA-6–PB blends at the following different rubber concentrations (vol%): (○) 0; (+)7.5; (▲) 15; (×) 22.5; (□) 30. From [12]; reproduced with permission from Elsevier Ltd.

For blends of a semiductile polymer with a rubber, it is now well accepted that the function of the elastomeric phase is to cavitate and thereby change the stress state of the matrix material in the neighborhood of the cavity [1–9]. Another effect of the cavitated particles is the stress field overlap with neighboring cavitated particles [4, 6, 10, 11]. As a result of cavitation, the plane-strain condition ahead of a notch is changed in the direction of the plane-stress state, and shear yielding of the matrix is now more likely to take place.

The deformation behavior of semicrystalline polymers, like that of some semiductile amorphous polymers, is strongly dependent on temperature (see Fig. 25.1) [12]. At low temperatures, these polymers fracture in a brittle manner with relatively low fracture energy. With increasing temperature, the fracture energy increases steadily, as a result of increasing deformation in the notch area. At a particular temperature, called the brittle–ductile transition, the fracture energy rises sharply. At this transition, one finds that some samples are broken in a brittle manner, while others give a ductile break. In the individual samples, combinations of partly brittle and partly ductile fracture are not observed. With rubber modification, the fracture energy at low temperatures is higher and the deformation in the notch at a particular temperature is greater. When rubber is added, the brittle–ductile transition temperature T_{bd} is considerably lower, and the transition is still discontinuous. Above T_{bd}, the fracture energy decreases with increasing temperature, and the level of fracture energy also

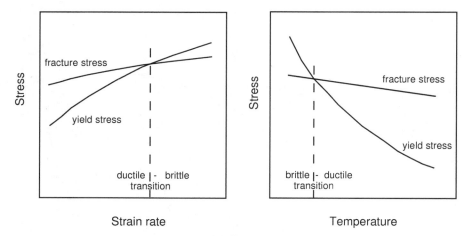

Strain rate   Temperature

Figure 25.2 Diagrams illustrating the Ludwig–Davidenkov–Orowan theories of brittle–ductile transition (a) strain rate, (b) temperature.

decreases with rubber concentration. This decrease in fracture energy in the ductile region with rubber concentration seems to be due to the decreasing amount of matrix material present [12].

The samples that show brittle fracture have a relatively low fracture energy (1–20 kJ/m^2 as measured with notched Izod), stress whitening in the notch, little visible deformation below the fracture plane (i.e., a thin stress-whitened zone), and a high fracture speed (> 200 m/s). The samples fractured in a ductile manner have a much higher fracture energy, a zone next to the fracture plane that has deformed plastically (i.e., a thick stress-whitened zone), and a low fracture speed (< 50 m/s). The level of impact strength in the ductile region corresponds to the thickness of the deformation (stress whitening) zone next to the fracture plane [13, 14]. The brittle–ductile transition is described by the Ludwig–Davidenkov–Orowan (LDO) criterion (see Fig. 25.2) [15, 16]. In this criterion, the brittle–ductile transition is where the yield stress and the fracture stress have the same value. The brittle–ductile transition temperature T_{bd} is very sensitive to changes in materials parameters and test conditions like test speed and test temperature.

The ductile fractured blends exhibit extensive plastic deformation. A large amount of the mechanical energy is dissipated as heat, and there is a marked temperature increase in the notch area (see Fig. 25.3) [12, 17]. If the plastic deformation takes place under adiabatic conditions (i.e., at high test speeds), an even greater temperature rise can be expected.

For these ductile materials, two generally accepted ways of studying the fracture behavior are the notched Charpy and the notched Izod method. Another very useful method for studying the fracture behavior of these materials is the single-edged notched tensile (SENT) test [18]. This instrumented method gives information on the fracture stress, the strain development before crack initiation, and the strain develop-

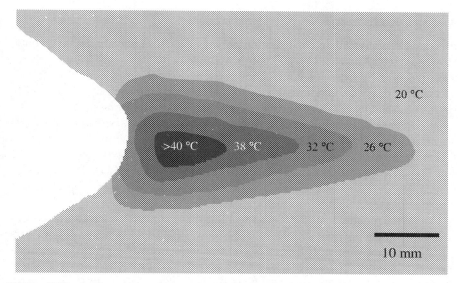

Figure 25.3 Surface temperatures of a 95/5 PP–EPDM blend ahead of a notch, as measured on SENT samples at 10^{-3} m/s and with an infrared camera with a spatial resolution of 0.130 mm. From [17]; reproduced with permission from Elsevier Ltd.

ment after crack initiation. For some blends, fracture energies have also been studied using linear elastic fracture mechanics (LEFM) [19–24].

II. BLEND FORMATION

The toughness of a material can be increased if crazing can be suppressed. A good balance of properties is obtained if the toughness is improved by using a dispersed rubber phase. Very critical in the deformation behavior of such materials is the morphology of the blends. It is now well accepted that the rubber in semicrystalline polymers has to be present as fine particles, with a particle size preferably less than 1 μm, but that very fine particles, such as polyamides smaller than 0.05 μm, do not take part in the toughening process [25–27]. The blends are mostly prepared by an extrusion compounding process. In special cases, such as for polypropylene [28] reactor processes are used. Blending may be carried out in a twin-screw extruder or a single-screw extruder with a mixing unit. It is also possible to carry out the blending during injection molding [29]. Parameters influencing the blending process are described in Chapters 16 and 17.

The rubbers most often used are EPDM, ethylene-propylene-rubber (EPR), ethylene-butylene-rubber (EBR), styrene-butadiene-styrene (SBS), styrene-ethylene-butylene-styrene (SEBS), styrene-butadiene-acrylonitrile (ABS), polybutadiene (PB), polyisoprene, polybutylacrylate, and polyurethane.

A. Extrusion Process

Blends of semicrystalline condensation polymers, such as polyamides (PA), poly-butylene terephthalate (PBT), and polyethylene terephthalate (PET), are nearly always prepared by extrusion compounding. Blends of polypropylene (PP) can also be made in this way, although this is not the method employed by industry [28]. Extrusion blending is a very flexible process that allows the variation of many material parameters. Blending with a rubber can be done with bulk rubber materials or preformed core–shell particles. In the compounding, the particle size of the rubber decreases as the molecular weight of the matrix increases (see Chapter 17 and Fig. 17.10a) [30, 31] and the molecular weight of the rubber decreases (see Fig. 25.4) [32]. Decreasing the rubber concentration also decreases the particle size (see Fig. 25.5) [32, 33].

One of the requirements for obtaining a fine dispersion is that the interfacial tension between the polymer and the rubber be low. PP–EPDM is such a system with a low interfacial tension. If the interfacial tension between the phases is too high, one has to modify the interface, by adding an adhesion promoter or by grafting at the interface in a reaction blending process. Particularly for PA systems, reactive extrusion with modified rubbers is a standard method. (See Chapter 17.) In reactive

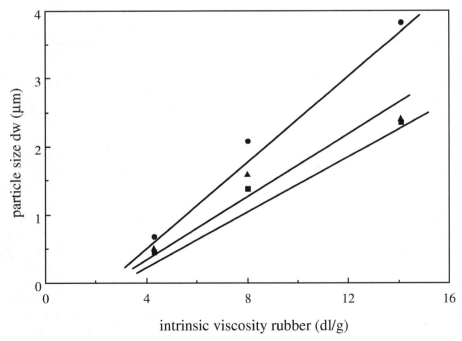

Figure 25.4 Influence of the molecular weight (intrinsic viscosity) of rubber on the weight-average particle diameter in PP–EPR blends at the following different concentrations (vol%): (■) 5; (▲) 10; (●) 20. From [32]; reproduced withpermission from Elsevier Ltd.

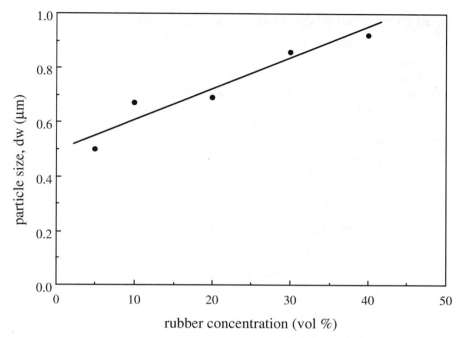

Figure 25.5 Influence of rubber concentration on the weight-average particle size in PP–EPDM blends. From [32]; reproduced with permission from Elsevier Ltd.

extrusion, rubbers grafted with maleic anhydride groups (-g-MA) are very frequently used [6, 31, 34–38].

B. Core–Shell Particles

Another way of obtaining the required fine phase structure is by starting with a core–shell particle impact modifier with a small particle size. Several core–shell systems consisting of a rubbery core and a hard shell (i.e., acrylate rubber–PMMA, polybutadiene–PMMA, and polybutadiene–SAN) have been developed. Extrusion of PA, PBT, and PET with core–shell impact modifiers gives materials with a good impact behavior [39–42]. The particle size of these core–shell modifiers is well defined and possibly even uniform. Core–shell materials are supplied as agglomerates, and these agglomerates, or clusters, have to be broken up. Without any specific interactions, this breaking up of the agglomerates is achieved only partially. The dispersion can be improved by adding an interfacial agent or by modifying the shell. The dispersion of core–shell particles, including those that are aggregated, is treated in detail in Chapter 24.

Core–shell particles can also be used in a reactor process. For the reactor process, the agglomerates have to be dispersed in a reaction mixture of low molar mass [43]. By starting from a latex instead of agglomerated particles, a well-dispersed system can easily be obtained [44].

III. BLENDS WITH SEMICRYSTALLINE POLYMERS

The deformation of semicrystalline polymers has been extensively reviewed (see also Chapter 6) [15, 45, 46]. The main properties of the most important semicrystalline polymers are given in Table 25.1. Most of these polymers do not break in an unnotched Izod test, but when notched they have a fairly low impact energy. The brittle–ductile transition temperature T_{bd} of most of these neat polymers, as measured by the notched Izod method, is well above room temperature. When these polymers blended with rubber, the T_{bd} can be lowered to well below room temperature in exchange for a loss in modulus and yield strength. The choice of the type of impact modifier depends on

- the ease of dispersing,
- the shift of T_{bd} that can be obtained,
- the fracture energy at room temperature,
- the thermal and UV stability of the rubber,
- the availability of the rubber,
- the melt rheological behavior of the blend, and
- the tensile properties of the blend.

The apolar olefinic rubbers seem to be the first choice for impact modifiers.

Table 25.1   Properties of Semicrystalline Polymers

	PA-6 (dry)	PA-66 (dry)	PBT	PET amorphous	PET crystalline	POM	PP	HDPE
T_m (°C)	215	265	220	-	253	175	170	130
T_g (°C)	60	60	45	70	80	0	5	-120
modulus (GPa)	2300	2300	2600	2500	2500	2800	1400	1000
yield stress (MPa)	75	75	60	60	62	65	35	28
Izod (kJ/m²)	n.b.[a]	n.b.	n.b.	n.b.	80	n.b.	n.b.	n.b.
notched Izod (kJ/m²)	4–5	4–5	6–8	3	2	8–9	4–6	n.b.
T_{bd} (°C) (notched Izod)	70	70	60	80	> 100	> 100	90	< R.T.[b]

[a] n.b. = not broken.
[b] R.T. = Room Temperature.

A. Polyamides

The most important polyamides are polyamide 6 and polyamide 66. They are about equally important as engineering plastics. Toughened grades of both polymers are on the market. The toughening of polyamide 6 (PA-6) has been studied extensively [3, 5, 10, 12, 14, 25, 27, 31, 34, 35, 37–39, 43, 47–69]. Unfortunately, a rubber for effective toughening is preferably apolar, and since PA-6 is rather polar, a fine dispersion of such rubbers is difficult to obtain. A good blend can be obtained with the aid of an interfacial agent or by interfacial reactions. (See Chapter 17.) The following dispersed phases have been used: maleic-anhydride-grafted (-g-MA) rubbers (e.g., EPDM-g-MA, EPR-g-MA, SEBS-g-MA, SBS-g-MA), mixtures of maleic anhydride rubbers with unmodified rubbers [6, 31, 35–38], maleic anhydride rubber with PP [60–62, 70–72], ABS, and other core–shell rubbers [39, 73–76], which are sometimes compatiblized with styrene–maleic anhydride copolymer (SMA). The mixtures of maleated and nonmaleated polymers with PA often give a broad particle-size distribution, with the -g-MA polymer at the interface [31]. The grafted maleic anhydride groups react readily with the amine and amide groups of the polyamide [31, 36, 37]. The graft polymer at the interface lowers the interfacial tension and stabilizes the dispersion against coalescence, which leads to a fine dispersion. Alternatively, with an adhesion promoter, a fine dispersion giving a good impact behavior can also be obtained [63, 64]. The adhesion promoter does not have to be a ductile material; even the very brittle SMA is effective [39, 73–76].

Rubber-modified polyamide 66 (PA-66) was the first marketed supertough engineering blend [5, 34, 57, 77–92]. Structure formation is more complex with PA-66, due to the bifunctionality (there are sometimes two amine end groups in one chain) of this polymer and the polyfunctionality of the -g-MA rubbers [26]. The rubbers used to toughen PA-66 are of the same type as for PA-6. With a given morphology, the deformation behavior is very similar to the PA-6–rubber blends.

B. Polyesters

Poly(butylene terephthalate) (PBT) and poly(ethylene terephthalate) (PET) are engineering thermoplastics that can also be toughened with rubbers [57, 93–95]. Both functionalized rubbers [40, 96–102] and core–shell rubbers [40, 41, 57, 103–105] are effective as impact modifiers for these polyesters. Core–shell rubbers are more effective than EPDM or PB rubbers [104, 105]. For reactive blending, epoxy groups are more effective than -g-MA functional groups [40]. Because the interfacial tension between acrylics and polyesters is high, the dispersion of the core–shell rubbers with acrylic shells is often not optimal [40]. The dispersion and impact properties can be improved considerably by adding an adhesion promoter. An efficient adhesion promoter for acrylic shell materials is PC [40, 41, 106–109]. ABS core–shell materials are also effective for toughening PBT [40, 110, 111]. The properties of the blends depend highly on the type of ABS used, as well as on the temperature of compounding and injection molding. It has been claimed that the modified rubber, EPDM-g-SAN, is useful for impact modification of PBT, PET, and PC [112].

PET is chemically and physically very similar to PBT. The toughening of PET has been described in the literature [41, 42, 113–119]. One of the primary ways in which PET differs from PBT is that its melting temperature T_m is higher, which gives PET blends the advantage of having a higher use temperature. However, the negative side of having a higher T_m is that the substance has a processing temperature that is high for some commonly employed toughening systems; also, the lower rate of crystallization of PET makes it more difficult to process. However, the dispersed particles can act as nucleation agents by inducing thermal stresses [115, 120]. Crystallization packages have been developed to compensate for the low crystallization rate of PET. Amorphous PET is far more ductile than semicrystalline PET and can also be toughened more easily [42]. For semicrystalline PET blends, high impact values are obtained only if the impact modifier is well dispersed. In blends of PET with acrylic shell rubbers, PC is an effective interfacial agent [42].

C. Polypropylene

Polypropylene (PP)–toughened blends have been studied extensively [17, 28, 30, 32, 33, 112, 121–154]. PP is most often toughened with EPR or EPDM. Satisfactory toughening is also obtained using SBS, SEBS, EBR, polybutadiene, or polyisoprene. Even though PP is not miscible with EPDM or EPR, there is a certain affinity that leads to good adhesion between the phases [155, 156]. With dynamically cross-linked EPDM, the particle size remains small during compounding, and at the same time, the impact behavior is improved [30, 126, 139, 152–154, 157]. For PP–polyisoprene blends, an optimum degree of cross-linking has been observed [158].

D. Polyoxymethylene

Owing to its high crystallinity and its lack of functional groups, polyacetal is more difficult to rubber modify. Polyoxymethylene (POM) blends have a high toughness [159–161]. A high toughness is obtained with a co-continuous network [162]. An intermeshed structure gave a supertough material with a notched Izod of 910 $J \cdot m^{-1}$; whereas discrete rubber particles gave a notched Izod impact strength of 200 $J \cdot m^{-1}$ [163]. Finally, POM can be impact modified effectively by blending it with polyurethanes [163, 164] or methyl methacrylate-butadiene-styrene (MBS) core–shell particles [165].

IV. MATRIX PARAMETERS

The yield and fracture behavior of semicrystalline polymers and their elementary deformation mechanisms have been reviewed [15, 23, 45, 166–169]. Yielding of semicrystalline polymers is accompanied by drastic rearrangement of the crystalline morphology. The spherulitic structure with lamellae is destroyed, and a fibrillar crystallite structure is the result. The deformation process inside the spherulites is strongly inhomogeneous, and the yielding of these materials often involves strain softening,

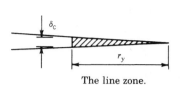

The line zone.

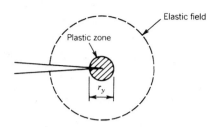

Elastic field and plastic zone

Figure 25.6 Possible shapes of the plastic zone ahead of sharp notch: (a) Line zone, (b) circular zone.

with the formation of a neck. At a high level of strain, a strain-hardening mechanism occurs. The strain hardening is important for stabilization of the deformed structure.

Because of the inhomogeneous deformation nature of the material and the notch in the sample, plastic deformation produces a Dugdale line zone (see Fig. 25.6a)—that is, a thin-layer zone (i.e., line zone) across the width of a sample, in which all the plastic deformation is concentrated. With a homogeneously deforming system ahead of a notch, the deformation zone has a cylindrical shape (i.e., it is a circular zone). (See Fig. 25.6b.)

The critical stress for yielding is dependent not only on material parameters, but also on the stress state in the sample. Ahead of a notch or crack, particularly for thick samples, a multiaxial stress state is present. To describe yielding under these conditions, the Tresca and the von Mises criteria can be used.

The fracture of semicrystalline polymers can occur by ductile tearing, like in rubbers, or by a crazing mechanism. Crazing is a dilatational process that yields the formation of cavities in the polymer ahead of an inhomogeneity or a notch. The craze stress increases with increasing entanglement density and molecular weight of the polymer [170].

Craze yielding and shear yielding are assumed to be independent processes, and the mechanism that requires the lowest stress is dominant. The yield stress is more sensitive to test parameters than is the craze stress. The yield stress increases with increasing strain rate and with decreasing temperature. The yield stress of a semicrystalline polymer also increases with crystallinity and lamellar thickness [171].

The mode of deformation is highly dependent on the stress state. As the stress state becomes more multiaxial, yielding becomes more difficult. (See Chapter 22.) The stress state can be altered by changing the sample dimensions or the notch size and by cavitation of a dispersed phase. Cavitation is a dilatation process and is very effective in changing the stress state. The cavitational behavior of a blend depends also on the thermal stresses in the sample [7, 172]. Lazzeri and Bucknall [11, 173] developed a yielding criterion for cavitated blends. (See also Chapter 22.)

The brittle–ductile transition temperatures of neat polymers, as measured by the notched Izod method, are usually well above room temperature. (See Table 25.1.) For ductile deformation at room temperature and below room temperature, a multiaxial stress state should be avoided. One way of avoiding a multiaxial stress state is to incorporate particles that cavitate before the craze stress is reached.

A. Molecular Weight

The molecular weight of the matrix has a strong effect on the dispersion of the rubber. (See Section II.A.) The molecular weight has also an effect on the tensile properties. The fracture stress and fracture strain of a polymer increase with the molecular weight of the polymer, while the yield stress remains constant. With increasing molecular weight (decreasing melt-flow index) of the neat polymer, the fracture stress increases relative to the yield stress and the brittle–ductile transition shifts to lower temperatures (see Fig. 25.7) [30]. The effect of the melt-flow index (MFI) for blends is similar to the effect observed with the neat polymer. Oshinski et al. [27] observed a shift in the curves of T_{bd} versus particle size with the molecular weight of PA. (See Fig. 25.8.) This means that at a particular particle size, the T_{bd} shifts to lower temperatures with increasing molecular weight. A T_{bd} shift with PA molecular weight has also been observed by Dijkstra et al. [58]. For PA blends, the fracture energy in the ductile region increases with the molecular weight of the PA [27, 58].

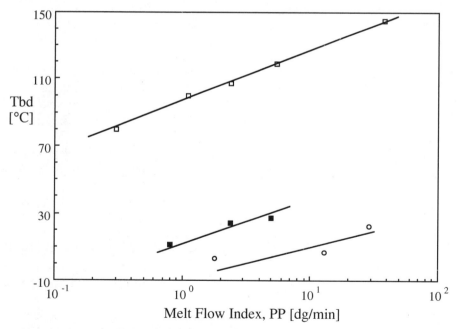

Figure 25.7   Influence of PP melt flow index (MFI) on T_{bd} in PP–EPDM blends, as measured by notched Izod tests: (□) PP; (■) 80/20 PP–EPDM ($d_w \approx 0.6 \ \mu$m); (○) 80/20 PP–EPDM prepared with dynamic cross-linking ($d_w \approx 0.6 \ \mu$m). From [30]; reproduced with permission from Elsevier Ltd.

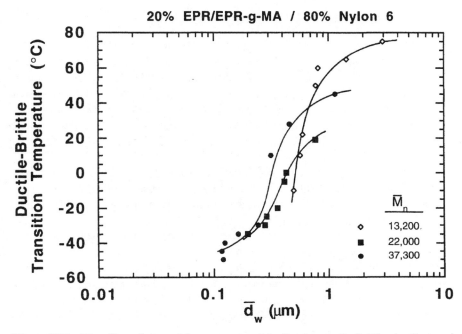

Figure 25.8   The effect of the weight-average particle diameter on the brittle–ductile transition temperatures of PA-6–EPR–EPR-g-MA, with the molecular weight of PA being varied. From [27]; reproduced with permission from Elsevier Ltd.

B. Crazing

As a sample is loaded in tensile or in bending, particularly ahead of a notch, a volume strain is built up in the sample. With increasing volume strain crazing or cavitation can take place. The ease of the crazing or cavitation depends on the interchain bonding and entanglement density of the polymer [174]. For many amorphous polymers, the craze initiation stress correlates well with the entanglement density [170]. However, with increasing entanglement density, the maximum draw ratio is reduced, which makes the polymer less ductile. If the cavities in the sample are of sufficient size, they can grow into a crack. The ease of cavitating of a polymer and the size of the cavities formed seem to be important parameters in this failure mechanism by crazing.

In blends, a competition between rubber cavitation and matrix crazing is expected if the cavitation stress of the matrix polymer is of the same order as that of the dispersed phase. In blends with semicrystalline matrix polymers, cavitation of the rubber is usually easier than crazing of the matrix. If large plastic deformation takes place, the deformation with semicrystalline matrix polymers nearly always occurs by shear yielding, and not by multiple crazing.

C. Crystallinity

The crystalline regions in semicrystalline polymers are the physical cross-links that hold the material together, particularly above the T_g of the material. With increasing crystallinity of the matrix, the modulus and the yield strength above the T_g increase [171]. Below the T_g, the effect of crystallinity on the modulus and yield strength is small. The number of crystalline cross-links below the T_g compared to the number of frozen-in entanglements is small. From these factors, it can be expected that the effect of crystallinity on the impact behavior is particularly strong above the T_g of a polymer.

The effect of crystallinity has been studied for PP (T_g at 5°C) [33]. The T_{bd}, as measured with notched Izod and SENT at 1 m/s, increased with increasing crystallinity. (See Fig. 25.9.) The increase in crystallinity of neat PP from 31 to 53% resulted in an increase in T_{bd} of 70°C. Also, in PET, a strong increase in T_{bd} with crystallinity was observed [42]. Thus, increasing crystallinity has a very strong negative effect on the brittle–ductile transition of the neat polymers.

In blends, the presence of the dispersed phase was found to have little effect on the degree of crystallinity of the matrix material [30], and the crystallization rate was hardly affected by the dispersed phase [87, 88]. However, the spherulitic structure of the matrix in the blends was distorted and much finer. In PA-66–EPDM-g-MA

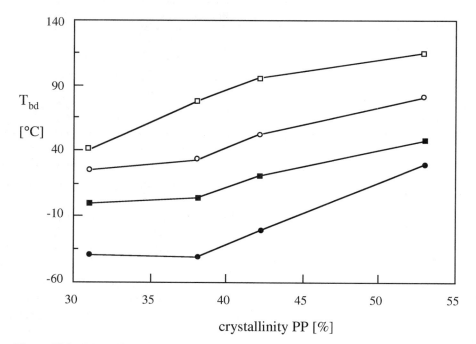

Figure 25.9   Thee effect of crystallinity in PP on the brittle–ductile transition of a PP–EPDM blends, as measured by notched Izod tests at the following EPDM concentrations (vol%): (□) 0; (○) 10; (■) 20; (●) 30. From [30]; reproduced with permission from Elsevier Ltd.

blends, a surface-crystallization layer around the rubber particles was observed [89]. This surface-crystalline structure has been proposed to facilitate the local yielding in between the particles.

The effect of crystallinity on the T_{bd} of the PP–EPDM blends has been studied. With increasing crystallinity of the PP, the T_{bd} of the blends shifted to higher temperatures in a similar way as for neat PP (see Fig. 25.9) [30]. It is noteworthy that at temperatures below the T_g of the PP, the increasing crystallinity has a negative effect on T_{bd}.

A balance of properties is of practical interest. It is important to have a high ductility combined with a high modulus and a high yield strength. A highly crystalline PP, which is more brittle, can also be modified to a system with low-temperature ductility by adding more rubber. (See Fig. 25.9.) For blends based on polymers with varying matrix crystallinity and rubber concentration, the T_{bd}–modulus relationship can be derived. (See Fig. 25.10.) Highly crystalline PP has a high modulus and a high T_{bd}; by increasing the rubber concentration, the modulus and the T_{bd} are lowered. The best balance of properties is obtained with the highly crystalline PP. The same principle holds for the relationship between yield strength and T_{bd}.

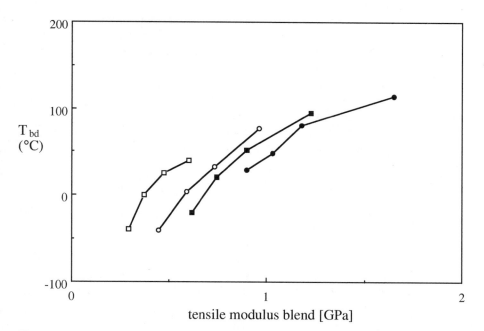

Figure 25.10   The modulus of PP–EPDM blends versus T_{bd} at different volume fractions of EPDM, with the following values for the crystallinity of PP (%): (□) 32; (○) 38; (■) 45; (●) 54. From [30]; reproduced with permission from Elsevier Ltd.

D. Matrix Glass Transition

The brittle–ductile transition can be described by the LDO criterion, or the ratio of yield stress to fracture stress [15]. The change in yield stress with temperature and test speed is therefore an important parameter. For semicrystalline polymers, the modulus drops at the T_g, as does the yield stress [175], while the Poisson's ratio increases. For PA and polyesters, the T_{bd}, as measured by Izod tests, coincides with the T_g. (See Table 25.1.) If the matrix is plasticized, the T_{bd} shifts together with the T_g [52]. For PP, however, the T_{bd} is well above the T_g (at 5°C). (See Fig. 25.7.) The T_{bd} is also heavily dependent on the crystallinity of the matrix. The strong change in the yield behavior near T_g is an important parameter that affects the T_{bd}, but other parameters, such as molecular weight, notch size, and sample dimensions, are also significant.

V. DISPERSED-PHASE PARAMETERS

Most semicrystalline engineering polymers are notch sensitive and when notched have a high brittle-to-ductile transition temperature. (See Table 25.1.) Upon blending of a rubber into a semicrystalline polymer, the T_{bd} is shifted to lower temperatures (see Fig. 25.1) [12]. The T_{bd} shift is dependent on rubber concentration, rubber particle size, particle dispersion, and morphology of the particles. The function of the rubber phase is to cavitate and thereby change the stress state of the matrix material in the neighborhood of the cavity and create an elastic stress-field overlap of the cavitated particles (see also Chapter 22) [1–8, 10, 11]. To what extent the properties of the rubber particles after cavitation are important, however, is still not very clear. Toughening with small holes (i.e., cavities without rubber) has been found to be as effective as toughening with cavities in the rubber [176, 177]; the compressive and shear strength of a system with holes is poor. Steenbrink et al. [178] calculated that the rubber properties of the cavitated rubber particles have an effect on the yield behavior. The higher the shear modulus of the dispersed phase, the more difficult it is for the cavities to grow.

The cavitational behavior of the dispersed rubber can be studied by volume strain experiments [82, 179–181], by observing the onset of stress whitening [7, 182], and by examining the cavitated structure after deformation. A common technique is to measure the volume strain with strain gauges [82, 179]. This method is accurate only at low test speeds. Measurement of the development of stress whitening is a non-sample-contacting method with a very short response time and can also be used at very high test speeds [7, 182]. It measures the change in transmittance of light due to stress whitening simultaneously with the tensile data. The onset of stress whitening is taken as the onset of cavitation.

A. Rubber Concentration

The moduli of the blends were found to decrease linearly with increasing rubber volume fraction; for a 40% blend, the decrease is approximately 55% [33, 49]. The

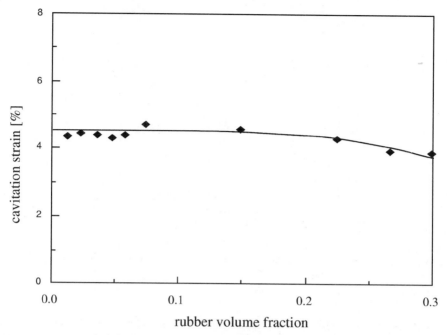

Figure 25.11   The onset of stress whitening strain (i.e., cavitation strain) as a function of the rubber content, as measured in a tensile test (at 10^{-2} s^{-1}) on a PA-6–PB blend. The results were obtained with a laser setup. From [7]; reproduced with permission from Chapman & Hall.

effect of rubber concentration on yield stress is more complex. As long as the rubber particles have not cavitated, they can bear a stress, but after cavitation the voided particles bear little stress. On cavitation, the stress field around particles changes, and there is greater stress-field overlap between cavities. The moment of cavitation is important for the yield stress of a blend. A particle distribution has an effect on the local concentration and on the local yielding [183]. In the regions with a high rubber concentration, the yielding is at a lower stress than in the regions with a low rubber concentration. The yielding is now a diffuse process and starts at a low stress.

   The neat polymers that shear yield show hardly any volume strain effects during tensile loading. In blends, the onset of rubber cavitation occurs at low strains (2–4%) and before the yield point of the neat polymer is reached [7, 82, 179]. Only at high concentrations does the rubber have an effect on the onset of cavitation (see Fig. 25.11) [7]. The volume strain at low concentrations increases linearly with the rubber concentration [179]. Model calculations indicate that the cavitational process at high concentrations is influenced by neighboring particles [10, 11]. Apparently, a few percent rubber is not sufficient to relieve all of the volume strain [179]. This means that for a sample that is deformed under conditions for which a multiaxial stress state can develop, e.g., ahead of a notch or a crack, this multiaxial stress state will be more difficult to develop if the sample has a high rubber concentration.

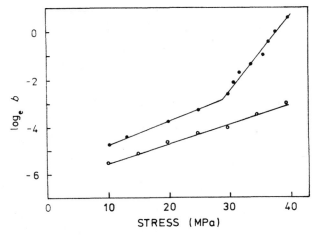

Figure 25.12 Creep behavior of PA-66 and a PA-66–EPDM blend, as shown by Eyring plots of $\log b$ against applied stress: (○) PA-66; (●) PA-66–EPDM blend. From [82]; reproduced with permission from Chapman & Hall.

The creep rate is found to increase considerably after cavitation of a blend (see Fig. 25.12) [114]. Thus, for limiting the creep in a blend, increased resistance to cavitation of the rubber is an advantage.

The notched Izod curves shift to lower temperatures with increasing rubber concentration at a constant particle size. (See Fig. 25.1.) A shift in the Izod curve of more than 100°C can be obtained by increasing the rubber concentration to 30% [12, 33, 49]. At the same time, the notched Izod impact strength at −35°C (in the brittle region) increases linearly with the rubber concentration. (See Fig. 25.13.) This increase in low-temperature impact strength is due to an increase in deformation in the notch area, resulting in a shift in the Izod curve to lower temperatures. At the high temperature of 80°C (in the ductile region), the impact strength decreased with increasing rubber concentration. (See Fig. 25.13.) This decrease in fracture energy is thought to be due to the decreasing amount of matrix polymer present with increasing rubber concentration. Most of the energy absorption in a sample comes from the shear deformation of the matrix material. Some workers have found that there is an optimum rubber concentration for room-temperature impact strength [12, 37]. Others have found that the impact strength in the ductile region increases with rubber concentration [13, 57]. A third effect that can sometimes, although not always, be observed in the Izod–temperature curve (see Fig. 25.1) in the ductile region is a slight decrease in fracture energy with increasing temperature. This decrease in fracture energy is thought to be due to the decrease in yield strength with increasing temperature.

The most important effect of an increase in rubber concentration is the shift in the brittle-to-ductile transition to lower temperatures. The decrease in T_{bd}, as measured with Izod and SENT at 1 m/s, is approximately linear with the increase in rubber concentration (see Fig. 25.14) [33, 49]. This factor suggests that the changes

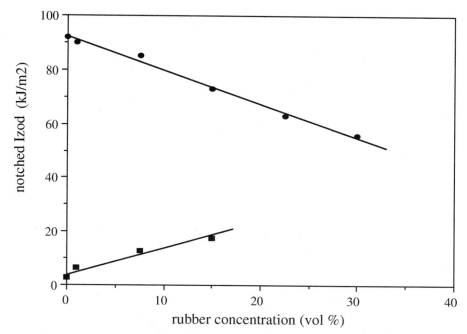

Figure 25.13 Notched Izod measurements as a function of rubber concentration for PA-6–PB blends at low and high temperature: (■) −35°C (in brittle region); (●) 80°C (in ductile region). From data in [12]; repoduced with permission from Elsevier Ltd.

in stress-state and stress-field overlap are gradual as the rubber concentration increases. At low test speeds, the T_{bd} decreases with increasing rubber concentration in a similar way as at high test speeds, but at a lower temperature.

B. Particle Size

The function of rubber particles is twofold; to cavitate and thereby change the stress state around the particles and to generate a local stress concentration [1–11]. However, the cavitated particles should not be initiation sites for the fracture process; therefore, the particles must be so small that they do not grow to a size at which they can initiate a crack.

Another requirement is that cavitation of the particles should not occur too early, as blends with cavitated particles have a higher creep rate (see Fig. 25.12) [114]. Just-in-time cavitation is what is wanted. It is expected that the larger particles will form larger cavities. However, this is true only if there is one cavity per particle. The cavitation of particles is a function of the rubber properties [6, 7, 178, 184], the stress state [7, 11, 178], and the particle size [11, 25–27, 185]. The cavitation of particles is a balance between the elastic energy in and around the particle and the energy for creating new surfaces [7, 11, 186]. If the particles are very small, cavitation becomes difficult [11, 25–27].

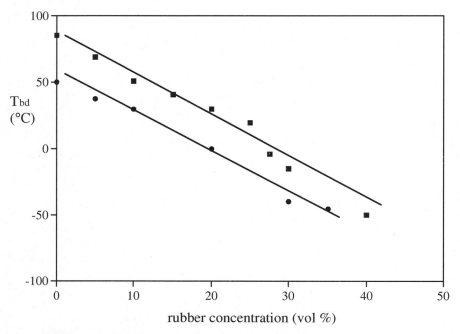

Figure 25.14   Brittle–ductile transition temperature as a function of rubber content, as measured by SENT, on a PP–EPDM blend at the following test speeds: (●) 1 mm/s; (■) 1 m/s. From [33]; reproduced with permission from Elsevier Ltd.

The effects of particle size have not been studied extensively [25–27, 32, 49], since it is difficult to change the particle size without changing other parameters. In an extrusion compounding process, the particle size is a function of shear rate, interfacial tension, matrix viscosity, dispersed-phase viscosity (see Fig. 25.4), and rubber concentration (see Fig. 25.5). (See also Chapters 16 and 17.) Changing the shear rate during compounding can change the particle size. On subsequent injection molding, however, the effect often is lost. Only if the structure is stabilized against coalescence can a reasonable particle-size variation in molded test specimen be obtained by changing the shear rate in compounding. The particles can be stabilized by a grafting reaction over the interface and also by dynamic cross-linking of the rubber during blending.

A less reliable method is to change the melt viscosity (i.e., the molecular weight) of the dispersed phase. One can use this method only if the change in the molecular weight of the rubber has little effect on its function. It has been reported that the effect of the molecular weight of the rubber phase on the impact strength for the PP–EPDM system is small [137]. Van der Wal et. al. [32] changed the particle size in a PP–EPR blend by changing the molecular weight of the EPR, while keeping the type of PP and the blending conditions constant. They and another group of workers found that the effect of particle size on the yield stress is small [32, 49], unless the particle size is very small (i.e., 0.14 μm) [7]. As the particle size of the rubber decreases, the onset

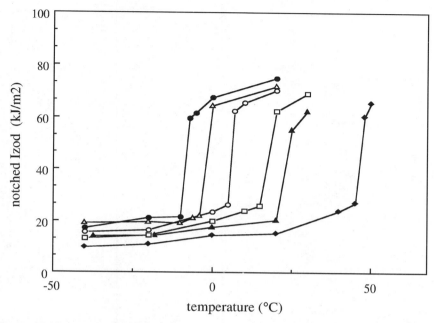

Figure 25.15 Notched Izod measurements as a function of temperature for PA-6 and a PA-6–EPDM-g-MA blend (26 vol%) at the following weight-average rubber particle sizes (µm): (◆) 1.59; (▲) 1.20; (□) 1.14; (○) 0.94; (△) 0.57; (●) 0.48. From [49]; reproduced with permission from Elsevier Ltd.

of cavitation occurs at higher strains, particularly for very small sizes. At the same time, the yield stress of the blend is changed [7].

The notched Izod energy curves are strongly influenced by particle size (see Fig. 25.15) [25–27, 32, 49]. With decreasing particle size, the ductile-to-brittle transition is shifted to lower temperatures. The Izod curves does not seem to change shape: The curves simply shift to lower temperatures. As the particle size of the rubber is reduced from 2 to 0.3 µm, this shift is about 50°C for both PA–EPDM and for PP–EPR systems. The effect of the particle size can best be correlated with the weight-average particle size d_w better than with the number-average particle size d_n [27, 32]:

$$d_n = \frac{\Sigma n_i d_i}{\Sigma n_i} \qquad d_w = \frac{\Sigma n_i d_i^2}{\Sigma n_i d_i}. \qquad (25.1)$$

In Eq. 25.1, n is the number of particles with size d.

Some PA-66–EPDM blends that had rubber particles of much larger sizes, with PA inclusions, gave impact behavior similar to that of a PA-6–EPDM blend with finely dispersed rubber particles [85, 187]. This PA-66 blend had a low T_{bd} in spite of its large particle size. The PA inclusions in the large particles are thought to suppress the coalescence of the cavities. The cavitation of large rubber particles is relatively

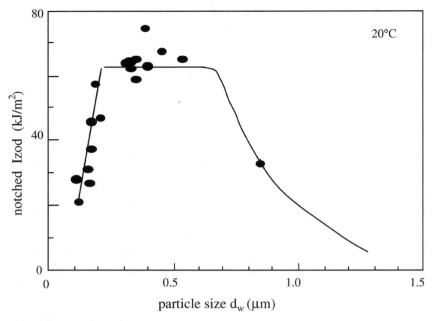

Figure 25.16 Notched Izod measuremnents as a function of the weight-average particle size for a PA-6–EPR blend (20 vol%). From [25]; reproduced with permission from Hüthig & Wepf.

insensitive to their size; thus, effects of particle size on the toughening of the blends must stem from another cause in this size region. The effect of particle size on the toughening of the blends in this region is not due to differences in ease of cavitation of the particles, but possibly to the ease of coalescence of cavities to large cavities that nucleate the fracture process.

Blends with very small particle sizes have poor impact behavior (see Fig. 25.16) [25, 27, 32, 67]. In fact, PA particles smaller than 50 nm apparently do not take part in the toughening process [27, 32]. This means that if the weight-average particle size falls below 200 nm, the impact strength at room temperature decreases and the T_{bd} increases. The reason for the lower limit in particle size seems to be that very fine particles are more difficult to cavitate. Furthermore, the fracture toughness at the lower limit is related to the degree of cavitation [25, 173, 178, 188, 190]. The fracture stress for blends with very small particles (i.e., about 50 nm) is much lower than for blends with larger particles [67]. This means that without cavitation, yielding in the notch is more difficult.

The effect of particle size at low test speeds is quite different from that at high test speeds. At low test speeds for PP–EPR and PA–EPR blends, the T_{bd} was found not to change much in the range of 0.2–2 μm (see Fig. 25.17) [32, 191]. If, however, very large particles are present, the fracture strain in a tensile test is considerably reduced (see Fig. 25.18) [192].

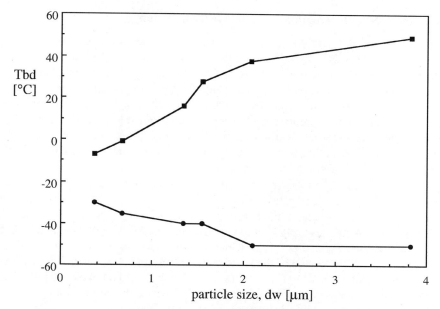

Figure 25.17 Brittle–ductile transition temperature of an 80/20 PP–EPDM blend as a function of particle size at the following test speeds: (■) Izod (high test speed); (●) SENT (1 mm/s; low test speed). From [32]; reproduced with permission from Elsevier Ltd.

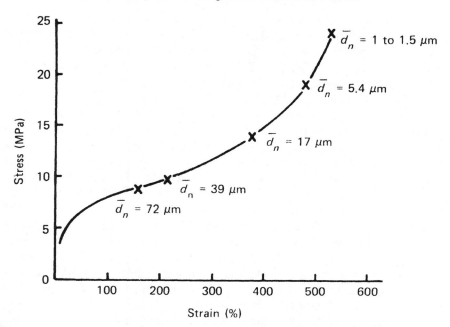

Figure 25.18 Stress-strain data of a 60/40 blend of PP and EPDM as a function of particle size. × denotes the fracture point. From [192]; reproduced with permission from the American Chemical Society.

C. Particle-Size Distribution and Particle Distribution

The particle-size distribution d_w/d_n of extrusion-blended systems is usually narrow, on the order of 1.4–2.1. The distribution with core–shell particles also can be very narrow. If the rubber core is a good rubber (i.e., it has a low T_g), like polybutadiene, and the particle size is small (e.g., 0.3 μm), then the toughening effect is excellent [12]. The particle-size distribution can be broadened if two rubbers are used, or if the rubber is not homogeneous in composition. For these broadly distributed blends, with a particle-size distribution of 4–7, the T_{bd} still correlates well with d_w [30]. However, a broad (i.e., bimodal) distribution of particle sizes has not been found to be advantageous with semiductile matrixes.

If the compound is not well blended, the distribution of particles within the matrix is poor. Particular aggregates of core–shell particles should be absent. It can be expected that if the particles are not homogeneously dispersed, then the deformation is less homogeneous, resulting in a lower fracture energy. With a PA–PB blend, it was found that the T_{bd} of a poorly distributed compound was higher than do that of a well-distributed compound of same blend material [12]. Systems with flocculated particles have lower impact strengths than do well-distributed blends [80]. The cavitation and the matrix yielding of blends with particles that are well dispersed in the matrix depend on local concentration fluctuations [11].

D. Composition Dispersed Phase

The function of the dispersed particles is to cavitate just in time, before the yield strength of the matrix is reached. The cavitational behavior of the dispersed phase depends on the type of elastomeric material. Model calculations of the particle cavitation have been made [173, 178, 188]. (See also Chapter 22.) These calculations indicate that in fully hydrostatic loading, the cavitation growth is directly related to the shear modulus of the rubber [7, 178].

Studies on the effect of the rubber showed that the type of rubber has little influence on the notched impact strength in the tough region and a strong effect on the T_{bd} (see Fig. 25.19) [6, 7, 27, 63]. The effect of the type of rubber on the T_{bd} could not be correlated with the tensile strength or the elongation at break of the elastomers. For the olefinic elastomers, a good correlation was found with the modulus (see Fig. 25.20) [25, 27]: The lower the modulus of the rubber, the lower the T_{bd} (at constant rubber concentration and rubber particle size). For elastomers like the polyetheresters, a better correlation is expected if the Poisson's ratio of the elastomer is also taken into account [25, 193]. The molecular weight of the rubber in the blend was found to have little effect [32, 137]. Also, it has been proposed that the optimal rubber must be cross-linked to some extent, so that after cavitation it bears stresses and prevents the formation of a fatal crack [1]. However, cross-linking a blend, in this case, HIPS, by radiation after the blend was formed did not improve the impact behavior of the blend [194]. With increasing cross-link density, the cavitation of the rubber becomes more difficult [184, 194].

Lightly cross-linked EPDM rubber gives a slightly lower T_{bd} (see Fig. 25.7) [32, 139]. In dynamic blending, a better adhesion was obtained, possibly because

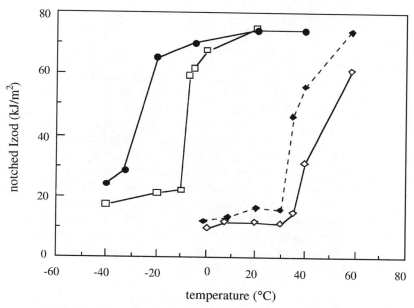

Figure 25.19 Notched Izod measurements as a function of temperature for PA-6 blends (26 vol%; particle size is 0.35–0.45 μm) with the following modifiers: (●) PB ($G' = 0.70$ MPa); (□) EPDM ($G' = 5.6$ MPa); (◆) LDPE ($G' = 130$ MPa); (◇) PE ($G' = 131$ Mpa).

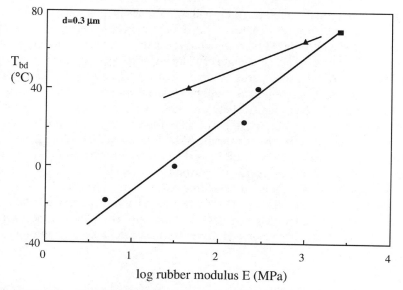

Figure 25.20 The effect of the rubber modulus on the brittle–ductile transition, as measured by notched Izod tests, for blends of PA-6 with the following rubbers (10 vol%; particle size is 0.30 μm): (●) olefinic polymers; (▲) polyetherester elastomers; (■) PA 6. From [25]; reproduced with permission from Hüthig & Wepf.

of the formation of a graft polymer at the interface. Cross-linking the rubber seems to prevent voids in the cavitated particles from coalescing to form large, unstable cavities [195].

The rubbers become effective near their glass transition temperature [27]. For polybutadiene, the lower temperature limit under impact conditions is $-60°C$; for EPDM, the limit is $-45°C$; for EPR and SEBS, it is -50 or $-20°C$, depending on their composition; and for butyl acrylate, it is $-20°C$.

Some rubbers give a higher impact strength in the ductile region, although their T_{bd} is not that low [27, 37]. This suggests that the impact strength and the T_{bd} are dependent on different parameters. Once a material fractures in a ductile mode, the crack speed is low, and considerable plastic deformation can take place next to the fracture plane. The fracture energy is related to the amount of plastic deformation and hence to the width of the deformation zone. The high impact values at room temperature are not related to the modulus of the rubber. One of the requirements seems to be a homogeneous distribution of the rubber particles in the blend. Formation of co-continuous microclusters of the particles seems beneficial for the fracture toughness [196].

E. Interfacial Effects

A low interfacial tension between components of a blend is important for obtaining a fine dispersion (see also Chapters 16 and 17) [197]. This condition is relevant not only for dispersing bulk rubber but also for dispersing aggregated core–shell materials. A low interfacial tension can be obtained through a specific interaction, through a grafting reaction at the interface, or by adding an adhesion promoter. To prevent delamination of the dispersed particles during deformation of the blend, the interfacial strength has to be only 1000 J·m^{-2}, which is the tearing strength of the rubber [79]. In fact, this level of interfacial strength can already have been obtained with van der Waals bonding. Acrylic shell particles are difficult to disperse in PBT and PET, but blends with good impact properties could be obtained with a high-molecular-weight matrix material, which has higher shear forces, or by adding a dispersion agent, like PC, to these systems [41, 42, 63, 64]. The adhesion promoter does not have to be a ductile material itself; even the very brittle SMA is effective [64, 73–76].

The T_{bd} of blends of PA with EPDM-g-MA (at a particular particle size) is independent of the amount of interfacial grafting [27, 36]. A lower amine end-group concentration in PA was found to have little effect on the dispersion of the rubber [31] or the T_{bd} of the blends [27]. Once a fine dispersion was obtained, the impact properties were good. Grafted MA polymers not only strengthened the interface, but also stabilized the dispersion and decreased the coalescence rate during further processing [12, 31].

VI. SAMPLE AND TEST PARAMETERS

It is important to recognize that in fabrication of samples for testing, particles size can increase, due to coalescence. The slow cooling in a compression molding process can result in larger particle sizes [198]. This effect is less likely to occur in injection molding, but injection molding can result in oriented particles, particularly at the surface of the sample.

A. Test Method

Generally accepted ways of studying the fracture behavior of ductile materials are the notched Charpy and the notched Izod methods. (See also Chapter 21.) For many nonengineering applications, the unnotched values are used for toughness.

In terms of their measurement of impact strength, the Izod and Charpy tests currently represent the easiest and fastest way of evaluating the mechanical behavior of blends at high strain rates. As for toughnes samples of superductile materials are often not fully broken, the fracture energy values in the ductile region obtained are not very reliable. Instrumented Charpy tests give more information on the force-deflection behavior both before and after crack initiation.

A very useful method for studying ductile blends is the instrumented single-edged notched tensile (SENT) test [18]. The stress displacement curves from this test give information about the deformation before and after crack initiation. Thus, failure behavior can be studied as a function of materials and test variables, and the samples are always fully broken. The samples used in SENT method are firmly gripped, which results in only small vibrations on the force signal. This method yields valuable information even at high test speeds.

Particularly critical in these ductile blends is the brittle–ductile transition. The brittle–ductile transition temperature approach has long been accepted as a means of studying ductility in steels [199]. In polymers, the change from brittle to ductile behavior is observed as a change in the structure of the deformation zone and as an increase in the fracture resistance G_c [166, 200]. The brittle–ductile transition is described by the Ludwig–Davidenkov–Orowan criterion (see Fig. 25.2) [15, 16]. This criterion is independent of the elastic energy in the sample. The brittle–ductile transition as measured with the notched Izod, notched Charpy, or the SENT method is, however, also a function of the elastic energy in the sample.

Better characterization of the fracture energy can be obtained using fracture mechanics methods, although no impact property is truly independent of geometry or impact conditions. Under plane-strain conditions (that is, in thick samples with sharp notches) in the brittle region (that is, the region with low temperatures), one can determine K_{Ic} and G_{Ic}. These parameters are thought to be true material parameters. (See also Chapters 20 and 21.) However, it has been observed that K_{Ic} and G_{Ic} change with test speed [21, 24, 166, 201]. If the materials fracture in a ductile manner, the linear elastic fracture methods (LEFMs) are not suitable. For ductile systems, with mixed-mode deformation, the J_{Ic} method can be used (see also Chapter 20) [202, 203]. The J_{Ic} can be determined at high test speeds. This method is also usable for studying semiductile materials [19, 20]. The J_{Ic} value for a PA-66–EPDM blend was found to be independent of test speed, but for an amorphous PA–EPDM blend, it increased with test speed. For PA-6–EPDM blends, the J_{Ic} and the dJ/da changed with test speed and test temperature [20]. The brittle–ductile

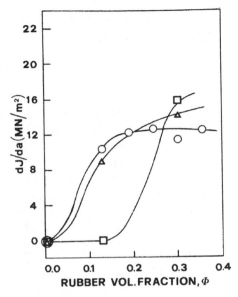

Figure 25.21 dJ/da as a function of temperature for PA-6–EPDM blends as a function of the volume fraction of EPDM at the following test speeds: (○) 5 mm/min; (△) 200 mm/min; (□) 0.5 m/s. From [20]; reproduced with permission from Hüthig & Wepf.

transition can be monitored by the change in dJ/da. (See Fig. 25.21.) The J_{Ic} values, however, are not independent of geometric factors [204]. A similar approach to studying the fracture toughness up to high test speeds is to measure fracture energies as functions of crack length [22, 67]. It is also possible to measure the crack initiation resistance G_C and the crack propagation resistance G_D [21, 205, 206]. G_D emerges as geometry independent, but crack-speed and temperature dependent. G_C is influenced by geometry and impact speed. In brittle fracture, there is an adiabatic thermal failure at the crack tip whereby a one-molecule-thick melt layer is formed [200].

B. Test Speed

The SENT test can be used to study the effect of clamp speed on the fracture stress and the fracture energy [17, 18]. Typical stress-displacement curves for 70/30 a ductile PP–EPDM blend at different test speeds are given in Fig. 25.22 [17]. In most cases, the maximum stress is the fracture stress. One can see that the fracture stress increases strongly with test speed. The ductile fracture at low test speeds occurs at a low fracture stress but a high fracture energy. At high test speeds, the total fracture energy seemed to be fairly constant or possibly even better at 0.75 m/s.

The tensile yield stress and the fracture stress for SENT samples of a PP–EPDM blend were studied as function of strain rate (see Fig. 25.23) [17]. The fracture stress

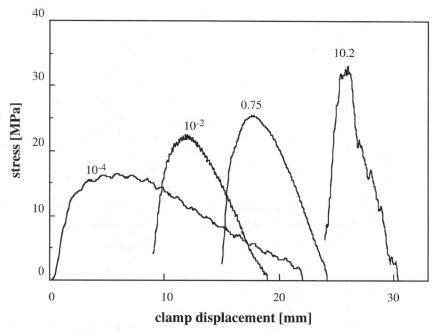

Figure 25.22   SENT stress-strain behavior of a PP–EPDM blend (30 vol%) at different test speeds (m/s). From [17]; reproduced with permission from Elsevier Ltd.

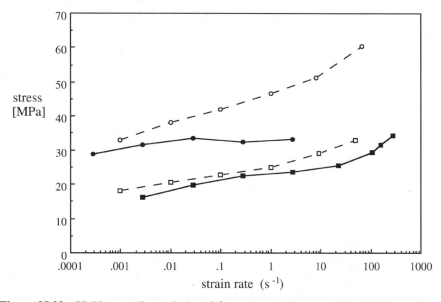

Figure 25.23   Yield stress (in tension) and fracture stress (measured by SENT) as a function of the strain rate on PP and a 70/30 PP–EPDM blend, under the following conditions: (○) PP tensile; (□) blend tensile; (●) PP SENT; (■) blend SENT. From [17]; reproduced with permission from Elsevier Ltd.

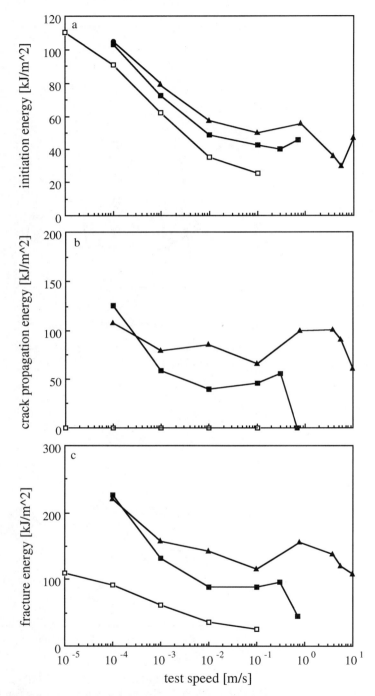

Figure 25.24   SENT fracture energy data as a function of test speed for the following PP and PP–EPDM blends: (□) PP; (■) 20% EPDM; (▲) 40% EPDM. From [17]; reproduced with permission from Elsevier Ltd.

in a notched sample is a function of the stress concentration at the notch. If a notch is blunted by local yielding, the fracture stress approaches that of a unnotched sample. The yield stress of both PP and the 70/30 PP–EPDM blend increased with strain rate. The fracture stress of PP SENT samples did not increase with strain rate. With strain rate, the fracture stress decreased relative to the yield stress. In the SENT test, the neat PP fractured in a brittle manner over the whole range of test speeds. The 30% EPDM blend showed an increase in fracture stress with strain rate, especially above 10 s^{-1}. The SENT fracture stress and the yield stress were close to each other, and the blend exhibited ductile fracture over the whole range of strain rates studied. The stress concentration factor of the blend must have been small.

The fracture energy, which is the integral of stress over displacement, is given for the PP–EPDM blends as a function of test speed in Fig. 25.24 [17]. The total fracture energy (see Fig. 25.24c); the energy up to the fracture initiation point, called the initiation energy (see Fig. 25.24a); and the fracture energy supplied to the sample after fracture initiation called the crack propagation energy (see Fig. 25.24b) are shown. At low test speeds, the fracture energy decreased with increasing test speed, as expected. (See Fig. 25.24c.) With increasing test speed (and decreasing test temperature), the yield stress increased, resulting in more brittle behavior. Above 10^{-2} m/s, the fracture energy was found to increase with increasing test speeds, but at very high test speeds, the fracture energy decreased again and the materials became brittle [3, 12, 18]. As yet, no accurate measurements of T_{bd} as a function of test speed have been made, but from the data available it seems that the deformation at low test speeds is a different process from that occurring at high test speeds. The brittle–ductile transition at low test speeds is a gradual process that takes place over 40°C, while at high test speeds the transition is a sharp, discontinuous transition (see Fig. 25.25) [33, 191].

C. Temperature Development

Blends that show ductile fracture have a large plastic deformation zone in the notch area and in the crack area. Most of this mechanical work is dissipated as heat [207]. As the test speed increases, the deformation becomes adiabatic, and with a strong plastic deformation a considerable temperature rise can be expected [18, 33, 47. 207]. As the temperature increases, the yield stress is lowered.

The temperature rise of the sample can be studied with an infrared camera which measures the surface temperature of blends (see Fig. 25.3) [17]. For example, the temperature in SENT tests at 10^{-3} m/s for a 95/5 PP–EPDM blend increases considerably. At higher rubber concentrations, e.g., a 30% blend, the temperature development is similar to that of the 5% blend, but the shape of the heated zone is more circular for the blends with more rubber [17].

With increasing test speed, the peak surface temperature increases steadily. (See Fig. 25.26.) Deviations from isothermal deformation start at about 10^{-5} m/s. Thus, with these ductile deforming systems, the fracture is isothermal only at very low speeds. At 10 m/s, the maximum measured surface temperature for PP–EPDM

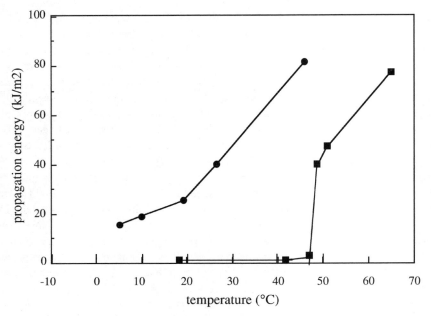

Figure 25.25   Fracture propagation energy in SENT on a PA–EPDM blend as a function of temperature at the following test speeds: (●) 1 mm/s; (■) 1 m/s. From data in [191]; reproduced with permission from the American Chemical Society.

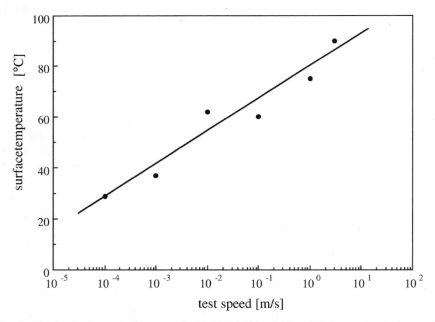

Figure 25.26   Surface temperatures of a PP–EPDM blend (30 vol%) 3 mm ahead of a notch, as measured on SENT samples with an infrared camera, as a function of test speed. From [17]; reproduced with permission from Elsevier Ltd.

bends is 90°C. Actually, the temperature in the fracture plane might well be considerably higher, as the spatial resolution of the IR camera used was only 130 μm, while the draw ratio λ increases strongly in the vicinity of 50 μm just ahead of the crack [89].

D. Inhomogeneous Deformation

Test speed also seems to influence the homogeneity of the deformation. Neat polymers, like PP, deform with necking in an inhomogeneous manner. The concentrated blends, however, deform without strain softening (see Fig. 25.27) [151]. With increasing test speed, cavitation seems to take place in bands (see Fig. 25.28) [3, 5, 11, 25, 47, 140, 151]. Thus, as the test speed increases (and as the temperature decreases), the deformation of blends is more inhomogeneous.

If the blend material has a layered structure, instead of a homogeneous composition, then the nature of the outer layer has a strong effect on the fracture energy [55, 208, 209]. A brittle outer layer, e.g., a brittle coating or weathered surface, gives a low fracture stress and a low fracture energy [208, 209]. A ductile layer on a brittle material gives a high fracture stress and a high fracture energy [55]. To increase the fracture energy of a brittle material or composite, adding a ductile outer layer is an option.

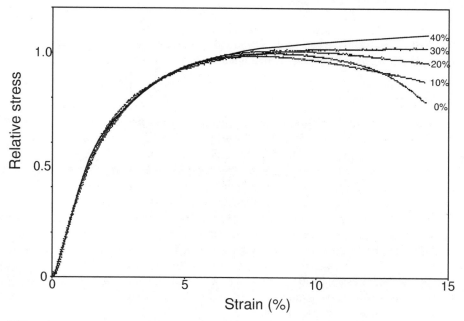

Figure 25.27   Strain-softening behavior of PP–EPDM blends at different rubber concentrations. (Strain rate $= 10^{-2}$ s^{-1}.)

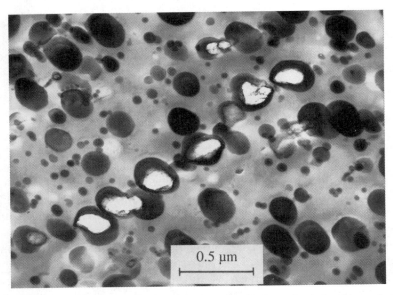

Figure 25.28   Transmission electron micrograph of an OsO_4-stained ultrathin section from a fractured Charpy specimen of a PA-6–PB blend; cavitation in a row can be seen. From [11]; reproduced with permission from Elsevier Ltd.

VII. FRACTURE GRAPHICS

Fracture graphics is widely used in postfailure analysis for studying the deformation mode of fracture and the role of the second phase. (See also Chapters 9 and 26.) The plastic deformation of a sample with a notch is localized ahead of the notch.

A. Stress Whitening

In blends, plastic deformation is accompanied by stress whitening of the sample, which is due to light scattering from crazes in the matrix or cavities in the rubber particles. On straining of a blend, a stress concentration first develops around the modifier particles [210]. Next, in blends of semicrystalline polymers, void formation takes place in the form of particle cavitation. On further straining, the voids elongate, due to the shear deformation of the matrix material next to the voids. A blend that fractures in a brittle manner may show strong stress whitening in the notch, but little stress whitening in the crack propagation region. A sample that fractures in a ductile manner often has a thick (1–2 mm) stress-whitened zone along the whole fracture surface. By observation of the stress-whitened zone developed, one can clearly determine whether the sample fractured in a brittle or ductile mode. At high test speeds, the transition from brittle to ductile is very sharp. At the transition, some samples are fully brittle, while others are fully ductile.

If the material deforms inhomogeneously, with strain-softening behavior, then a Dugdalelike line zone is formed ahead of the notch. (See Fig. 25.6.) The Dugdale

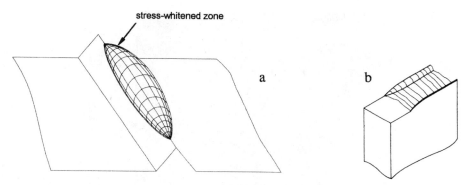

Figure 25.29   (a) Schematic of the development of a stress-whitened zone before a notch and (b) the formation of a shear lip as seen on a fractured sample.

zone can be observed as both a stress-whitened zone and a zone in which the temperature is increased. (See Fig. 25.3.) A Dugdale zone is evident for most amorphous polymers and can also be seen in some semicrystalline materials, like PP. With PA, a Dugdalelike deformation zone is less evident. The PP blends with a high rubber content, however, deform with little or no strain softening (see Fig. 25.27), and the deformation zone is more circular.

Since the stress state at the sample wall is a state of plane stress, there is a gradual buildup to a multiaxial stress state through the sample's thickness, which results in a wider yield zone at the walls. The wall material does not show stress whitening, evidently because there is no cavitation there (see Fig. 25.29a), and on the fracture surface, shear lips can be seen (see Fig. 25.29b). For PP and PA, the thickness of the noncavitated layer is on the order of 0.3 mm. For brittle materials, this layer thickness is much smaller.

B. Fracture Surface

The fracture surface of a sample that fractures in a brittle mode often shows first a smooth, semicircular (mirror) zone and next a rippled zone. This change in texture is due to an increase in fracture speed as the crack front progresses [15, 87, 88, 210]. At high fracture speeds, the fracture plane is smooth. At very high fracture speeds, crack branching takes place. In this region, rumples can be seen in the fracture plane, and the periodicity corresponds to the length of the crack branches. On a smaller scale, one can see a patchy surface (see Fig. 25.30a) [88]. For samples that fracture in a ductile manner, a rumpled surface accompanying cavitation can be seen. (See Fig. 25.30b.) The rumples lie parallel to the notch and give rise to tufts of highly drawn material. The rumples have been explained as being due to considerable drawing ahead of the crack tip before unstable fracture sets in. The extensive deformation of the cavities ahead of the crack tip gives rise to these structures [14].

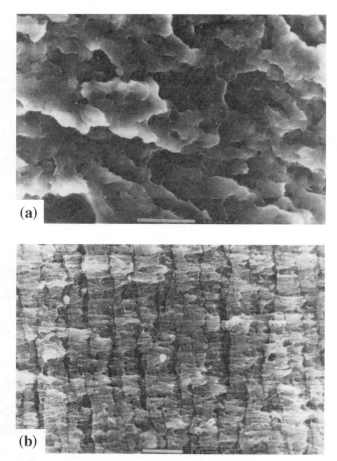

Figure 25.30   Fracture surfaces of a PA–EPDM-g-MA blend: (a) brittle fracture, (b) ductile fracture. From [88]; reproduced with permission from Elsevier Ltd.

C. Structure Below the Fracture Surface

The structure of the deformation zone in the sample perpendicular to the fracture surface has been studied by a number of authors [3, 9, 14, 18, 64, 87–89]. In brittle fractured samples, there is only a thin cavitated layer [151]. In materials that show ductile fracture, there is a stress-whitened layer with a thickness of 1–2 mm in which the rubber particles have cavitated but are still spherical in shape (see Fig. 25.31) [151]. Close to the fracture surface, the cavities are strongly deformed, which suggests a strong plastic deformation of the matrix. PP–EPR blends show at low test speeds a 70-μm-thick deformation layer with strongly deformed cavities. (See Fig. 25.31a.) Approaching the fracture surface increases the stretch ratio of the cavities, with a ratio of 10–20 at the fracture surface. The stretch ratio of the cavities is the draw ratio of the matrix material next to the cavities. At high test speeds in PA-66–EPDM blends,

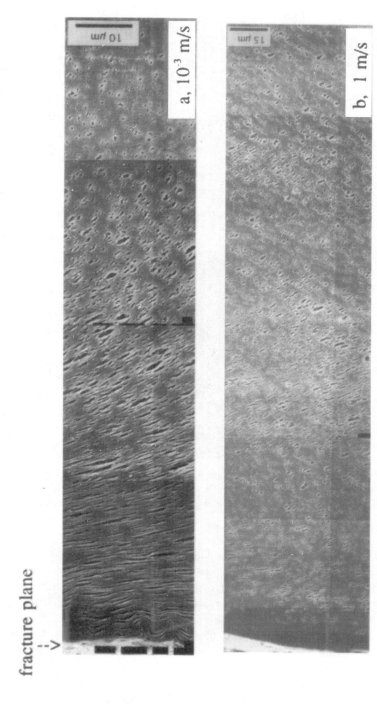

fracture plane

Figure 25.31 The structure of a stress-whitened zone below the fracture plane on a PP–EPR blend (20 vol%), for ductile fractured samples tested at the following speeds: (a) 10^{-3} m/s (low test speed); (b) 1 m/s (high test speed). From [151]; reproduced with permission from Elsevier Ltd.

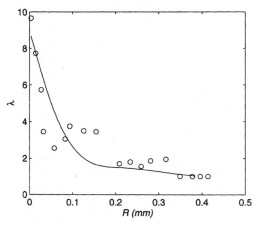

Figure 25.32 Draw ratios ahead of the crack front in the plane of the crack in an Izod sample of a PA–66 blend. From [89]; reproduced with permission from Elsevier Ltd.

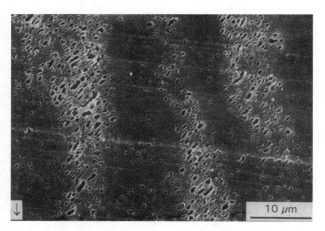

Figure 25.33 Structure of a PA–EPR (20 vol%) sample tested at a very high test speed (12.5 m/s). The cavitation and deformation are in broad bands. The arrow indicates the direction of crack growth. From [47]; reproduced with permission from Chapman & Hall.

the matrix deformation in this zone can have a draw ratio as large as 10–12 next to the fracture surface (see Fig. 25.32) [87, 88]. Surprisingly, in PA-6 blends and PP blends at high test speeds (SENT at 0.1–10 m/s), the cavities near the fracture surface seem to be less stretched, and at the fracture surface there is a thin layer (5–20 μm) without cavities. (See Fig. 25.31b.) The formation of this layer without cavities has been explained as being due to a relaxation of the matrix material [118, 151]. Relaxation of orientation in semicrystalline polymers can take place at high temperatures. The formation of this relaxation layer is accompanied by an increase in fracture energy (see Fig. 25.24), possibly caused by the crack blunting effect of the elastic layer. In PA-66–EPDM blends, the relaxation layer was not observed [120, 121].

With polarized light microscopy, one can see that samples tested at high test speeds have shear bands [3]. This deformation in bands is also observed by using electron microscopy [11, 151, 211]. At very high test speeds, the deformation occurs in broad bands (see Fig. 25.33) [18].

VIII. SUMMARY

Unnotched samples of semicrystalline polymers often fracture in a ductile manner; however, with a notch or defect they nearly all show brittle fracture. Semicrystalline polymers can be made tough by rubber modification. The function of the rubber is to cavitate and thereby relieve the multiaxial stress state the near the defect (see also Chapter 22) [1–7, 9, 11]. The energy absorption caused by cavitation is only a small part of the total fracture energy. In notched tests, cavitation of the rubber particles is a prerequisite for the enhanced ductile deformation for which matrix shear yielding is the principal energy-absorbing mechanism. It has been found that micromechanical deformation processes in particle-filled thermoplastics are very similar to the processes in rubber-modified polymers [9]. Important parameters in rubber toughening are as follows:

- the stress state of the material
- the cavitational behavior of the rubber
- the stability of the cavitated structure
- the deformation behavior of the matrix.

The stress state in a sample ahead of a notch depends on the material's properties, the sample dimensions, the notch dimensions, and the test speed. Possibly, thermal blunting takes place as a result of strong plastic deformation under adiabatic conditions. The amount of rubber seems to have a direct effect on the stress state after cavitation. The T_{bd} changes gradually with the rubber concentration, which suggests that the stress state also changes gradually with the rubber concentration.

The ease of cavitation of the rubber in a particular stress state depends on the properties of the rubber, the local rubber concentration, and the particle size. The cavitation of olefinic rubbers is easier than that of other rubbers, like polyetherester elastomers [6]. Thus, a low cohesive strength of the rubber is an advantage. Crosslinking of the rubber does not improve toughening, although the effect of light crosslinking is not yet clear. As the particle size decreases, cavitation takes place at a higher strain. Very small particles do not cavitate. In PA, the lower limit on particle size for toughening is about 50 nm [26].

Studies on the stability of the cavitated structure are few. It is known that multiphase systems with large particles have a low fracture strain and poor impact strength. If the cavities grow or coalesce, they can become unstable and initiate fracture. Since the ductile–brittle transition temperature is correlated with d_w, this suggests that a few large particles (cavities) determine T_{bd}. If the particles have a complex morphology with inclusions, the relationship between d_w and T_{bd} seems to be different [187]. This is probably a result of the fact that inclusions stabilize the cavitated structure. Coalescence of voids also seems to be depressed by light cross-linking of the rubber phase.

Matrix deformation is strongly dependent on temperature, test speed, and stress state. The molecular parameters that influence matrix deformation are the cohesive properties, the crystallinity, and the molecular weight. The cohesive strength and molecular weight of the matrix polymer have a strong effect on the craze strength [170], while the crystallinity increases the yield strength.

Since DuPont first introduced supertough polyamides on the market in 1976, the explanation for the toughening behavior of these polyamides has changed considerably. In blends based on semicrystalline polymers, the main energy-absorbing mechanism is shear yielding [4, 18, 82, 87, 88, 179, 212, 213]. Multiple crazing has not been observed in these systems. Local crazing can take place as part of the fracture process, ahead of a notch or running crack, but is not an important energy-absorbing mechanism.

The fracture behavior of supertough polymers is characterized as having an S-type fracture energy behavior as a function of temperature. (See Fig. 25.1.) In particular, the brittle–ductile transition temperature is a critical parameter. With the shift in the T_{bd} to lower temperatures, the fracture energies at low temperatures increase. These fracture energies are due mainly to deformation (in the notch area) before the crack is initiated. It has been suggested that T_{bd} is the temperature at which the fracture energy has attained a critical amount of deformation [13, 14]. After plastic deformation during initiation, a sharp crack is initiated, but the conditions are such that yielding ahead of the crack is still possible. The rubber concentration has a strong effect on the ease of yielding under these conditions. Three main types of deformation have been observed: a brittle fracture process, a ductile fracture process at low test speeds, and a ductile fracture process at high test speeds.

A brittle fracture process for blends of semicrystalline polymers is characterized by a very high fracture rate, on the order of 400–500 m/s for PA and PP blends. In this process, the chains fracture, and it has been proposed that a one-molecule-thick melt layer is formed [21]. Next to the fracture surface is a thin layer of cavitated particles that are visible, as schematically presented in Fig. 25.34a.

Ductile fracture at low test speeds develops gradually as the test speed increases. During the ductile stage at a loading rate of 10^{-4} m/s, crack speeds can be very low, that is, less than 10^{-2} m/s. The average crack speed is a function of the test speed. As shown in Fig. 25.34b, the deformation of the sample is concentrated into a thick layer that has cavitated (I) and a layer next to the fracture surface that is highly deformed (II). In the layer next to the fracture surface, a layer that is 50 μm thick, draw ratios are possible. Although the tests are carried out at low speeds, the process is not isothermal. At test speeds, higher than 10^{-5} m/s, a considerable temperature increase in the fracture zone, 3 mm away from the fracture surface, was observed with infrared thermography (see Fig. 25.26) [17]. It can be expected that the temperatures in the 50-μm layer next to the fracture surface are appreciably higher. At high test speeds, the brittle–ductile transition is discontinuous, meaning that the samples break in either a brittle or a ductile mode. The cavitation zone is thick, and three

brittle   ductile, low speed   ductile, high speed

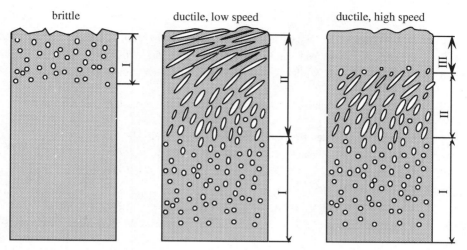

Figure 25.34 Schematic presentation of the structure of the stress-whitened zone for the following conditions: (a) brittle fracture, (b) ductile fracture at low test speeds, (c) ductile fracture at high test speeds.

layers can be observed (see Fig. 25.34c): A 1–2 mm-thick layer with round cavitated particles (I); a layer that is strongly deformed (II); and a layer next to the fracture surface, where relaxation of the deformed structure has taken place (III) [18]. Relaxation of highly oriented material takes place at high temperatures. The relaxation layer has a thickness of 3–5 μm [59] for PA and 10–25 μm for PP [151]. The formation of a relaxation layer is accompanied by an increase in fracture energy. Ductile deformation at high test speeds is a process with a thermal blunting process.

The T_{bd} of the blends depends on both the particle size and the rubber concentration. For PA–rubber blends, a good correlation is found between the T_{bd} and the interparticle distance as measured by notched Izod tests [13, 49]. However, this correlation does not hold for PA–rubber blends containing rubber particles with inclusions [187]. For PA and PP blends at low test speeds, the T_{bd} is not dependent on particle size and thus also is not correlated with the interparticle distance [32, 191]. The interparticle-distance relationship as measured by notched Izod tests is not valid for PP–EPR blends as well [32]. Thus, it is clear that with the interparticle-distance model, only a few of the observed particle-size effects can be described.

IX. REFERENCES

1. C. B. Bucknall, *Adv. Polym. Sci.*, **27**, 121 (1978).
2. A. M. Donald and E. J. Kramer, *J. Mat. Sci.*, **17**, 1765 (1982).
3. F. Ramsteiner and W. Heckmann, *Polym. Commun.*, **26**, 199 (1985).
4. A. F. Yee and R. A. Pearson, *J. Mat. Sci.*, **21**, 2462 (1986).

5. S. Wu, *J. Appl. Polym. Sci.*, **27**, 335 (1987).

6. R. J. M. Borggreve, R. J. Gaymans, and J. Schuijer, *Polymer*, **30**, 71 (1989).

7. K. Dijkstra, A. van der Wal, and R. J. Gaymans, *J. Mat. Sci.*, **29**, 3489 (1994).

8. Y. Ijichi, T. Kojima, Y. Susuki, T. Nishio, M. Kakugo, and Y. Amemiya, *Macromol.*, **26**, 829 (1993).

9. G.-M. Kim, G. H. Michler, M. Gahleitner, and J. Fiebig, *J. Appl. Polym. Sci.*, **60**, 1391 (1996).

10. K. Dijkstra and G. H. ten Bolsher, *J. Mat. Sci.*, **29**, 4286 (1994).

11. A. Lazzeri and C. B. Bucknall, *Polymer*, **36**, 2895 (1995).

12. K. Dijkstra, H. Wevers, and R. J. Gaymans, *Polymer*, **35**, 323 (1994).

13. S. Wu, *Polymer*, **26**, 1855 (1985).

14. F. Speroni, E. Castoldi, P. Fabbri, and T. Casiraghi, *J. Mat. Sci.*, **24**, 2165 (1989).

15. I. M. Ward, *Mechanical Properties of Solid Polymers*, John Wiley & Sons, New York, 1983.

16. E. Orwan, *Rept. Prog. Phys.*, **12**, 185 (1949).

17. A. Van der Wal and R. J. Gaymans, "PP–EPDM Blends No. 3," *Polymer*, **40**, 6045 (1999).

18. K. Dijkstra, J. ter Laak, and R. J. Gaymans, *Polymer*, **35**, 315 (1994).

19. B. A. Crouch and D. D. Huang, *J. Mat. Sci.*, **29**, 861 (1994).

20. T. Ricco and A. Pavan, *Angew. Makromol. Chemie*, **201**, 23 (1992).

21. P. S. Leevers, *Polym. Eng. Sci.*, **36**, 2296 (1996).

22. T. Vu-Khahn, *Polymer*, **29**, 979 (1988).

23. J. G. Williams, *Fracture Mechanics of Polymers*, Ellis-Horwood, Chichester, England, 1984.

24. D. D. Huang and J. G. Williams, *J. Mat. Sci.*, **22**, 2503 (1987).

25. R. J. Gaymans, R. J. M. Borggreve, and A. J. Oostenbrink, *Makromol. Chem., Macromol. Symp.*, **38**, 125 (1990).

26. A. J. Oshinski, H. Keskkula, and D. R. Paul, *Polymer*, **33**, 268 (1992).

27. A. J. Oshinski, H. Keskkula, and D. R. Paul, *Polymer*, **37**, 4919 (1996).

28. P. Gali and J. C. Haylock, *Macromol. Chem., Macromol. Symp.*, **63**, 19 (1992).

29. A. J. Oostenbrink, L. J. Molenaar, and R. J. Gaymans, *PRI Polymer Blends*, Plastics and Rubber Institute, London, Cambridge, 1990, p. 3.

30. A. Van der Wal, J. J. Mulder, J. Oderkerk, and R. J. Gaymans, *Polymer*, **39**, 6781 (1998).

31. A. J. Oshinski, H. Keskkular, and D. R. Paul, *Polymer*, **37**, 4891 (1996).

32. A. van der Wal, A. J. J. Verheul, and R. J. Gaymans, "PP–EPDM Blend 4," *Polymer*, **40**, 6057 (1999).

33. A. van der Wal, J. J. Mulder, R. Nijhof, and R. J. Gaymans, "PP–EPDM Blend 2," *Polymer*, **40**, 6031 (1999).

34. B. N. Epstein, U.S. Patent 4,174,358 (to E. I. duPont) (1979).

35. M. Abbate, V. Di Liello, E. Martuscelli, P. Musto, G. Ragosta, and G. Scarinzi, *Polymer*, **33**, 2940 (1992).

36. R. J. M. Borggreve and R. J. Gaymans, *Polymer*, **30**, 63 (1989).

37. P. Maréchal, G. Coppens, R. Legras, and J.-M. Dekoninck, *J. Polym. Sci.*, **33**, 757 (1995).

38. Y. Takeda, H. Keskkula, and D. R. Paul, *Polymer*, **33**, 3394 (1992).

39. B. Majumdar, H. Keskkula, and D. R. Paul, *J. Polym. Sci.*, **32**, 1386 (1994).

40. E. Hage, W. Hale, H. Keskkula, and D. R. Paul, *Polymer*, **38**, 3237 (1997).

41. A. J. Brady, H. Keskkula, and D. R. Paul, *Polymer*, **35**, 3665 (1994).

42. F. S. Mazer, Ph.D. Dissertation, Cranfield Inst. of Technology, 1991.

43. R. J. Gaymans and J. W. van der Werff, *Polymer*, **35**, 3658 (1994).

44. K. Udipi, *J. Appl. Polym. Sci.*, **36**, 117 (1988).

45. A. J. Kinloch and R. J. Young, *Fracture Behaviour of Polymers*, Elsevier Applied Science, London, 1985.

46. K. Friedrich, *Adv. Polym. Sci.*, **91–92**, 353 (1990).

47. K. Dijkstra and R. J. Gaymans, *J. Mat. Sci.*, **29**, 3231 (1994).

48. S. Cimmino, F. Coppola, L. D'Orazio, R. Greco, G. Maglio, M. Malinconinco, C. Mancarella, E. Martuscelli, and G. Ragosta, *Polymer*, **27**, 1874 (1986).

49. R. J. M. Borggreve, R. J. Gaymans, J. Schuijer, and J. F. Ingen Housz, *Polymer*, **28**, 1489, (1987).

50. R. J. M. Borggreve, R. J. Gaymans, and A. R. Luttmer, *Macromol. Symp.*, **16**, 195, (1988).

51. R. J. M. Borggreve and R. J. Gaymans, *Polymer*, **29**, 1441 (1988).

52. R. J. Gaymans, R. J. M. Borggreve, and A. B. Spoelstra, *J. Polym. Sci.*, **37**, 479 (1989).

53. R. J. M. Borggreve and R. J. Gaymans, *Polymer*, **30**, 78 (1989).

54. R. J. Gaymans and R. J. M. Borggreve, "Toughening Mechanism in Polyamide-Rubber Blends," in *Contemporary Topics of Polymer Science*, Vol. 6, B. M. Culbertson (ed.), Plenum, New York, 1989, p. 461.

55. R. J. Gaymans and K. Dijkstra, *Polymer*, **31**, 971 (1990).

56. B. Majumdar, H. Keskkula, and D. R. Paul, *J. Polym. Sci.*, **32**, 2127 (1994).

57. D. Neuray and K.-H. Ott, *Angew. Makromol. Chemie.*, **98**, 213 (1981).

58. K. Dijkstra and R. J. Gaymans, *Polymer*, **35**, 332 (1994).

59. H. Janik, R. J. Gaymans, and K. Dijkstra, *Polymer*, **36**, 4203 (1995).

60. A. González-Montiel, H. Keskkula, and D. R. Paul, *Polymer*, **36**, 4587 (1995).

61. A. González-Montiel, H. Keskkula, and D. R. Paul, *Polymer*, **36**, 4605 (1995).

62. A. González-Montiel, H. Keskkula, and D. R. Paul, *Polymer*, **36**, 4621 (1995).

63. M. Lu, H. Keskkula, and D. R. Paul, *J. Appl. Polym. Sci.*, **59**, 1467 (1996).

64. M. Lu, H. Keskkula, and D. R. Paul, *J. Appl. Polym. Sci.*, **58**, 1175 (1995).

65. M. Lu, H. Keskkula, and D. R. Paul, *Polymer*, **34**, 1874 (1993).

66. A. J. Oshinski, H. Keskkular, and D. R. Paul, *Polymer*, **37**, 4909 (1996).

67. Y. Kayano, H. Keskkular, and D. R. Paul, *Polymer*, **38**, 1885 (1997).

68. N. Billon and J. M. Haudin, *Polym. Eng. Sci.*, **37**, 1761 (1997).

69. A. Mamat, T. Vu-Khanh, P. Cigana, and B. D. Favis, *J. Polym. Sci.*, **35**, 2583 (1997).

70. R. Hösti-Miettinen and J. Seppala, *J. Polym. Eng. Sci.*, **32**, 868 (1992).

71. A. González-Montiel, H. Keskkula, and D. R. Paul, *J. Polym. Sci., Polym. Phys. Ed.*, **33**, 1751 (1995).

72. S. N. Sathe, S. Devi, G. S. Rao, and K. V. Rao, *J. Appl. Polym. Sci.*, **61**, 97 (1996).

73. D. M. Otterson, B. H. Kim, and R. E. Lavengood, *J. Mat. Sci.*, **26**, 1478 (1991).

74. V. J. Triacca, S. Ziaee, J. W. Barlow, H. Keskkula, and D. R. Paul, *Polymer*, **32**, 1401 (1991).

75. B. K. Kim and S. J. Park, *J. Appl. Polym. Sci.*, **43**, 357 (1991).

76. J. C. Angola, Y. Fujita, T. Sakai, and T. Inoue, *J. Polym. Sci., Polym. Phys. Ed.*, **26**, 807 (1988).

77. B. N. Epstein, U.S. Patent 4,172,895 (to E. I. duPont) (1979).

78. E. A. Flexman, *Polym. Eng. Sci.*, **19**, 564 (1979).

79. S. Wu, *J. Polym. Sci., Polym. Phys. Ed.*, **21**, 699 (1983).

80. S. Wu, *J. Appl. Polym. Sci.*, **35**, 549 (1988).

81. D. W. Gilmore and M. J. Modic, *J. Soc. Plast. Eng. ANTEC*, **47**, 1371 (1989).

82. C. B. Bucknall, P. S. Heather, and A. Lazzeri, *J. Mat. Sci.*, **16**, 2255 (1989).

83. D. F. Lawson, W. L. Hergenrother, and M. G. Matlock, *J. Appl. Polym. Sci.*, **39**, 2331 (1990).

84. M. J. Modic and L. A. Pottick, *Plast. Eng.*, **7**, 39 (1991).

85. Y. Takeda, H. Keskkula, and D. R. Paul, *Polymer*, **33**, 3173 (1992).

86. C. J. Wu, J. F. Kuo, C. Y. Chen, and E. Woo, *J. Appl. Polym. Sci.*, **52**, 1695 (1994).

87. O. K. Muratoglu, A. S. Argon, R. E. Cohen, and M. Weinberg, *Polymer*, **36**, 921 (1995).

88. O. K. Muratoglu, A. S. Argon, R. E. Cohen, and M. Weinberg, *Polymer*, **36**, 4771 (1995).

89. O. K. Muratoglu, A. S. Argon, R. E. Cohen, and M. Weinberg, *Polymer*, **36**, 4787 (1995).

90. C. E. Scott and C. W. Macosko, *Int. Polym. Process.*, **10**, 36 (1995).

91. S. V. Nair, A. Subramaniam, and L. A. Goettler, *J. Mat. Sci.*, **32**, 5335 (1997).

92. S. V. Nair, A. Subramaniam, and L. A. Goettler, *J. Mat. Sci.*, **32**, 5347 (1997).

93. D. J. Hourston and S. Lane, "Toughened Polyester and Polycarbonates," in *Rubber Toughened Engineering Plastics*, A. A. Collyer (ed.), Chapman & Hall, London, 1994, Ch. 8.

94. M. Okamoto, K. Shiomi, and T. Inoue, *Polymer*, **35**, 4618 (1994).

95. F. Polato, *J. Mat. Sci.*, **20**, 1455 (1985).

96. D. J. Hourston, S. Lane, and H. X. Zhang, *Polymer*, **36**, 3051 (1995).

97. T.-K. Kang, Y. Kim, G. Kim, W.-J. Cho, and C.-S. Ha, *Polym. Eng. Sci.*, **37**, 603 (1997).

98. M. Hert, *Angew. Makromol. Chemie.*, **196**, 89 (1992).

99. R. M. Hosti-Miettinen, M. T. Heino, and J. V. Sappala, *J. Appl. Polym. Sci.*, **57**, 573 (1995).

100. P. Laurienzo, M. Malinconico, E. Martuscelli, and G. Volpe, *Polymer*, **30**, 835 (1989).

101. A. Cecer, R. Greco, G. Ragosta, and G. Scarinzi, *Polymer*, **31**, 1239 (1990).

102. H. Kanai, A. Auerbach, and A. Sullivan, *J. Appl. Polym. Sci.*, **53**, 527 (1994).

103. C. J. G. Plummer, *Macromol. Phys.*, **197**, 2047 (1996).

104. I. A. Abu-Isa, C. B. Jaynes, and J. F. O'Gara, *J. Appl. Polym. Sci.*, **59**, 1957 (1996).

105. M. Penco, M. A. Pastorino, E. Occhiello, F. Garbassi, R. Braglia, and G. Giannotta, *J. Appl. Polym. Sci.*, **57**, 329 (1995).

106. S. Y. Hobbs, M. E. J. Dekker, and V. H. Watkins, *J. Mat. Sci.*, **23**, 1219 (1988).

107. M. E. J. Dekker, S. Y. Hobbs, and V. H. Watkins, *J. Mat. Sci.*, **23**, 1225 (1988).

108. M. Okamoto, Y. Shimodu, T. Kojima, and T. Inoue, *Polymer*, **34**, 4868 (1993).

109. J. Wu and Y.-W. Mai, *J. Mat. Sci.*, **28**, 6167 (1993).

110. C. M. Benson and R. P. Burford, *J. Mat. Sci.*, **30**, 573 (1995).

111. P.-C. Lee, W.-F. Kuo, and F.-C. Chang, *Polymer*, **35**, 5641 (1994).

112. C. J. Chou, K. Vijayan, D. Kirby, A. Hiltner, and E. Baer, *J. Mat. Sci.*, **23**, 2521 (1988).

113. J.-P. Meyer, *Kunststoffe*, **85**, 9 (1995).

114. M. Penco, M. A. Pastorino, E. Occhiello, F. Garbassi, R. Braglia, and G. Giannotta, *J. Appl. Polym. Sci.*, **57**, 329 (1995).

115. V. Tanrattanakul, W. G. Perkins, F. L. Massey, A. Moet, A. Hiltner, and E. Baer, *J. Mat. Sci.*, **32**, 4749 (1997).

116. W. D. Cook, T. Zhang, G. Moad, G. Van Deipen, F. Cser, B. Fox, and M. O'Shea, *J. Appl. Polym. Sci.*, **62**, 1699 (1996).

117. W. D. Cook, G. Moad, B. Fox, G. Van Deipen, T. Zhang, F. Cser, and L. McCarthy, *J. Appl. Polym. Sci.*, **62**, 1709 (1996).

118. S. S. Morye and D. D. Kale, *J. Polym. Mat.*, **13**, 217 (1996).

119. K. Yokoyama, Y. Fujita, N. Higashida, and T. Inoue, *Makromol. Chemie*, **83**, 157 (1994).

120. D. L. Wilfong, A. Hiltner, and E. Baer, *J. Mat. Sci.*, **30**, 587 (1995).

121. J. Karger-Kocsis, *Polypropylene Structure and Properties*, Vol. 2, Chapman Hall, London, 1995.

122. T. Laus, *Angew. Makromol. Chemie*, **60/61**, 87 (1977).

123. F. Ramsteiner, *Polymer*, **20**, 839 (1979).

124. J. Karger-Kocsis, A. Kalló A. Szafner, G. Bodor, and Zs. Sényei, *Polymer*, **20**, 37 (1979).

125. J. Karger-Kocsis, L. Kiss, and V. N. Kuleznev, *Acta Polymer.*, **33**, 14 (1982).

126. K. C. Dao, *J. Appl. Polym. Sci.*, **27**, 4799 (1982).

127. L. D'Orazio, R.Greco, C. Mancarella, E. Martuscelli, G. Ragosta, and C. Silvestre, *Polym. Eng. Sci.*, **22**, 536 (1982).

128. B. Z. Jang, R. D. Uhlmann, and J. B. Vander Sande, *J. Appl. Polym. Sci.*, **30**, 2485 (1985).

129. B. Z. Jang, R. D. Uhlmann, and J. B. Vander Sande, *J. Appl. Polym. Sci.*, **29**, 3409 (1984).

130. B. Z. Jang, R. D. Uhlmann, and J. B. Vander Sande, *J. Appl. Polym. Sci.*, **29**, 4377 (1984).

131. R. Greco, C. Mancarella, E. Martuscelli, G. Ragosta, and Y. Jinghua, *Polymer*, **28**, 1929 (1987).

132. D. L. Faulkner, *J. Appl. Polym. Sci.*, **36**, 467 (1988).

133. B. Pukánszky, F. Tüdös, A. Kalló, and G. Bodor, *Polymer*, **30**, 1407 (1989).

134. V. Choudhary, H. S. Varma, and I. K. Varma, *Polymer*, **32**, 2534 (1991).

135. F. Ramsteiner, *Acta Polym.*, **42**, 584 (1991).

136. W.-Y. Chiang, W.-D. Yang, and B. Pukánszky, *Polym. Eng. Sci.*, **32**, 641 (1992).

137. L. D'Orazio, C. Mancarella, E. Martuscelli, G. Sticotti, and P. Massari, *Polymer*, **34**, 3671 (1993).

138. K. Hayashi, T. Morioka, and S. Toki, *J. Appl. Polym. Sci.*, **48**, 411 (1993).

139. M. Ishikawa, M. Sugimoto, and T. Inoue, *J. Appl. Polym. Sci.*, **62**, 1495 (1996).

140. G. H. Michler, *Kuststof-Mechanik: Morphologie, Deformations, und Bruchmechanismen*, Carl Hanser Verlag, Munich, 1992.

141. F. M. Mirabella, Jr., *J. Appl. Polym. Sci.*, **32**, 1205 (1994).

142. S. Norzalia, B. Surani, and M. Y. Ahmad Fuad, *J. Elastom. Plast.*, **26**, 183 (1994).

143. G. Snooppy, J. Reethamma, T. Sabu, and K. T. Varughese, *Polymer*, **36**, 4405 (1995).

144. R. C. Cieslinski, H. C. Silvis, and D. J. Murray, *Polymer*, **36**, 1827 (1995).

145. T. Nomura, T. Nishio, T. Fujii, J. Sakai, M. Yamamoto, A. Uemura, and M. Kakugo, *Polym. Eng. Sci.*, **35**, 1261 (1995).

146. E. P. Moore, Jr. and O. H. Cincinnati, in *Polypropylene Handbook*, Vol. XIX, C. Jaar (ed.), Hanser-Gardner Publications, Munchen, Germany, 1996, p. 419.

147. Y. Yokoyama and T. Ricco, *Plastics, Rubber, and Composites*, **25**, 417 (1996).

148. S. C. Tjong, W. D. Li, and R. K. Y. Li, *Polym. Bull.*, **38**, 721 (1997).

149. Z. Zhang, D. S. Chiu, and G. G. Siu, *J. Reinf. Plast. and Comp.*, **16**, 603 (1997).

150. L. Li, L. Chen, P. Bruin, and M. A. Winnik, *J. Polym. Sci.*, **35**, 979 (1997).

151. A van der Wal and R. J. Gaymans, "Polypropylene–Rubber Blends 5," *Polymer*, **40**, 6067 (1999).

152. T. Inoue, *J. Appl. Polym. Sci.*, **54**, 723 (1994).

153. T. Inoue and T. Suzuki, *J. Appl. Polym. Sci.*, **56**, 1113 (1995).

154. T. Inoue and T. Suzuki, *J. Appl. Polym. Sci.*, **59**, 1443 (1996).

155. D. J. Lohse, L. J. Fetters, M. J. Doyle, H.-C. Wang, and C. Kow, *Macromolecules*, **26**, 3444 (1993).

156. D. Yank, B. Zhang, Y. Yang, Z. Fang, G. Sun, and Z. Feng, *Polym. Eng. Sci.*, **24**, 612 (1984).

157. S. C. Tjong, W. D. Li, and R. K. Y. Li, *Polym. Bull.*, **38**, 721 (1997).

158. A. J. Thinker, *Sagamore Army Res. Conf. Proc. 155, CA*, **107**, 2414OZ (1987).

159. E. A. Flexman, *ACS Polym. Prep.*, **29**, 189 (1988).

160. H.-Q. Xie, D.-S. Feng, and J.-S. Guo, *J. Appl. Polym. Sci.*, **64**, 329 (1997).

161. F. Kloos, *Angew. Makromol. Chemie*, **133**, 1 (1985).

162. C. J. G. Plummer, Ph. Béguelin, and H.-H. Kausch, "High Strain-Rate Mechanical Properties of Polyoxymethylene," in *Impact and Dynamic Fracture of Polymers and Composites*, J. G. Williams and A. Pavan (eds.), ESIS Publication 19, Mechanical Engineering Publication Limited, London, 1995, p. 265.

163. L. H. Wadhwa, T. L. Dolce, and H. L. La Nieve, *SPE Retec, Columbus, Ohio*, December 12 (1985).

164. E. A. Flexman, Jr., U.S. Patent 4,804,716 (to DuPont) (1989).

165. K. Kurz and K. Witar, *Kunststoffe*, **86**, 1852 (1996).

166. W. Browstow and R. P. Corneliussen, *Failure of Plastics*, Hanser Verlag, Berlin, 1987.

167. J.-H. Haudin, in *Plastic Deformation of Amorphous and Semicrystalline Materials*, B. Escaig and C. G'Sell (eds.), Publ. Les Editions de Physique, Les Ulis, France, 1982.

168. H. H. Kausch and J. G. Williams, "Fracture and Fatigue," in *Encyclopedia of Polymer Science and Technology,* 2nd Ed., Vol. 7, John Wiley & Sons, New York. 1985, p. 328.

169. D. W, Hadley and I. M. Ward, "Mechanical Properties," in *Encyclopedia of Polymer Science and Technology*, 2nd Ed., Vol. 2, John Wiley & Sons, New York, 1985, p. 379.

170. S. Wu, *Polym. Eng. Sci.*, **30**, 753 (1990).

171. R. J. Young, *Mater. Forum*, **11**, 210 (1988).

172. S. Wu, C. P. Bosnyak, and K. Seheanobish, *J. Appl. Polym. Sci.*, **65**, 2209 (1997).

173. A. Lazzeri and C. B. Bucknall, *J. Mat Sci.*, **28**, 6799 (1993).

174. E. J. Kramer and L. L. Berger, *Adv. Polym. Sci.*, **91/92**, 1 (1990).

175. B. Hartmann and R. F. Cole, *Polym. Eng. Sci.*, **25**, 65 (1985).

176. R. Bagheri and R. A. Pearson, *Polymer*, **36**, 4883 (1995).

177. D. I. Collias, D. G. Baird, and R. J. M. Borggreve, *Polymer*, **35**, 3978 (1994).

178. A. C. Steenbrink and E. Van der Giessen, *J. Eng. Mat. Technol.*, **119**, 256 (1997).

179. R. J. M. Borggreve, R. J. Gaymans, and H. M. Eichenwald, *Polymer*, **30**, 78 (1989).

180. S. Y. Hobbs and M. E. J. Dekker, *J. Mat. Sci.*, **24**, 1316 (1989).

181. H.-J. Sue and A. F. Yee, *J. Mat. Sci.*, **24**, 1447 (1989).

182. H. Breuer, F. Haaf, and J. Stabenow, *J. Maromol. Sci., Phys.*, **B14**, 387 (1977).

183. R. J. M. Smit, "Toughness of Hetrogeneous Polymeric Systems," Ph.D. Thesis, Technical University of Eindhoven, 1998.

184. A. C. Steenbrink, V. M. Litvinov, and R. J. Gaymans, *Polymer*, **39**, 4817 (1998).

185. D. Dompas and G. Groeninkx, *Polymer*, **35**, 4743 (1994).

186. D. Dompas, G. Groeninckx, M. Isogawa, T. Hasegawa, and M. Kadokura, *Polymer*, **35**, 4750 (1994).

187. A. J. Oshinski, H. Keskkula, and D. R. Paul, *Polymer*, **33**, 284 (1992).

188. A. N. Gent and D. Tompins, *J. Polym. Sci.*, **A-2**, 1483 (1969).

189. A. N. Gent, *Rubber Chem. Technol.*, **63**, 949 (1990).

190. D. D. Huang and B. A. Wood, *ACS PMSE*, **67**, 629 (1992).

191. R. J. Gaymans, K. Dijkstra, and M. H. ten Dam, "Ductile-to-Brittle Transitions in Blends of Polyamide-6 and Rubber," in *Toughening of Plastics II: ACS Advances in Chemistry*, C. K. Riew and A. J. Kinloch (eds.), No. 252, American Chemical Society, Washington, DC, 1996, Ch. 20, p. 303.

192. A. Y. Coran and R. Patel, *Rubber Chem. Technol.*, **53**, 141 (1980).

193. H. Liang, W. Jiang, J. Zang, and B. Jiang, *J. Appl. Polym. Sci.*, **59**, 505 (1996).

194. C. B. Bucknall, H. H. Yang, and X. C. Zhang, "Effects of Rubber Cross-Link Density on Deformation Kinetics in HIPS," in *Polymat '94, Mat. Inst.*, London, September 1994, p. 673.

195. R. A. Bubeck, D. J. Buckley, Jr., E. J. Kramer, and H. R. Brown, *J. Mat Science*, **26**, 6249 (1991).

196. R. A. Pearson, H. R. Azimi, R. Bagheri, J. Qian, and Y.-C. Huang, *Deformation and Yield and Fracture of Polymers*, Materials Institute, Cambridge, 1997, p. 175.

197. S. Datta and D. J. Lohse, *Polymer Compatiblizers*, Soho Press, Inc., New York, 1996.

198. P. H. M. Ellemans, J. G. M. Van Grisbergen, and H. E. H. Meijer, "Morphology of the Model System Polystyrene/Polyethylene," in *Integration in Polymer Science and Technology*, Vol. 2, P. J. Lemstra and L. A. Kleintjes (eds.), Elsevier, London, 1988, p. 261.

199. J. F. Knott, *Fundamentals of Fracture Mechanics*, Butterworth-Heinemann, London, 1979, p. 1.

200. P. S. Levers, M. Douglas, M. Chong, and J. G. Williams, *Deformation and Yield and Fracture of Polymers*, Materials Institute, Cambridge, 1997, p. 106.

201. Ph. Béguelin and H. H. Kauch, "A technique for Studying the Fracture of Polymers from Low to High Loading Rates," in *Impact and Dynamic Fracture of Polymers and Composites*, J. G. Williams and A. Pavan (eds.), ESIS Publication 19, Mechanical Engineering Publication Limited, London, 1995, p. 3.

202. J. G. Williams and S. Hashemi, *Polymer Blends and Mixtures*, Kluwer Academic, Amsterdam, 1995.

203. B. A. Crouch and D. D. Huang, "Impact Testing of Polymers Using the J-Integral Technique," in *Impact and Dynamic Fracture of Polymers and Composites*, J. G. Williams and A. Pavan (eds.), ESIS Publication 19, Mechanical Engineering Publication Limited, London, 1995, p. 225.

204. S. G. Larsson and A. J. Carlsson, *J. Mech. Phys. Solids*, **21**, 263 (1973).

205. P. S. Leevers and R. E. Morgan, *Eng. Fract. Mech.*, **52**, 999 (1995).

206. P. S. Leevers, *Int. J. of Fract.*, **73**, 109 (1995).

207. Y. K. Godovsky, *Thermal Properties of Polymers*, Springer, Berlin, 1992, Chs 7 and 8.

208. C. B. Bucknall and D. G. Street, *J. Appl. Polym. Sci.*, **12**, 1311 (1986).

209. C. Verpy, J.-L. Gacougnolle, A. Dragon, B. Bourchet, A. Chesneau, F. Cozze, and A. Vanlerberghe, "The Origin of the Embrittlement of a Ductile Polymer by a Brittle Thin Coating," in *Impact and Dynamic Fracture of Polymers and Composites*, J. G. Williams and A. Pavan (eds.), ESIS Publication 19, Mechanical Engineering Publication Limited, London, 1995, p. 233.

210. E. A. Flexman, "Impact Behavior of Nylon-66 Compositions: Ductile–Brittle Transitions," *Proc. Int. Conf. on Toughening of Plastics*, PRI, London, July 1978, p. 14.

211. G.-M. Kim and G. H. Michler, *Deformation and Yield and Fracture of Polymers*, Materials Institute, Cambridge, 1997, p. 486.

212. R. A. Pearson and A. F. Yee, *J. Mat. Sci.*, **21**, 2475 (1986).

213. D. S. Parker, H. J. Sue, J. Huang, and A. F. Yee, *Polymer*, **31**, 2267 (1990).

26 Toughening of Epoxies

A. F. YEE AND J. DU

Department of Materials Science and Engineering
The University of Michigan
Ann Arbor, MI 48109

M. D. THOULESS

Department of Mechanical Engineering and Applied Mechanics
The University of Michigan
Ann Arbor, MI 48109

Polymer Blends, Volume 2: Performance. Edited by D. R. Paul and C. B. Bucknall.
ISBN 0-471-35280-2. © 2000 John Wiley & Sons, Inc.

I. INTRODUCTION

Thermosets are highly cross-linked polymers with a three-dimensional molecular structure. The network structure gives rise to high stiffness, high strength, and good heat resistance and solvent resistance. These excellent properties mean that thermosets are frequently the engineering polymers of choice for structural applications. For example, thermosets are widely used as adhesives and as matrix materials in various composites that range in use from aerospace structures to dental fillers. However, one major drawback of thermosets, which also results from their network structure, is their poor resistance to impact and to crack initiation. Consequently, enormous efforts have been made to improve the fracture toughness of thermosets while still maintaining their desirable properties. One approach to toughening thermosets is the addition of second-phase polymeric particles, such as rubbers and thermoplastics. Notable successes have been achieved in this area over the past few decades.

Epoxy resins are among the strongest thermosets. They are formed from low-molecular-weight compounds, also called epoxy resins, having two or more epoxide groups (oxirane rings) in their molecular structure. Cross-linked epoxies are produced by a "curing reaction." In this process, the epoxide groups of the liquid precursor may react with a suitable curing agent to achieve a three-dimensional structure. Curing agents are organic compounds with between two and four functional groups that act by opening the oxirane rings. Alternatively, ring opening is achieved by adding a catalyst. Owing to the diversity of available epoxy resins, curing agents, and curing procedures that can be used in different combinations, a great number of epoxies with a variety of properties can be obtained. This versatility results in the epoxies being the most extensively used class of thermosets in structural applications. Consequently, most investigations into the toughening of thermosets have focused on epoxies [1–50]. A relatively good understanding has been achieved in this area, and several toughening mechanisms have been proposed.

Rubber toughening is the most thoroughly studied route to improved properties. Rubber-toughened epoxies are generally made by mechanically mixing liquid rubber and liquid epoxy resin together, so that a homogeneous solution is formed. When this solution is cured, rubber particles precipitate as a second phase. It has been observed that dramatic improvements in toughness, sometimes by an order of magnitude, can be obtained by adding only a small amount of rubber. It is now generally agreed that, in the case of epoxies with a fair amount of intrinsic ductility, the primary toughening mechanism is shear yielding of the matrix, which is facilitated by the rubber particles. However, the rubber-modified epoxies often suffer from a relatively significant loss in modulus and sometimes in strength. Furthermore, rubber toughening is effective only for epoxies with low cross-link density and, therefore, with a relatively high level of intrinsic ductility.

Relatively recently, it has been found that a significant enhancement in toughness can be obtained by adding a suitable rigid thermoplastic to an epoxy. This approach, which has great potential for further development, was originally motivated by a desire to toughen highly cross-linked epoxies, which are relatively difficult to toughen, while retaining the strength and the modulus of the original epoxies. To process these epoxies, the thermoplastic is dissolved in the liquid epoxy resin. It then separates during curing, through nucleation and growth or spinodal decomposition to form second-phase particles or a co-continuous structure. It is also possible simply to mix the thermoplastic in a particulate form into the epoxy. A crack-bridging mechanism involving plastic deformation of the second phase appears to be the main toughening mechanism. However, it should be noted that the co-continuous structures that usually result from the addition of large amounts of the thermoplastic modifier can produce significant decreases in some of the other desirable properties, such as the modulus and the solvent resistance, even as the toughness exhibits dramatic improvements.

Previous work on the two aforementioned types of toughening is comprehensively reviewed in this chapter, while Chapter 13 reviews issues of the formulation of such blends. The review starts with a description of the deformation and fracture of unmodified epoxies. This description is followed by detailed reviews of the basic deformation and fracture behavior of modified epoxies, qualitative descriptions of the toughening mechanisms proposed for modified epoxies, and a description of some quantitative toughening models. Details on some recent developments in the field of rubber-modified epoxies are also presented in this chapter.

II. DEFORMATION AND FRACTURE BEHAVIOR OF UNMODIFIED EPOXIES

As mentioned previously, epoxy resins are extensively used in structural engineering applications, owing to their excellent properties. However, the vast majority of epoxies are brittle materials that need to be toughened. A comprehensive knowledge of the deformation and fracture behavior of unmodified epoxies is essential to an understanding of toughening mechanisms and is very important in achieving the optimal behavior of the toughened epoxies.

A. Deformation Behavior of Unmodified Epoxies

In tensile tests of moderate testing rates, pure epoxies at relatively low temperatures fail at very low strains prior to yielding, while at temperatures approaching their glass transition temperatures, T_g, they yield before failure, but without necking and subsequent cold drawing [51]. Crazing was originally believed to be the underlying mechanism for tensile fracture at low temperatures [52]. However, doubts as to the existence of crazes in epoxies were first advanced by Kramer [53]. He demonstrated that the occurrence of crazing depends on entanglement density. Specifically, crazing is suppressed in polymers with high entanglement densities. Therefore, it is virtually impossible for crazes to form in epoxies, because of their relatively high cross-link densities. It is now commonly accepted that cracking is the reason for tensile fracture. However, we note that crazelike structures have recently been observed in thermosets with relatively low cross-link densities [54].

In contrast, unmodified epoxies usually exhibit yielding in compression tests, even at relatively low temperatures [55, 56]. Therefore, the yield behavior of pure epoxies is generally studied under compression. A schematic example of a typical stress–strain curve obtained in uniaxial compression is shown in Fig. 26.1. It can be seen that the stress–strain behavior of unmodified epoxies is very similar to that of other glassy polymers. Generally, there is an initial linear response, followed by yielding. Further deformation is accompanied by strain softening before strain hardening sets in. The yield stress is often defined as the maximum point at which plastic deformation is so extensive that subsequent deformation can occur without further increase in stress. The form of yielding may vary from sharp shear banding to diffuse shear yielding, depending on the testing rate, temperature, and thermal history of the specimen. Localized shear banding usually dominates at relatively low temperatures or high rates.

The yield behavior of pure epoxies is usually pressure sensitive; that is, the yield stress decreases with hydrostatic tension and increases with hydrostatic compression. This dependence is often described by the pressure-modified Von Mises yield criterion [57]. Furthermore, the yield behavior of unmodified epoxies is frequently rate and temperature dependent as well; specifically, the yield stress increases as the testing rate increases or as the testing temperature decreases [55].

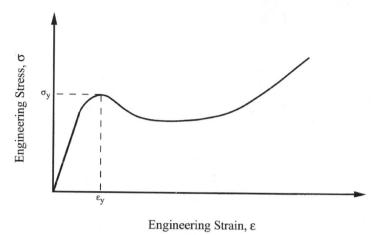

Engineering Strain, ε

Figure 26.1 A schematic of a typical stress–strain curve for an epoxy under uniaxial compression.

B. Fracture Behavior of Unmodified Epoxies

The fracture behavior of a material can be described effectively through the application of fracture mechanics. A more detailed treatment of this subject is given in Chapter 20. Since pure epoxies are generally very brittle, exhibiting little plastic deformation during fracture, their fracture behavior is usually analyzed using linear elastic fracture mechanics (LEFM). In this case, the resistance to fracture, i.e., fracture toughness, is often quantified in terms of a critical strain-energy release rate G_C or a critical stress-intensity factor K_C under plane-strain conditions. Furthermore, the fracture toughness of materials is usually measured for Mode I (opening mode) cracking, because this mode is technically the most important and causes the material to behave in the most brittle manner.

The fracture toughness of unmodified epoxies is comparatively low at room temperature; that is, K_{Ic} is well below 1.0 MPa$\sqrt{m}$ or G_{Ic} is on the order of 100 J/m^2 [58, 59]. Although pure epoxies are very brittle, inelastic deformation in the form of viscoelastic and/or plastic deformation still occurs during the fracture process. This attribute was originally inferred from the fact that the measured fracture toughness greatly exceeds the theoretical estimate for purely brittle fracture [60]. Later, this finding was confirmed by fractographic observations of tearing lines randomly emanating from the initial crack front [61].

Two modes of crack propagation are usually observed in unmodified epoxies tested under stable methods, such as the compact tension (CT) test or the double-cantilever-beam (DCB) test: (1) unstable, "stick–slip" crack propagation and (2) stable, continuous crack propagation [55, 58, 59]. A schematic example of the typical load-displacement curves obtained in the two modes of crack propagation is shown in Fig. 26.2. The stick–slip mode is often observed at relatively low testing rates or high testing temperatures. This factor is attributed to the occurrence of crack-tip

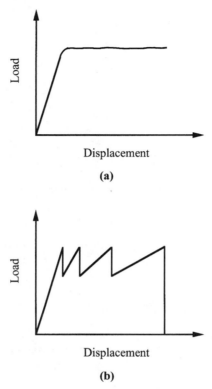

Figure 26.2 A schematic of the typical load-displacement curves for crack propagation in epoxies under tapered DCB tests. (a) stable, continuous propagation, (b) unstable, "stick–slip" propagation.

blunting prior to crack propagation, caused by localized plastic stretching. In contrast, the continuous mode is usually observed at relatively high testing rates or low testing temperatures. This factor is attributed to the suppression of crack-tip blunting and the intrinsically stable nature of fracture in the specimens used.

The stability of cracking in pure epoxies has been extensively investigated by several researchers [55, 62–66]. One approach to understanding this issue is to study the effects of testing rate and temperature on crack stability [55, 63, 66]. Yamini and Young [55] showed that the difference between the initiation and arrest toughness decreased as the rate increased or as the temperature decreased. Hence, as mentioned previously, crack propagation in the continuous mode usually occurs at relatively high rates or temperatures, at which the initiation and arrest toughness become nearly equal. The other approach is to study the effects of crack velocity on crack stability [62, 64, 65]. Mai and Atkins [62] and Andrews and Stevenson [64] showed that the fracture toughness of certain epoxies decreased with increasing crack velocity. The negative rate of change of fracture toughness with crack velocity was considered as the cause of crack instability. However, in later work, Gledhill and Kinloch [65] showed that fracture toughness increased slightly with increasing crack velocity during stable crack propagation. Hence, it seems that a controversy about the cause of crack instability exists in the literature.

C. Summary

In sum, the compressive yield behavior of unmodified epoxies is generally rate and temperature dependent and pressure sensitive. Pure epoxies are usually brittle, undergoing very slight, highly localized inelastic deformation during fracture. Both unstable, "stick–slip" and stable, continuous crack propagations can be observed, depending on the testing rates or temperatures. Crack instability is probably related to the intrinsic connection between fracture toughness and crack velocity. Unfortunately, this relationship has not yet been unambiguously established.

III. RUBBER TOUGHENING OF EPOXIES

The technology of rubber toughening was first utilized in the toughening of thermoplastics in the 1940s [67]. The best known example is perhaps the toughening of high-impact polystyrene (HIPS). Substantial improvement in fracture toughness, as compared to that of the unmodified polystyrene, can be achieved by the incorporation of a rubbery phase. Bucknall and Smith [68] concluded from their optical microscopic observations that the role of the uniformly dispersed rubber particles was to initiate multiple crazing in the polystyrene matrix. Furthermore, they suggested that rubber particles might also toughen cross-linked epoxies by a similar approach. Their astute suggestion encouraged subsequent studies on the rubber toughening of epoxies, although the rubber particles were later shown to initiate massive shear yielding in the matrix, rather than multiple crazing. Rubber toughening of epoxies was initiated by researchers at B. F. Goodrich Company and first studied by McGarry and his coworkers [1–3] in the late 1960s and early 1970s. Since then, extensive investigations [4–30] have been conducted to elucidate the toughening mechanisms in rubber-modified epoxies.

A. General Overview

In the initial investigation of toughening mechanisms by McGarry et al. [1–3], a liquid carboxyl-terminated butadiene acrylonitrile (CTBN) copolymer was used to modify a diglycidyl ether of bisphenol A (DGEBA) epoxy resin cured with a 2,4,6-tri(dimethylaminoethyl)phenol (DMP). Although other types of functionalized rubber were studied, CTBN was found to be the most effective, and so this system was the most extensively studied. CTBN rubber is soluble in DGEBA epoxy, and simple rubber particles of the size of microns are precipitated out before gelation during curing. (See Fig. 26.3.) "Salami-type" rubber particles are also formed under some curing conditions. The fracture toughness significantly increased by an order of magnitude when the rubber content was increased (see Fig. 26.4) and when the

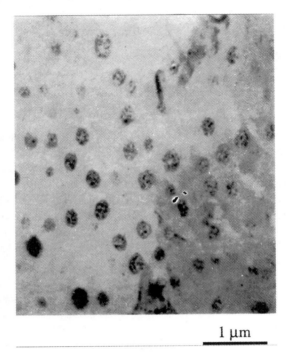

1 μm

Figure 26.3   Transmission electron micrograph of a microtomed, OsO_4-stained section from a CTBN rubber-modified DGEBA epoxy [1].

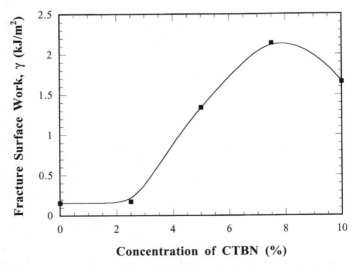

Figure 26.4   Fracture toughness plotted as a function of rubber content for a CTBN rubber-modified DGEBA epoxy [1].

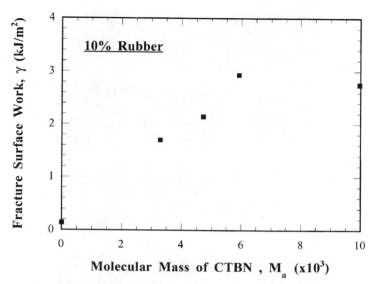

Figure 26.5 Fracture toughness plotted as a function of the molecular weight of rubber for a CTBN rubber-modified DGEBA epoxy [1].

molecular weight of the rubber was increased (see Fig. 26.5). This effect, however, was achieved only for large rubber particles (i.e., particles no less than 0.3 μm in diameter) that possessed good adhesion with the epoxy matrix. The toughening effect was thought by McGarry et al. to be caused by the crazing of the epoxy matrix, based on the stress whitening observed around the tip of the slowly growing crack [1]. They also demonstrated that the toughness of modified epoxies increased as the testing temperature increased or as the testing rate decreased. Furthermore, they showed that the toughening effect could persist to lower temperatures when the butadiene content of the rubber was increased. They also noted that CTBN rubber is not effective in toughening epoxies with high cross-link densities.

The deformation mechanisms of toughened epoxies were also studied under uniaxial and biaxial stress states [3]. It was proposed that the yield behavior of modified epoxies could be represented by a pressure-sensitive Von Mises criterion. McGarry et al. suggested that shear yielding is the main deformation mechanism for epoxies with small particles, while crazing is the dominant flow mechanism for epoxies with large particles. Microcavitation and shear banding were observed on the fracture surfaces of epoxies with large particles. McGarry et al. also noted that the size of the cavities is larger than the particle size of the undeformed epoxies. However, this finding was incorrectly taken as the evidence for crazing of the matrix. Although Bucknall [67] later proposed shear flow as an additional toughening mechanism, crazing was considered to be the main toughening mechanism [1–5] until doubt was cast by Yee et al. [10–15].

In contrast to the matrix crazing mechanism, Kunz-Douglass et al. [6, 7] emphasized the role of the rubber particles, rather than that of the epoxy matrix, in

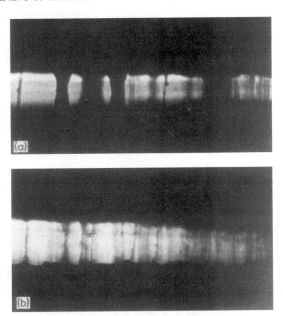

Figure 26.6   Optical micrographs of (a) rubber stretching and (b) tearing for a CTBN rubber-modified DGEBA epoxy [6].

toughening. In their studies, a CTBN rubber was used to toughen a DGEBA epoxy cured with 4,4'-diaminodiphenylmethane (DDM). They observed by optical microscopy that stretched rubber particles spanned the crack and experienced stable tearing and rupture as the crack was continuously opened. (See Fig. 26.6.) Cavities and particle protrusion observed on the fracture surfaces were taken as further indication of rubber stretching and tearing. Based on these observations, Kunz-Douglass et al. proposed crack bridging by rubber particles as the main toughening mechanism. Furthermore, they developed a quantitative toughening model based on the idea that energy was irreversibly dissipated during particle tearing. They predicted that the toughness should increase with the volume fraction and the tearing energy of the particles and depend weakly on particle size. Good agreement with their experimental results was obtained [6, 7]. However, this model predicts only the small amount of toughness improvement that Kunz-Douglass et al. obtained, rather than the order of magnitude achieved by other researchers. Furthermore, it can explain neither the extensive stress-whitening phenomenon observed in most rubber-modified epoxies nor the high toughness values at high temperatures. These failings of the model arise because it completely neglects any toughening effect from the matrix, which was later shown to be the most important factor in toughening. Furthermore, it should be noted that the particles are relatively large in the model material that Kunz-Douglass et al. investigated, so rubber bridging was perhaps unavoidably observed. However, in practical systems, the particles are small compared to the size of the process zone, so significant rubber bridging rarely occurs.

At around the same time that Kunz-Douglass et al. performed their study, Bascom et al. [8, 9] investigated the fracture behavior of rubber-modified epoxies, both in bulk and as adhesives. In their investigations, the system studied was a CTBN and/or solid rubber-modified DGEBA epoxy cured with piperidine (PIP). In all cases, the toughness was increased dramatically by increasing the rubber concentration. Continuous, stable crack growth was observed for these rubber-modified epoxies in tapered-DCB tests, while stick–slip, unstable crack propagation occurred for the unmodified epoxies. Nearly spherical holes were observed on the fracture surfaces of stable-crack-growth (usually stress-whitened) regions. (See Fig. 26.7a.) On the basis of scanning electron micrographs, they surmised that there was shear yielding of the surrounding matrix, although no conclusive evidence was offered. The formation of these cavities was attributed to the dilatation of the rubber particles in the presence of the triaxial stress state around the crack tip. In contrast, little evidence of hole formation was observed on the fracture surfaces of fast-crack-growth (usually smooth and featureless) regions. (See Fig. 26.7b.) This was attributed to the high strain rate of fast cracking, which prevented significant plastic flow from occurring. Furthermore, a significant increase in the size of the critical deformation zone was observed upon the addition of rubber.

Based on these observations, Bascom et al. proposed rubber cavitation and plastic-zone formation as the main toughening mechanisms. The fracture toughness of these modified epoxies was also shown to be rate dependent; that is, the fracture toughness decreased as the testing rate increased. This factor was attributed to the decreasing capability of the epoxy matrix to undergo viscoelastic and/or plastic deformation with increasing strain rate. Interestingly, a bimodal distribution of particles was obtained by modifying an epoxy with both liquid and solid rubber, and the toughness could be double that obtained using either rubber individually. This suggests that a synergistic toughening effect might be achieved in the presence of a bimodal distribution of particles. It should be noted that Bascom et al. were the first to suggest shear yielding of the matrix as an energy-dissipating process in rubber-modified epoxies. Unfortunately, they did not explain the role of rubber cavitation in the toughening process. The importance of rubber cavitation was not fully realized until the work of Yee et al. [10–15] was published.

Yee and Pearson [10–12] investigated the toughening mechanisms in a CTBN rubber-modified DGEBA epoxy cured with PIP. Complete phase separation was obtained after curing. They found no evidence for the plasticization of the epoxy by residual, dissolved rubber. The particle size decreased as the acrylonitrile content of the CTBN rubber was increased; meanwhile, the particle-size distribution varied from monodisperse to polydisperse. The fracture toughness increased significantly with increasing rubber concentration, but did not depend on the rubber type or the particle size within the range studied. Furthermore, the toughness could be directly correlated with the size of the plastic zone (i.e., the stress-whitened zone) ahead of the crack tip. The fracture surfaces of this zone were shown to be characterized by numerous cavities and appreciable plastic deformation in the surrounding matrix. Yee and Pearson showed that the cavity walls were actually lined with a layer of rubber and that the cavities in this region were larger than those in the fast-crack-

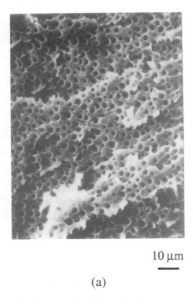

10 µm

(a)

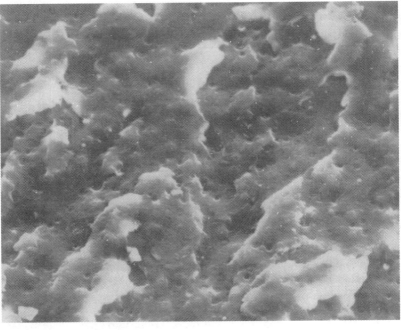

10 µm

(b)

Figure 26.7   Scanning electron micrographs of the fracture surface of a tapered DCB specimen for a CTBN rubber-modified DGEBA epoxy. (a) Stable-crack-growth region, (b) fast-crack-growth region [8].

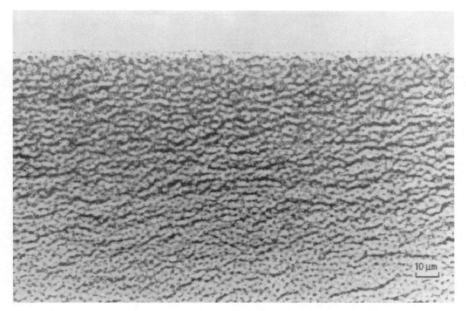

Figure 26.8   Transmission optical micrograph of a thin section perpendicular to the fracture surface of a single-edge-notched three-point-bending specimen for a CTBN rubber-modified DGEBA epoxy [12].

growth region. (The cavities in the fast-crack-growth region were of approximately the same size as the undeformed particles.) This indicated that these cavities were essentially formed by cavitation of the rubber particles and subsequent dilatation of the surrounding matrix. Furthermore, massive shear bands connecting the cavitated particles were observed in the subsurface damage zone. (See Fig. 26.8.) This zone was often embedded within another zone consisting solely of cavitated particles without shear banding. This attribute suggested that rubber cavitation preceded and enhanced shear banding of the matrix.

Based on these observations, Yee and Pearson proposed the cavitation of rubber particles followed by void growth and induced shearing banding of the matrix as the main toughening mechanism. This mechanism was confirmed by the independent work of Kinloch et al. [16–21], which will be discussed later. However, it should be noted that rubber cavitation also degrades the matrix proportionally to the volume fraction of the rubbery phase. Therefore, an optimal toughening effect usually exists through the addition of between 10 and 15 phr (parts per hundred resins by weight) rubber.

The effect of cross-link density on the toughenability of epoxies was examined in a CTBN rubber-modified DGEBA epoxy cured with 4,4′-diaminodiphenyl sulphone (DDS) [13]. Various DGEBA epoxies with different epoxide equivalent weights were used to control the molecular weight between cross-links, effectively

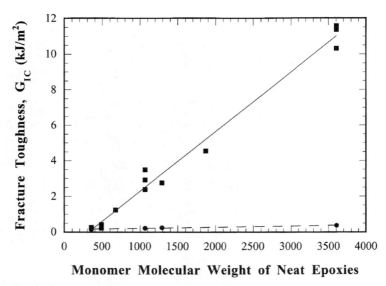

Figure 26.9 Fracture toughness plotted as a function of monomer molecular weight for both elastomer-modified (■) and neat (●) epoxy resins [13].

the cross-link density. It was shown that the cross-link density usually decreased with increasing the monomer molecular weight of the epoxy. It was also shown that the fracture toughness of pure epoxies was only very weakly dependent on the cross-link density, while that of rubber-modified epoxies increased dramatically with decreasing cross-link density. (See Fig. 26.9.) This observation suggested that the toughenability of pure epoxies increased with decreasing cross-link density. The relatively poor fracture resistance of epoxies with high cross-link densities was shown to correlate with a low extent of plastic deformation observed in microscopic examinations.

Later, Yee and Pearson also investigated the effect of particle size and its distribution on toughening mechanisms in rubber-modified epoxies [14]. In this investigation, both liquid (CTBN and CTB) and solid core–shell (methacrylated butadiene-styrene (MBS)) rubber were used to modify a DGEBA epoxy cured with PIP. Yee and Pearson showed that the fracture toughness increased with decreasing particle size. This relationship was attributed to the role played by the particles, which was governed by the size of the process zone. Yee and Pearson claimed that small particles induced extensive shear yielding of the matrix by the constraint relief effect of their cavitation, while large particles only bridged and/or deflected the crack. It was shown that the core–shell MBS rubber with a comparatively high cavitational stress provided the greatest toughening effect. Consequently, Yee and Pearson suggested that an appropriately great cavitational resistance would be desirable, because relatively high strain energy could be stored prior to the cavitation of the particles; hence, a relatively large plastic zone could be formed before cleavage fracture occurred. Furthermore, a bimodal distribution of particles was obtained in epoxies modified by both CTBN and CTB rubber, but no synergistic toughening was observed. Therefore, they claimed that no synergistic effects would occur between bridging particles and cavitating particles. The importance of rubber cavitation was further studied by Yee et al. [15] using tests with different degrees of constraint. They suggested that the triaxial tension at the crack tip must be relieved before the crack propagates, so that the deviatoric stress can reach a critical value for yielding to occur. They demonstrated that this could be achieved by rubber cavitation in terms of the relief of local constraint—that is, by changing the stress state of the surrounding matrix from dilatational dominated (highly triaxial) to distortional dominated (highly deviatoric).

Kinloch et al. [16–19] also studied the deformation and fracture behavior of a CTBN rubber-toughened DGEBA epoxy cured with PIP. The CTBN rubber completely precipitated out, as deduced from the observation that T_g remained unchanged in the modified epoxy. Continuous, stable crack propagation was observed in CT tests at relatively low rates or high temperatures, while stick–slip, unstable crack propagation occurred at relatively high rates or low temperatures. A significant improvement in fracture toughness was obtained by the addition of 15 phr rubber, and the increase in toughness was also shown to be rate and temperature dependent; that is, the toughness increased with decreasing testing rate or increasing testing temperature. Cavities and evidence for shear yielding of the surrounding matrix were observed using SEM on the fracture surfaces of stable-crack-growth regions. Correlation was also found between the fracture toughness and the size of the plastic zone ahead of the crack tip. Based on these observations, Kinloch et al. proposed voiding and shear yielding of the matrix as the main toughening mechanisms. They considered shear yielding of the matrix initiated by the stress concentration of the particles as the primary source of energy dissipation, while plastic void growth owing to internal cavitation or interfacial debonding of the particles was of secondary importance. In fact, they suggested that cavitation was neither necessary nor desirable, since solid rubber alone was sufficient for inducing stress concentrations.

Recently, Guild and Kinloch [20] demonstrated, using three-dimensional finite element analysis (FEA), that the relatively high Poisson's ratio (close to 0.5) of the rubber particles would induce cavitation first. Cavitation would result in a considerable increase in the Von Mises stress and hence promote extensive shear yielding and plastic void growth in the matrix. This concept is apparently consistent with the notion proposed by Yee et al. that rubber cavitation precedes and enhances shear banding of the matrix. Interestingly, Huang and Kinloch [21] showed that preexisting microvoids of the size of microns could also provide a significant increase in the toughness of epoxies. They suggested that a strong interface is not strictly required for significant toughening effect to be achieved. Quite recently, Bagheri and Pearson [22] showed that both rubber particles and preexisting microvoids could toughen epoxies by inducing massive shear yielding of the matrix. They also claimed that cavitation resistance of the rubber particles might not play an important role in toughening of the epoxies. However, it should be noted that the epoxies studied were those with relatively low cross-link density, and hence with relatively high intrinsic ductility. Therefore, the stress concentrations associated with the preexisting microvoids might induce extensive shear banding in the epoxy matrix, thereby substantially improving the toughness. However, for epoxies with relatively high cross-link density and hence with relatively low intrinsic ductility, the stress concentrations associated with preexisting microvoids might cause cohesive failure of the matrix instead, without the matrix undergoing significant plastic deformation. In contrast, a delayed cavitation of rubber particles with relatively high cavitational strength can relieve the triaxial tension ahead of the crack tip; this triaxial tension governs the cohesive failure of the epoxy matrix. Furthermore, the stress concentration associated with cavitated particles may induce massive shear banding of the matrix and hence significantly enhance the toughness. An excellent example illustrating this issue is provided by the core–shell rubber-modified epoxies that are discussed next.

Recently, preformed core–shell rubbers were employed to toughen epoxies, with notable success [23–25]. This technique was motivated by the requirement to toughen epoxies with relatively high cross-link densities, for which liquid rubber modification was shown to be ineffective. Furthermore, the morphology of modified epoxies can easily be controlled by varying the shell chemistry, without changing the composition and size of the rubbery core. Sue [23] studied a grafted-rubber concentrate (GRC)–modified DGEBA epoxy cured with DDS. The GRC rubber used had a butadiene-styrene core with a multicomponent shell. The size of rubber particles (about 100 nm) is determined by their latex synthesis and hence is independent of curing procedure. A significant improvement in fracture toughness (about $0.4 \text{ MPa}\sqrt{\text{m}}$, or 300 J/m^2) was obtained by adding 10 wt% of core–shell GRC rubber. This improvement was attributed to rubber cavitation and matrix shear yielding (up to 60% plastic strain) in the vicinity of the crack, which were observed by microscopic examinations. Sue emphasized that the cavitation resistance of rubber particles should be sufficiently high to permit the shear yielding of the surrounding matrix. Therefore, the effectiveness of this rubber, as compared to liquid rubbers, was attributed to its relatively high cavitational strength. Furthermore, minor toughening mechanisms, e.g., crack bifurcation and deflection, were also observed in this system.

Later, Sue et al. [24] showed that the spatial distribution of the rubber particles could be altered by changing the composition and the thickness of the shell. A transition from a random, multimodal distribution to a local clustering or even a large-scale segregation of the GRC particles was obtained by decreasing the glycidyl methacrylate (GMA) content, the acrylonitrile (AN) content, and the thickness of the shell. An increase in fracture toughness was observed when the dispersion of the particles was varied from random to locally clustered. This increase was attributed to the synergistic effect of matrix shear yielding and crack deflection. However, a decrease in fracture toughness was observed when an enormous segregation of the particles occurred. This decrease was attributed to the observation that only crack deflection operated in this case. In other words, large clusters of small rubber particles behaved as if they were large particles of the size of the clusters.

Quite recently, Bagheri and Pearson [25] investigated the role of blend morphology in an MBS or MBS-COOH core–shell rubber-modified DGEBA epoxy cured with PIP or aminoethyl piperazine (AEP). They showed that the morphology of modified epoxies depends strongly on the shell composition, the curing agents, and the extent of agitation prior to casting. Whereas MBS-COOH-modified epoxies had a uniform dispersion of particles, MBS-modified epoxies had a continuous morphology characterized by highly aggregated particles. The marked aggregation of MBS particles was attributed to the immiscibility of the poly(methyl methacrylate) (PMMA) shell in cured DGEBA epoxy, while the uniform dispersion of MBS-COOH particles was attributed to the compatibility conferred by the COOH groups that reacted with epoxy and suppressed the clustering of particles. Bagheri and Pearson also showed that MBS particles were significantly well dispersed in AEP-cured epoxies, as compared to PIP-cured epoxies. This factor was attributed to the relatively short gelation time for AEP-cured epoxies. Furthermore, they showed that proper agitation prior to gelation could suppress clustering of MBS particles and hence result in a relatively uniform dispersion of particles. The fracture toughness of epoxies with a connected morphology was significantly higher than that of epoxies with a uniform particulate morphology. This difference was attributed to the relatively large crack-tip plastic zone observed in epoxies with a connected microstructure. However, a significant decrease in uniaxial tensile yield strength was observed in the epoxies with a connected morphology.

The aforementioned studies demonstrate that the issues of rubber-particle cavitation resistance and particle clustering effects remain unresolved.

B. Qualitative Toughening Mechanisms

Along with the development of rubber toughening of epoxies, several toughening mechanisms have been proposed to explain the enhancement in fracture toughness of rubber-modified epoxies. These mechanisms include crazing of the matrix, rubber-particle bridging, formation of a plastic zone in the matrix, rubber cavitation and induced shear deformation of the matrix, and crack bifurcation and/or deflection by rubber particles. The optimal toughening mechanism for a particular system is strongly dependent on the material properties, such as the intrinsic ductility of the matrix and the size of the rubber particles. Furthermore, several mechanisms may be promoted independently or cooperatively under certain suitable circumstances (e.g., a bimodal distribution of particle size). Therefore, a comprehensive understanding of these mechanisms and their relative effectiveness is essential for the fundamental physics of toughening to be understood.

1. *Crazing of the Matrix*   Rubber particles act not only as stress concentrators, initiating crazes in the surrounding matrix, but also as craze terminators, preventing uninhibited growth of the crazes, which would result in premature failure. Craze stabilization allows additional crazes to grow. The multiple-crazing process consumes significant amounts of energy and hence increases the fracture toughness. This mechanism was proposed on the basis of interpretations of several observations, such as stress whitening ahead of the crack tip, and on the basis of the strong dependence on the particle size of fracture toughness and pressure sensitivity of yielding. Theoretically speaking, the relatively high cross-link density of epoxies should prevent crazing from occurring. Moreover, crazes have never been observed in epoxies with high cross-link densities. Consequently, it is now agreed that toughening by crazing is generally not operative in rubber-modified epoxies.

2. *Rubber-Particle Bridging*   Rubber particles span the crack and exert closure tractions on the crack surfaces. This process effectively reduces the applied stress-intensity factor at the crack tip. Alternatively, rubber particles are stretched and torn to fracture, which consumes additional fracture energy. This mechanism was proposed on the basis of microscopic observations of the crack profiles and the fracture surfaces. However, some inconsistencies exist between this mechanism and experimental observations. For example, the mechanism can predict only a two-fold improvement in fracture toughness and cannot explain the stress-whitening phenomenon usually observed on fracture surfaces. Furthermore, the mechanism can not account for the dependence of fracture toughness on temperature. These inconsistencies apparently arise from the fact that the bridging mechanism totally neglects the primary toughening contribution from the matrix. Therefore, it is commonly accepted that rubber-particle bridging is a weak, secondary toughening mechanism in rubber-modified epoxies. However, rubber particle bridging has been shown to be important for highly cross-linked epoxies modified by relatively large rubber particles. Generally, good interfacial adhesion is a prerequisite for this mechanism to be effective.

3. *Formation of a Plastic Zone in the Matrix*   Rubber particles increase the size of the plastic zone ahead of the crack tip and hence increase the fracture toughness. This factor was attributed to the possible suppression of microcrack formation in the presence of a highly flexibilized interface, even though hole formation was observed on the fracture surface, owing to the triaxial dilatation of the rubber particles. Although the role of rubber particles was later shown to be misinterpreted in this manner and the observation of plastic deformation was based only on SEM of the fracture surface, this mechanism unambiguously suggests that extensive plastic deformation in the matrix is the major energy-dissipative process in rubber-modified epoxies.

4. *Rubber Cavitation and Induced Shear Deformation in the Matrix*   Rubber particles cavitate because of the triaxial tension around the crack tip. This cavitational process relieves the plane-strain constraint ahead of the crack tip, allowing the stress concentrations associated with cavitated particles to activate extensive shear deformation of the matrix in the form of dilatational void growth, discrete shear bands between cavitated particles, or even diffuse shear yielding. It was proposed on the basis of microscopic observations of the fracture surfaces and the subsurface damage zone. It is now generally accepted as the primary toughening mechanism for rubber-modified epoxies, especially those with relatively low cross-link densities and thus relatively high toughenability.

5. *Crack Bifurcation and/or Deflection by Rubber Particles* Rubber particles cause the main crack to fork into many secondary cracks, distributing the local stress intensity of the main crack to multiple cracks and/or deviating the crack off the principal plane of propagation, which increases the area of the crack surface and induces a mixed mode (containing Modes I and II) of crack propagation. This mechanism was proposed on the basis of microscopic observations of the profiles of a subcritical crack. The mechanism was shown to be ineffective unless it interacts with other mechanisms. Therefore, it is usually considered to be a secondary toughening mechanism in rubber-modified epoxies. However, crack bifurcation and/or deflection by particles might be important for highly cross-linked epoxies modified with relatively large particles, especially rigid particles.

C. Quantitative Toughening Models

A qualitative understanding of toughening mechanisms may be valuable; however, it is highly desirable to develop quantitative models that describe the experimental observations using the fundamental microstructural parameters of the toughened system. Furthermore, predictive models might provide some valuable suggestions for the appropriate selection of an optimal toughening strategy for a new system to be modified.

1. *Rubber-Particle Bridging Model* This model was first proposed by Kunz-Douglass et al. [6] and is based on the concept that the increase in fracture toughness is solely contributed from the energy irreversibly dissipated during tearing of the rubber particles. The predicted improvement in toughness, ΔG_{Ic}, is given by

$$\Delta G_{\text{Ic}} = \left(1 - \frac{6}{\lambda_t^2 + \lambda_t + 4}\right) 4 \Gamma_t f, \qquad (26.1)$$

where λ_t is the extension ratio of the rubber particles at tearing, Γ_t is the tearing energy of the rubber particles, and f is the volume fraction of the rubber particles. This model predicts a linear dependence of the increase in fracture toughness on the volume fraction of the rubber particles. Furthermore, this model predicts only a weak dependence of the improvement in fracture toughness on the particle size.

Later, Ahmad et al. [26] proposed an alternative model by considering the crack shielding effect. This model was based on the reduction of local stress intensity at the crack tip caused by the compressive tractions on the crack surfaces induced by the rubber particles. The predicted fracture toughness K_{Ic} is given by

$$K_{\text{Ic}} = (1 - f) K_{\text{Ic}}^m \frac{f E_m^* \Gamma_t}{2(1 - f) K_{\text{Ic}}^m}, \qquad (26.2)$$

where f is the volume fraction of the rubber particles, K_{Ic}^m is the fracture toughness of the pure epoxy, E_m^* is the effective Young's modulus of the pure epoxy (i.e., E_m for plane stress and $E_m/(1 - \nu_m^2)$ for plane strain, where E_m and ν_m are the Young's modulus and Poisson's ratio of the pure epoxy, respectively), and Γ_t is the tearing energy of the rubber particles. This model predicts that the fracture toughness increases with increasing rubber concentration and rubber particle size.

Both models can predict only a two- to threefold increase in the fracture toughness, because they neglect any toughening contribution from the matrix, which is often the most significant contribution of all. Furthermore, both models incorrectly predict the dependence of the fracture toughness on temperature.

2. Synergistic Rubber Stretching, Plastic Dilatation, and Shear Banding Model

Evans et al. [27] developed this model by considering both the rubber stretching that occurs along the crack plane and the plastic dilatation and shear banding of the matrix that occur in a process zone. The model is based on a general method of toughening analysis that relates the total enhancement in fracture toughness to the total energy dissipation per unit crack area within a process zone as the crack propagates through the system. The resultant energy change usually consists of contributions from residual deformation in the wake, plastic energy consumed in the crack-tip process zone, and the energy of newly created free surfaces. This energy change can be determined quantitatively with a knowledge of the stress–strain behavior experienced by the material within a process zone during crack propagation.

The toughening due to rubber stretching, ΔJ_{rs}, was evaluated using the first rubber particle bridging model described previously. Toughening due to plastic dilatation, ΔJ_{pd}, was associated with the nonlinear mean stress-volumetric strain behavior of the material within a process zone and the attendant hysteresis loss in the wake (alternatively, the compressive tractions on the crack surfaces exerted by the expanding process zone) and is given by

$$\Delta J_{pd} = \frac{2\phi f}{(z-1)(\sigma_y^P \varepsilon_y^P l)^{z-1}} J_{pd}^z, \tag{26.3}$$

where ϕ and z are coefficients that are dependent only upon the work-hardening exponent of the epoxy matrix; f is the volume fraction of the rubber particles; σ_y^P and ε_y^P are the yield stress and strain of the epoxy matrix, respectively; l is the rubber-particle spacing, and J_{pd} is the net toughness owing to plastic dilatation. It can be seen that plastic-dilatation toughening depends on several intrinsic properties of the epoxy matrix (e.g., the yield characteristics and the strain-hardening coefficient) and the interparticle distance. Toughening due to shear banding, ΔJ_{sb}, was provided by the nonlinear shear stress–shear strain behavior of the material within a process zone and the resultant hysteresis loss in the wake and is given by

$$\Delta J_{sb} = \frac{0.4\gamma_b f_b J_{sb} E}{\tau_c}, \tag{26.4}$$

where γ_b is the net shear strain experienced by an individual shear band, τ_c is the critical shear stress at which γ_b occurs, f_b is the volume fraction of shear-banded epoxy, J_{sb} is the net toughness owing to shear banding, and E is the Young's modulus of the epoxy matrix. It can be seen that the shear-banding toughening is dependent on the deformation characteristics of shear bands (e.g., shear strain and stress), the volume fraction of shear-banded epoxy, and the elastic modulus of the epoxy matrix.

Evans et al. claimed that the toughening phenomena associated with the process zone, such as the plastic dilatation and the shear banding, are strictly additive. On the other hand, the toughening effect of each phenomenon is multiplied by the rubber-particle bridging across the crack surface, because the toughening effect scales with the size of the process zone. Therefore, a synergism is expected when the three mechanisms operate simultaneously. It was also emphasized that the predicted toughness is valid only when a steady-state process zone is present. Furthermore, it was suggested that the toughening effect derives primarily from the crack wake, while a loading zone ahead of the crack tip provides no toughening. Consequently, an R-curve behavior is expected in which the toughness increases with crack growth until a steady state is reached. However, no experimental verification was provided for this model. Furthermore, the plastic dilatation and the shear banding were considered to be independent in this model. This condition is apparently inappropriate, because the shear yielding of epoxy is strongly dependent on the mean stress that governs the plastic dilatation of epoxy.

3. Crack-Tip Shielding Model   Argon [28] proposed a model by considering the crack-tip shielding induced by plastic deformation in an inelastic zone surrounding the crack. He claimed that the toughening effect in rubber-modified epoxies resulted primarily from an effective reduction of the plastic resistance, i.e., the tensile yield strength, of the epoxies, caused by the cavitating particles. The shielding, ΔJ, was evaluated using a similar approach employed by Evans et al. [27] and is given by

$$\Delta J = J_m \left[\frac{n A_n}{(n+1)} \frac{(\sigma_c/\sigma_y)^{n-1}}{(1-f)^{n-2}(\tilde{\sigma}_{\theta\theta}(0))^{n+1}\pi(1-\nu^2)} - f \right],\qquad(26.5)$$

where J_m is the intrinsic toughness of the epoxy matrix, n is the strain-hardening exponent of the epoxy matrix, f is the volume fraction of the rubber particles, σ_c is the cohesive strength of the epoxy matrix, σ_y is the tensile yield strength of the epoxy matrix, $\tilde{\sigma}_{\theta\theta}$ is the angular function of the stress distribution, A_n is a coefficient dependent upon the previous five parameters, and ν is the Poisson's ratio of the epoxy matrix. This model significantly underestimates the improvement in fracture toughness for a rubber-modified epoxy. This was attributed to the additional shielding from rubber cavitation, which was not taken into account.

4. Additive Localized Shear Banding, Plastic Void Growth, and Rubber Bridging Model   Huang and Kinloch [19] developed this model by considering the localized shear banding and the plastic void growth that occur in a crack-tip process zone and the rubber-particle bridging that occurs along the crack plane. An approach similar to that used by Evans et al. [27] was employed in this model. Furthermore, it was assumed that the overall increase in fracture toughness has additive contributions from the localized shear banding, the plastic void growth, and the rubber-particle bridging.

The contribution from the shear banding, ΔG_s, was determined by integrating the dissipated strain-energy density for the shear banding estimated over a plastic zone of predicted size and is given by

$$\Delta G_s = 0.5 \left(1 + \frac{\mu_m}{\sqrt{3}}\right)^2 \left[\left(\frac{4\pi}{3f}\right)^{1/3} - \frac{54}{35}\right] f \sigma_{yc} \gamma_f r_{yu} K_{vm}^2, \qquad (26.6)$$

where μ_m is the pressure sensitivity of yielding of the epoxy matrix, f is the volume fraction of the rubber particles or voids, σ_{yc} is the compressive yield stress of the epoxy matrix, γ_f is the shear fracture strain of the epoxy matrix, r_{yu} is the radius of the crack-tip plastic zone, and K_{vm} is the maximum stress-concentration factor of the Von Mises stress in the epoxy matrix.

The contribution from the plastic void growth, ΔG_v, was calculated by integrating the consumed strain-energy density estimated for the plastic void growth over the same plastic zone and is given by

$$\Delta G_v = \left(1 - \frac{\mu_m^2}{3}\right)(f_v - f_r)\sigma_{yc} r_{yu} K_{vm}^2, \qquad (26.7)$$

where f_v and f_r are the volume fractions of rubber particles and voids, respectively. The contribution from rubber bridging, ΔG_r, was evaluated using the rubber-particle bridging model proposed by Kunz-Douglass et al. [6].

It was shown that the predicted values of fracture toughness were in good agreement with those measured experimentally, in terms of the effects of testing temperature and rate on the toughness. Furthermore, it was shown that localized shear banding is usually the main toughening mechanism. Plastic void growth becomes dominant at elevated temperatures, while rubber bridging plays an important role at relatively low temperatures. However, this model focused only on cavitation and flow in a process zone associated with a stationary crack and hence neglected any toughening contribution from the crack wake. Actually, the method of toughening analysis used in this model is valid only when a fully developed, steady-state process zone is present. Furthermore, shear banding and plastic void growth were also inappropriately considered to be independent in this model.

5. *Dilatational Yielding Model* Lazzeri and Bucknall [29, 30] proposed a model for dilatational yielding in terms of the formation of cavitated shear bands (dilatational bands) in rubber-toughened polymers. This model is treated in some detail in Chapter 22. In this model, it was emphasized that rubber cavitation reduces the yield stress of the matrix by increasing the pressure sensitivity of yielding and hence enables the matrix to shear yield through the formation of dilatational bands. Furthermore, it was shown that the predicted band angles are in good agreement with experimental values.

D. Some New Developments in the Field of Rubber-Modified Epoxies

Some new developments in the field of rubber-modified epoxies are presented in this section to illustrate the importance of the interaction between crack propagation and evolution of process zone in polymer toughening [69].

Previous work on rubber-modified epoxies has focused on exploring the effects of cavitation and flow in a process zone situated ahead of a stationary crack. However, it has been established in the mechanics literature [70–72] that the strengths of toughened brittle materials depend critically on how a nonlinear, irreversible process zone interacts with a propagating crack. Generally, a continual development of the process zone with crack growth gives rise to an R-curve behavior in which the toughness depends on the extent of crack advance [71]. Typically, no enhancement of toughness above the intrinsic toughness of the matrix is associated with the initial process zone that forms ahead of a stationary crack. It is only once the crack has begun to extend that the toughness increases [71]. Provided that the specimen boundaries are sufficiently remote from the process zone, a fully developed, steady-state process zone of uniform thickness eventually evolves in the wake of the crack, and the toughness rises to a fully toughened value that is independent of crack length [70–72].

Exploration of wake effects in rubber-modified epoxies appears to have been limited to cases in which bridging by rubber particles was thought to be the main toughening mechanism [6]. However, while the contribution of a nonbridging crack wake to toughness is clear from mechanics analyses [70–72], it is highly desirable that this contribution be demonstrated in polymers, as such a demonstration is crucial for establishing the legitimacy of adapting the well-developed mechanics literature on toughening [70–72] to describe engineering polymers. Therefore, the effects of the crack wake on toughness and on R-curve behavior are investigated for a CTBN rubber-modified DGEBA epoxy cured with PIP [69]. No bridging mechanism was shown to operate, owing to the fine dispersion of the rubber particles [14].

A significant R-curve behavior was observed for a 10-phr rubber-modified epoxy (see Fig. 26.10) using DCB tests. Upon a small crack growth of less than 1 mm, an initial fracture resistance of about 2 kJ/m^2 was obtained. The toughness then rose rapidly to an approximately steady-state value of about 7 kJ/m^2 after a crack growth of about 20 mm. These values should be compared with the toughness value of 0.2 kJ/m^2 for the unmodified epoxy [14, 15]. The initial toughness measured is in good agreement with the previous value of about 1.7 kJ/m^2, which was determined from single-edge notched specimens tested in three-point-bending geometry [14, 15].

The features of the experimental R-curves, which show a rise in toughness until an approximate steady state is reached, can be directly correlated to the evolution of the process zone obtained from Transmission Optical Microscopy (TOM) examinations. (See Fig. 26.11.) As the crack grew, the process zone moved with it and left a trace of its boundary in the crack wake. During the early stage of crack growth, there was an apparent fanning out of the crack wake. This observation is consistent with prior TOM observations of the process zone developing ahead of the crack tip [14]. Eventually, the wake reached a uniform thickness, which implies a steady state with a process zone of constant size.

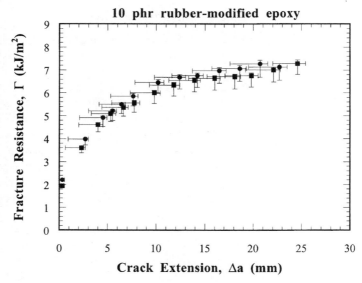

Figure 26.10   Experimental *R*-curves for a 10-phr CTBN rubber-modified DGEBA epoxy. The results for two separate specimens are shown and respectively designated by (■) and (●).

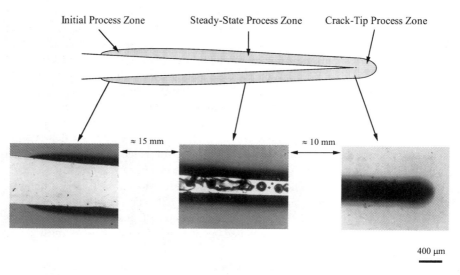

Figure 26.11   Transmission optical micrographs of the process zone for a 10-phr CTBN ruber-modified DGEBA epoxy.

The contribution of the crack wake to the toughness was investigated using *R*-curve measurements in which a portion of the crack wake was removed after an approximate steady state was reached. (See Fig. 26.12.) After the crack wake had

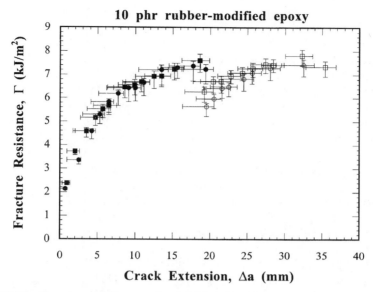

Figure 26.12 Experimental R-curves for a 10-phr CTBN rubber-modified DGEBA epoxy. The solid symbols (■ and ●) represent the data points obtained before the crack wake was removed. The open symbols (□ and ○) represent the data points obtained after the crack wake was removed.

been removed, the crack-growth resistance dropped by about 2 kJ/m^2. Further crack growth resulted in the toughness increasing back up to the original steady-state value. This investigation demonstrated unambiguously that the crack wake contributed to the toughness, despite the absence of any bridging in this material. This behavior is expected, because unloading of material in the process zone as it passes into the wake results in hysteresis losses [27, 73]. (Alternatively, it is possible to consider the dilatational characteristics of the cavitation and subsequent void growth within the wake as inducing closure tractions on the crack tip [70].) The removal of this residual deformation (or compression) in the crack wake results in a decrease in the fracture resistance. The subsequent rise in toughness can be attributed only to the redevelopment of a wake of uniform thickness, because the process zone was fully developed when the wake was removed.

The aforementioned behavior is consistent with several analyses of the toughening of rubber-modified polymers in which the toughness is assumed to derive only from the nonlinear, irreversible deformation of material as it passes through the process zone and into the crack wake [27, 73]. Furthermore, it is consistent with the notion that a successful toughening strategy would significantly improve the strength only if a substantial portion of the R-curve can be traversed during failure [74, 75]. Less than optimal strengths usually result from premature, unstable crack growth caused by loading geometry or rate effects.

E. Summary

In sum, epoxy resins can be significantly toughened by the incorporation of a rubbery phase. The fracture toughness of rubber-modified epoxies increases with increasing rubber concentration up to 10–15 phr and depends strongly on the size of the rubber particles. Generally, small particles of the size of microns are more effective in toughening than large particles of the size of tens of microns. The enhanced toughness is rate and temperature dependent; that is, it increases with decreasing testing rate or increasing testing temperature. It is commonly agreed that cavitation of the rubber particles and subsequent hole growth by matrix shear deformation is the most important toughening mechanism. However, secondary toughening mechanisms, e.g., rubber-particle bridging and crack bifurcation and/or deflection, may play a dominant role when particle sizes are relatively large. Rubber-modified epoxies usually exhibit a significant R-curve behavior. This behavior supports the notion that successful toughening strategies for engineering materials should include ways of ensuring that the full potential of the R-curves can be realized. The R-curve behavior of rubber-modified epoxies derives primarily from the nonlinear, irreversible deformation of the material within the wake of the crack. This indicates that a full understanding of the mechanics of toughening requires a detailed knowledge of how the crack tip interacts with the material in the process zone as it passes into the crack wake.

IV. THERMOPLASTIC TOUGHENING OF EPOXIES

Although epoxy resins can be substantially toughened by the addition of a rubbery phase, the improvement in toughness is inevitably accompanied by a significant loss in elastic modulus and yield strength. Furthermore, rubber toughening of highly cross-linked epoxies is shown to be inefficient, owing to the limited ability of these epoxies to undergo shear deformation. Therefore, an alternative approach has recently emerged in which epoxies are toughened by the incorporation of rigid, thermoplastic particles [31–50]. Generally, the thermoplastics blended are those possessing a high modulus, a high glass transition temperature, and a high level of toughness. Thus, the aim is to achieve an enhancement in fracture toughness for the modified epoxies, while still retaining other desirable properties.

A. General Overview

Little toughening effect was achieved in the first work in this area, by Bucknall and Partridge [31]. In their investigation, poly(ether sulfone) (PES) was employed as the thermoplastic modifier. Phase separation of PES was generally observed (see Fig. 26.13), and the morphology of the PES-rich phase was varied by changing the epoxy resins (trifunctional and tetrafunctional) and the curing agents (DDS and dicyanodiamide (Dicy)). However, only slight improvements in fracture toughness (no more than 0.3 MPa$\sqrt{\text{m}}$) were obtained, irrespective of the degree of phase separation or the morphology. The toughening was attributed to crack deflection by the

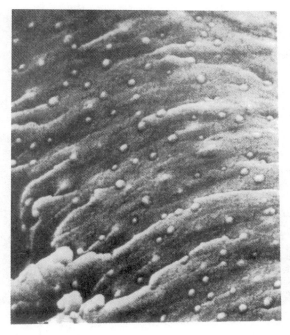

1 µm

Figure 26.13 Scanning electron micrograph of the fracture surface for a PES-modified epoxy [31].

PES-rich phase. In later work, a PES–epoxy (tetrafunctional) blend cured with an aromatic anhydride was studied by Raghava [32]. A two-phase morphology with a bimodal particle distribution was observed in the modified epoxies. Again, only marginal enhancements in fracture toughness were obtained. This was attributed to the relatively low elongations of PES at failure and the limited ability of the brittle matrix to undergo shear deformation. Furthermore, it was noted that interfacial adhesion seemed to be poor, owing to the relatively high solubility parameter of the PES.

It was not until the work by Hedrick et al. [33] that impressive successes were made. In their study, phenolic hydroxyl-terminated polysulfone (PSF) oligomers were used to modify a DGEBA epoxy. The PSF molecules were capped by DGEBA resin prior to the addition of a stoichiometric amount of DDS curing agent; hence, minimal chain extension was obtained. A particulate morphology was observed in the modified epoxies (see Fig. 26.14), and the particle size increased with increasing molecular weight of the PSF. Furthermore, good interfacial adhesion was obtained, owing to the presence of the reactive group in the PSF. The fracture toughness considerably increased (up to 0.7 MPa$\sqrt{m}$) with increasing PSF content, without a significant loss in flexural modulus. The toughening effect was attributed to ductile drawing of the PSF particles and induced plastic deformation of the surrounding matrix.

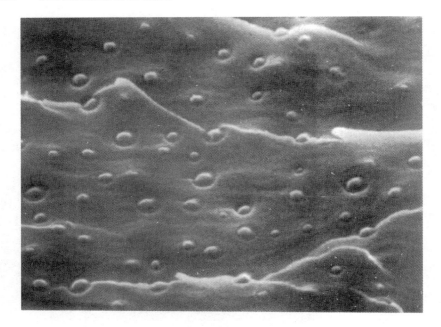

1 μm

Figure 26.14 Scanning electron micrograph of the fracture surface for a hydroxyl-terminated PSF-modified epoxy [33].

Furthermore, the fracture toughness was shown to increase with increasing molecular weight of the PSF, owing to the enhanced ductility of the PSF. It was particularly emphasized that good interfacial adhesion is extremely important in achieving a substantial enhancement in the toughness of thermoplastic-modified epoxies. This work was the first to demonstrate the effectiveness of the functionalization of the thermoplastic modifier in promoting a good interface, even though it was intuitively thought to be advantageous for good adhesion to exist.

Similar approaches were extensively used by researchers in the subsequent development of this area [34–41]. Most of these investigations were again on PES- or PSF-modified epoxies, but they generally focused on incorporating thermoplastics with relatively high molecular weights. Cecere and McGrath [34] studied DDS-cured epoxies modified by an amine-terminated PSF and a poly(ether ketone) (PEK). Dramatic improvements in fracture toughness were obtained, with negligible loss in flexural modulus. Those improvements were also attributed to the ductile tearing of the modifier. Furthermore, the fracture toughness was shown to increase with increasing molecular weight and amount of modifier. A particulate morphology was observed at relatively low contents of the modifier, while a co-continuous structure occurred at high concentrations of the modifier. (See Fig. 26.15.) Although the presence of the co-continuous network might provide relatively high toughness, it may

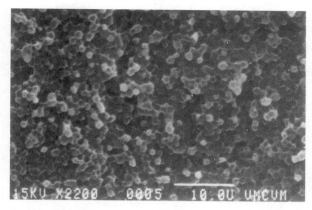

Figure 26.15   Scanning electron micrograph of the fracture surface for an amine-terminated PSF-modified epoxy [34].

also make the control of morphology very difficult and possibly result in a reduction in some desirable mechanical, chemical, and thermal properties of the modified epoxies.

Kim and Brown [35] investigated the toughening mechanisms in a resorcinol-based, DDS-cured epoxy modified by a glassy second component. A particulate morphology was also observed, with phase inversion occurring at relatively high concentrations of the modifier. The fracture toughness increased monotonically with increasing modifier content, even though an apparent transition from brittle to ductile behavior was observed on fracture surfaces, owing to the phase inversion occurred at relatively high concentrations of the modifier. Kim and Brown claimed that at low modifier contents, the toughening occurred via the local yielding of the epoxy matrix initiated by yielding of the modifier particles, while at high modifier contents the toughening resulted from the considerable deformation of the modifier matrix. It should be noted that most of the work prior to Kim and Brown's study attributed improvements in toughness to the deformation of thermoplastic-rich particles. This work appears to be the first to emphasize the possible importance of the deformation of an epoxy matrix with relatively low cross-link density, even though the work was based mainly on observations of thin-film deformation in which plane-stress conditions prevailed.

Kinloch et al. [40] recently studied toughening mechanisms in DDS-cured epoxies modified by a reactively-terminated PES copolymer. The morphology of the modified epoxies was varied from single phase to particulate to co-continuous and finally to phase inverted by steadily increasing the modifier content. (See Fig. 26.16.) The fracture toughness significantly increased with increasing modifier concentration only when phase separation occurred. (See Fig. 26.17.) The toughening effect was attributed to crack deflection by the microstructural components and the relatively high toughness of the PES-rich phase. Furthermore, the toughness was also shown to increase continuously, even though the microstructure changed from particulate to co-continuous and then to phase inverted. It was particularly emphasized that

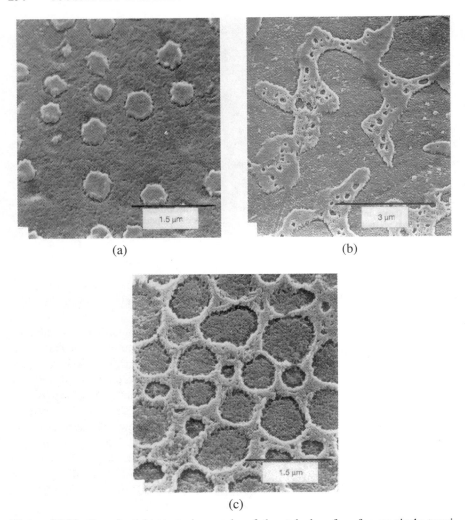

(a)   (b)

(c)

Figure 26.16 Scanning electron micrographs of the etched surface for reactively terminated PSF-modified epoxies. (a) Particulate microstructure, (b) co-continuous microstructure,
(c) phase-inverted microstructure [40].

no indications of cavitation or debonding of the particulate phase or plastic yielding
of the matrix phase were observed on the fracture surfaces. In later work by Kubotera
[41], the toughness improvement of a PES–epoxy system was attributed to bridging
and pinning by particles for the particulate morphology and to the microcracking
of the epoxy-rich phase and the plastic drawing of the PES-rich phase for the co-
continuous morphology.

   In addition to work on the modification of epoxies using thermoplastics with reactive groups, extensive investigations on physically blended systems have also been

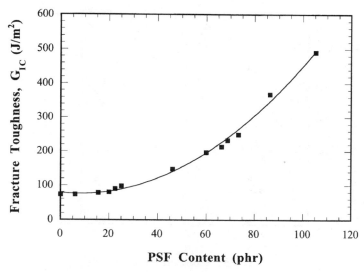

Figure 26.17 Fracture toughness plotted as a function of PSF content for reactively termi-nated PSF-modified epoxies [40].

reported. These systems include modifications of epoxy resins by poly(ether imides) (PEI) [42, 43], polyphenyleneoxide (PPO) [44], poly(methyl methacrylate) (PMMA) [45], and polycarbonate (PC) [46]. Notable successes have been reported in this area. Bucknall and Gilbert [42] studied a DDS-cured, tetrafunctional epoxy modified by PEI. A particulate morphology was observed for the modified epoxies in terms of the phase separation of PEI. Good interfacial adhesion was obtained, although no reactive groups were introduced into the modifier. The fracture toughness substan-tially increased (up to 400 J/m^2) with increasing PEI content. (See Fig. 26.18.) The toughening effect was attributed to ductile drawing of PEI particles through crack bridging. This work demonstrated that a significant enhancement in toughness can also be obtained for a physically blended system, provided that a sufficiently strong interface is present. Pearson and Yee [44] investigated a PIP-cured DGEBA epoxy modified by PPO. Phase separation of PPO was observed and was stabilized by the addition of a styrene–maleic anhydride copolymer. A particulate morphology devel-oped at relatively low PPO concentrations, while a co-continuous structure formed at relatively high PPO concentrations. The fracture toughness significantly increased with increasing PPO concentration, while the tensile yield strength was almost inde-pendent of the PPO content. (See Fig. 26.19.) The toughening effect was primarily attributed to the formation of massive microcracking in the matrix ahead of the crack tip (see Fig. 26.20), while the ductile tearing of crack-bridging PPO particles might also have contributed to the overall toughness. It was emphasized that the promotion of profuse microcracking over particle bridging might be very important in achieving relatively high toughening efficiency in highly cross-linked epoxies.

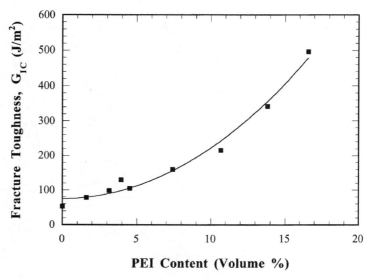

Figure 26.18   Fracture toughness plotted as a function of PEI content for PEI-modified epoxies [42].

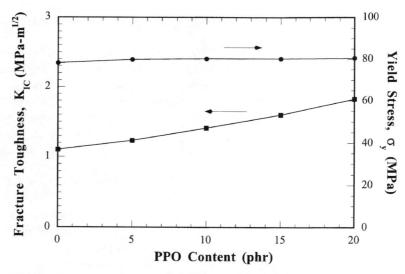

Figure 26.19   Fracture toughness and yield stress plotted as functions of PPO content for PPO-modified DGEBA epoxies [44].

Recently, Kim and Robertson [47] used three semicrystalline thermoplastics—poly(butylene terephthalate) (PBT), nylon 6, and poly(vinylidene fluoride) (PVDF)—to toughen a DGEBA epoxy cured with a mixture of aromatic amines.

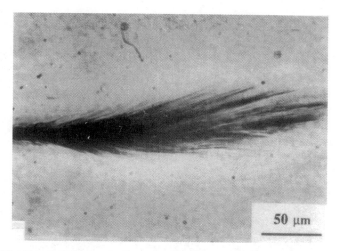

Figure 26.20   Transmission optical micrograph of the damaged zone of a cracked single-edge-notched specimen for a PPO-modified DGEBA epoxy [44].

A particulate morphology was observed in all three types of modified epoxies. The fracture toughness dramatically increased with increasing concentration of modifiers, without a significant loss in modulus and strength. The toughening effect was attributed to a combination of energy-dissipating processes induced by thermoplastic-rich particles, including crack bridging, crack pinning (or bowing), crack bifurcation and/or deflection, ductile tearing, and phase transformation. The first four mechanisms were confirmed for epoxies toughened by PBT, while the last was claimed to provide the extra increase in toughness observed. The first four were also confirmed for epoxies modified by nylon 6, while the first three were confirmed for epoxy–PVDF blends. Furthermore, the toughness of epoxies toughened by PBT was shown to be independent of particle size, while the toughness of epoxies modified by nylon 6 decreased with increasing particle size.

The toughening of epoxies by employing both thermoplastic and elastomeric phases has also received considerable attention, and very promising results have been obtained [44, 48–50]. Pearson and Yee [44] employed styrene-butadiene-styrene (SBS) and, in a second experiment, CTBN rubber to modify a PPO–epoxy blend. SBS rubber partitioned into the PPO phase, while CTBN rubber concentrated in the epoxy matrix. A significant increase in fracture toughness was obtained by adding only a small amount of rubber. For DGEBA–PPO–SBS blends, the toughening effect was attributed to the increased density of microcracks in the matrix initiated by the composite particles. (See Fig. 26.21.) For DGEBA–PPO–CTBN blends, the toughening effect was attributed to the suppression of microcracking and the promotion of shear banding in the matrix induced by the triaxial constraint relief of rubber cavitation. (See Fig. 26.22.) Kishi et al. [50] studied the effect of PES content on the toughenability of a highly cross-linked epoxy resin (a DGEBA epoxy cured with

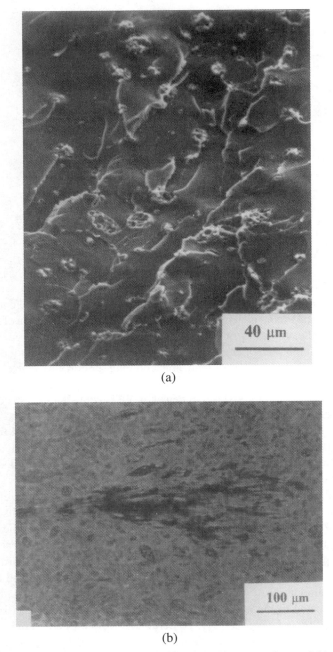

(a)

(b)

Figure 26.21 (a) Scanning electron micrograph of the fracture surface and (b) transmission optical micrograph of the damage zone of single-edge-notched specimens of an SBS rubber–PPO–epoxy blend [44].

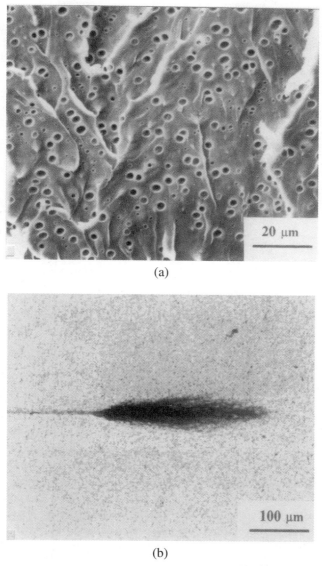

(a)

(b)

Figure 26.22   (a) Scanning electron micrograph of the fracture surface and (b) transmission optical micrograph of the damage zone of single-edge-notched specimens of an CTBN rubber–PPO–epoxy blend [44].

DDS). Although the shear ductility of the epoxy was enhanced by the incorporation of the PES, the fracture toughness dramatically increased with increasing PES concentration only when CTBN rubber was present. (See Fig. 26.23.) This was attributed to the fact that the enhanced ductility of the matrix was activated only when the triaxial constraint ahead of the crack tip was dissipated by cavitation of the rubber particles.

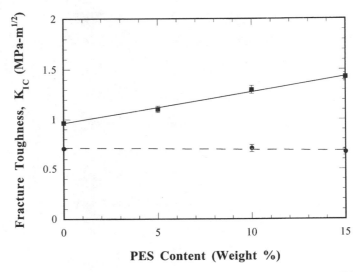

Figure 26.23   Fracture toughness plotted as a function of PES content for modified epoxies with (■) and without (●) CTBN rubber [50].

B. Qualitative Toughening Mechanisms

Along with the development of processes that toughen epoxies by using rigid thermoplastic particles, several qualitative toughening mechanisms were proposed to account for the enhancement in fracture toughness. These mechanisms include crack bridging by thermoplastic particles, crack pinning (or bowing) by thermoplastic particles, crack deflection and/or bifurcation by thermoplastic particles, shear banding of the matrix, massive microcracking of the matrix, and transformation of thermoplastic particles. Generally, several mechanisms operate simultaneously to produce the overall toughening effect. Furthermore, the optimal toughening mechanism is strongly dependent on the intrinsic properties of the material (e.g., type, size, and ductility of the thermoplastic particles and the ability of the matrix to undergo shear deformation and microcracking). Therefore, a complete understanding of these mechanisms and their relative effectiveness is essential for the optimal design of the thermoplastic modifier and the epoxy matrix.

1. *Thermoplastic-Particle Bridging*   Thermoplastic particles span the crack, which necessitates their ductile stretching and tearing. This mechanism provides a closure traction to the crack surfaces and effectively reduces the local stress-intensity factor at the crack tip. It was proposed on the basis of microscopic observations of the fracture surface that showed tearing of the thermoplastic particles. It may induce a significant toughening effect, especially for highly cross-linked epoxies, owing to the inherently high yield stress of the thermoplastic modifier. However, it is effective only when relatively large particles with strong interfaces are present.

2. *Crack Pinning or Bowing by Thermoplastic Particles*   Rigid, thermoplastic particles act as impenetrable objects and effectively pin the advancing crack. The pinned crack front bows out, which consumes additional energy. This mechanism was proposed on the basis of microscopic examinations of fracture surfaces that showed "tails" near the particles and usually interrelates with the thermoplastic-particle bridging mechanism. Furthermore, it is effective only when there is good interfacial adhesion.

3. *Crack Deflection and/or Bifurcation by Thermoplastic Particles*   Thermoplastic particles change the crack path by causing the crack to deviate from its principal plane of propagation and/or causing a crack to split into several secondary cracks. The deflection of the crack path increases the total area of crack surface and forces crack propagation into a mixed mode (a mixture of Modes I and II), thereby resulting in greater energy dissipation. The bifurcation of a crack reduces the local stress-intensity factor at the tip by distributing it over multiple cracks. This mechanism, which was proposed on the basis of microscopic observations of the profiles of subcritical cracks, may provide only a modest toughening effect and usually accompanies other toughening mechanisms.

4. *Shear Banding of the Matrix*   Thermoplastic particles behave as stress concentrators, owing to the rather significant mismatch in modulus between the particle and the matrix. This stress concentration causes the matrix to undergo extensive shear deformation, generally in the form of massive shear banding between the particles. The formation of the shear bands absorbs considerable energy, thereby enhancing the fracture toughness. This mechanism was proposed on the basis of microscopic examinations of the fracture surface and the crack-tip process zone. It may provide a significant improvement in toughness, provided that the intrinsic ductility of the matrix can be fully exploited.

5. *Microcracking of the Matrix*   Thermoplastic particles cause stress concentrations and initiate massive microcracks in the surrounding matrix, dissipating extra fracture energy and increasing the fracture toughness. This mechanism was proposed on the basis of microscopic observations of the crack-tip process zone and could account for a modest improvement in fracture toughness. It appears to be effective for highly cross-linked epoxies when the particles are relatively rigid and capable of debonding.

6. *Transformation of Thermoplastic Particles*   Particles of semicrystalline thermoplastics may undergo a stress-induced phase transformation in crystal structure. The dilatation usually associated with this transformation reduces the tensile stresses ahead of the crack tip, thus effectively lowering the local stress-intensity factor. This mechanism was proposed on the basis of microscopic examinations of fracture surfaces and is valid only for epoxies toughened by semicrystalline thermoplastics.

C. Quantitative Toughening Models

No quantitative toughening models have been proposed for thermoplastic-modified epoxies, probably because of the novelty of this approach. However, reasonable modification of several predictive models, which were proposed for toughened metals or ceramics, might well elucidate the experimental observations for thermoplastic-modified epoxies. Therefore, it is very useful to discuss these models in some detail so that their fundamental physics can be fully understood.

1. Ductile Particle-Bridging and Particle-Pinning Models   Przystupa and Courtney [76] developed a model for a brittle metal toughened with ductile particles by considering both particle bridging and particle pinning. Particle bridging reduced the stress intensity at the crack tip by exerting closure tractions on the crack surfaces, while particle pinning increased the critical stress intensity for crack propagation in the matrix by forcing the crack front to bow out. The predicted apparent toughness K_c is given by

$$K_c = AK_c^m + \frac{E_m^* f W_p l_p}{2AK_c^m},\tag{26.8}$$

where A is a coefficient accounting for crack bowing, K_c^m is the critical stress intensity of the matrix, E_m^* is the effective Young's modulus of the matrix (i.e., E_m for plane stress and $E_m/(1 - v_m^2)$ for plane strain, where E_m and v_m are the Young's modulus and Poisson's ratio of the matrix, respectively), f is the volume fraction of the particles, W_p is the plastic work to failure of the particles, and l_p is the length of particles between the fracture surfaces that is subject to plastic strain. This model was shown to be consistent with experimental studies in a two-phase alloy. In particular, this model was shown to be in good agreement with experimental results on nylon-12-modified epoxies [48].

2. Crack Deflection Model   Faber and Evans [77] developed a model to predict the improvement in fracture toughness owing to crack deflection around second-phase particles. Crack deflection resulted in a nonplanar crack and thus induced a reduction in the driving force of the crack. The predicted toughening increment ΔG_c for spherical particles is given by

$$\Delta G_c = 0.87 f G_c^m,\tag{26.9}$$

where f is the volume fraction of the particles and G_c^m is the intrinsic toughness of the matrix. This model indicates that the increase in toughness associated with crack deflection is invariant with particle size, but depends on particle shape and volume fraction of particles. Experimental results in a ceramic were directly correlated with this model. Unfortunately, no attempts have been made to apply this model to thermoplastic-modified epoxies.

3. *Microcracking Model*   Evans and Faber [72] proposed a model for brittle materials toughened by microcracking. Microcracking resulted in both a decrease in elastic modulus and an attendant dilatation in the wake of the process zone, thereby providing an additional contribution to crack shielding. The predicted increase in toughness, ΔK_c, is given by

$$\Delta K_c = 0.4 f K_c^m + 0.25 E_m \sqrt{h\theta}, \tag{26.10}$$

where f is the volume fraction of the microcracks, K_c^m is the intrinsic toughness of the matrix, E_m is the Young's modulus of the matrix, h is the thickness of a fully developed crack wake, and θ is the total dilatation caused by microcracking. This model suggests that the dilatational contribution is relatively substantial and is dependent on the size of the process zone. It was shown to be consistent with experimental results for several ceramics. Unfortunately, no efforts have been made to apply this model to thermoplastic-modified epoxies.

D. Summary

In sum, epoxies with relatively high cross-link densities, in which rubber toughening is usually not effective, can be significantly toughened by the addition of a rigid, ductile thermoplastic resin. Generally, phase separation of the thermoplastic modifier in the form of either a particulate or a co-continuous morphology is a prerequisite for substantial improvements in toughness to be obtained. Good interfacial adhesion owing to either chemical or physical bonding is a further requirement for successful toughening. The fracture toughness of thermoplastic-modified epoxies continuously increases with increasing content of the modifier, even though the microstructure may change abruptly from particulate to co-continuous. The co-continuous structure, as compared to the particulate morphology, usually provides a strong toughening effect. However, some desirable properties of the modified epoxies may also be degraded.

Thermoplastic-particle bridging is generally the main toughening mechanism, while the optimal toughening efficiency is usually obtained when several mechanisms cooperate. Furthermore, the incorporation of a rubbery phase might further enhance the fracture toughness, because extensive shear deformation may be induced in the matrix through the constraint relief effect of rubber cavitation. Several other factors, such as the mechanical properties of the modifier and the matrix, the morphology of the modified epoxies, and the distribution and size of the particles, are also very important in achieving appreciable toughening effects. However, a detailed, quantitative explanation of the effects of these parameters on toughening mechanisms and toughening efficiency has yet to be unambiguously established. This is because the independent control of these factors is impossible to achieve using current techniques for the formulation of modified epoxies. Therefore, investigations both on suitable approaches to controlling these parameters independently and on their influence on the toughening process are essential for further development.

V. CONCLUDING REMARKS

Epoxies are highly cross-linked polymers with many excellent properties; hence, they are usually the engineering polymers of choice for an enormous range of structural applications. However, most of the epoxies are brittle materials that need to be toughened. One successful approach to toughening epoxies is the incorporation of second-phase polymeric particles, such as rubbers and thermoplastics.

Rubber toughening is effective for epoxies with relatively low cross-link densities. The fracture toughness of rubber-modified epoxies increases with increasing rubber content up to 10–15 phr. Relatively small particles usually cavitate, owing to the triaxial stress state ahead of the crack tip, which relieves the plane-strain constraint associated with the crack. This induces extensive shear deformation in the matrix, which consumes a significant amount of fracture energy. Relatively large particles usually bridge the crack and exert closure tractions on the crack surfaces. This reduces the stress intensity at the crack tip and provides a modest degree of toughening.

Thermoplastic toughening is effective for epoxies with relatively high cross-link densities, provided that a strong interface is obtained through either chemical bonding or physical blending. The fracture toughness of thermoplastic-modified epoxies increases continuously with increasing thermoplastic concentration, even though phase inversion may occur at a relatively high modifier content. Crack bridging by thermoplastic particles is often the main energy-dissipating process, while optimal toughening effects are obtained usually when several toughening mechanisms cooperate.

It should be noted that while this chapter addresses only the toughening of epoxies, similar toughening mechanisms operate in many other thermosets. These mechanisms are addressed in other chapters of this book. The subject of thermoset toughening by rigid, inorganic particles is left out entirely, owing to space limitations. However, it should be mentioned that a surprising amount of toughening is possible through the use of rigid, inorganic fillers in the case of intrinsically brittle thermosets.

VI. ACKNOWLEDGMENTS

The authors are thankful to Dr. J. Harcup for his valuable comments on the manuscript. This work was supported by a grant from the National Science Foundation, Grant #CMS9523078.

VII. REFERENCES

1. F. J. McGarry, *Proc. Roy. Soc. London*, **A319**, 59 (1970).
2. J. N. Sultan, R. C. Laible, and F. J. McGarry, *Appl. Polym. Symp.*, **16**, 127 (1971).
3. J. N. Sultan and F. J. McGarry, *Polym. Eng. Sci.*, **13**, 29 (1973).
4. A. C. Meeks, *Polymer*, **15**, 675 (1974).

5. C. K. Riew, E. H. Rowe, and A. Siebert, "Rubber Toughened Thermosets," in *Advances in Chemistry* **154**, R. D. Deanin and A. H. Crugnola, Eds., American Chemical Society, Washington, DC, 1976.

6. S. Kunz-Douglass, P. W. R. Beaumont, and M. F. Ashby, *J. Mater. Sci.*, **15**, 1109 (1980).

7. S. C. Kunz and P. W. R. Beaumont, *J. Mater. Sci.*, **16**, 3141 (1981).

8. W. D. Bascom, R. L. Cottingham, R. L. Jones, and P. Peyser, *J. Appl. Polym. Sci.*, **19**, 2545 (1975).

9. W. D. Bascom, R. Y. Ting, R. J. Moulton, C. K. Riew, and A. R. Siebert, *J. Mater. Sci.*, **16**, 2657 (1981).

10. A. F. Yee and R. A. Pearson, NASA Contractor Report 3718, U.S. Government Rainting Office, Washington DC, August 1983.

11. A. F. Yee and R. A. Pearson, *J. Mater. Sci.*, **21**, 2462 (1986).

12. R. A. Pearson and A. F. Yee, *J. Mater. Sci.*, **21**, 2475 (1986).

13. R. A. Pearson and A. F. Yee, *J. Mater. Sci.*, **24**, 2571 (1989).

14. R. A. Pearson and A. F. Yee, *J. Mater. Sci.*, **26**, 3828 (1991).

15. A. F. Yee, D. Li, and X. Li, *J. Mater. Sci.*, **28**, 6392 (1993).

16. A. J. Kinloch, S. J. Shaw, D. A. Tod, and D. L. Hunston, *Polymer*, **24**, 1341 (1983).

17. A. J. Kinloch, S. J. Shaw, and D. L. Hunston, *Polymer*, **24**, 1355 (1983).

18. Y. Huang and A. J. Kinloch, *J. Mater. Sci.*, **27**, 2753 (1992).

19. Y. Huang and A. J. Kinloch, *J. Mater. Sci.*, **27**, 2763 (1992).

20. F. J. Guild and A. J. Kinloch, *J. Mater. Sci.*, **30**, 1689 (1995).

21. Y. Huang and Kinloch, *Polymer*, **33**, 1330 (1992).

22. R. Bagheri and R. A. Pearson, *Polymer*, **37**, 4529 (1996).

23. H. J. Sue, *Polym. Eng. Sci.*, **31**, 275 (1991).

24. H. J. Sue, E. I. Garcia-Meitin, D. M. Pickelman, and P. C. Yang, " Optimization of Model-I Fracture Toughness of High-Performance Epoxies by Using Designed Core-Shell Rubber Particles," in *Advances in Chemistry* **233**, C. K. Riew and A. J. Kinloch (eds.), American Chemical Society, Washington, DC, 1993.

25. R. Bagheri and R. A. Pearson, *J. Mater. Sci.*, **31**, 3945 (1996).

26. Z. B. Ahmad, M. F. Ashby, and P. W. R. Beaumont, *Scripta Metall.*, **20**, 843 (1986).

27. A. G. Evans, Z. B. Ahmad, D. G. Gilbert, and P. W. R. Beaumont, *Acta. Metall.*, **34**, 79 (1986).

28. A. S. Argon, "Sources of Toughness in Polymers," in *Advances in Fracture Research*, Vol. 4, K. Samala, K. Ravi-Chander, D. M. R. Taplin, and P. Rama Rao (eds.), Pergamon Press, New York, 1989.

29. A. Lazzeri and C. B. Bucknall, *J. Mater. Sci.*, **28**, 6799 (1993).

30. A. Lazzeri and C. B. Bucknall, *Polymer*, **36**, 2895 (1995).

31. C. B. Bucknall and I. K. Partridge, *Polymer*, **24**, 639 (1983).

32. R. S. Raghava, *J. Polym. Sci. Polym. Phys. Ed.*, **25**, 1017 (1987).

33. J. L. Hedrick, I. Yilgör, G. L. Wilkes, and J. E. McGrath, *Polym. Bull.*, **13**, 201 (1985).

34. J. A. Cecere and J. E. McGrath, *Polym. Prepr.*, **27(1)**, 299 (1986).

35. S. C. Kim and H. R. Brown, *J. Mater. Sci.*, **22**, 2589 (1987).

36. R. S. Raghava, *J. Polym. Sci. Polym. Phys. Ed.*, **26**, 65 (1988).

37. Z. Fu and Y. Sun, *Polym. Prepr.*, **29**, 177 (1988).

38. J. L. Hedrick, I. Yilgör, M. Jurek, J. C. Hedrick, G. L. Wilkes, and J. E. McGrath, *Polymer*, **32**, 2020 (1991).

39. A. J. MacKinnon, S. D. Jenkins, P. T. McGrail, and R. A. Pethrick, *Macromolecules*, **25**, 3492 (1992).

40. A. J. Kinloch, M. L. Yuen, and S. D. Jenkins, *J. Mater. Sci.*, **29**, 3781 (1994).

41. K. Kubotera,"Morphological and Fracture Studies of Alloys of Thermosets and Thermoplastics," Ph.D. Thesis, The University of Michigan, Ann Arbor, MI, 1995.

42. C. B. Bucknall and A. H. Gilbert, *Polymer*, **30**, 213 (1989).

43. D. J. Hourston and J. M. Lane, *Polymer*, **33**, 1379 (1992).

44. R. A. Pearson and A. F. Yee, *Polymer*, **34**, 3658 (1993).

45. B. Das, D. Chakraborty, A. K. Hajra, and S. Sinha, *J. Appl. Polym. Sci.*, **53**, 1491 (1994).

46. C. B. Bucknall, I. K. Partridge, L. Jayle, I. Nozue, A. Fernyhough, and J. N. Hay, *Polym. Prepr.*, **33**, 378 (1992).

47. J. Y. Kim and R. E. Robertson, *J. Mater. Sci.*, **27**, 161 (1992).

48. B. J. Cardwell, "Toughening of Polymer Blends Through Thermoplastic Additions," Ph.D. Thesis, The University of Michigan, Ann Arbor, MI, 1994.

49. S. T. Kim, J. K. Kim, C. R. Choe, and S. I. Hong, *J. Mater. Sci.*, **31**, 3523 (1996).

50. H. Kishi, Y. B. Shi, J. Huang, and A. F. Yee, *J. Mater. Sci.*, **32**, 761 (1997).

51. E. Pink and J. D. Campbell, *Mater. Sci. Eng.*, **15**, 187 (1974).

52. R. J. Morgan and J. E. O'Neal, *J. Mater. Sci.*, **12**, 1966 (1977).

53. E. J. Kramer, "Microscopic and Molecular Fundamentals of Crazing," in *Advances in Polymer Science*, Vol. 52/53, H. H. Kausch (ed.), Springer-Verlag, Berlin, 1983.

54. P. C. Yang, H. J. Sue, and M. T. Bishop, "Crazing in Intrinsically Tough-High-Performance Thermosets Resins," in *Polymer Toughening*, C. B. Arends (ed.), Marcel Dekker, New York, 1996.

55. S. Yamini and R. J. Young, *J. Mater. Sci.*, **15**, 1814 (1980).

56. B. W. Cherry and K. W. Thomson, *J. Mater. Sci.*, **16**, 1913 (1981).

57. A. J. Kinloch and R. J. Young, *Fracture Behaviour of Polymers*, Elsevier Applied Science, London, 1983.

58. L. J. Broutman and F. J. McGarry, *J. Appl. Polym. Sci.*, **9**, 609 (1965).

59. S. Mostovoy and E. J. Ripling, *J. Appl. Polym. Sci.*, **10**, 1351 (1966).

60. J. B. Berry, "Brittle Behavior of Polymeric Solids," in *Fracture Process in Polymeric Solids*, B. Rosen (ed.), Interscience, New York, 1964.

61. R. L. Patrick, "The Use of Scanning Electron Microscopy," in *Treatise on Adhesion and Adhesives*, Vol. 3, R. L. Patrick (ed.), Dekker, New York, 1973.

62. Y. W. Mai and A. G. Atkins, *J. Mater. Sci. Lett.*, **10**, 2000 (1975).

63. R. J. Young and P. W. R. Beaumont, *J. Mater. Sci. Lett.*, **11**, 766 (1976).

64. E. H. Andrews and A. Stevenson, *J. Mater. Sci.*, **13**, 1680 (1978).

65. R. A. Gledhill and A. J. Kinloch, *Polym. Eng. Sci.*, **19**, 82 (1979).

66. A. J. Kinloch and J. G. Williams, *J. Mater. Sci.*, **15**, 987 (1980).

67. C. B. Bucknall, *Toughened Plastics*, Applied Science, London, 1977.

68. C. B. Bucknall and R. R. Smith, *Polymer*, **6**, 437 (1965).

69. J. Du, M. D. Thouless, and A. F. Yee, *Int. J. Fract.*, **92**, 271 (1998).

70. R. M. McMeeking and A. G. Evans, *J. Am. Ceram. Soc.*, **65**, 242 (1982).

71. B. Budiansky, J. W. Hutchinson, and J. C. Lambrapolous, *Int. J. Solids Struct.*, **19**, 337 (1983).

72. A. G. Evans and K. T. Faber, *J. Am. Ceram. Soc.*, **67**, 255 (1984).

73. W. Yang, *Acta Mech. Sin.*, **7**, 131 (1991).

74. Y. W. Mai and B. R. Lawn, *Ann. Rev. Mater. Sci.*, **16**, 415 (1986).

75. R. F. Cook and D. R. Clarke, *Acta Metall.*, **36**, 555 (1988).

76. M. A. Przystupa and T. H. Courtney, *Metall. Trans.*, **A13**, 881 (1982).

77. K. T. Faber and A. G. Evans, *Acta Metall.*, **31**, 565 (1983).

27 Fatigue-Crack Propagation in Polymer Blends

R. A. PEARSON*

Department of Materials Science and Engineering
Lehigh University
Bethlehem, PA 18015

L. PRUITT

Department of Mechanical Engineering
University of California at Berkeley
Berkeley, CA 94720

* Completed while on sabbatical leave at Sandia National Laboratories, Albuquerque, NM 87185-0958.
Polymer Blends, Volume 2: Performance. Edited by D. R. Paul and C. B. Bucknall.
ISBN 0-471-35280-2. © 2000 John Wiley & Sons, Inc.

I. INTRODUCTION

Fracture mechanics concepts are highly effective in describing crack-tip stresses and can be used to determine the residual strength of flawed components. The use of fracture mechanics for fatigue design is based on the assumption that many structural components are intrinsically flawed, but are capable of sustaining a large degree of stable crack growth prior to fracture. The fatigue life of a component based on the defect-tolerant approach is dictated by the number of loading cycles needed to propagate a crack of an initial size to a critical dimension. This philosophy is distinct from the total-life approach, which relates fatigue life to the initiation and subsequent growth of a flaw. Over the last few decades, numerous researchers have provided detailed reviews [1–8] of fatigue behavior in polymers, based both on total-life and on fracture mechanics approaches. A brief overview of fatigue crack propagation in polymers is given next.

The use of fracture mechanics is widely applicable to the characterization of fatigue-crack-propagation behavior in polymer blends, as the majority of these polymeric materials are capable of sustaining a large amount of subcritical crack growth prior to fracture. Further, many of these polymers are used in safety-critical applications, for which defect-tolerant life estimates are essential. Characterizing crack-growth behavior in polymer blends is complicated by the fact that fatigue cracks are known to propagate at different rates, depending on the near-tip micromechanisms, mean stress, frequency, crack geometry, or test environment. These factors are of considerable interest and practicality to the safe design of structural polymeric components subjected to repetitive loading.

The aim of this chapter is to provide an overview of near-tip micromechanisms associated with fatigue crack propagation in rubber-toughened polymers. The importance of rubber particle size, blend morphology, and the interaction between rubber particles and inorganic fillers in resisting crack propagation is reviewed. Emphasis is placed on crack-tip shielding mechanisms, but mean stress effects, variable amplitude loading, and issues relating to notches and short cracks are also discussed.

A. Review of Crack Growth Under Monotonic Loading

The stress intensity parameter K, derived from linear elastic fracture mechanics, can be used to describe the magnitude of the stresses, strains, and displacements in the region ahead of the crack tip. As discussed in Chapter 20, there are three distinct ways or modes whereby a crack can be loaded. Mode I is the normal, or opening, mode, which is usually considered to be the harshest mode. Mode II is the in-plane shear mode, and mode III is the out-of-plane shear mode. The linear elastic solution for the normal stress in the y-direction (σ_{yy}) for the mode I loading case incorporates K_I as a scaling parameter and is written as a function of distance r and angle θ away from the crack tip [9]:

$$\sigma_{yy} = \frac{K_I}{\sqrt{2\pi r}} \cdot \cos\frac{\theta}{2}\left\{1 + \sin\frac{\theta}{2}\sin\frac{3\theta}{2}\right\}. \tag{27.1}$$

Here, K_I is the mode I (opening mode) stress intensity factor, which incorporates the boundary conditions of the cracked body and is a function of loading, crack length, and specimen geometry. The stress intensity factor can be found for a wide range of specimen geometries and is used to scale the effect of the far-field load, crack length, and geometry of the flawed component [10]:

$$K_I = \sigma^\infty\sqrt{\pi a} \cdot F\left(\frac{a}{W}\right). \tag{27.2}$$

Here, σ^∞ is the remote far-field stress and F is the geometric factor for the specimen or component geometry. F is a function of the crack length a normalized by the specimen width W.

The efficacy of the stress intensity parameter derives from its ability to correlate the inception and growth of cracks contained in different specimens through a concept known as similitude. This concept states that two cracks with the same stress intensity factor experience the same driving force for crack propagation. The similitude concept enables fracture toughness data taken from laboratory specimens to be used for strength predictions of load-bearing components. Therefore, the conditions for crack growth under monotonic load conditions for specimens scaled to laboratory size or for large engineering structures is given by

$$K_I \geq K_{Ic}, \tag{27.3}$$

where K_{Ic} is the fracture toughness of the material containing the crack and is a true material property. Next, we will examine the mechanisms that determine the magnitude of K_{Ic}.

B. Crack-Tip Shielding Mechanisms Under Monotonic Loading

Crack-tip shielding occurs when the local stress intensity K_{local} is less than the remote stress intensity K_{remote} calculated using σ^∞ in Eq. 27.2, such that the crack feels an effectively lower driving force. This phenomenon can come about through micromechanisms activated ahead of or behind the crack tip. These micromechanisms can improve the toughness of the material through energy dissipation in the process zone at the crack tip or through crack-bridging mechanisms in the crack wake. There are several well-established extrinsic crack-tip shielding mechanisms found in engineering solids [11, 12]. These mechanisms include crack deflection; zone shielding through phase transformations, localized microcracking, or void nu-

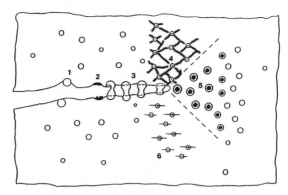

Figure 27.1   Crack-tip shielding mechanisms under monotonic loading include (1) crack deflection, (2) rubber tearing, (3) bridging by fibers or second-phase particles, (4) massive shear banding, (5) rubber particle cavitation and plastic void growth, and (6) particle-induced multiple craze formation or microcracking.

cleation; and contact shielding from surface asperity contact, ligament or fiber bridging, or plasticity-induced closure. Figure 27.1 schematically illustrates several of the shielding mechanisms that may be activated in polymer blends, including (1) crack deflection, (2) particle tearing, (3) bridging by fibers or second-phase particles, (4) massive shear banding, (5) plastic void growth, and (6) multiple craze formation or microcracking.

C. Modeling the Sources of Toughness

Toughening mechanisms are often divided into two categories: (a) those involving the process zone ahead of the crack tip and (b) crack-wake mechanisms, which occur behind the crack tip, as seen in Fig. 27.1. Some polymeric structures, such as rubber-toughened polymers, are capable of activating both mechanisms to obtain synergistic toughening [13, 14]. The process-zone mechanism of toughening is typically described using crack-tip energetics [15–17]. The energy release rate G characterizes the release in potential energy with the advancement of the crack plane. The energy release rate is related to the stress intensity parameter under linear elastic conditions (i.e., plane stress) and can be written as $G = K^2/E$, where E is the elastic modulus. If one considers a polymeric material that generates a process zone of width $2w$ at its crack tip, then as the crack advances through the material, it leaves a wake behind the crack tip. (See Fig. 27.2a.) The change in toughness due to energy changes in the strip is given as

$$\Delta G_C = 2 \int_0^W \left\{ \int_0^{\varepsilon_{ij}} \sigma_{ij}\varepsilon_{ij} \right\} dy, \tag{27.4}$$

where the integral in braces represents the strain-energy density. The critical energy-release rate for propagation is then equal to the work required to advance the crack

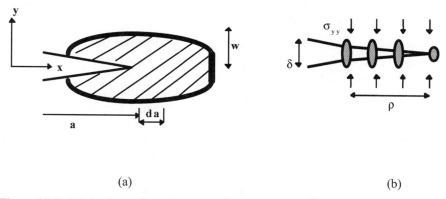

(a)   (b)

Figure 27.2 Mechanisms of toughening include the development of (a) a process zone of width $2w$ ahead of the crack tip and (b) bridging tractions in the wake of the crack tip.

from a to $(a + da)$ and is normalized by the new crack area da for unit thickness:

$$G_C = \Delta G_C + G_i = 2 \int_0^w \left\{ \int_0^{\varepsilon_{ij}} \sigma_{ij}\varepsilon_{ij} \right\} dy + 2\gamma_e. \tag{27.5}$$

In this equation, G_i is the intrinsic toughness and γ_e is the energy to generate new crack surfaces under elastic conditions.

Many toughened polymer systems utilize second-phase particles or rubber reinforcements capable of nonlinear deformation. An analogous form for G can be developed for nonlinear deformations by using the J-integral form of Eq. 27.5. $J_c = \Delta J_c + J_i$. If the reinforcing particles are responsible for all of the energy dissipation, one can assume that the strain energy does not depend on y (the direction normal to the crack plane). For f particles per unit volume, one can write the toughness as

$$G_C = 2wf \int_0^{\varepsilon_{ij}} \sigma_{ij}\varepsilon_{ij} + 2\gamma_e. \tag{27.6}$$

Particle bridging provides another mechanism of toughening. In this case, an advancing crack leaves second-phase particles intact behind the crack tip as it advances through the material. (See Fig. 27.2b.) This results in a traction force across the crack faces. These tractions have been modeled by Evans et al. [13], who showed that rubber-particle bridging and tearing could lead to improvements in toughening. Unfortunately, rubber particles are very compliant, and the amount of toughening from this mechanism is considered to be small.

Many of these toughening mechanisms are activated in rubber-toughened polymers through rubber-particle bridging and cavitation as well as through matrix plasticity, resulting in the nucleation of multiple crazes and shear bands. (See Chapter 22.) Numerous theories have been postulated on the predominance and ordering of these mechanisms [13, 18, 19].

D. Review of Crack Growth Under Cyclic Loading

The velocity of an advancing fatigue crack subjected to a constant stress amplitude loading is determined from the change in crack length a as a function of the number of loading cycles N. The fatigue-crack-propagation rate per cycle, da/dN, is found from experimentally generated curves, which plot a as a function of N. For constant stress amplitude loading, the rate of crack growth increases as the crack grows longer. Paris, Gomez, and Anderson [20] suggested that the range of the stress intensity factor, $\Delta K = K_{max} - K_{min}$, which itself captures the far-field stress, crack length, and geometry, should be the characteristic driving parameter for fatigue crack propagation for metals.

Many relationships have been proposed to describe the fatigue-crack growth in polymeric solids. However, the most widely used formulation is that proposed by Paris, Gomez, and Anderson [20]. This formulation is known as the Paris Law, and it states that da/dN scales with ΔK through the power-law relationship, such that

$$\frac{da}{dN} = C \cdot (\Delta K)^m, \qquad (27.7)$$

where C and m are material constants. It can be seen from Fig. 27.3 that the Paris equation (27.7) is valid for intermediate ΔK levels, spanning crack-propagation rates

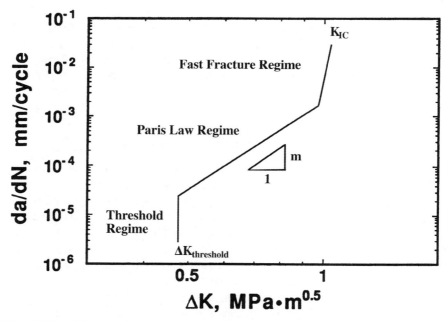

Figure 27.3 Schematic diagram representing regimes of stable crack propagation. The linear relationship between the logarithm of the crack-growth rate and the logarithm of the driving force ΔK is termed the Paris regime.

from 10^{-6} to 10^{-4} mm/cycle. The material constants C and m can be strongly affected by the morphology, test frequency, and load ratio of the fatigue cycle. The load, or stress, ratio R is defined as the ratio of the minimum stress to the maximum stress of the fatigue cycle. Both the stress intensity range and the stress ratio during the loading cycle must be provided in order to specify fully the conditions of a fatigue test.

The Paris Law is a useful tool for fatigue-life prediction. It is implied in this defect-tolerant approach that all structural components are intrinsically flawed. If no initial flaw is detected, the limit of resolution of the nondestructive evaluation technique is used as the initial crack size a_i. Assuming that the fatigue loading is performed under constant stress amplitude conditions, that the geometric factor F does not change within the limits of integration (i.e., that an embedded penny-shaped crack remains an embedded penny-shaped crack), and that fracture occurs when the crack reaches a critical value a_c (dictated by the fracture toughness of the polymer), one can integrate Eq. 27.8 in order to predict the fatigue life of the component:

$$N_f = \frac{2}{(m-2)CF^m(\Delta\sigma)^m\pi^{m/2}}\left[\frac{1}{a_i^{(m-2)/2}} - \frac{1}{a_c^{(m-2)/2}}\right], \qquad m \geq 2. \quad (27.8)$$

Another powerful feature of the Paris Law is that it offers general applicability to metals, ceramics, polymers, and composites [15–17]. This is not to say that the fatigue-crack-propagation micromechanisms in polymers are the same as those observed in metals or ceramics. In fact, there are several factors that play an important role in polymer fatigue, but are of little concern in fatigue characterization of inorganic solids. For example, many polymeric materials are highly sensitive to the frequency and waveform of the fatigue test. These mechanical factors play an important role in the viscous micromechanisms, including creep and chain disentanglement, of the polymer.

E. Issues in Modeling Crack-Growth Behavior

As in the case for monotonic loading, the crack driving force near a fatigue crack tip, ΔK_{tip}, will be lower than the "applied" crack driving force ΔK_a when extrinsic toughening mechanisms are present. The presence of extrinsic toughening mechanisms shields the crack-tip, thereby decreasing the crack driving force and hence the crack-growth rate. The extrinsic crack-tip shielding effect has been expressed by Ritchie [21] as

$$\Delta K_{tip} = \Delta K_a - K_s, \qquad (27.9)$$

where K_s is the stress intensity factor due to shielding. Under monotonic loading conditions, extrinsic toughening mechanisms shield the crack by reducing the applied stress intensity at the crack tip. However, under cyclic loading conditions, the type of extrinsic toughening mechanisms determines the nature of the shielding effect. (See Fig. 27.4.)

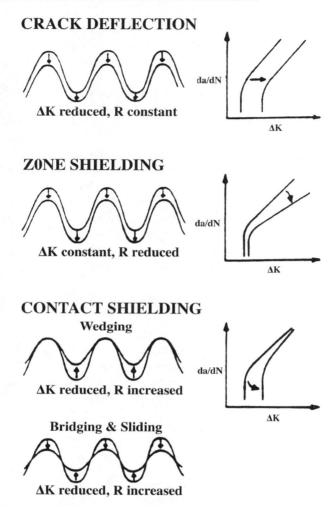

Figure 27.4   Schematic illustration of the shielding mechanisms that may operate under cyclic loading conditions. (Modeled after Ritchie.)

There are three types of shielding mechanisms that can operate in cyclic loading situations: crack deflection, zone shielding, and contact shielding. Shielding due to crack-path deflection results in improvements in FCP behavior over all ranges of ΔK. By contrast, process-zone shielding mechanisms operate more effectively at high ΔK levels, whereas contact shielding mechanisms are more effective at low ΔK levels. Although the model shown in Fig. 27.4 is generic and must be refined for each material system, it is quite useful, in a qualitative sense, for identifying shielding mechanisms in FCP studies.

The amount of shielding due to crack-path deflection has been modeled by Suresh, who derived the effective fatigue-crack driving force and subsequent crack-growth

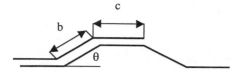

Figure 27.5   Geometry of a deflected crack modeled by Suresh.

rates by analyzing a small segment of crack with periodic tilts. For the crack deflection geometry shown in Fig. 27.5, the model predicts that

$$\Delta K_{\text{tip}} = \frac{b \cos^2(\theta/2) + c}{b + c} \Delta K_a \tag{27.10}$$

$$\frac{da}{dN} = \frac{b \cos(\theta) + c}{b + c} \left(\frac{da}{dN} \right)_n, \tag{27.11}$$

where θ is the deflection angle, b is the deflected distance, and c is the undeflected distance.

It is important to note that unloading can occur due to premature closure and that plastic deformation can accompany the deflected crack. Such complications will increase the amount of shielding above that predicted by Eq. 27.10. Interestingly, crack-path deflection has been observed in particulate-filled polymers [22].

The amount of shielding due to process-zone mechanisms depends on the nature of the plastic deformation, i.e., massive crazing or shear banding [15–17]. The yielding in front of the crack due to far-field tensile loading (monotonic) results in the formation of a plastic zone. For tension fatigue loading, the crack will be continuously loaded and unloaded, and a zone of reversed plastic zone will be generated within the monotonic plastic zone. The size of the cyclic plastic zone, r_c, can be estimated as [15]

$$r_c \approx \frac{1}{\pi} \left(\frac{\Delta K_a}{2\sigma_y} \right)^2, \tag{27.12}$$

where σ_y is the macroscopic yield strength of the material. For an elastic, perfectly plastic material, the residual stress acting in the cyclic plastic zone is equal to the flow that of stress, and the size of the plastic zone is one fourth that of the monotonic plastic zone for zero-tension, zero-loading conditions. Cyclic plastic zones have been observed in several amorphous polymer systems and have been shown to be very important in the inception of cracks under cyclic compression loading [15]. Qualitatively, it is easy to see that the size of the plastic zone increases with ΔK. Therefore, process-zone shielding mechanisms are effective at high ΔK levels.

Contact shielding involves physical contact between mating crack surfaces due to the presence of aspirates, second-phase particles, and/or fibers. Premature contact between the crack surfaces occurs during unloading at a stress intensity level known as K_{cl}, which is often referred to as the closure stress intensity. Therefore, one can calculate the shielding due to closure effects as

$$\Delta K_{tip} = K_{max} - K_{cl}, \tag{27.13}$$

where K_{max} is the maximum stress intensity reached during cyclic loading. As mentioned previously, contact shielding can arise from contact between asperitities, as in the case of rough surfaces, but perhaps the most potent contact shielding mechanism is fiber bridging. Fiber bridging has been shown to be a potent shielding mechanism in short-fiber composites by Lange and his coworkers [23].

In sum, resistance to fatigue crack propagation can be improved by extrinsic shielding mechanisms. The active shielding mechanism depends on the type of additive or modifier used. The crack-tip shielding concept can also be applied to ceramics and metals. The following discussion focuses on phenomena that are specific to polymers.

1. Mean-Stress Effects   The response of a fatigue crack propagating in a polymeric material is highly sensitive to the mean stress or stress ratio of the fatigue cycle. Polymers are unlike metals and ceramics in that there is a well-established dichotomy between different polymers in the fatigue response to an increase in mean-stress ratio, or R-ratio. Within the available research literature on engineering polymers, there are two distinct responses to an increase in mean stress: For a given stress intensity range ΔK, some polymers exhibit an increase in crack-propagation rate, while others show a decrease in crack-growth rate. The mean stress is defined as the average of the applied stresses in the fatigue cycle and can be written in terms of the R-ratio as

$$\sigma_{mean} = \frac{\sigma_{min} + \sigma_{max}}{2} = \frac{\sigma_{max}(R+1)}{2}. \tag{27.14}$$

The published research on the effects of mean stress and the R-ratio covers a broad range of polymer classes, including amorphous, semicrystalline, cross-linked, and rubber-modified polymers. (See Table 27.1.) Increasing crack growth rates associated with an increased stress ratio or mean stress are observed in epoxy resin [24], poly(methyl methacrylate) [25,26], high-density polyethylene [27] and high-density polyethylene copolymers [28], polystyrene [29, 30], poly(vinyl chloride), and nylon [4]. This trend of increasing crack propagation with increased stress ratio or mean stress is similar to the well-documented behavior reported in the metals literature. (See the work of Hertzberg [16], Suresh [15], or Vasudevan [32] for a review.) This response is associated with heightened levels of K_{max} that enhance the likelihood of bond rupture. That is, the crack accelerates more readily for the same stress intensity amplitude.

A number of different explanations and relationships have been proposed in an attempt to rationalize the effect of mean stress on fatigue crack propagation. Arad et al. [32] suggested that the fatigue-crack-growth rates could be scaled to the stress

Table 27.1   The Effect of Increasing the Mean Stress or _R_-Ratio (0.1 to 0.5) on the Fatigue-Crack-Propagation Rate of Several Polymer Systems

Investigator	Polymer System
(a) Crack-propagation rate increases as the mean stress or _R_-ratio is increased at constant ·K	
Bucknall and Dumpleton [27]	High-density-polyethylene
Hertzberg and Manson [5]	Nylon
Murakami et al. [67]	Fiber-reinforced polycarbonate
Mai and Williams [29]	Polystyrene
Pruitt [30]	
Mukherjee and Burns [25]	Poly(methyl methacrylate)
Clark et al. [26]	
Sutton [24]	Epoxy
Zhou and Brown[28]	Polyethylene copolymer
(b) Crack-propagation rate decreases as the mean stress or _R_-ratio is increased at constant ΔK	
Arad, Radon, and Culver [32]	Polycarbonate
Bucknall and Dumpleton [27]	Rubber-toughened poly(methyl methacrylate)
Hertzberg and Manson [5]	Acrylonitrile-butadiene-styrene (ABS)
Hertzberg and Manson [5]	Low-density polyethylene
Manson et al. [35]	Polycarbonate
Mills and Walker [39]	Poly(vinyl chloride)
Moskala [40]	Toughened polycarbonate–copolyester blends
Pruitt [30]	Polycarbonate
Pruitt and Rondinone [37]	
Pruitt [30]	Rubber-modified polystyrene
Takemori [38]	Polycarbonate

intensity factor with the following relationship:

$$\frac{da}{dN} = \beta\lambda^n = \beta\left(K_{max}^2 - K_{min}^2\right)^n. \tag{27.15}$$

In this equation, β depends on the loading environment, frequency, and material properties and n is a material constant. $\lambda = K_{max}^2 - K_{min}^2$ can be rewritten as $2\Delta K \cdot K_{mean}$, making Eq. 27.15 similar in form to the relationship proposed by Mukherjee et al. [25] for the effect of mean stress on fatigue-crack propagation in commercial PMMA:

$$\frac{da}{dN} = C \cdot K_{mean}^{213} \cdot \Delta K^{239}. \tag{27.16}$$

Equations 27.15–27.16 capture an enhanced crack-growth rate for a nominally constant ΔK level due to an increase in the mean stress or R-ratio (from $R = 0.1 - 0.5$). The parameter λ was found by many researchers to be an acceptable normalizing parameter for increased crack-propagation rates in polymeric solids.

A micromechanistic explanation for this response to an increase in stress ratio or mean stress is also possible. Polymers that become more prone to fracture with increasing mean stress are most likely to be affected by the monotonic fracture process associated with the maximum portion of the loading cycle as it approaches a critical stress intensity level. In general, polymers are susceptible to crazing, chain scission, or cross-link rupture. For example, a crazeable polymer, such as polystyrene, will nucleate a single craze ahead of its crack tip when subjected to fatigue loading. (See Kramer et al. [33] and Argon et al. [34] for details on craze initiation mechanisms.) Cyclic loading of the crazeable polymer results in cavitation and damage accumulation, fibrillation of the polymer, breakdown of the primary load-supporting fibrils, and subsequent crack advance. As the maximum stress of the fatigue cycle approaches the strength or critical extension of the craze fibril, the leading fibril or fibrils will fail and subsequently transfer load to the adjacent fibrils. An increased mean stress on this type of polymer raises the stresses on the fibrils and enhances the likelihood of fracture. Increased mean stresses are also detrimental to polymers with restricted chain mobility. For example, Clark et al. [26] have shown that chain scission is promoted in high-molecular-weight PMMA when it is subjected to increased R-ratios. This degradation is linked to the hindered chain mobility associated with greater chain entanglement density. The restriction of chain mobility promotes stress concentrations in polymer chains near the crack tip, and crack propagation occurs through bond rupture or scission of the chains. Bond rupture at the fatigue-crack tip is also favored in chemically cross-linked polymers, such as epoxy resin [24] and cross-linked PMMA [26]. The cross-linked structure hinders chain mobility and the ability to dissipate elastic energy ahead of the crack tip. For these polymer types, an increase in mean stress results in faster crack-propagation rates.

Remarkably, numerous polymer blends offer improved resistance to crack propagation as the mean stress is increased. These polymers include ABS (acrylonitrile-butadiene-styrene) polymers [5], polycarbonate [30, 35–38], extruded poly(vinyl chloride) [39], rubber-modified polystyrene [30], rubber-toughened poly(methyl methacrylate) [27], and low-density polyethylene [5]. For polycarbonate and polycarbonate–copolyester blends, studies [40] indicate that crack-growth rates when plotted against λ decrease by two to three orders of magnitude as R is increased from 0 to 0.6.

The notion of increased resistance to crack propagation as the mean stress is increased is not without foundation. For over 50 years, it has been known that rubbers can undergo strain-induced crystallization and that this structural reorganization can extend the fatigue life of the material [41]. This benefit has been attributed to the strength of the crystallites nucleated at an enhanced level of strain on increasing the mean stress of the fatigue cycle. It has been proposed by Hertzberg and Manson [5] that the mean-stress effects observed in polymeric materials depend on opposing mechanisms. While it is intuitive that the increased mean stress would be detrimental to all materials, certain polymeric materials can undergo strain- or stress-induced micromechanisms that render them more resistant to fatigue-crack propagation. Hertzberg and Manson [5] have found that many of these same polymeric materials demonstrate marked improvement of fracture toughness with increased mean stress. Hertzberg postulated that the strain energy normally available for crack extension is consumed through deformation or structural reorganization ahead of the crack tip. The use of strain energetics to describe fracture processes in polymers, especially rubbers and elastomers, is well founded. Andrews [1] utilized a fracture energy parameter T analogous to the energy release rate proposed by Griffith [42] and Irwin [43] for elastic solids. T is given as

$$T = T_o \left[\frac{C}{C - f(\beta)} \right], \tag{27.17}$$

where T is the total energy used by the solid to create a new unit area of surface through crack advance, T_o is the energy expended for an ideal elastic solid, C is a material constant that depends on strain, and β is the hysteresis ratio. β reflects the energy lost due to inelastic energy expenditure. If β is large, the amount of energy needed to cause fracture increases, and hence it is expected that the crack-growth rate will be reduced as β increases. Andrews demonstrated that β is amplified in elastomeric solids as the strain is increased. Analogously, it is expected that an increased R-ratio would require a greater level of ΔT in order to propagate the crack at the same rate, owing to the increased level of strain sustained in the fatigue cycle.

Hertzberg and Manson [5] proposed that the effect of mean stress on the fatigue-crack-propagation resistance of polymeric materials is directly linked to the parameter β. For an ideally elastic solid, β is zero, and thus an increase in the R-ratio is more likely to produce a critical stress intensity concomitant with fracture. Thus, polymeric materials with a structure that makes them less susceptible to hysteretic losses are more likely to exhibit greater crack-growth rates at a given ΔK as R is increased, as discussed previously. Conversely, polymers capable of structural reorganization or inelastic losses are likely to be more resistant to fatigue-crack propagation as the mean stress is increased. This theory is in good agreement with the published literature summarized in Table 27.1.

Polymers exhibiting a decrease in crack-growth rate on increase of the mean stress have near-tip processes that dissipate elastic energy ahead of the crack tip. Such toughening mechanisms include particle or rubber toughening, orientation hardening, chain slip, and shear banding. Rubber-toughening mechanisms are reviewed in Chapter 22. A critical step in most toughened polymers is the formation of voids in the rubber phase. The cavitated rubber particles act as stress concentrators, increase shear plasticity of the matrix, and aid in the nucleation of multiple crazes in the matrix material. The compliant rubber particles are also capable of stretching to accommodate the extending and of strain hardening under applied stress. As the fibrillated rubber strain hardens, it continuously supports more of the applied stress, until it eventually fails. These mechanisms result in a process zone ahead of the crack tip that diverts energy from the advancing crack tip and accounts for the decreased crack-propagation rates with increasing R-ratios observed in high-impact polystyrenes [30]. Takemori [44] has found that polycarbonate is susceptible to the dual deformation mechanisms of crazing and shear banding. When the two mechanisms are activated simultaneously, they form an epsilon-shaped zone ahead of the crack tip. This zone formation results in an increased time to fibril breakdown in the craze and a concomitant reduction in the crack-propagation rate. Dissipative mechanisms, such as chain slip, can also be beneficial at high mean stresses. Clark et al. [26] found that lower molecular-weight PMMA resins were better able to resist fatigue-crack propagation at higher mean stresses, as compared to their higher molecular-weight and cross-linked counterparts. While molecular weight is generally known to increase fatigue resistance, it also increases chain entanglement density, which at critical levels can restrict chain slip and at high mean stresses can promote scission or rupture.

In summary two distinct classes of crack-propagation response are triggered by an increased mean stress at fixed ΔK. In one class of polymers, the effect of increasing the mean stress or R-ratio is a reduction in the driving force necessary for crack advance at a given propagation rate. This situation is similar to the behavior observed in metals and ceramics: as R is increased, the maximum stress intensity of the fatigue cycle approaches the critical stress intensity or fracture toughness of the material. The local fracture criterion is satisfied within the cyclic damage zone, and the fatigue crack advances more readily with increased mean stress. For the second class of polymers, an increase in mean stress requires an increased level of driving force ΔK to sustain a given crack-propagation rate. In these instances, there is a structural evolution within the cyclic damage zone that is more resistant to fatigue-crack propagation. These responses to increased mean stresses have important consequences for materials subjected to overloads and variable-amplitude fatigue.

2. Variable Amplitude Fatigue   Variable-amplitude fatigue plays an important role in the design of polymeric components subjected to fluctuations in the load cycle or in components that may experience tensile or compressive overloads. Traditionally, variable-amplitude loading is modeled using the concept of cumulative damage. While this concept has strength in crack initiation models, it cannot capture the role of overload type or order in the loading sequence, which is important in predicting the crack-propagation life of the component. It is well known that the application of a single tensile overload can extend the life of a cracked component by retarding the rate of crack advance [15]. The initial overload typically results in a temporary increase in crack advance followed by a prolonged regime of retarded crack velocity. This transient crack-propagation behavior is often controlled by several mechanisms, including crack closure [45], residual compressive stresses upon unloading [46, 47], and crack-tip blunting [48]. The Elber crack closure concept explains the retardation of crack velocity in terms of residual compressive stresses left in the plastically deformed wake of the advancing crack. The plastic wake results in contact between the crack faces while the specimen is still in the tensile portion of the fatigue cycle. The closure load, denoted by $P_{closure}$, effectively reduces the stress intensity range from $\Delta K = K_{max} - K_{min}$ to $\Delta K_{eff} = K_{max} - K_{closure}$, where $K_{closure}$ is the stress intensity corresponding to the closure load. This crack closure mechanism has been proposed by Pitoniak et al. [49] to describe crack retardation in PMMA and by Murakami [50] for polycarbonate. Pitoniak used interferometric methods to determine the closure loads in PMMA experimentally and found that significant tensile loads were needed to separate the crack faces following tensile overloads.

While there is experimental evidence for a small amount of crack closure in polymeric solids, many researchers have found little evidence of crack closure in their systems and have suggested other mechanisms. Blunting has been proposed by Banasiak, Grant, and Montulli [52] to describe the reduced crack velocity following tensile overloads in polycarbonate. Local plastic flow activated during the overload was suggested as the mechanism of crack-tip blunting. Although crack-tip blunting can affect the crack velocity subsequent to the overload it does not provide a basis for the prolonged regime of crack retardation. The zone of residual compressive stresses sustained at the crack tip upon unloading is known to increase in size and magnitude as the far-field tensile load is increased. This has been validated experimentally using X-ray methods [52, 53] and through photoelastic and interferometric techniques [36]. These residual compressive stresses sustained at the crack tip are believed to decrease the crack-propagation rate following the tensile overload. The crack has to grow through this zone of enhanced residual compression before it can return to its initial crack-propagation rate for the ΔK sustained prior to overload. This trend has been observed in polycarbonate and acrylate systems [36]. Many current life-prediction models are formulated on the basis of residual compressive stresses for the rationalization of crack retardation.

Compressive overloads also play a critical role in the life of a component subjected to variable loading conditions. A crack propagating at a constant rate due to the application of a constant level of ΔK will experience an enhancement of crack velocity if subjected to a compressive overload. As in the case of a tensile overload, a few mechanistic explanations are offered. One is that the compressive overload can sharpen a crack that has been blunted in the tensile regime or that surface asperities contributing to closure can be flattened in the compressive cycle. While this explanation offers a physical justification, it does not account for the transient effect of the crack acceleration following the compressive overload. In recent years, it has been shown that the application of fully compressive cyclic loads results in the inception and growth of fatigue cracks ahead of stress concentrations and notches [30, 54]. The source of this crack growth is the generation of a zone of residual tensile stresses upon unloading from far-field compression. It is well known that for an elastic, perfectly plastic material that contains a nonclosing flaw and that is subjected to fully compressive cyclic stresses, a zone of residual tension (of magnitude equal to the flow strength in tension) is created over a distance one fourth the size of the monotonic plastic zone created on application of a monotonic far-field compressive stress [48]. If instead of plastic flow, the material exhibits a different mode of permanent deformation, the development of residual tensile stresses can still result from the creation of a cyclic damage zone. In the case of polymeric solids, permanent deformation ahead of the notch tip can be induced by crazing, shear flow, chain reorientation, or a combination of these processes. (See Fig 27.4c.) Unloading from far-field compression, therefore, results in residual tensile stresses in much the same way as for fully crystalline solids. Pruitt et al. [54] have shown that fully compressive cyclic loads result in crack inception and growth for polyethylene, polystyrene, rubber-toughened polystyrene, polycarbonate, and CR39. Further, Bucknall and Dumpleton [27] have found that the application of a compressive load intermittently during tension fatigue or fatigue tests (i.e., using negative R-ratios) results in enhanced crack propagation in polyethylene and toughened poly(methyl methacrylate). Pruitt and Suresh [47] have shown through experimental stress analysis on model polymer systems that the zone of residual tension increases in magnitude and size with increasing far-field compressive load thus providing an explanation for the enhancement of crack-growth rates following compressive overloads.

3. *The Effect of Notches and Short Cracks*   Notches are important in fatigue because they serve as a stress concentration sites and play an important role in the inception of propagating cracks. For a slender notch of root-radius ρ, the variation of normal stress on the plane of the notch, σ_{yy}, with distance r ahead of the notch tip is given as [55]

$$\sigma_{yy} = \frac{K_I}{\sqrt{2\pi r}} \frac{\rho}{2r} + \frac{K_I}{\sqrt{2\pi r}} + \sigma^\infty, \tag{27.18}$$

where the origin of the coordinate system is located at a distance of $\rho/2$ behind the crack tip. Substituting the for of the stress intensity factor (using the notch length a') into Eq. 27.18 yields an expression for the elastic stress concentration factor K_t due to the presence of the notch:

$$K_t = \frac{\sigma_{yy}}{\sigma^\infty} = \frac{F\rho}{2r}\sqrt{\frac{a'}{2r}} + F\sqrt{\frac{a'}{2r}} + 1. \tag{27.19}$$

The application of a far-field cyclic stress of amplitude $\Delta\sigma^\infty$ results in a maximum stress intensification at the notch tip, given as

$$\Delta\sigma_{max} = K_I(\Delta\sigma^\infty) = \frac{2}{\sqrt{\pi}} \frac{\Delta K_I}{\sqrt{\rho}}, \tag{27.20}$$

where ΔK_I is the stress intensity range at the tip of a fatigue crack subjected to the same far-field range of load and of the same length as the notch. This expression has been found to offer validity under small-scale yielding conditions for small notch-tip radii and for notch lengths significantly greater than the root radius [56]. The forms of these expressions indicate that the crack will be initiated in a number of cycles that scales with the stress intensity range normalized by the square root of the root radius of the notch:

$$N_{init} = \alpha\Delta K/\sqrt{\rho}. \tag{27.21}$$

Moreover, notches play a critical role in the inception of sharp mode I fatigue cracks in many semicrystalline, amorphous, and blended polymeric solids, as discussed previously [47, 54].

Monitoring the crack inception from notches is difficult, and subsequently characterizing the propagation behavior of short cracks is complex. Short cracks often propagate at different rates for identical stress intensity conditions, owing to differences in crack-face contact, notch effects, and micromechanistic processes. This "short-crack problem" leads to a breakdown in similitude and is known to exist in metals [57], ceramics [15], and polymers [38]. It is particularly evident when the microstructure offers bridging mechanisms behind the crack tip, which cannot be realized in the shorter crack. For such cases, crack-propagation rates become significantly greater for shorter cracks. Suresh and Ritchie [57] defined a crack to be microstructurally or mechanically small if the fatigue crack is short in comparison to the characteristic microstructural dimension or if the size of the plastic zone is comparable to the crack length, respectively. Short cracks have important implications for particulate-reinforced polymers and rubber-toughened polymers when the cracks are smaller than the average spacing of the second phase. In such cases, the cracks cannot utilize the bridging mechanism of the particulate or rubber phase across the crack boundary. Short cracks are also important when the fatigue crack is smaller than the craze or shear band zone ahead of the crack [44]. In such cases, the cracks can propagate at faster rates than a longer crack can for equivalent stress intensity ranges. These discrepancies can result in an underestimation of fatigue life if predictions are based on experimental data obtained using long cracks.

II. FATIGUE-CRACK PROPAGATION OF RUBBER-TOUGHENED POLYMERS

Successful application of rubber-toughened polymers in structural components exposed to cyclic stresses requires an understanding of damage accumulation and subcritical crack growth. Surprisingly, there are relatively few papers that address micromechanical deformation mechanisms under cyclic loading as compared to the number of papers that address monotonic loading situations. Interesting, the rare papers on this subject have shown that fracture surfaces generated under stable fatigue-crack propagation (FCP) conditions are significantly different from crack surfaces generated by monotonic loading.

For example, the fracture surfaces for polycarbonate, an amorphous thermoplastic polymer, tested under FCP conditions have been shown to be significantly different than those produced under monotonic loading. Specifically, the fracture surface produced under FCP conditions shows signs of discontinuous growth bands, whereas monotonic loading gives an essentially featureless surface. Examination of the subsurface damage reveals a characteristic epsilon-shaped plastic zone in the case of fatigue-crack propagation. Such evidence clearly illustrates that the micromechanical deformation mechanisms operating under FCP conditions are significantly different from those observed under monotonic loading. Often, ductile thermoplastic polymers that deform by massive shear banding at the crack tip under monotonic loading conditions tend to deform by crazing under FCP conditions.

Hertzberg and Manson [6] pointed out that the effect of an added rubbery phase on the fatigue behavior of a rigid polymer is difficult to predict. The addition of rubber particles usually causes decreases in the modulus, yield strength, and creep resistance. Thermal effects due to either an increase in the loss modulus E'' or simply the increase of stress amplitude required to grow the crack are often more intense. Note that, for different cases, such thermal effects can be either beneficial or deleterious, and it is not surprising to find exceptional diversity of FCP behavior in rubber-modified systems.

Since the purpose of this chapter is not to review the fatigue-crack propagation of every rubber-modified polymer ever studied, but instead to discuss issues generic to all polymers, we will focus on relatively simple rubber-modified polymers, viz., rubber-toughened epoxies. Epoxy polymers are thermosetting materials containing cross-links that inhibit the formation of crazes. Therefore, neat epoxies display neither discontinuous growth bands nor epsilon-shaped plastic zones. We focus on the modification of model epoxy polymers with rubber particles or glass fillers or both.

The purpose of Section II is to (a) review the early correlations between monotonic fracture toughness and FCP behavior, (b) examine crack-tip shielding mechanisms in the near-threshold regime, (c) investigate the effect of rubber particle size on the crack-growth rate, and (d) speculate on the role of blend morphology at low growth rates.

A. Correlating Fracture Toughness and FCP Behavior

Early investigations on the FCP behavior of rubber-toughened blends were often performed at high crack-growth rates ($> 10^{-5}$ mm/cycle). The focus on high crack growth rates was perhaps due to the restricted capabilities of the equipment of that era, the cost associated with slow crack-growth testing, or the need to supply FCP data in a short time. Whatever the reason for the focus on high crack-growth rates, such work did provide an interesting correlation between monotonic fracture toughness and FCP behavior: The addition of rubber not only improved static fracture toughness, but also retarded subcritical crack-growth.

The correlation between static fracture toughness and FCP behavior was perhaps best illustrated by Hertzberg and coworkers [58], who developed the concept of K^*, the stress intensity range at a crack growth rate of 7.5×10^{-4} mm/cycle. By plotting K^* versus K_{Ic}, a direct correlation between fracture toughness and K^* and hence FCP behavior was observed for many rubber-toughened blends. One possible conclusion to be drawn from the correlation between K^* and K_{Ic} is that cheaper static fracture-toughness tests can be used to develop fatigue-resistant materials. However, such a conclusion would be misguided: Azimi [59] and coworkers [60, 61] showed that K^* correlates well with K_{Ic} at 7.5×10^{-4} mm/cycle, but that K^* is not a strong function of K_{Ic} at low growth rates. (See Fig. 27.6.)

B. Slow Crack-Growth Behavior

Many studies have verified a linear relationship between log da/dN and log ΔK, i. e., the so-called Paris-Law behavior. Some polymers exhibit vanishing low values of crack-growth rates at low values of ΔK. Such limiting values of ΔK are often described as threshold values ΔK_{thres}. The crack-growth rates are essentially zero at

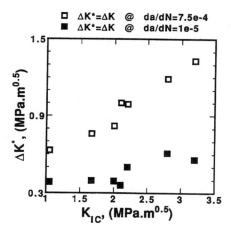

Figure 27.6 K^* is a strong function of K_{Ic} at a crack-growth rate of 7.5×10^{-4} mm/cycle, but is almost independent of K_{Ic} at a crack growth rate of 1.0×10^{-5} mm/cycle. [From Ref. 61.]

ΔK levels lower than ΔK_{thres} for these polymers. However, most polymers do not exhibit a practical ΔK_{thres} regime, and the crack-growth rates simply decrease more slowly as ΔK decreases in this near-threshold regime. Since the majority of the total lifetime of a cracked component is spent at low levels of ΔK, the near-threshold behavior may be the life-controlling phenomenon in many load-bearing situations. Therefore, it is important that we understand the effect of adding rubber on slow crack-growth behavior in rubber-toughened polymers.

The role of crack-tip shielding mechanisms on the crack growth-rate regime has been discussed by Ritchie [21] and has been reviewed in the introduction to this chapter. The addition of rubber particles to a ductile epoxy polymer is known to trigger a process-zone shielding mechanism involving massive shear banding of the epoxy matrix. According to Ritchie's model, the occurrence of a process-zone shielding mechanism should change the slope m in the Paris-Law region of the FCP curve, but should not change the crack-growth behavior at low ΔK levels. Experimental support of Ritchie's model has been given by Azimi et al. [60, 61] and is reviewed next.

The importance of interactions between rubber particles and the plastic zone has been discussed by Azimi et al. [60, 61], who studied the FCP behavior of rubber-modified epoxies. Note that these polymers did not display ΔK_{thres} behavior. (See Fig. 27.7.) The addition of rubber clearly decreased the slope m at high crack-growth rates; that is, the rubber particles toughened the epoxy matrix and, hence, retarded crack-growth. Similar behavior has been reported by Karger-Kocsis and Friedrich [62]. Interestingly, at low values of ΔK, the crack-growth rates for the rubber-toughened epoxies were nearly identical to those of the neat resin. The change in crack-growth behavior was observed at the so-called transition level ΔK_T, which is the value of ΔK at which the crack-growth rate of the rubber-modified epoxies equals that of the neat resin.

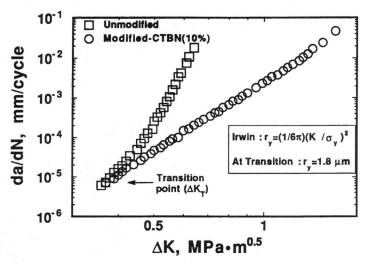

Figure 27.7   At low ΔK levels, the crack-growth rates for the rubber-toughened epoxies were no greater than that of the unmodified matrix. [From Ref. 61.]

The occurrence of a transition in crack growth-rate behavior can be rationalized by considering the interaction between the rubber particles and the crack-tip process zone. At low ΔK levels, the process zone in front of the crack tip is small and the rubber particles are not highly stressed. Thus, the crack grows with minimal plasticity. On the other hand, at high ΔK levels, the process zone is much larger than the rubber particles and the rubber particles are highly stressed. Subsequent rubber-particle cavitation causes significant additional plasticity in the matrix. Therefore, the matrix undergoes significant plasticity as the crack advances. Experimental evidence for this explanation and a schematic representation of this hypothesis are shown in Fig. 27.8.

C. Effect of Rubber Particle Size

It was previously mentioned that rubber-modified epoxies exhibit a transition in FCP behavior at a value of ΔK called ΔK_T. When tested at ΔK values below ΔK_T, the crack-growth rates of rubber-toughened epoxies are nearly identical to those for the neat resin; that is, the rubber particles no longer provide a significant crack-tip shielding mechanism. Azimi et al. have shown that such a transition in crack-growth behavior will occur when the size of the plastic zone at the crack tip is on the order of the size of the particles. Further investigation by Azimi et al. [61] clearly showed that particle size—and not interparticle distance—controlled ΔK_T. Their investigation is discussed next.

Azimi et al. [61] prepared rubber-modified epoxies with different CTBN rubber contents in order to determine whether interparticle distance controls ΔK_T values. Figure 27.9 shows that rubber-modified epoxies with CTBN contents of 1, 5, and

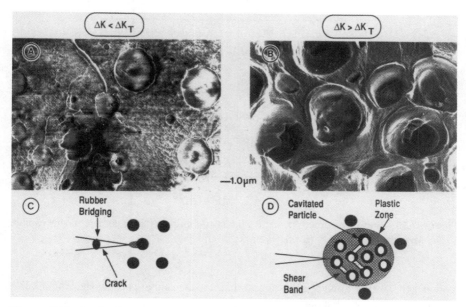

Figure 27.8 Lack of toughening at low values of ΔK has been attributed to a transition between fracture mechanisms: (a) At low ΔK levels, the process zone is smaller than the rubber particles, and the crack advances with little plasticity; (b) at high ΔK levels, the process zone is much larger than the rubber particles, and the particles initiate plasticity in the matrix while the crack advances. [From Ref. 61.]

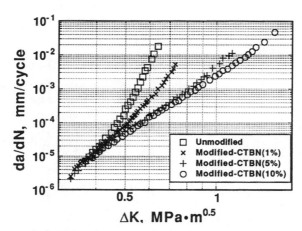

Figure 27.9 Plot of da/dN versus ΔK for several rubber contents, revealing that ΔK_T is independent of interparticle distance. [From Ref. 61.]

10 vol% displayed the same ΔK_T value. Note that for CTBN-modified epoxies, the particle size increased slightly with increasing rubber content, but was nominally 1.5 microns in diameter. The surface-to-surface interparticle distance was found to range

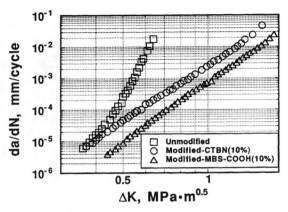

Figure 27.10   Rubber-modified epoxies containing 0.4-micron rubber particles do not exhibit a ΔK_T over the range of ΔK tested. [From Ref. 61.]

from 4.1 to 0.7 μm. Such results suggest that it is the rubber particle size and not the interparticle distance that determines ΔK_T values.

The rationale for this effect is as follows: The rubber particle size has to be smaller than the plastic zone of the neat epoxy resin in order to interact with the stress field surrounding the crack tip. Such interactions produce a process zone in which massive shear banding can occur, thereby shielding the crack tip and slowing down crack growth. To further illustrate their point, Azimi et al. prepared a rubber-modified epoxy using small (0.4 μm in diameter) MBS particles, as the toughening agent. As predicted, MBS-modified epoxy polymer did not display a ΔK_T region, and the crack-growth rates were up to seven times slower than in the neat resin at low values of ΔK. (See Fig. 27.10.)

In sum zone-shielding mechanisms arising from rubber-particle addition may not retard crack-growth at low values of ΔK. The occurrence of a ΔK_T depends on the size of the rubber particles and the size of the plastic zone of the epoxy matrix. To avoid the ΔK_T phenomenon and to reduce the crack-growth rate, submicron rubber particles should be used. However, further improvements can be obtained by controlling the dispersion of the rubber particles, as discussed next.

D. Effect of Blend Morphology

Changes in blend morphology can produce strong effects in the near-threshold region but have no effect on the slope m in the Paris-law regime. For example, Azimi et al. [61] studied the FCP behavior of epoxy modified with MBS and MBS-COOH particles. MBS-COOH particles can be dispersed uniformly in the epoxy matrix, whereas MBS particles tend to cluster on the microscopic scale [63, 64]. The MBS-COOH-modified epoxies exhibited crack-growth rates seven times lower than CTBN-modified epoxies at low ΔK values, owing to the smaller size of the MBS particles. An additional reduction in the crack-growth rate, to a rate five times lower

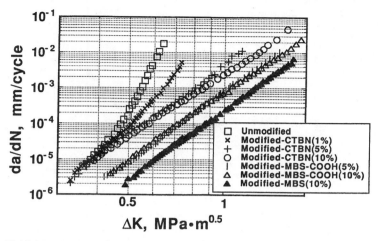

Figure 27.11 Rubber-modified epoxies containing microclustered 0.4-micron rubber particles (the solid triangles) exhibit lower crack-growth rates. [From Ref. 61.]

than that of CTBN-modified epoxies, was found for MBS particles that microclustered. (See Fig. 27.11.) Therefore, both rubber particle size and blend morphology affect the FCP behavior by altering the preexponential value A.

III. FATIGUE-CRACK-PROPAGATION BEHAVIOR OF REINFORCED, RUBBER-TOUGHENED BLENDS

In some cases, rubber-toughened blends can be further improved by the addition of a few percent of rigid fillers. Such materials are often referred to as hybrid particulate composites since they possess two types of toughening particles. The preparation of hybrid particulate composites allows the study of interactions between the toughening mechanisms triggered by each type of toughening particle. Soft rubber particles can trigger massive shear banding at the crack tip, whereas hard inorganic fillers often pin the crack tip or bridge the crack wake. Under some circumstances, the interaction between these various types of toughening can be synergistic. Before discussing the conditions for synergism, it is useful to discuss briefly the shielding mechanisms in filled epoxies.

A. Brief Overview of the FCP Behavior of Filled Polymers

Figure 27.12 describes the FCP behavior of epoxies containing hollow glass spheres (HGS), treated solid glass spheres (TSGS), and solid glass fibers (SGF). From the features of these curves, one could hypothesize that spherical fillers shield the fatigue crack via a zone-shielding or crack-deflection mechanism, but the amount of shielding is rather modest. By contrast, short glass fibers provide a significant amount of contact shielding. The contact shielding is probably due to bridging and sliding of

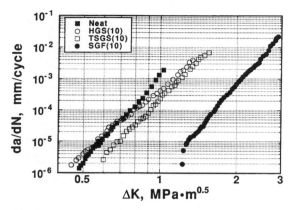

Figure 27.12   FCP behavior of various filled epoxies. [From Ref. 59.]

the glass fibers in the crack wake. The shapes of these curves can be compared to the generic FCP curves described in Fig. 27.4.

The purpose of this section is to explore when and if synergistic interactions occur between (a) solid glass spheres and rubber particles, (b) hollow glass spheres and rubber particles, and (c) short glass fibers and rubber particles when tested under FCP conditions. In essence we are examining the interaction between the zone-shielding mechanisms induced by rubber particles and the contact-shielding and crack-deflection mechanisms induced by rigid fillers.

B. The FCP Behavior of Hybrid Composites

Blending rubber-toughened polymers with a small amount of inorganic filler can improve FCP resistance. Such blends are often referred to as hybrid composites. Hybrid composites can also vary by the type of filler particles used. Hybrid composites containing either hollow glass spheres; solid glass spheres; or short, solid glass fibers are discussed next.

1. *Rubber-Toughened Blends Containing Hollow Glass Spheres*   Mixing various amounts of rubber and hollow glass spheres into a ductile polymer matrix can also lead to blends with improved FCP resistance. An example of these types of glass-filled, rubber-toughened blends has been studied by Azimi et al. [65], who examined the effect of adding hollow glass spheres to a rubber-toughened epoxy. Their work showed that the FCP response is maximized for the 7.5/2.5 CTBN–HGS formulation. (See Fig. 27.13.) The underlying cause of the synergism was believed to be the interaction between the hollow glass filler at the crack tip and the plastic zone triggered by the rubber particles. The presence of hollow glass spheres also caused the plastic zone to branch and stretch. (See Fig. 27.14.)

2. *Rubber-Toughened Blends Containing Solid Glass Spheres*   Mixing various amounts of rubber and solid glass spheres into a ductile polymer matrix can also

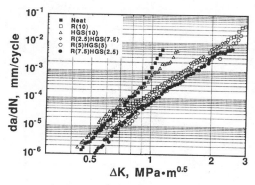

Figure 27.13 FCPs responce of HGS/CTBN hybrid composites. Note that the formulation containing 2.5 vol% HGS and 7.5 vol% CTBN rubber is more fatigue resistant than blends with 10 vol% CTBN or 10 vol% HGS. [From Ref. 65.]

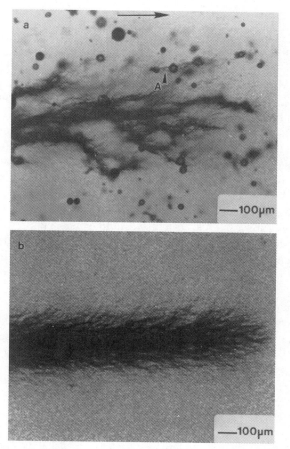

Figure 27.14 Plastic-zone stretching and branching mechanism in an epoxy polymer containing 2.5 vol% HGS and 7.5 vol% CTBN rubber. [From Ref. 65.]

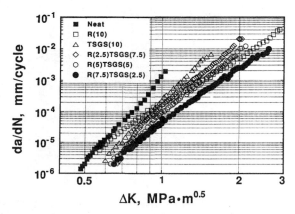

Figure 27.15 Hybrid epoxy composites containing CTBN rubber and treated, solid glass spheres (TSGS) can display synergistic toughening in fatigue. [From Ref. 66.]

lead to blends with improved FCP resistance. Azimi et al. [66] studied the effect of adding solid glass spheres to a rubber-toughened epoxy and found that the FCP response is maximized for the 7.5/2.5 CTBN–SGS formulation. These results are similar to those found in the study of hollow glass sphere. (See Fig. 27.15.)

As in the blends containing hollow glass spheres, the underlying cause of the synergism was believed to occur from the interaction between the rigid filler at the crack tip and the plastic zone triggered by the rubber particles. The presence of glass spheres caused the plastic zone to branch and stretch, i.e., a mechanism very similar to that of the HGS–CTBN-modified epoxies. Such interactions do not occur at higher concentrations of glass spheres, which tend to retard the growth of the plastic zone.

3. Rubber-Toughened Blends Containing Solid Glass Fibers   Adding solid glass fibers (SGF) and rubber particles can also lead to blends with improved FCP resistance [59], but the behavior of such blends is significantly different from that of those blends containing spherical fillers. The FCP response is influenced by a competition of toughening mechanisms rather than a positive interaction between toughening mechanisms, and a lack of synergism is observed. (See Fig. 27.16.) Solid glass fibers are much more effective in bridging the crack wake than spherical fillers. One possible cause for the lack of synergism is the rubber modifier's ability to shield the fiber–matrix interface, thereby retarding the fiber-sliding mechanism. Indeed, SEM examination of fracture surfaces shows less sliding and more fiber breakage when rubber is added. (See Fig. 27.17.)

IV. CONCLUDING REMARKS

The use of fracture mechanics for fatigue design assumes that flawed structural components are capable of sustaining a large degree of stable crack growth prior to fracture. In general, homopolymers possess a limited range of stable fatigue-crack

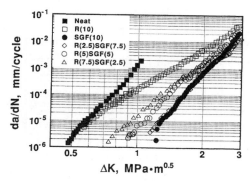

Figure 27.16 The addition of 2.5 vol% short glass fibers significantly improves the fatigue-crack-propagation resistance at low crack-growth rates. [From Ref. 59.]

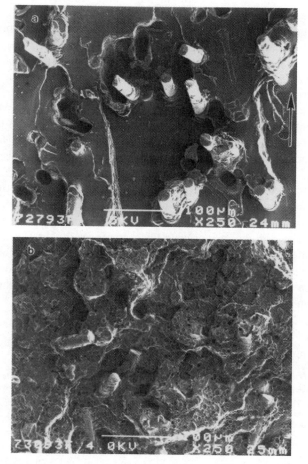

Figure 27.17 SEM of fracture surfaces reveals more fiber breakage in blends containing rubber particles. [From Ref. 59.]

growth. However, the range of stable fatigue-crack growth can be increased by blending rubber particles and/or organic fillers into the polymer matrix. Rubber particles provide an extrinsic crack-tip mechanism involving rubber-particle cavitation and concomitant matrix plasticity. Such a process-zone mechanism slows the rate of crack growth as well as increases the critical stress intensity value for unstable fracture, thereby increasing fatigue life. Rubber particle size and blend morphology have been shown to influence the effectiveness of this extrinsic toughening mechanism. In contrast, inorganic fillers, such as glass fibers, can elicit an extrinsic toughening mechanism involving bridging of the crack wake. Such crack-wake mechanisms increase the critical stress intensity factor and increase the fatigue-crack-propagation threshold. Again, the net result is improved fatigue performance. In some cases, blending both rubber particles and inorganic fillers can produce synergistic toughening. However, the interactions between extrinsic toughening mechanisms, such as plastic-zone yielding and fiber bridging, are poorly understood and merit further investigation.

The crack-growth rate can be affected by a variety of external parameters, including a change in mean stress, a variation in stress amplitude, and the presence of a notch. Increasing the mean stress often causes an increase in the crack-growth rate, but some polymers exhibit a reduction in the crack-growth rate instead. The application of a stress overload can lead to an improved FCP resistance that has been attributed to crack closure resulting from asperities formed in the overload cycle. The presence of a nearby notch can result in an overlap of stress fields that results in higher growth rates. Although micromechanisms have been proposed to explain the effects of external variables on crack-growth rates, these mechanisms are not fully understood.

In summary fracture mechanics provides a design approach for predicting the life of structural components under cyclic loading conditions. The threshold, Paris-law, and fast-fracture regimes can be improved by the use of inorganic fibers and/or rubber particles. The addition of inorganic fibers improves the threshold performance by a fiber-bridging mechanism. The addition of rubber particles slows down the crack-growth rates in the Paris-law regime by a plastic-zone mechanism. Both micromechanisms increase the critical stress intensity factor for fracture, thus improving performance in the fast-fracture regime.

V. REFERENCES

1. E. H. Andrews, "Fatigue in Polymers," in *Testing of Polymers*, W. Brown (ed.), Wiley, New York, 1969, p. 237.

2. P. Beardmore and S. Rabinowitz, *Treat. Mat. Sci. and Tech.*, **6**, 267 (1975).

3. J. A. Sauer and G. C. Richardson, *Int. J. Fract.*, **16**, 499 (1980).

4. R. W. Hertzberg, M. D. Skibo, and J. A. Manson, "Fatigue Fracture Micromechanisms in Engineering Plastics," in *Fatigue Mechanisms*, ASTM STP 675, Philadelphia, 1979, p. 471.

5. R. W. Hertzberg and J.A. Manson, *Fatigue of Engineering Plastics*, Academic Press, New York, 1980.

6. R. W. Hertzberg and J. A. Manson, in *Encyclopedia of Polymer Science and Engineering*, Wiley, New York, 1986, p. 378.

7. H. H. Kausch and J. G. Williams, "Fracture," in *Encyclopedia of Polymer Science and Engineering*, Wiley, New York, 1986, p. 341.

8. J. A. Sauer and M. Hara, "Effects of Molecular Weights on Crazing and Fatigue of Polymers," in *Advances in Polymer Science 91/92*, H.H. Kausch (ed.), Springer-Verlag, Berlin, 1990, p. 71.

9. M. L.Williams, *J. Appl. Mech.*, **24**, 109 (1957).

10. G. C. Sih, *Handbook of Stress Intensity Factors*, Lehigh University Press, Bethlehem, 1973.

11. R. O. Ritchie, "Extrinsic Toughening Mechanism in Solid," in *Mechanical Behaviour of Materials: Proceedings of the 5th International Conference*, M. G. Yan, S. H. Zhang, and Z. M. Zheng (eds.), Pergamon, Oxford, 1988, p. 5.

12. B. Lawn, *Fracture of Brittle Solids*, 2nd Ed., Cambridge Solid State Sciences Series, Cambridge University Press, Cambridge, 1993.

13. A. G. Evans, Z. B. Ahmad, D. G. Gilbert, and P. W. R. Beaumont, *Acta Metall.*, **34**, 79 (1986).

14. C. B. Bucknall and W. W. Stevens, "Thoughened Plastics," in *Toughening of Plastics*, Plastics and Rubber Institute, London, 1978, p. 24.

15. S. Suresh, *Fatigue of Materials*, 2nd Ed., Cambridge University Press, Cambridge, 1998.

16. R. W. Hertzberg, *Deformation and Fracture Mechanics of Engineering Materials*, 4th Ed., Wiley, New York, 1996.

17. T. L. Anderson, *Fracture Mechanics: Fundamentals and Applications*, CRC Press, Ann Arbor, MI, 1995.

18. A. S. Argon, "Sources of Tougheness in Polymers," *Proceedings of 7th International Conference on Advances in Fracture Research*, K. Samala, K. Ravi-Chander, D. M. R. Taplin, and P. Rama Rao (eds.), Pergamon Press, New York, 1989.

19. A. J. Kinloch and F. J. Guild, "A Rational Analytical Theory of Fatigue," in *Toughened Plastics II: Novel Approaches in Science and Engineering*, Advances in Chemistry Series 252, American Chemical Society, Washington, DC, 1996, p. 1.

20. P. C. Paris, M. P. Gomez, and W. P. Anderson, *Trend in Engin.*, **13**, 9 (1961).

21. R. O. Ritchie, Weikay Yu, and R. J. Bucci, *Eng. Fract. Mech.*, **32**, 361 (1989).

22. H. R. Azimi, R. A. Pearson, and R. W. Hertzberg, *J. Mater. Sci.* (1996).

23. R. W. Lang, J. A. Manson, R. W. Hertzberg, and R. Schirrer, *Polym. Eng. Sci.*, **24**, 833 (1984).

24. S. A. Sutton, *Eng. Fract. Mech.*, **6**, 587 (1974).

25. B. Mukherjee and D. J. Burns, *Mech. Eng. Sci.*, **11**, 433 (1971).

26. T. R. Clark, R. W. Hertzberg, and N. Nohammadi, "Fatigue Mechanisms in Polymers at Throuthold Effects of MW and Mealy Sheet," in *8th International Conference on Deformation, Yield and Fracture of Polymers*, Plastics and Rubber Institute, London, 1991, p. 31/1.

27. C. B. Bucknall and P. Dumpleton, *Poly. Eng. Sci.*, **27**, 110 (1987).

28. Y.-Q. Zhou, and N. Brown, *J. Polym. Sci.: B: Polym. Phys.*, **30**, 477 (1992).

29. Y. W. Mai and J. G. Williams, *J. Mat. Sci.*, **14**, 1933 (1979).

30. L. Pruitt, "Cyclic Damage Aheat of Fatigue Cracks in Amorphous Solids: Theory, Experiments and Implication," Ph.D. Thesis, Providence, RI, Brown University, 1993.

31. A. K. Vasudevan and K. Sadananda, *Met. Mat.Trans. A*, **28**, 1221 (1995).

32. S. Arad, J. C. Radon, and L. E. Culver, *J. Mech. Eng. Sci.*, **13**, 75 (1971).

33. E. J. Kramer and L. L. Berger, "Fundamental Processes of Crack Grouth and Fracture," in *Advances in Polymer Science 91/92*, H. H. Kausch (ed.), Springer-Verlag, Berlin, 1990, p. 1.

34. A. S. Argon and R. E. Cohen, in *Advances in Polymer Science 91/92*, H. H. Kausch (ed.), Springer-Verlag, Berlin, 1990, p. 300.

35. J. A. Manson, R. W. Hertzberg, and P. E. Bretz, "Normalization of Fatigue Cracks Propagations Behavior in Polymers," in *Advances in Fracture Research*, D. Francois (ed.), Pergamon Press, Oxford, 1981, p. 443.

36. L. Pruitt and S. Suresh, *Polymer*, **35**, 3221 (1994).

37. L. Pruitt and D. Rondinone, *Polym. Eng. Sci.*, **36**, 1300 (1996).

38. M. T. Takemori, *Polym. Eng. Sci.*, **22**, 937 (1982).

39. N. J. Mills and N. Walker, *Polymer*, **17**, 335, (1976).

40. E. J. Moskala, "Fatigue Crack Propagation in Toughened Polycarbomide Copolymer Blends," in *8th International Conference on Deformation, Yield and Fracture of Polymers*, Plastics and Rubber Institute, London, 1991, p. 51/1.

41. S. M. Cadwell, R. A. Merrill, C. M. Sloman, and F. L. Yost, *Ind. Eng. Chem. Anal. Ed.*, **12**, 19 (1940).

42. A. A. Griffith, "The Phenomenon of Rupture and Flow in Solids," in *Philosophical Transactions of the Royal Society*, **A221**, London, 1921, p. 163.

43. G. R. Irwin, "Plastic Zone Near a Crack and Fracture Tougheness," in *Proceedings of the 7th Sagamore Ordinance Materials Conference* IV, Syracuse University, New York, 1960, p. 63.

44. M. Takemori, "Copetition Between Cracking and Shear Flow During Fatigue," in *Advances in Polymer Science 91/92*, H. H. Kausch (ed.), Springer-Verlag, Berlin, 1990, p. 263.

45. W. Elber, *Eng. Fract. Mech.*, **2**, 37 (1970).

46. S. Suresh, *Eng. Fract. Mech.*, **18**, 577 (1983).

47. L. Pruitt and S. Suresh, *Phil. Mag. A*, **67**, 1219 (1993).

48. J. R. Rice, "Mechanics of Crack Tip Determination and Extension by Fatigue," in *Fatigue Crack Propagation*, ASTM STP 415, Philadelphia, American Society for Testing and Materials, 1967, p. 247.

49. F. J. Pitoniak, A. F. Grandt, L. T. Montulli, and P. F. Packman, *Eng. Fract. Mech.*, **6**, 663 (1974).

50. R. Murakami, S. Noguchi, K. Akizono, and W. G. Ferguson, *J. Fract. Eng. Mater. Struct.*, **6**, 461 (1987).

51. D. H. Banasiak, A. F. Grandt, and L. T. Montulli, *J. App. Polym. Sci.*, **21**, 1297 (1977).

52. J. E. Allison, "Measurement of Crack Tip Stress Distracturers by X-ray Diffraction," in *Fracture Mechanics*, ASTM STP 677, Philadelphia, American Society for Testing and Materials, 1979, p. 550.

53. S. Taira and K. Tanaka, "Local Residual Stress Near Fatigue Crack Tip," *Trans. of the Iron and Steel Inst. of Japan*, **19**, 411 (1979).

54. L. Pruitt, R. Herman, and S. Suresh, *J. Mat. Sci.*, **27**, 1608 (1992).

55. H. Tada, P. C. Paris, and G. R. Irwin, *Stress Analysis of Cracks Handbook*, Del Research Corporation, Hellertown, PA, 1973.

56. J. M. Barson and S. T. Rolfe, *Fracture and Fatigue Control in Structures*, 2nd Ed., Prentice Hall, Englewood Cliffs, NJ, 1987.

57. S. Suresh and R. O. Ritchie, *Int. Met. Rev.*, **29**, 445 (1984).

58. J. F. Hwang, J. A. Manson and R. W. Hertzberg, *Polym. Sci. Eng.*, **29**, 1477 (1989).

59. H. R. Azimi, "Toughend Epoxy Polymers Fatigue Crash Properties Mechanism," Ph.D. Dissertation, Lehigh University, Bethlehem, PA, 1994.

60. H. R. Azimi, R. A. Pearson, and R. W. Hertzberg, *J. Mater. Sci. Lett.*, **13**, 1460 (1994).

61. H. R. Azimi, R. A. Pearson, and R. W. Hertzberg, *J. Mater. Sci.*, **31**, 3777 (1996).

62. J. Karger-Kocsis and K. Friedrich, *Colloid. Polym. Sci.*, **270**, 549 (1992).

63. R. Bagheri and R. A. Pearson, *J. Mater. Sci.*, **31**, 3945 (1996).

64. J. Y. Qian, R. A. Pearson, V. L. Dimonie, O. L. Shaffer, and M. S. El-Aasser, *Polymer*, **38**, 21 (1997).

65. H. R. Azimi, R. A. Pearson, and R. W. Hertzberg, *Polym. Eng. & Sci.*, **36**, 2352 (1996).

66. H. R. Azimi, R. A. Pearson, and R. W. Hertzberg, *J. Appl. Polym. Sci.*, **58**, 449 (1995).

67. R. Murakami, S. Noguchi, K. Akizono, and W. G. Ferguson, *J. Fract. Eng. Mat. Struct.*, **6**, 461 (1987).

28   Transmission and Reflection of Light in Multiphase Media

ROBERTO ALEXANDER-KATZ

Departamento de Física
Universidad Autónoma Metropolitana-Iztapalapa
México D.F. México

I. INTRODUCTION

Most of the technologically interesting polymers are commonly multiphase inhomogeneous materials. Optical inhomogeneity refers to the difference in refractive index between the different domains within a material. The limit between an inhomogeneous and a homogeneous material depends on the length scale of the probe used. As far as optical properties are concerned, the probe used is light and therefore, the relevant length scale is the wavelength of the light employed. A highly dispersed multiphase material can be homogeneous from a mechanical point of view, but opaque or turbid to the eye, manifesting optical heterogeneities on a scale of the wavelength of the incident light. Surface roughness can also be a source of turbidity, as long as this roughness is significant on a wavelength scale. Therefore, in a transparent material (i.e., optically homogeneous) the refractive indexes of the domains are matched or

Polymer Blends, Volume 2: Performance. Edited by D. R. Paul and C. B. Bucknall.
ISBN 0-471-35280-2. © 2000 John Wiley & Sons, Inc.

else the size of the domains and of the surface roughness is very small in comparison to the wavelength used.

Besides chemical composition, another element that introduces heterogeneities in the material is the processing itself; the strain is not homogeneous within a manufactured item, leading to zones with different mechanical and optical properties. In the case of multiphase materials, local stresses give rise to different morphologies within the processed object itself, and in particular, surface roughness is very sensitive to processing conditions. This relationship between processing conditions and surface roughness has been studied extensively, for example, in processes such as tubular film extrusion [1–3], wherein the limiting factor in obtaining an optically clear film is related primarily to surface roughness. The elongational flow that is present in almost all processing operations is particularly intense, for example, close to the surface of an extrudate or at the advancing front in an injection-molding process [4]. This intense flow field induces the migration of filler particles or other phases to the surface, giving rise to surface roughness [5, 6]. In the case of injection-molded parts, the presence of fillers or other phases can couple with process variables, like the mold-filling rate, and have a strong effect on the gloss of the part [7].

We shall limit our study to the distortions introduced by the heterogeneous nature of the material in the image formed by light passing through or reflected from the specimen. There are two important aspects involved in appearance, one related to the colors transmitted or reflected and the other associated with the resolution and contrast of the images seen through the specimen. The first of these aspects refers to the wavelength distribution of the light transmitted or reflected by the object and is connected to the absorbing and scattering properties of the medium. The second aspect has to do with the directional distribution of light produced by the specimen and is mainly related to its scattering properties, although the presence of an absorbing medium can have an effect on the angular distribution of the scattered light. In this chapter, we shall be concerned only with the second aspect and consider the absorption of light only when it is relevant to the discussion of transparency, gloss, and translucency. The reader interested in science and color technology is referred to books dedicated to this subject [8, 9].

This chapter stresses the basic concepts of light scattering by random media and their relation to the basic attributes that are commonly measured in the assessment of transparency, gloss, and translucency. However, we do not discuss the methods of measuring these attributes, as these methods have been thoroughly reviewed by Willmouth [10]. We refer only to the quantities that are being measured. Special attention is given to the relationship between microstructure and optical properties, since this topic becomes crucial in controlling the appearance of industrial products.

In the second section, we introduce some basic concepts of the interaction of light with a random medium and define the fundamental lengths associated with scattering phenomena by a turbid medium or rough surfaces. The third section analyzes in more detail the relationship between microstructure and the attributes measured in the assessment of transparency, gloss, and translucency.

II. BASIC CONCEPTS

When we see an object through a polymeric specimen, there are several processes that occur as light goes through the specimen. If the specimen has perfectly smooth incoming and exit surfaces and the bulk is optically homogeneous, with no fluctuations in the refractive index, then the laws governing the reflection and transmission of light will be given by Fresnel equations [11]. Any ray of light that crosses such a medium suffers phase changes, which, in the absence of diffraction effects, are determined by the laws of reflection or refraction, and the image formed will depend on the refractive index of the medium and its geometric form. To simplify matters, we shall adopt throughout this chapter a simple geometry for our polymer specimen, that of an infinite slab with parallel entrance and exit faces. We shall also assume that we illuminate a large area, so that diffraction at the boundaries can be neglected.

Two scattering sources are involved in the emergence of diffuse light when we illuminate a translucent plastic. One originates in the bulk, wherein a multiple scattering process takes place as the beam traverses a random optically inhomogeneous medium, and the other is caused by scattering at the surface due to surface roughness. The incident beam will be partly scattered in all directions, and only a fraction of light will be transmitted and reflected in the direction given by the laws of reflection and refraction. We shall call this fraction of light the transmitted or reflected coherent component, due to its predictable and constant phase relative to the incident beam. The scattered field produced by surface roughness or by many scattering events in the bulk is usually called the diffuse or incoherent component, owing to its wide angular spread and the lack of phase relationship with the incident field.

Let us consider a collimated incident beam perpendicular to the entrance face. In this case, the undeviated transmitted, or specularly reflected, light corresponds to the coherent component. The incident coherent field is partially converted by scattering into a diffuse field or else is absorbed and transformed into heat.

To understand how the microstructure of a multiphase polymer specimen influences its appearance, it is important to develop some concepts that relate the transmittance, or reflectance, with the internal structure of the material. For that purpose, we shall divide the discussion in two parts. First, we will take into account only the inhomogeneneous nature of the bulk, assuming that the surfaces are smooth, and secondly, we will specifically address the appearance problems associated with scattering from rough surfaces.

In order to visualize the interaction of light with an optically inhomogeneous random medium, one could assume that photons diffuse following random-walk paths. As we shall see, this similarity has been well established from a theoretical as well as an experimental basis, as long as the number of scattering events is large [12–23]. The mean free path Λ of a photon is, in analogy with the kinetic theory of gases, equal to $1/(NC_{\text{ext}})$, where N is the number of heterogeneities (e.g., particles) per unit volume and C_{ext} is the extinction cross section of the inhomogeneity (e.g., domain, particle, etc.), which is a measure of the total loss of the incident light beam caused by scattering and absorption by this single inhomogeneity. C_{ext} can be written in terms of its scattering and absorption contributions as

$$C_{ext} = C_{sca} + C_{abs}, \tag{28.1}$$

where C_{sca} and C_{abs} are the scattering and absorption cross sections, respectively.

The probability of a photon being scattered (or absorbed) after traveling a thickness $d\ell$ is $d\ell/\Lambda$. On the other hand, let us assume that a collimated beam of intensity I transverses a thickness $d\ell$ and that the loss in the transmitted intensity is dI; therefore, the probability of a photon of being scattered (or absorbed) is dI/I. Consequently,

$$\frac{dI}{I} = -\frac{d\ell}{\Lambda}. \tag{28.2}$$

Hence, after a depth ℓ, the final undeviated beam intensity is given by

$$\frac{I}{I_o} = \exp\left(-\frac{\ell}{\Lambda}\right), \tag{28.3}$$

where I_o is the incident intensity. After substituting the value for $\Lambda = (NC_{ext})^{-1}$ in Eq. 28.3 we arrive at the well-known Beer–Lambert law for coherent transmittance. This means that the probability that a photon is not deviated after traveling a distance ℓ through the material is $\exp(-\ell/\Lambda)$. Λ gives us the distance over which we lose $1/e$ of the incident coherent component. This relationship implies that a plastic should be much thinner than Λ to form high-resolution images. Therefore, all scattering phenomena that limit perfect transparency, such as haziness and lack of clarity, involve mainly single scattering events. Haziness corresponds to a loss in contrast while clarity measures the ability of the specimen to transmit fine details [24]. Λ depends on the size of the inhomogeneities as well as on their index of refraction and their volume fraction.

Figure 28.1 shows Λ, calculated by means of Mie's theory [25–27], as a function of particle size for spherical inclusions for three indexes of refraction. The first index corresponds to rubber particles, the second to air bubbles, and the third to TiO_2, which is a strong scatterer. All of these particles are in a polystyrene (PS) matrix at a volume fraction Φ of 10%. To correct for concentration effects, we follow Gate's [22] procedure and assume that the index of refraction of the medium is that of the composite medium, PS plus particles, and use a volume-fraction mixing rule. For TiO_2 particles, Λ drops sharply with particle size down to its minimum value at a particle diameter D of approximately 0.6λ. For larger particle sizes, Λ tends to exhibit an asymptotic behavior close to $\Lambda \approx 3D$. As the indexes of refraction of the dispersed phase are brought closer to that of the matrix, the decay of Λ with particle size becomes smoother. For instance, for air bubbles in PS, the minimum Λ is at $D \approx \lambda$ and is roughly four times larger than the corresponding minimum value of Λ for TiO_2. For PB particles in PS, the minimum value of Λ corresponds to a particle diameter of about 3.5 μm, which is already in the asymptotic zone for

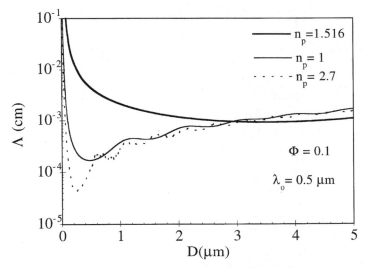

Figure 28.1   Calculated mean free path for spherical particles of different indexes of refraction as a function of particle size at a volume fraction $\Phi = 0.1$.

TiO_2, where $\Lambda \approx 3D$. At this concentration, for a small PB particle, say of around 0.2 μm, the mean free path is on the order of 145 μm, and a film with a thickness less than this value should be reasonably transparent since a large fraction of the incident beam remains undeviated. On the other hand, for TiO_2 of the same particle size, Λ is only 0.52 μm; therefore, for a film thickness of 145 μm, there are on average about 280 scattering events and the light that finally emerges is diffuse. Actually, the transmitted light loses memory of the incident angular dependence after roughly 20 successive scattering events [12–14].

Unless the phases are index matched, in many multiphase materials the number of scattering events is large for the common thickness of plastic goods. If this is the case, as mentioned previously, then one can visualize the transmission of light through a multiple-scattering medium as a diffusion process.

If the particles or domains are on the order of or larger than the wavelength of light, the scattering will be anisotropic, as illustrated in Fig. 28.2, where we show the scattering pattern of a PB particle in a PS medium. As Fig. 28.2 shows for radii of 0.25 μm and 0.5 μm and a wavelength (in vacuum) $\lambda_o = 0.5$ μm, the ratios between the forward- and back-scattering intensities are 1,122 and 20,592, respectively.

Scattering anisotropy introduces a correlation between the pre- and postscattering directions. The extension of this correlation is the so-called transport mean free path Λ_{eff}, which is equivalent to the persistence length in polymer conformation theory. That is, as a photon travels through a multiple-scattering medium, it follows a random path with steps of an average length Λ, as shown schematically in Fig. 28.3. Λ_{eff} is the average of the sum of projections in the initial direction. If the scattering by the heterogeneities has a cylindrical symmetry, then the ensemble of the photon's paths

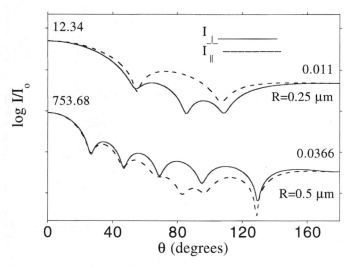

Figure 28.2   Scattering patterns according to Mie's theory of PB ($n = 1.52$) for particles of radius 0.25 µm and 0.5 µm in a matrix of PS ($n = 1.59$) for an incident wavelength (in vacuum) of $\lambda_o = 0.5$ µm. $I_\perp$ and $I_\parallel$ are the scattering intensities for an incident light polarized perpendicular and parallel to the plane of incidence, respectively.

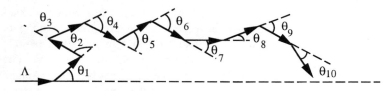

Figure 28.3   Schematic diagram of a photon path in a multiple-scattering medium.

will be equivalent to the conformations followed by a freely rotating chain. In this case, for a large number of steps, Λ_{eff} can be approximated by its asymptotic form as

$$\Lambda_{\text{eff}} = \Lambda\left(1 + \overline{\cos\theta} + \overline{\cos\theta}^2 + \cdots\right) = \frac{\Lambda}{1 - \overline{\cos\theta}}, \qquad (28.4)$$

where $\Lambda(\overline{\cos\theta})^j$ is the average projection of the jth step in the initial direction.

After a distance greater than Λ_{eff}, pre- and postscattering directions are uncorrelated. Therefore, in the case of anisotropic scattering, the paths followed by photons have the statistical properties of a random walk wherein the elementary step has been renormalized to Λ_{eff}.

For small particle sizes—that is, when $R \ll \lambda$, where R is the radius of the particle—$C_{\text{ext}} \approx R^6$, while $N \approx R^{-3}$, and therefore, $NC_{\text{ext}} \approx R^3$. However, for small particle radii, the scattering is isotropic, and therefore $(1 - \overline{\cos\theta})$ is al-

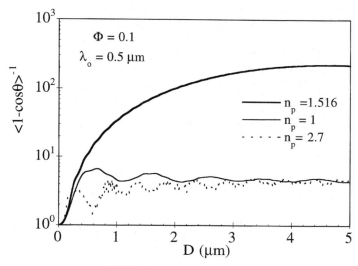

Figure 28.4   Calculated $(1 - \overline{\cos\theta})^{-1}$ versus R by Mie's theory, for spheres with indexes of refraction n_p of 1.516, 1, and 2.7 at a volume fraction of $\Phi = 0.1$ and a wavelength (in vacuum) $\lambda_o = 0.5$ μm.

most unity. This means that the initial slope of log Λ_{eff} vs. log R must be approximately -3. For larger particle sizes $(1 - \overline{\cos\theta})^{-1}$ starts growing, and when $R \gg \lambda$, it oscillates around a constant plateau, as shown in Fig. 28.4. Also, C_{ext} oscillates around $2\pi R^2$ when $R \gg \lambda$. Therefore, in this limit, Λ_{eff} should increase linearly with R. This asymptotic behavior is shown in Fig. 28.5 in which we plot Λ_{eff} versus particle size for the spherical scatterers considered in Fig. 28.1 at the same volume fraction $\Phi = 0.1$. In the interval around $R \approx \lambda$, there is a competition between scattering efficiency and anisotropy of scattering.

For $R \ll \lambda$, Λ_{eff} is large, because the scattering cross sections are very small, which implies that a large part of the incident light remains undeviated and the material will be almost transparent. By contrast, if $R \gg \lambda$ the scattering cross section is quite large, but most of the scattering lies in the forward direction and Λ_{eff} can also become large. Thus in this last case, the material will transmit a substantial part of the incident beam, although the resolution of the images will be limited. Finally, when $R \approx \lambda/2$, the scattering cross section is reasonably large, while the scattering anisotropy is moderate and Λ_{eff} is minimized, leading to a marked decrease in the transmittance. This means that as far as scattering in the bulk is concerned, there are two important parameters: Λ and Λ_{eff}. The first defines the fraction of the incident beam that is transmitted coherently and therefore rules transparency. The second determines how diffusive the material is. All other parameters, such as size and morphology of the heterogeneities, indexes of refraction, and so on, influence the contribution of scattering to transmittance via Λ and Λ_{eff}. When the medium absorbs or there are absorbing pigment particles, the absorbance plays an important role as well. As we shall see, popular theories that have been applied to the study of transport

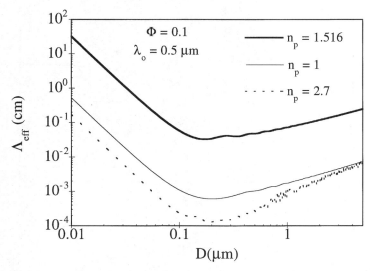

Figure 28.5   Calculated Λ_{eff} as a function of R for the same parameters as in Fig. 28.4.

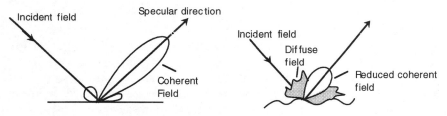

Figure 28.6   Schematic representation of the scattered coherent and diffuse fields.

of diffuse light in an optically heterogeneous material, such as the Kubelka–Munk two-flux theory [28, 29], involve empirical parameters that can be written in terms of Λ and Λ_{eff} and the absorbing properties of the medium.

Up to this point, we have discussed only the bulk scattering effects on transmittance. However, as we mentioned previously, transmittance losses can also be ascribed to surface roughness. In many multiphase plastics when the specimen is very thin, such as polyethylene blown films, the surface contribution to scattering is dominant over that from the bulk. On the other hand, the surface roughness of opaque polymer materials defines attributes such as gloss. If the surface is smooth, then the light will be scattered preferentially in the specular direction, as shown schematically in Fig. 28.6a. As the illuminated area becomes larger, diffraction effects can be neglected, and light will be scattered only in the specular direction. The specularly reflected field is called the coherent field. When a surface is rough, a fraction of the incident beam will be scattered in all directions, reducing the specular component, as shown schematically in Fig. 28.6b. Finally, for a very rough surface, all of the collimated incident beam becomes diffuse after a reflection.

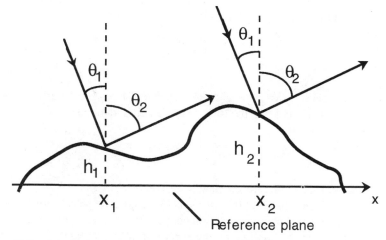

Figure 28.7   Schematic representation of the difference in optical paths between two rays.

The diffuse component reflected (or transmitted) from a rough surface is produced by interference due to changes in phase as the light is reflected from the surface. To illustrate this concept, let us consider two incident parallel rays at an angle θ_1 relative to the average normal to the surface, as shown in Fig. 28.7. Their phase difference is [30, 31]:

$$\Delta\phi = k\big[(h_1 - h_2)(\cos\theta_1 + \cos\theta_2) + (x_1 - x_2)(\sin\theta_1 - \sin\theta_2)\big], \qquad (28.5)$$

where $k = 2\pi/\lambda$ is the magnitude of the wave vector; the h's and x's are the heights and lateral positions, respectively, of the protuberances, as shown in Fig. 28.7, and θ_2 is the scattering angle. In the specular direction ($\theta_1 = \theta_2$), the phase difference depends only on the difference in heights; that is,

$$\Delta\phi = 4\pi\cos\theta_1\,\Delta h/\lambda. \qquad (28.6)$$

Let us now visualize the scattered field as the resultant of the interference between the different secondary wavelets emanating from the scattering surface. For a perfectly smooth surface, the phase difference is due exclusively to the difference in position across the surface ($x_1 \neq x_2$). However, in the specular direction, $\Delta\phi = 0$, and all of these wavelets interfere constructively, giving rise to a strong specularly scattered field. Assuming that the area illuminated has an infinite extension, there is destructive interference for any other direction, since $x_1 - x_2 \gg \lambda/2$. When the surface is rough, the difference in phase in the specular direction is given by Eq. 28.6, and there is some destructive interference along this direction, reducing the amplitude of the scattered coherent field. As we shall see for a Gaussian height distribution, the ratio between the coherently reflected intensity and that corresponding to a smooth surface of the same material is given by [32]

$$\frac{R_{\text{coh}}}{R_F} = \left(\overline{e^{-(\Delta\phi)}} \right)^2 = e^{-g_s}, \tag{28.7}$$

where $g_s = (4\pi\sigma \cos\theta_1/\lambda)^2$; R_F is the reflectance of the smooth surface given by Fresnel equations [11]; and σ is the RMS average height defined as

$$\sigma = \sqrt{\overline{h^2}}, \tag{28.8}$$

where we have chosen the reference plane such that $\bar{h} = 0$.

For scattering angles different from the specular angle, the phase varies randomly from 0 to 2π, and there is no phase relation between the incident and scattered wave. The phase changes are determined by the surface profile, and light scattered in the nonspecular direction depends on both the height distribution and the lateral correlation between heights, the latter represented by the autocorrelation function of heights

$$C(r) = \frac{\overline{h(r)h(0)}}{\sigma^2}, \tag{28.9}$$

which is the normalized autocovariance function. As will be discussed later, it seems that most surfaces have a height correlation function close to an exponential $C(r) = \exp(-r/L_c)$ [33–35], where L_c is the correlation length, defined as the lateral distance at which the correlation function between heights decays by $1/e$. One could say that, in most cases, the correlation length gives a measure of the average extent of the "hills and ravines" of a rough surface.

To illustrate this heuristic definition, Fig. 28.8a shows two mathematically generated random surfaces with an exponential autocorrelation function that have the same Gaussian height distribution function (i.e., same σ), but different correlation lengths. It is evident that as L_c increases, the extent of the bumps also increases. This implies that L_c has an important effect on the angular distribution of the light scattered around the specular direction. Fig. 28.8b shows the scattering pattern of the diffuse component around the specular direction for the two surfaces shown in Fig. 28.8a. As L_c/λ increases, the diffuse pattern concentrates more around the specular angle, and consequently, L_c, as well as σ, influence image resolution and haziness.

The same concepts also apply to the transmission of light through a rough surface, except that the phase differences in transmission are much smaller than in reflection, and therefore, the fraction of coherent light is much larger in transmission than in reflection.

From a purely heuristic point of view, σ/L_c somehow represents a sort of average slope of the surface bumps: the steeper these bumps are, the more they "reflect" light at a wider angle, and as these become smoother they scatter light closer the specular direction, as shown schematically in Fig. 28.9.

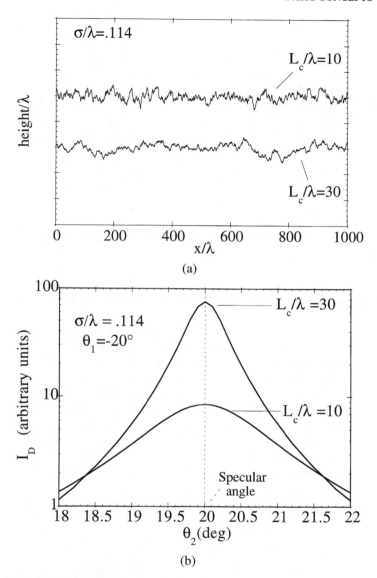

Figure 28.8 (a) Mathematically generated surface profile for a Gaussian distribution of heights with $\sigma/\lambda = 0.114$ for two values of L_c/λ. (b) Corresponding diffuse scattering pattern calculated by means of Kirchhoff scalar theory. Taken and reedited form [84], with permission from John Wiley & Sons.

As in scattering in a random medium, the main features of the light scattered by a rough surface are ruled by two parameters: σ/λ and L_c/λ. The first parameter determines how "strongly" the surface scatters, and the second influences the degree of anisotropy of the scattered light in the same way as Λ and Λ_{eff} affect transmission

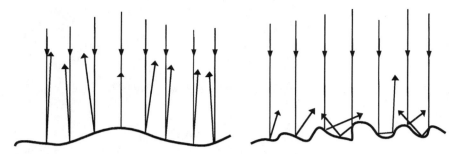

Figure 28.9 Schematic representation of the difference between light scattered by a soft-slope vs. a sharp-slope rough surface.

in a random medium. When we come to the discussion on gloss, we will see that the correlation length plays an important role.

III. MODELS FOR TRANSPARENCY, GLOSS, AND TRANSLUCENCY

In this section, we discuss the theories and models used to relate microstructure to the basic attributes of appearance. Some of these topics have been already reviewed in the literature, and in those cases, we refer the reader to the corresponding review. However for the sake of completeness, we do give a concise exposition of those topics that have been thoroughly treated elsewhere. Such is the case of transparency, for which one can find excellent general reviews, such as that by Willmouth [10], or else specialized ones, like that by White and Cakmak [3] addressing films produced by tubular film extrusion. Also, the field of optics of semicrystalline polymers was developed in the 1960s and 1970s and has already been reviewed by Haudin [36]; nevertheless, we refer to some of the classical papers on the subject, as the discussion of the effects of morphology on appearance requires it.

By contrast, the importance of surface roughness in the optical properties of polymers was recognized a long time ago, yet the theory of light scattering by rough surfaces has not permeated the polymer literature, and therefore, we particularly stress this aspect. In this section, we discuss a model for gloss based on the well-known Kirchhoff scalar approximation in light scattering by rough surfaces and point out the extension of these ideas to the problem of transmission through specimens with rough surfaces.

Finally, we examine the theory of transmission and reflection of light in translucent plastics and describe the main models used in predicting total transmittance and reflectance by a turbid medium.

A. Transparency

Transparency is perhaps the topic that has drawn most of the attention out of the plastics literature concerned with appearance. Perfect transparency of multiphase materials is normally limited by the presence of haziness or a reduced clarity, which arises primarily from failure to perfectly match the indexes of refraction of the polymer phases (or other phases present in the material) and surface roughness. Haziness refers to a loss in contrast when we observe an object through a specimen. This means that a light source seen through a hazy specimen looks like a less intense source surrounded by an illuminated background, because a hazy specimen scatters in all directions.

The standard quantitative assessment of haziness is haze [37], which is defined as the fraction of the transmitted light that deviates from the directly transmitted beam by more than 2 1/2°. This means that haze measures forward-rather than back-scattering haziness. As mentioned previously, the specimen must have a thickness less than that or a photon's mean free path in order to be close to transparency, which implies that any small departure from perfect transparency is related to single-scattering events. Particles that are very large, as compared to the wavelength of light, scatter mostly at low angles in the forward direction, mainly within the 2 1/2° of the incident direction, and consequently have a small effect on haze [38]. By contrast, heterogeneities that are very small as compared to λ scatter uniformly in all directions; however, their scattering cross sections are very small (e.g., for spheres of radius $R \ll \lambda$, $C_{sca} \approx R^6$), and therefore their contribution to haze also is not very important. The size that maximizes haze depends on the refractive index and corresponds to a balance between scattering anisotropy and scattering strength.

The image of a point source seen through a specimen with optical heterogeneities that are large compared to the wavelength appears to be surrounded by a bright halo rather than a uniform illuminated background. This halo limits the resolution of the image of an object seen through the polymer sample and, therefore, its clarity. This means that a quantitative measure of clarity involves the determination of light scattered at very low angles. There are several methods designed to quantify clarity, and they are discussed in depth by Willmouth [10]. However, the method most frequently used in this respect, and one that is part of an ASTM standard [39], is a measure of the direct transmittance DT, defined as the fraction of the incident beam transmitted without deviation. Nevertheless, there is an uncertainty in this method, since wide, and small-angle scattering as well as the absorption of the medium contribute to the attenuation of the directly transmitted beam.

Similarly, haziness arising from surface roughness is related not only to the RMS average height σ, but also to the average slope of the broken topography, as discussed previously. That is, σ together with the correlation length of heights define the scattering strength and the directionality of the scattered light, and therefore haziness and image resolution depend upon whether we have "sharp" or "smooth" roughness. Actually, for high-gloss surfaces, there are equivalent attributes, such as reflection haze and distinctness-of-image gloss [40]. For a specular angle of 30°, the first attribute measures the light scattered at 2° (narrow-angle reflection haze) and 5° (wide-angle reflection haze) from the specular direction, while the second measures the scattering at 0.3° from the specular angle.

To illustrate the effects of size and index of refraction of the optical heterogeneities on haze and direct transmittance, let us consider a parallel-sided, smooth-surfaced slab of material of thickness ℓ containing randomly distributed microdomains. The matrix material has an index of refraction n_m with a number density N of optical heterogeneities of index of refraction n_p. Since we want to study small deviations from transparency, the mean free path must satisfy the condition that $\Lambda \gg \ell$, and therefore, $N C_{ext} \ell \ll 1$. Thus, in the upcoming discussion, we choose N, ℓ, or C_{ext} to be sufficiently small that $\Lambda \gg \ell$ is always satisfied.

Using the well-established theory of scattering by single particles [25–27], Willmouth expressed the transmission parameters used in the characterization of transparency in terms of the scattering amplitudes of the optical heterogeneities. Assuming real and isotropic refractive indexes, smooth slab surfaces, and only single reflections at the entrance and exit walls, he got for an unpolarized incident beam

$$\text{direct transmittance} = \frac{F_t}{F_o} = \frac{16 n_m^2 \exp(-N C_{ext} \ell)}{(n_m + 1)^4}, \tag{28.10}$$

where F_t is the undeviated transmitted flux and F_o is the incident flux. Equation 28.10 is the Beer–Lambert law corrected by a factor $[4 n_m / (n_m + 1)^2]^2$, corresponding to the loss by reflection at the two slab faces. By definition,

$$\text{haze} = \frac{(F_s)_{2.5}^{90}}{F_t + (F_s)_0^{90}}, \tag{28.11}$$

where $(F_s)_{\alpha_1}^{\alpha_2}$ is the scattered flux emerging between angles α_1 and α_2 and is given by

$$(F_s)_{\alpha_1}^{\alpha_2} = \frac{4 n_m F_o (C^*)_{\alpha_1}^{\alpha_2}}{(n_m + 1)^2 C_{ext}} [1 - \exp(-N C_{ext} \ell)], \tag{28.12}$$

where $(C^*)_{\alpha_1}^{\alpha_2}$ can be written as

$$(C^*)_{\alpha_1}^{\alpha_2} = \frac{\lambda^2}{4\pi} \int_{\sin^{-1}(\sin \alpha_1 / n_3)}^{\sin^{-1}(\sin \alpha_2 / n_3)} \left[|S_1(\theta)|^2 T_\perp(\theta) + |S_2(\theta)|^2 T_\parallel(\theta) \right] \sin \theta \, d\theta. \tag{28.13}$$

S_1 and S_2 are the amplitudes of the scattering light polarized with the electric vector perpendicular and parallel to the scattering plane (i.e., the plane defined by the incident and scattering directions), respectively. $T_\perp$ and $T_\parallel$ are the Fresnel transmittances for light polarized perpendicular and parallel to the scattering plane, respectively [11]. In the limit of $\ell/\Lambda \ll 1$, haze symply reduces to

$$\text{haze} \approx \frac{(n_m + 1)^2}{4 n_m} (C^*)_{2.5}^{90} N \ell. \tag{28.14}$$

$(C^*)_{\alpha_1}^{\alpha_2}$ depends on the structure of the scattering domains. For spheres and cylinders, there is an exact analytical theory given by Mie's theory [25–27]. When the domains almost match the index of refraction of the matrix, the anomalous diffraction approximation [25–27] applies, which simplifies the evaluation of $(C^*)_{\alpha_1}^{\alpha_2}$. Additionally, if the phase shifts induced by the particles are small, then the Rayleigh–Gans–Debye approximation [25–27] simplifies the calculation even more. For nonspherical and noncylindrical particles, regular or irregular, a number of approximate methods have been developed, such as perturbation methods [41, 42], the Purcell–Pennypacker method [43], the T matrix method [44, 45]. More recently, fractal concepts have been introduced into the study of the optical properties of aggregates clusters [46–48].

Figure 28.10 illustrates the main effect of size and relative index of refraction on haze and direct transmittance for spherical particles, where in Mie's theory was used in the calculation of $(C^*)_{\alpha_1}^{\alpha_2}$. As discussed previously, haze decreases for large and for very small particle size relative to λ, while direct transmittance does the opposite. As the refractive index of the particle gets closer to that of the matrix, the particle size that gives the maximum haze increases. As shown in Fig. 28.10, for PB particles in PS, maximum haze is obtained at $D \approx 2.6$ μm, while for TiO_2 in the same matrix, the maximum is at $D \approx 0.24$ μm.

In theory, for homogeneous particles or domains, as we index match the domains with the matrix, haze tends to zero and the direct transmission to 100%. However, in multicomponent blends, the scattering domains are not necessarily homogeneous. In most cases, besides the internal structure of the scattering domains, there is a compatibilizing interface. Under such circumstances, one can only match up to an index of refraction that minimizes scattering, rather than nullifying it. However, there is no unique way of minimizing scattering. For instance, one can minimize the particle-scattering cross section, which implies that the direct transmittance is maximized, or else one can match the refractive index so that haze is minimized. Nevertheless, the effective particle refractive index that minimizes haze does not coincide with the one that maximizes direct transmittance [49]. Theoretically, for PMMA–PS core–shell particles in an index-matched medium, the difference in composition between maximum direct transmittance and minimum haze can be as large as 2% or more, depending on the wavelength used [49]. In fact, the author of this chapter has shown that this can occur even in the case of small particles [49] and that the origin of this is that in the neighborhood of transparency even very small composite particles can have strongly anisotropic scattering patterns [50]. An interesting conclusion is that controlling the morphology of the particle can lead to a very small haze equivalent to the one obtained with a highly index-matched homogeneous particle of the same size [49].

If the domains are anisotropic in nature, such as for spherulites, anisotropic rods, and so on, then the dielectric properties of the domains are represented by a dielectric tensor rather than a scalar. Stein and coworkers [51–54] and others [55] have extended the theory, within the Raylaigh–Gans–Debye approximation, to include the anisotropic properties of spherulites and rods and their work constitutes the basis of most light scattering studies on the morphology of semicrystalline materials. These authors have included morphological features such as truncation [56–58], incomplete growth [59], internal disorder [60–63], deformation [64] and interference effects between scattering particles [65–68]. On the other hand, Meeten [69] has extended

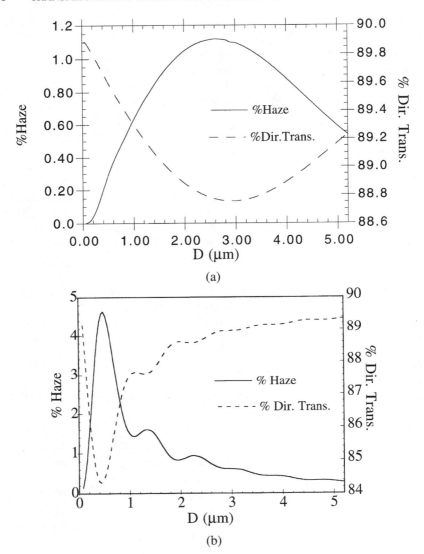

Figure 28.10   Haze and direct transmittance vs. particle diameter for $n_m = 1.59$, $\Phi = 10^{-4}$, $\ell = 0.1$ cm, and $\lambda_o = 589$ nm. (a) $n_p = 1.51$, (b) $n_p = 1$, and (c) $n_p = 2.7$.

the anomalous diffraction approximation to include incident polarized light and anisotropy in the scattering elements. This field has been thoroughly reviewed by Haudin [36].

So far, we have discussed scattering by the bulk; however, in many applications, surface roughness is the dominant scattering source that limits transparency. Such is the case, for instance, for blown-polyethylene (PE) and blown-polypropylene (PP) films. Probably, blown-PE film has been the system most studied in this respect.

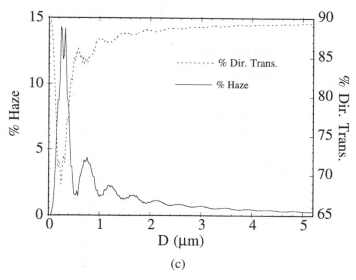

(c)

Figure 28.10   (Continued).

Since the classical study of Huck and Clegg [2], which examined the effect of extrusion conditions on haze of blown low-density polyethylene (LDPE), there have been numerous papers that have tried to establish on a scientific basis many of the conjectures put forward by Huck and Clegg. Stehling et al. [70] showed experimentally that the principal source of haze was surface roughness. Their visual interpretation of scanning electron microscope (SEM) and light micrographs of blown LDPE surfaces together with a small-angle light-scattering (SALS) study led them to relate the increment of surface roughness to an increase in haze. Later on, Ashizawa et al. [71] corroborated Stehling et al.'s observations by using a profilometer and quantifying the degree of roughness by means of the RMS average height σ. However, Stehling et al. characterized haze according to the ASTM standard mentioned previously, while Ashizawa and collaborators monitored direct transmittance to determine the clarity of the film. Stehling and coworkers concentrated on the effects on haze of molecular-weight distribution, chain microstructure, melt elasticity, and mechanical treatment, trying to establish the mechanisms involved and the origin of haze (extrusion haze and crystallization haze) and summarizing in a consistent picture most of the work up to that time. Ashizawa et al. [71] have studied extensively the relationship between processing parameters in tubular film extrusion and direct transmittance and found a correlation between surface direct transmittance and σ. They also found that within the range of different processing conditions studied, these had little effect on crystalline orientation.

A similar study for PP was made by Bheda and Spruiell [72]. They found that the dominant cause of haze in PP tubular film was surface scattering, and they also found a correlation between the direct surface transmittance and the surface roughness. White et al. [73] also came to similar conclusions. Both groups of researchers established that surface roughness is related to the crystallization process. An extensive quantitative study on orientation development in tubular film extrusion of polyethylene can be found in the work of Choi et al. [74]. The development of orientation in polypropylene tubular film was investigated by Shimomura et al. [75]. White et al. [73] compared roughness and the contribution of the surface to the transmittance of a range of tubular films and showed that, to a first approximation, surface roughness correlates with percent crystallinity developed in the process.

The first investigations on the morphology of oriented polyethylene films date from the 1950s with the classical work by Holmes, Palmer, Miller, and Bunn [76] and Keller [77], and since then, it has become an important topic in polymer physics. White and Cakmak [3] reviewed this subject in the context of tubular film extrusion. Recently, Smith et al. [78] studied haze and surface morphology with atomic force microscopy (AFM) on blown-polyethylene films processed under various conditions. The AFM images showed the presence of surface structures that resembled twisted lamellae, which are predominantly aligned transverse to the machine direction. This finding confirmed Keller and Machin's [79] model to explain the development of crystalline orientation with uniaxial stress and the subsequent findings of other authors. With regard to haze, Smith et al. arrived at similar conclusions as those of Stehling et al. and Ashizawa et al., namely that haze is due mainly to surface roughness and that the latter is related to surface crystallinity.

On the other hand, Sung et al. [80] studied the structure and the properties of oriented polymer surfaces for three series of uniaxially oriented films of PP, polyethylene terephthalate (PET), and PS. They characterized relative surface crystallinity, surface orientation, topology, and surface roughness by FT–IR–ATR dichroism, microscopy, and profilometry. The relative surface crystallinity of PP was found to increase modestly with draw ratio. For drawn PET film surfaces, the molecular orientation of the *trans* ethylene glycol unit, at 975 cm^{-1} is highest. Sung et al. monitored the relative amount of surface *trans* conformer, which increased with drawing, indicating that the relative crystallinity also increases slightly. For uniaxially drawn PP, the surface becomes rougher as we increase the draw ratio and becomes anisotropic with the peaks and valleys elongated along the draw direction. By contrast, Sung et al. found that the surfaces of uniaxially drawn PET and PS are smooth. White et al. [73] also found that the films of PS produced by tubular extrusion are smooth.

Although the theory of light scattering by rough surfaces is a well-established field [30, 31], the polymer literature does not register its application, in spite of its relevance in understanding the final optical properties of films and plastics in general. The aforementioned literature at most characterizes the statistical nature of the surface by a single parameter—the RMS average height σ or another height average. However, as explained in the previous section, σ governs the scattering strength— that is, how much of the coherent beam becomes diffuse—but cannot determine uniquely the angular distribution of the diffuse component. The latter depends also on L_c. Haze actually depends on the angular distribution of scattered light. As well as coherent reflectance, direct transmittance through a surface with a Gaussian distribution of heights has the same functional dependence on σ—that is, $\log(DT) \approx -\sigma^2$—

except that the proportionality constant is smaller in transmittance than in reflection [81]. This means that experimental data on the surface contribution to the attenuation of direct transmittance in films with Gaussian height distribution should conform to this law.

Information on height autocorrelation functions for polymer surfaces is lacking in spite of the growing interest in AFM studies of polymers [82]. Among the few papers that characterize the correlation function of polymer surfaces are those of Méndez et al. [33] in the case of poly(acrylonitrile-butadiene-styrene) (ABS) and those of Lettieri et al. [34] for glossy coating on paper. In both cases, the Kirchhoff scalar approximation with a Gaussian height distribution was used to calculate the angle-resolved light scattering. Méndez et al. [33] showed that for different processing conditions, an exponential height correlation function describes accurately the angle-resolved scattering in the vicinity of the specular angle. For glossy paper, Lettieri et al. [34] found that a quasiexponential correlation function fitted best the angle-resolved light-scattering data. In a more general context, Bennett and Mattsson [35] have pointed out that for a wide variety of surfaces, the surface height correlation function is, in general, closer to an exponential. This implies that a simple Kirchhoff scalar theory with a quasiexponential correlation function seems to be a reasonable model in explaining the surface scattering features of many complex systems encountered in industrial applications. In the next subsection, a theory of gloss is derived and the importance of surface height correlations is particularly stressed.

B. Kirchhoff Scalar Theory of Gloss

Specular gloss, according to the ASTM D523-89 standard [83], is the relative luminous reflectance factor of a specimen in the mirror direction. The relative luminous reflectance factor is defined as the ratio of the luminous flux reflected from a specimen to that of a standard surface under the same geometric conditions. For the purpose of measuring specular gloss, the standard surface is assumed to be a polished glass. The relative specular reflectance around a solid angle Ω_1 can be written as

$$\rho(\Omega) = \frac{1}{P_N} \int_{\Omega_1}^{\Omega_1 + \Omega_D} \left(\frac{dP}{d\Omega}\right) d\Omega, \tag{28.15}$$

where $dP/d\Omega$ is the power scattered by the surface per solid angle, Ω_D is the solid angle defined by the receiving optics of the detector, and the subscript 1 refers to the specular direction. P_N is the total power scattered by a smooth surface, received by the same detector-collecting system, at the specular angle θ_1; that is,

$$P_N = \int_{\Omega_1}^{\Omega_1 + \Omega_D} \left(\frac{dP_o}{d\Omega}\right) d\Omega, \tag{28.16}$$

where $dP_o/d\Omega$ is the power scattered per solid angle by a smooth surface.

Standard glossmeters use large receiving-field apertures ($\approx 2°$–$4°$) and illuminate a large sample area (≈ 1–2 cm^2). From the discussion in Section II, it is clear that in addition to the coherent reflected beam, part of the diffuse light enters the detector receiving optics. Therefore, the contribution of the incoherent component to gloss depends on the correlation length and σ. In order to calculate the coherent and incoherent contribution to $dP/d\Omega$, Alexander–Katz and Barrera [84] used Kirchhoff scalar theory [30, 31].

Kirchhoff theory is the most widely used approximation in the study of wave scattering from rough surfaces; the name derives from the analogy with Kirchhoff's theory of light diffraction by an aperture. The Kirchhoff approximation assumes that the field at any particular point on the surface is the same as the field reflected by an infinite plane tangent to the surface at that point. This approximation is accurate when $L_c > \lambda$ and $L_c \gg \sigma$, as long as the incidence angle (with respect to the normal to the surface) is not too large. A full discussion on the accuracy of the Kirchhoff approximation is given in Sections 4.2 and 4.3 of Ogilvy's book [31]. These conditions are well satisfied in the case of injection-molded ABS [33] and paper coatings [34], mentioned previously and probably in many other polymer applications, although, as pointed out before, there is very little published work on correlation functions for polymer surfaces.

Here, it is assumed that the incident beam is monochromatic and that there is no cross-polarized surface scattering, as is the case for the two examples discussed previously, which validates the use of Kirchhoff scalar theory. Within this approximation, the coherent contribution to the specular relative reflectance $\rho_s^{(c)}$ for a Gaussian distribution of heights is given by

$$\rho_s^{(c)} = \frac{|r|^2}{|r_o|^2} \exp(-g), \qquad (28.17)$$

where $|r|^2$ and $|r_o|^2$ are the reflectance in the specular direction of a smooth specimen surface and of the standard surface, respectively, as given by the usual Fresnel formulas and $g = k^2\sigma^2(\cos\theta_1 + \cos\theta_2)^2$, where $k = 2\pi/\lambda$ is the magnitude of the wave vector of the incident beam, σ is the surface RMS height, and θ_1 and θ_2 are the incident and scattering angles, respectively, as shown in Fig. 28.7. Given the angular range seen by the detector, g can be approximated by $g \approx g_s = (2k\sigma\cos\theta_1)^2$, and therefore $\rho_s^{(c)}$ reduces to the Bennett and Porteus [32] expression

$$\rho_s^{(c)} \cong \frac{|r|^2}{|r_o|^2} \exp(-g_s), \qquad (28.18)$$

which is the same as Eq. 28.7 when the reflectance of a smooth specimen surface is the same as that of the reference standard polished glass.

For the range of angular receiving apertures typical of a glossmeter, Alexander-Katz and Barrera [84] showed that, for a surface that is both isotropic and stationary, and that has Gaussian height statistics, the diffuse contribution to the relative specular reflection $\rho_s^{(d)}$ can be approximated by

$$\rho_s^{(d)} = \frac{2}{\pi} \frac{|r|^2}{|r_o|^2} e^{-g_s} \int_0^{Y_D} \int_0^{Y_D \cos\theta_1} \left[\int_0^\infty J_o\left(\sqrt{\alpha^2 + \beta^2}\, x\right)\left(e^{g_s C(x)} - 1\right) x\, dx \right] d\alpha\, d\beta.$$
(28.19)

In this equation, $Y_D = kL_c|(\delta\theta)_D|$, L_c is the correlation length and $(\delta\theta)_D$ is the detector-collecting half-angle; J_o is a Bessel function of zero order and $C(x)$ is the autocorrelation function. Here, it must be pointed out that the normalization used in this expression corresponds to that of a large-aperture detector; that is, $\Omega_D \gg \lambda^2/A_M$, where A_M is the illuminated area. This means that the detector sees almost all the specularly scattered diffraction pattern. Therefore, the normalization applied in Eq. 28.19 differs from that used by Beckmann [30].

One can observe that for a given angle of incidence, the incoherent contribution to gloss depends on only two parameters: g_s and Y_D. This implies that $\rho_s^{(d)}$ varies with L_c in the same way as with the detector-collecting angle. Equation 28.19 is valid for all correlation functions with a single characteristic length. The gloss, G can then be written as

$$G = \rho_s^{(c)} + \rho_s^{(d)}.$$
(28.20)

For an exponential correlation function $C(R) = \exp(-R/L_c)$, G can be written as the following series [84]:

For $\theta_1 > 0$,

$$G_{(\exp)} = \frac{|r|^2}{|r_o|^2} \exp(-g_s)\left\{1 + \frac{2}{\pi} \sum_{n=1}^\infty \left[\frac{g_s^n}{n!}\right] \arctan\left\{\frac{Y_D^2 \cos\theta_1}{n\sqrt{n^2 + Y_D^2(1 + \cos^2\theta_1)}}\right\}\right\}.$$
(28.21)

For normal incidence,

$$G_{(\exp)}(\theta_1 = 0) = \frac{|r|^2}{|r_o|^2} \exp(-g_s)\left\{\exp(g_s) - \sum_{n=1}^\infty \frac{g_s^n}{(n-1)!(n^2 + Y_D^2)^{1/2}}\right\}.$$
(28.22)

Figure 28.11 shows a plot of gloss as a function of Y_D for several values of σ/λ for an exponential correlation function. As expected, for very small values of σ/λ, gloss should be highly insensitive to Y_D, since the coherent component is dominant and the incoherent contribution represents a small percentage of the total specularly reflected light. In this limit, Eqs. 28.21 and 28.22 reduce to the Bennett and Porteus expression [32] given by Eq. 28.18. It is also observed that for very large σ/λ, gloss is once again insensitive to changes in Y_D. However, the reason for this insensitivity is that for such large values of g_s (≈ 22), the light scattered diffusely is almost angle independent within the angular interval defined by the detector-receiving system. Although, in this limit, $G \approx (Y_D)^2$, yet the proportionality constant is very small.

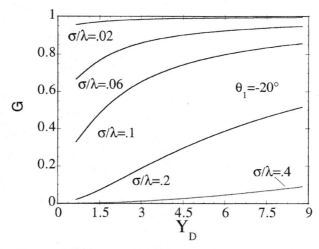

Figure 28.11   Gloss vs. Y_D at an incidence angle of 20° for an exponential correlation function for different values of σ/λ. $(\delta\theta)_D = 1°$. From [84]; reprinted with permission from John Wiley & Sons.

Table 28.1   Contribution of the Diffuse Field to Gloss for Some Values of σ/λ and L_c/λ^a

σ/λ	L_c/λ	$G(\times 100)$	$G_i/G(\%)$
0.06	40	89.9	32.9
0.1	40	73.8	66.6
0.1	10	41.3	40.4

$^a \theta_1 = 20°, (\delta\theta)_D = 0.017$ rad.

In many applications, surfaces are neither extremely glossy nor fully matte; instead, they are fairly glossy, with moderate values of g_s (≈ 1). It is in this intermediate region that gloss has the strongest sensitivity to Y_D. For a moderate value of σ/λ, as L_c/λ increases, the diffuse scattering tends to concentrate more within the detector's acceptance angle, as shown in Fig. 28.8b, and as a consequence, gloss is increased by the contribution of the incoherently scattered field. Table 28.1 shows the fraction of gloss coming from the diffuse component for some selected values.

As one can see, the incoherent fraction of gloss can be quite substantial for moderately glossy surfaces or even highly glossy surfaces as long as L_c/λ is large. The proportion of the incoherent contribution can be enhanced by increasing the receiving aperture, since the detector will register a larger portion of the diffuse field. In conclusion, for moderately glossy surfaces, gloss, as usually measured, can be strongly incoherent in nature.

Alexander–Katz and Barrera [84] also discussed the case of quasiexponential correlation functions and a Gaussian correlation function. The two types of quasiexponential correlation functions considered were of the form $C(R) = \exp(-|R/L_c|^\alpha)$

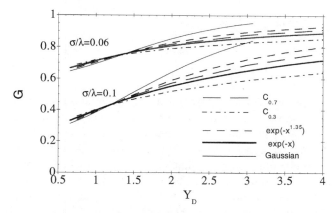

Figure 28.12   Gloss vs. Y_D at an incidence angle of 20° for various correlation functions for two values of σ/λ. $(\delta\theta)_D = 1°$. From [84]; reprinted with permission from John Wiley & Sons.

and

$$C_\nu(R) = \frac{(p_\nu R/L_c)^\nu}{2^{\nu-1}\Gamma(\nu)} K_\nu(p_\nu R/L_c),\qquad(28.23)$$

where K_ν is the modified Bessel function of order ν, Γ is the gamma function, and p_ν is a scalar whose value is such that for $R = L_c$, $C_\nu(L_c) = e^{-1}$. The first type of function was used by Lettieri et al. [34] to fit their angle-resolved scattering data, and the second type was proposed by Hoenders et al. [85] in the context of a model of a fluctuating facet-scattering surface. Figure 28.12 shows the effect that the different correlation functions have on the functional dependence of gloss on Y_D and σ. The fact that all of the functions cross at almost the same point suggests a simple method for measuring the correlation length from gloss measurements, even if we do not know the surface correlation function.

Kirchhoff theory can be extended to transmission through rough surfaces. Welford [81] has discussed the case of the transmission through a phase screen—that is, a screen that introduces a random phase in the transmitted wave—as a model of a rough screen. This approach can be translated to the transmission of light through thin rough films, as discussed previously.

During the last few decades, there has been considerable effort to control and predict the morphology of multiphase media (see Chapter 15) and the effects of processing on morphology (see Chapter 16). However, although surface roughness plays a major role in the optical properties of films, and notwithstanding the large amount of work done on PP and PE films in establishing relations between material structure, process variables, and surface roughness, the author is not aware of any work predicting on a theoretical basis the statistical nature of surface roughness of such films in terms of material and process variables. In this sense, in the case of plastics, only a limited number of references can be found [5, 6, 86–90] wherein an effort was made to model the formation of protuberances at interfaces due to the presence of fillers.

C. Translucency

In terms of the transmission of light, a translucent material is somewhere in-between an opaque material, which transmits no light, and a transparent material. In a highly transparent specimen, the transmitted coherent component is fully dominant over the diffuse one, while in a translucent material, the diffuse field is predominant and the objects seen through it are poorly visible or not visible at all. The degree of visibility depends on the ratio between the coherent and diffuse transmitted components. In some applications, it is important to maximize the total transmitted light and, at the same time, partially obscure the objects beyond, or else to maximize the back-scattered light so that the appearance is as bright as possible.

Among the parameters commonly measured are the total transmittance and total reflectance—that is, the fraction of the incident light that is transmitted or reflected, respectively. Another parameter usually employed is the contrast ratio, defined as $C = R_o/(R)_{R_b}$, where R_o is the reflectance in the absence of a backing or else with a perfectly absorbing backing and $(R)_{R_b}$ is the reflectance when the specimen is placed against a backing with a reflectance R_b. In the limit when the backing reflectance is unity, the contrast ratio tends to the ideal contrast ratio.

The application of the scattering theory to translucency is a subject rarely found in the plastics literature, in spite of the fact that most engineering plastics are translucent. However, the testing methods and instruments used to assess the aforementioned parameters and others associated with color technology apply flux theories in their operation. By contrast, the paint and paper industries have applied radiation transport theories in the design of their products for a long time, and have been a drive in the development of this field from a theoretical as well as an experimental point of view. Still, as is usual, the first transport theory [91] originated in astrophysics, and since then, it has been a primary source of theoretical and experimental studies in this field [92]. For this reason, there are presently a number of theories with a reasonable experimental basis that have been translated into other areas.

There are two approaches to the study of the interaction of an electromagnetic field with a random medium. One approach may be called the analytical theory and the other the transport theory, better known as radiative transfer theory. The former starts from the basic equations of electromagnetism, and from the scattering and absorption properties of the particles constructs the differential or integral equations for the statistical properties of the fields. Within this framework, one can include multiple scattering, diffraction, and interference effects. However, in practice, there are very few problems that can be solved via this approach. The transport theory, on the other hand, is more phenomenological and is concerned only with the energy transport through a medium filled with scattering domains. Although it includes the scattering and absorption of the single scatterers, it does not consider diffraction effects. Intensity is assumed to be additive, and therefore no correlation between fields is considered. If the scattering mean free path Λ is larger than the wavelength

λ and the thickness of the specimen is larger than the transport mean free path Λ_{eff}, then the transfer equation can be approximated by a diffusion equation [15].

Some of the most often-used simplifications of the transport theory are the n-flux theories. The most popular theory in this class is the Kubelka–Munk two-flux theory [28, 29] for diffuse illumination, which is equivalent to Schuster's [91] early approach to the study of radiation through a foggy atmosphere. That is, Kubelka and Munt divided the light crossing a specimen (or any other multiple-scattering medium) into two fluxes flowing in directions opposite to each other and coupled by linear differential equations. Later on, Kubelka [29] gave useful expressions for the reflectance and transmittance for different experimental situations in terms of the parameters of the theory.

In order to include a collimated illumination, it was necessary to introduce, in addition to the opposite diffusing fluxes, two more fluxes for the collimated forward and backward directions. The collimated beams are continuously converted into the backward and forward diffuse fluxes, leading to four coupled linear differential equations [15, 93]. The basic balance equations are

$$\frac{dF_{c+}}{dz} = -(k^* + S_1^* + S_2^*)F_{c+}, \tag{28.24}$$

$$\frac{dF_{c-}}{dz} = (k^* + S_1^* + S_2^*)F_{c-}, \tag{28.25}$$

$$\frac{dF_{d+}}{dz} = S_1^* F_{c+} - (K + S)F_{d+} + SF_{d-} + S_2^* F_{c-}, \tag{28.26}$$

and

$$\frac{dF_{d-}}{dz} = -S_2^* F_{c+} - SF_{d+} + (K + S)F_{d-} - S_1^* F_{c-}, \tag{28.27}$$

where F_{c+} and F_{c-} are the forward and backward collimated fluxes, respectively, and F_{d+} and F_{d-} are the forward and backward diffuse fluxes, respectively. S is the scattering coupling parameter between the diffuse back and forward fluxes, which represents the fraction lost by scattering per unit length of a diffuse flux in favor of the opposite flux, and K is the fraction lost of F_{d+} or F_{d-} per unit length due to absorption. S_1^* and S_2^* are the fraction lost by scattering per unit length of the collimated beam by forward and back scattering, respectively, and k^* is the contribution to the extinction coefficient due to absorption. Also,

$$S_1^* + S_2^* = s, \tag{28.28}$$

where s is the scattering contribution to the extinction coefficient.

In addition to these coefficients, the boundary conditions introduce the diffuse and coherent reflectance of the backing and of the external and internal surfaces of the slab. We shall denote the coherent reflectance and the diffuse internal and external reflectance of the slab boundaries by r_c, r_d^i, and r_d^e, respectively. Also, we shall call r_c^b and r_d^b the coherent and diffuse reflectance of the backing, respectively, and τ_c and τ_d the corresponding coherent and diffuse transmittance of the backing, respectively. We can write the coherent transmittance τ_{cc} and the coherent reflectance R_{cc} in terms of the fluxes as

$$\tau_{cc} = \frac{(1 - r_c)(1 - r_c^b)\tau_c}{1 - r_c r_c^b} \frac{F_{c+}(\ell)}{F_c^o + F_d^o} \tag{28.29}$$

and

$$R_{cc} = \frac{r_c F_c^o + (1 - r_c)F_{c-}(0)}{F_c^o + F_d^o}, \tag{28.30}$$

where F_c^o and F_d^o are the incident coherent and diffuse fluxes, respectively. The total diffuse transmittance τ_{dt} and total diffuse reflectance R_{dt} can be written as

$$\tau_{dt} = \frac{(1 - r_d^b)(1 - r_d^i)\tau_d}{1 - r_d^e r_d^b} \frac{F_{d+}(\ell)}{F_c^o + F_d^o} \tag{28.31}$$

and

$$R_{dt} = \frac{r_d^e F_d^o + (1 - r_d^i)F_{d-}(0)}{F_c^o + F_d^o}. \tag{28.32}$$

Applying the radiative transfer theory, Ishimaru (see Chapter 10 in [15]) has shown that the four coupling parameters S, K, S_1^*, and S_2^* are constant only in the case that the incoherent fields are isotropic. In this case, Ishimaru proved that these parameters can be written in terms of the differential scattering cross section σ_s and the absorption cross section of the scattering elements. For a granular type internal structure, it is common to assume that the grains (e.g., particles, domains, etc.) are the scattering elements and σ_s is their differential cross section. In many cases, σ_s can be approximated by the Henyey–Greenstein [94] formula

$$\sigma_s(\mu) = \frac{C_{sca} W_o}{4\pi}(1 - \bar{\mu}^2)[1 + \bar{\mu}^2 - 2\bar{\mu}\mu]^{-3/2}, \tag{28.33}$$

where $\mu = \cos\theta$ and $\bar{\mu} = \overline{\cos\theta}$, where θ is the scattering angle and $W_o = C_{sca}/C_{ext}$ is the albedo. By definition, $\overline{\cos\theta} = (1 - \Lambda/\Lambda_{eff})$. Therefore, in a nonabsorbing system, all of the coupling constants can be expressed in terms of Λ and Λ_{eff}, as stated in Section II.

Mudgett and Richards [93], following the work of Chandrasekhar [92], used the discrete-ordinate method of radiative transfer calculations and provided a general framework for many-flux theories. Instead of deriving from a multiflux analysis the structure of the coupling constants of a two-flux model, these authors looked for the best values of S and K in a two-flux theory that could give the same results as a multiflux calculation. In all of the cases analyzed, the optical depth (slab thickness/Λ) was 50, which implies that the authors only examined cases for which the fluxes could be considered isotropic and for which the diffusion approximation is valid. They found that a good approximation for S and K that accurately reproduce the results of a multiflux theory are, in our terminology,

$$S = \frac{3}{4} \Lambda_{eff}^{-1} \tag{28.34}$$

and

$$K = 2k^*. \tag{28.35}$$

Brinkworth [16–19] and Gate [20–23] arrived at similar conclusions by an independent line of thought. They applied the photon diffusion concept to study the transport of the incoherent field through a diffusing medium. They assumed that the photons diffuse through the material, under steady-state conditions with a light source of constant strength, in a manner described by Fick's law. In the case of isotropic scattering, the diffusion coefficient $\mathcal{D}$ is given in analogy with the kinetic theory of gases as $\mathcal{D} = c\Lambda/3$, where c is the velocity of light and Λ is the mean free path of photon in the diffusing medium. The theory was generalized to the case of anisotropic scattering by replacing Λ with Λ_{eff}. Basically, what Brinkworth and Gate assume is that photons follow random paths with a great number of steps, so that their statistical properties are well represented by a diffusion equation. This approximation is valid for $\Lambda/\lambda \gg 1$ and distances greater than Λ_{eff}. Brinkworth [17] generalized Eq. 28.34 for the case of a mildly absorbing ($k^*/s < 0.01$) material to $S = 3/4\Lambda_{eff}^{-1} - k^*$. For a very instructive review of the early photon diffusion theory and its experimental verification, the reader is referred to Chapter 6 in Meeten's book [95]. A comprehensive discussion of this approximation in the context of the general transport theory can be found in Ishimaru's book. (See Chapter 9 in [15].)

Maheu et al. [96, 97] have given useful general analytical expressions for the total coherent and diffuse transmittance and reflectance in terms of the parameters involved in the theory and the optical properties of the backings. However, the S and K parameters are written in terms of two extra parameters ε and ζ_d, to allow for non-isotropic incoherent beams. That is, $S = \varepsilon(1-\zeta_d)s$ and $K = \varepsilon k^*$. Here, ε determines the average path length traveled by the diffuse beam as compared to that traveled by collimated one, and ζ_d is the ratio of the forward scattering to total scattering for the diffuse beams. All of the parameters in the theory, except for ε and ζ_d, can be obtained from single-scattering calculations. However, ε and ζ_d are known in the extreme cases of isotropic and totally anisotropic scattering. In the limit of isotropic scattering, $\varepsilon = 2$ and $\zeta_d = 1/2 = \zeta_c$, where ζ_c is the ratio of forward scattering to total scattering of the collimated (coherent) beam. For total-scattering anisotropy, wherein the light is scattered completely in the forward incoming direction, $\varepsilon = 1$ and $\zeta_d = \zeta_c$. For other cases, ε and ζ_d have not been reduced to single-scattering calculations.

Niklasson [98] compared Maheu et al.'s results to exact multiple-scattering calculations tabulated by van de Hulst [99]. For isotropic scattering, he found good agreement between Maheu et al.'s results with $\varepsilon = 2$ and $\zeta_d = \zeta_c$ and the exact multiple-scattering results. This was also the case for moderately anisotropic scattering. For strongly anisotropic scattering, though, some fitting must be made to adjust the van de Hulst data. However, for nonabsorbing media, the four-flux calculation with $\varepsilon = 2$ and $\zeta_d = \zeta_c$ is fairly close to that of van de Hulst's curves of transmittance versus optical depth, even for large scattering anisotropy and small optical depths (e.g., 0.6). For a strongly absorbing medium ($k^* > 1.5$ s) and pronounced scattering anisotropy, the choice of $\varepsilon = 1$ and $\zeta_d = \zeta_c$ adjusts better the van de Hulst data. Therefore, for nonabsorbing or strongly absorbing materials, if we know the transmittances and reflectances of the backings and of the entrance and exit surface of the specimen, Maheu et al.'s approach [96, 97] can be reduced to single-scattering calculations. The author of this chapter and coworkers have used this approach to study the role of the incoherent and coherent fields in the UV-degradation patterns in filled and unfilled polytetrafluoroethylene (PTFE) [100, 101].

At this point, it is important to stress that although, as we have shown, the main coupling parameters can be written in terms of the scattering properties of the single optical heterogeneities, the effect of the surrounding composite medium on the optical properties of the single scatterers is taken into account commonly in the spirit of an effective field approach, assuming that the index of refraction of the medium surrounding the particle is that of the composite medium. For instance, in his experimental study with PS latex particles to test the validity of the diffusion theory, Gate [22] used for the medium an effective index of refraction for the composite system of water–latex particles, calculated by means of a volume-fraction mixing rule. This was substituted further on in the calculations C_{sca} and $\overline{\cos\theta}$, in order to compare the diffusion theory with the experimental data. On the other hand, Ishimaru et al. [14], in their own experimental test of the diffusion theory, measured the equivalent extinction coefficient and used this value in the theoretical calculations of the diffusion field, which is equivalent to Gate's procedure, although in this case the mixing rule is given experimentally. This procedure implies that Λ and Λ_{eff} have a nonlinear concentration dependence. Fitzwater and Hook, III, [102] used a different approach to take into account the concentration effects at high concentrations based on the concept of dependent scattering, which implies a screening effect of the surrounding particles.

Commonly, all other parameters in the theory can be calculated or measured, except the internal diffuse reflectance for which we must know the angular distribution of the diffuse light, and therefore, in principle, a multiflux theory is required. Even if the incoherent flux is isotropic, the isotropy is broken near the slab boundaries. (See Chapter 9 in [15].) In recent years, there has been a renewed interest in photon diffusion theories associated with the development of diffusing wave spectroscopy (DWS) [103–108]. One aspect that has been studied is the consequences of internal reflectivity at the boundaries of a disordered medium on multiply scattered light. To account more accurately for the behavior near the boundaries, photon diffusion theories chose boundary conditions that uniquely specify the diffuse photon density, for a given geometry and source term, to compensate for the shortcomings of the diffusion approximation [109, 110]. The boundary condition chosen is such that the photon density becomes zero at a distance $z_e \Lambda_{eff}$ outside the sample. z_e is a phenomenological parameter called the extrapolation length ratio, and for consistency, it is taken as

$$z_e = \frac{2}{3} \frac{1 + R_2}{1 - R_1},\tag{28.36}$$

where the R's are moments of the total reflectivity $R(\mu)$ of a photon striking a boundary at an angle $\cos^{-1}(\mu)$ and are defined as

$$R_n = \int_0^1 (n + 1)\mu^n R(\mu)d\mu.\tag{28.37}$$

Based on the diffusion approximation and using the aforementioned boundary conditions, Vera and Durian [110] recently found that the probability $P(\mu_e)$ of a photon being transmitted between angles $\cos^{-1}(\mu_e)$ and $\cos^{-1}(\mu_e + d\mu_e)$ from the exterior normal is given by

$$\frac{P(\mu_e)}{\mu_e} = \frac{3}{2}\left(\frac{n_e}{n_i}\right)^2 (z_e + \mu_i)[1 - R(\mu_i)],\tag{28.38}$$

where $R(\mu_i)$ is the total reflectivity for a photon striking the boundary at $\cos^{-1}(\mu_i)$ from the interior normal. This shows that the angular distribution of the light transmitted depends exclusively on the internal reflectivity, giving rise to a method to determine this. Vera and Durian [110] compared their expression, and its further generalizations to polarization dependence, with both random-walk computer simulations and experiments on glass frits, colloidal suspensions, and foams. They could confirm that, under the conditions for which the diffusion approximation is valid, the polarization effects come exclusively from photon reflections at the boundary and that the transmitted angular dependence is hardly affected by scattering anisotropy. In the case of the colloidal suspensions (PS latex particles), the exterior and interior boundaries and their respective refractive indexes, are given, so by varying these values, they could verify Eq. 28.38. The experimental data confirmed the theory remarkably well. For the glass frits, the boundary is rough and not specified, yet the experimental data conformed accurately to the μ dependence of the theory. By fitting the experimental data with Eq. 28.38 they could provide an estimate of the reflectivity. Finally, in the case of aqueous foams, again the theory reproduced the major features of the data.

Based on the relationship of the time-dependent autocorrelation function with the internal reflectivity of the boundaries, other techniques have been proposed to measure the internal reflectivity. Zhu et al. [109] obtained explicit expressions for the angular autocorrelation functions of a specimen that is rotated and showed that the angular autocorrelation functions depended on sample thickness, absorption, and internal reflection, giving rise to a new method to determine internal reflectivity.

The development of DWS in this last decade has brought new insight into the understanding of the interaction of light with a random medium. It provides information about the dynamics of turbid media and has been applied to numerous multiple-scattering systems, including colloidal suspensions [107, 111] , emulsions [112], foams [113], and porous media [107]. Very recently, Stark and Lubensky [114] provided a general framework for the study of the propagation of light in anisotropic random media—that is, media with dielectric anisotropy. In particular, they discussed in detail light diffusion in nematic liquid crystals and the generalization of DWS to random anisotropic media. DWS measures the fluctuations in the transmitted light, expressed in terms of the normalized electric-field autocorrelation function, as in common dynamic light scattering (DLS). However, because the measured intensity fluctuations arise from phase shifts that come from many scattering events, the time scale measured in DWS is much shorter than in DLS.

From the previous discussion, we see that the theory of light scattering by random media has reached a degree of maturity and can provide useful information about the structure and dynamics of complex systems. In spite of this fact, the author has not found any reference to the application of this field to the study of polymer blends with complex morphologies.

IV. CONCLUSIONS

In this chapter, we discussed how the structure of a material influences the directional distribution of the light transmitted or reflected by a specimen. For nonabsorbing materials, it was shown that two basic lengths define the transport of light in random media. In the bulk, these lengths are the photon mean free path and the transport mean free path. The first defines the fraction of the incident beam that is transmitted coherently and therefore rules transparency, and the second determines how diffusive the material is. It was shown that for an almost transparent specimen, basic appearance parameters such as haze and direct transmittance could be related to the optical properties of single-scattering elements. It was indicated in the case of translucent materials how the main parameters of a four-flux theory, and therefore the total transmittance and reflectance, could be also reduced to effective single elements (e.g., particles, domains, etc.) in scattering calculations. When the scattering is produced by surface roughness, the RMS average height and the height correlation length define the strength and directionality of the diffuse field. A theory of gloss was proposed in terms of these parameters, and, in particular, the effect of surface height correlations on gloss was emphasized. It was also pointed out that there is insufficient experimental information about the statistical nature of polymeric surfaces and that, with the exception of a few cases, there is no theoretical framework relating surface roughness to material and process variables. Finally, the relation between the transmitted or reflected light and the transport mean free path gives us new methods to study the structure of a multiphase material by using Λ_{eff} as a probe of the medium structure. New techniques, such as DWS and other multiple-light-scattering techniques, could, in principle, provide useful information about the structure and dynamics of multiphase polymeric materials.

V. REFERENCES

1. P. L. Clegg and N. D. Huck, *Plastics*, **26** ,114 (1961).

2. N. D. Huck and P. L. Clegg, *SPE Trans*, **1**, 121 (1961).

3. J. L. White and M. Cakmak, *Advances in Polymer Technology*, **8(1)**, 27 (1988).

4. Z. Tadmor and C. G. Gogos, *Principles of Polymer Processing*, John Wiley & Sons, New York, 1979.

5. R. L. Hoffman, "Bump Generation by rigid spheres in an extensional flow field near a liquid-fluid interface," in *Advances in Rheology, 2. Fluids, Proceedings of the IX International Congress on Rheology*, B. Mena, A. García Rejón and C. Rangel Nefaile (eds.), UNAM. México D.F., 565 (1984).

6. R. L. Hoffman, *J. Rheol.*, **29**, 579 (1985).

7. L. Fritch, *Plast. Eng.*, **35(5)**, 68 (1979).

8. F. W. Billmeyer and M. Saltzmann, *Principles of Color Technology*, 2nd Ed., John Wiley & Sons, New York, 1981.

9. R. S. Hunter, *The Measurement of Appearance*, Wiley-Interscience, New York, 1975.

10. F. M. Willmouth, "Transparency, Translucency and Gloss," in *Optical Properties of Polymers*, Meeten, G. H. (ed.) Elsevier Applied Science Publishers, London, 1986, p. 265–333.

11. E. Hecht, *Optics*, 2nd Ed., Addison-Wesley Publishing Company, New York, 1987.

12. C. Smart, R. Jacobsen, M. Kerker, J. P. Kratohvil, and E. J. Matijevic, *J. Opt. Soc. Am.*, **55**, 947 (1965).

13. D. H. Woodward, *J. Opt. Soc. Am.*, **54**, 1325 (1964).

14. A. Ishimaru, Y. Kuga, R. L.-T. Cheung, and K. Shimidzu, *J. Opt. Soc. Am.*, **73**, 131 (1983).

15. A. Ishimaru, *Wave Propagation and Scattering in Random Media*, Vols 1 and 2, Academic Press, New York, 1978.

16. B. J. Brinkworth, *Br. J. Appl. Phys.*, **15**, 733 (1964).

17. B. J. Brinkworth, *J. Phys. D, Appl. Phys.*, **4**, 1105 (1971).

18. B. J. Brinkworth, *J. Phys. D, Appl. Phys.*, **5**, 476 (1972).

19. B. J. Brinkworth, *Appl. Optics*, **11**, 1434 (1972).

20. L. F. Gate, *J. Phys. D, Appl. Phys.*, **4**, 1049 (1971).

21. L. F. Gate, *J. Phys. D, Appl. Phys.*, **5**, 837 (1972).

22. L. F. Gate, *J. Opt. Soc. Am.*, **68**, 312 (1973).

23. L. F. Gate, *Appl. Optics*, **13**, 236 (1974).

24. For an excellent discussion on this subject see [10].

25. H. C. Van de Hulst, *Light Scattering by Small Particles*, Wiley, New York, 1957.

26. M. Kerker, *Scattering of Light and Other Electromagnetic Radiation*, Academic, New York, 1969.

27. C. F. Bohren and D. R. Huffman, *Absorption and Scattering by Small Particles*, John Wiley & Sons, New York, 1983.

28. P. Kubelka and F. Munk, *Z. Tech. Phys.*, **12**, 593 (1931).

29. P. Kubelka, *J. Opt. Soc. Am.*, **38**, 448 (1948).

30. P. Beckmann and A. Spizzichino, *The Scattering of Electromagnetic Waves from Rough Surfaces*, Pergamon, Oxford, 1963.

31. J. A. Ogilvy, *Theory of Wave Scattering from Random Rough Surfaces*, Adam Hilger, Bristol, UK, 1991.

32. H. E. Bennett and J. O. Porteus, *J. Opt. Soc. Am.*, **51**, 123 (1961).

33. E. R. Méndez, R. G. Barrera, and R. Alexander-Katz, *Physica A*, **207**, 137 (1994).

34. T. R. Lettieri, E. Marx; Jun-Feng Song, and T. V. Vorburger, *Appl. Opt.*, **30**, 4439 (1991).

35. J. M. Bennett and L. Mattsson, *Introduction to Surface Roughness and Scattering*, Optical Society of America, Washington, DC, (1989). See especially p. 52.

36. J. M. Haudin, "Optical Studies of Polymer Morphology," in *Optical Properties of Polymers*, G. H. Meeten (ed.), Elsevier Applied Science Publishers, London, 1986, p. 167–264.

37. Standard Test Method for Haze and Luminous Transmittance of Transparent Plastics, ASTM Standard D1003, 1961.

38. R. J. Tabar, C. T. Murray, and R. S. Stein, *J. Polym. Sci. Polym., Phys. Ed.*, **21**, 831 (1983).

39. Standard Test Method for Transparency of Plastic Sheeting, ASTM Standard D 1746, 1970.

40. Standard Test Methods for Measurement of Gloss of High-Gloss Surfaces by Goniophotometry, ASTM Standard E 430-91, 1991.

41. C. Yeh, *Phys. Rev.*, **135A**, 1193 (1964).

42. V. A. Erma, *Phys. Rev.*, **179**, 1238 (1969).

43. E. M. Purcell and C. R. Pennypacker, *Astrophys. J.*, **186**, 705 (1973).

44. S. Strom, *Am. J. Phys.*, **43**, 1060 (1975).

45. C. Yeh and K. K. Mei, "On the Scattering from Arbitrary Shaped Inhomogeneous Particles—Exact Solutions," in *Light Scattering by Irregularly Shaped Particles*, D. Schuerman (ed.), Plenum, New York, 1980, p. 201–206.

46. Z. Chen, P. Sheng, D. A. Weitz, H. M. Lindsay, M. Y. Lin, and P. Meakin, *Phys. Rev. B.*, **37**, 5232 (1988).

47. M. Y. Lin, H. M. Lindsay, D. A. Weitz, R. C. Ball, R. Klein, and P. Meakin, *Nature*, **339**, 360 (1989).

48. M. Y. Lin, H. M. Lindsay, D. A. Weitz, R. C. Ball, R. Klein, and P. Meakin, *Proc. R. Soc. Lond. A*, **423**, 71 (1989).

49. R. Alexander-Katz, *J. Polym. Sci., Polym. Phys. Ed.*, **31**, 663 (1993).

50. R. Alexander-Katz, *Phys. Rev. A*, **42(11)**, 6816 (1990).

51. R. S. Stein and M. B. Rhodes, *J. Appl. Phys.*, **31**, 1873 (1960).

52. S. Clough, J. Van Aartsen, and R. S. Stein, *J. Appl. Phys.*, **36**, 3072 (1965).

53. S. Clough, R. S. Stein, and C. Picot, *J. Polym. Sci., Part A-2*, **9**, 1147 (1971).

54. D. Y. Yoon and R. S. Stein, *J. Polym. Sci., Polym. Phys. Ed.*, **12**, 735 (1974).

55. R. J. Samuels, *J. Polym. Sci., Polym. Phys. Ed.*, **12**, 1417 (1974).

56. R. S. Stein and C. Picot, *J. Polym. Sci., Part A-2*, **8**, 2127 (1970).

57. R. E. Prud'homme and R. S. Stein, *J. Polym. Sci., Polym Phys. Ed.*, **11**, 1683 (1973).

58. R. J. Tabar, A. Wasiak, S. D. Hong, T. Yusa, and R. S. Stein, *J. Polym. Sci., Polym. Phys. Ed.*, **19**, 49 (1981).

59. C. Picot, R. S. Stein, M. Montegi, and H. Kawai, *J. Polym. Sci., Part A-2*, **8**, 2115 (1970).

60. A. E. M. Keijzers, J. J. Van Aartsen, and W. Prins, *J. Am. Chem. Soc.*, **90**, 3107 (1968).

61. R. S. Stein and W. Chu, *J. Polym. Sci., Part A-2*, **8**, 1137 (1970).

62. T. Hashimoto and R. S. Stein, *J. Polym. Sci., Part A-2*, **9**, 1747 (1971).

63. D. Y. Yoon and R. S. Stein, *J. Polym. Sci., Polym. Phys. Ed.*, **12**, 763 (1974).

64. R. J. Samuels, *J. Polym. Sci., Part C*, **13**, 37 (1966).

65. T. Hashimoto, S. Ebisu, and H. Kawai, *J. Polym. Sci., Polym. Phys. Ed.*, **19**, 59 (1981).

66. R. S. Stein and C. Picot, *J. Polym. Sci., Part A-2*, **8**, 1955 (1970).

67. R. E. Prud'homme and R. S. Stein, *J. Polym. Sci, Polym. Phys. Ed.*, **11**, 1357 (1973).

68. T. Hashimoto, A. Todo, and H. Kawai, *Polym. J. (Japan)*, **10**, 521 (1978).

69. G. H. Meeten, *Optica Acta*, **29**, 759 (1982).

70. F. C. Stehling, C. S. Speed, and L. Westerman, *Macromolecules*, **14**, 698 (1981).

71. H. Ashizawa, J. E. Spruiell, and J. L. White, *Polym. Eng Sci.*, **24**, 1035 (1984).

72. J. Bheda and J. E. Spruiell, *Polym. Eng. Sci.*, **26**, 736 (1986).

73. J. L. White, Y. Matsukura, H. J. Kang, and H. Yamane, *Int. Polym. Proc.*, **1**, 83 (1987).

74. K. J. Choi, J. E. Spruiell, and J. L. White, *J. Polym. Sci., Polym. Phys. Ed.*, **20**, 27 (1982).

75. Y. Shimomura, J. E. Spruiell, and J. L. White, *J. Appl. Polym. Sci.*, **27**, 2663 (1982).

76. D. R. Holmes, R. P. Palmer, R. G. Miller, and C. W. Bunn, *Nature*, **171**, 1104 (1953).

77. A. Keller, *Nature*, **174**, 826 (1954).

78. P. F. Smith, I. Chun, G. Liu, D. Dimitrievich, J. Rabsurn, and G. J. Vancso, *Polym. Eng. Sci.*, **36**, 2129 (1996).

79. A. Keller and M. J. Machin, *J. Macromol. Sci. Phys.*, **B1**, 41 (1967).

80. N. H. Sung, H. Y. Lee, P. Yuan, and C. S. P. Sung, *Polym. Eng. Sci.*, **29**, 791 (1989).

81. W. T. Welford, *Opt. Quant. Elec.*, **9**, 269 (1977).

82. S. N. Magonov and D. H. Reneker, *Annu. Rev. Mater. Sci.*, **27**, 175 (1997).

83. Standard Test Method for Specular Gloss, ASTM Standard D 523–589, 1989.

84. R. Alexander-Katz and R. G. Barrera, *J. Polym. Sci.: Part B: Polym. Phys.*, **36**, 1321 (1998).

85. B. J. Hoenders, E. Jakeman, H. P. Baltes, and B. Steinle, *Optica Acta*, **26**, 1307 (1979).

86. R. Alexander-Katz, "Optical Properties of Polymers under Process," in *Polymers: From Polymerization to Properties*, J.-Y. Cavaillé, M. García-Ramírez and G. Vigier (eds.), Polythecnica, Paris, 1996, p. 291–298.

87. R. López, R. Alexander-Katz, J.-Y. Cavillé, and J. Castro "Efectos de la Rigidez de las Partículas en el Brillo de un ABS: Determinación de las Propiedades Mecánicas de la Fase Dispersa," in *Avances en Ingeniería Química*, AMIDIQ, México, 1994, p. 151.

88. M. B. Khan and C. Keener, *Polym. Engl. Sci.*, **36**, 1290 (1996).

89. M. B. Khan, *Polym. Plast. Technol. Eng.*, **34**, 989 (1995).

90. M. B. Khan, B. J. Briscoe, and S. M. Richardson, *Polym. Plast. Technol. Eng.*, **33**, 295 (1994).

91. A. Schuster, *Astrophys. J.*, **21**, 1 (1905).

92. S. Chandrasekhar, *Radiative Transfer*, Oxford University Press, London and New York, 1950, republished by Dover Publications, Inc., New York, 1960.

93. P. S. Mudgett and L. W. Richards, *Appl. Opt.*, **10**, 1485 (1971).

94. L. Henyey and J. Greenstein, *Astophys. J.*, **93**, 70 (1941).

95. G. H. Meeten, "Polymer Latex Optics," in *Optical Properties of Polymers*, G. H. Meeten (ed.), Elsevier Applied Science Publishers, London, 1986, p. 335–392.

96. B. Maheu, J. N. Letoulouzan, and G. Gouesbet, *Appl. Opt.*, **23**, 3353 (1984).

97. B. Maheu and G. Gouesbet, *Appl. Opt.*, **25**, 1122 (1986).

98. G. A. Niklasson, *Appl. Opt.*, **26**, 4034 (1987).

99. H. C. Van de Hulst, *Multiple Scattering. Tables, Formulas and Applications*, Vol. 1 and 2, Academic, New York, 1980.

100. L. Ferry, G. Vigier, R. Alexander-Katz, and C. Garapon, *J. Polymer Sci.: Part B: Polym. Phys.*, **36**, 2057 (1998).

101. R. Alexander-Katz, L. Ferry, and G. Vigier, *J. Polymer Sci.: Part B: Polym. Phys.*, **36**, 2069 (1998).

102. S. Fitzwater and J. W. Hook III, *J. Coat. Technol.*, **57**, 39 (1985).

103. A. A. Golubenntsev, *Zh. Eksp. Teor. Fiz.*, **86**, 47 (1984). [*Sov. Phys. JETP*, **59**, 26 (1984).]

104. G. Maret and P. E. Wolf, *Z. Phys. B.* **65**, 409 (1987).

105. D. J. Pine, D. A. Weitz, P. M. Chaikin, and E. Herbolzheimer, *Phys. Rev. Lett.*, **60**, 1134 (1988).

106. M. J. Stephen, *Phys. Rev. B.* **37**, 1 (1988).

107. D. J. Pine, D. A. Weitz, G. Maret, P. E. Wolf, E. Herbolzheimer, and P. M. Chaikin, "Dynamical Correlations of Multiply Scattered Ligh," in *Scattering and Localization of Classical Waves in Random Media*, Vol. 8 of *Directions in Condensed Matter Physics*, P. Sheng (ed.), World Scientific, Singapore, 1990, p. 312–372.

108. D. A. Weitz and D. J. Pine, "Diffussing-Wave Spectroscopy," in *Dynamic Light Scattering: The Method and Some Applications*, W. Brown (ed.), Claredon, Oxford, 1993, p. 652.

109. J. X. Zhu, D. J. Pine, and D. A. Weitz, *Phys. Rev. A*, **44**, 3948 (1991).

110. M. U. Vera and D. J. Durian, *Phys. Rev. E*, **53**, 3215 (1996).

111. X. L. Wu, D. J. Pine, J. S. Huang, P. M. Chaikin, and D. A. Weitz, *J. Opt. Soc. Am. B*, **7**, 15 (1990).

112. H. Gang, A. H. Krall, and D. A. Weitz, *Phys. Rev. Lett.*, **73**, 3435 (1995).

113. D. J. Durian, D. A. Weitz, and D. J. Pine, *Science*, **252**, 686 (1991).

114. H. Stark and T. C. Lubensky, *Phys. Rev. E*, **55**, 514 (1997).

29　Thermomechanical Performance of Polymer Blends

J. J. SCOBBO, JR.

GE Plastics
General Electric Company
Selkirk, NY 12158

I. INTRODUCTION

One of the key aspects of engineering performance is the ability of a material system to withstand loads at elevated temperatures. For thermoplastic resin systems, load-bearing ability decreases with increasing temperature and can deteriorate further through glass transitions and melting transitions. One way to enhance engineering performance at elevated temperatures is through the blending of polymers.

Polymer Blends, Volume 2: Performance. Edited by D. R. Paul and C. B. Bucknall.
ISBN 0-471-35280-2.　© 2000 John Wiley & Sons, Inc.

End-use applications dictate the target performance when one is either designing or specifying an engineering blend. This chapter discusses the blending strategies that one can employ to design a polymer blend system for use at temperatures for which single thermoplastics are insufficient. Various examples are used to illustrate amorphous–amorphous and crystalline–amorphous blends.

One of the most important technological advances in the field of polymer blends has been that of reactive compatibilization. In this chapter, several blends are used as model systems to illustrate some of the basic concepts of crystalline–amorphous blends and the effects that compatibilization has on high-temperature modulus behavior. Specifically, the role of morphology in determining thermal performance is described.

This chapter focuses on modulus–temperature and modulus–temperature–time relationships. To that end, the experimental framework for discussion is based on relationships between common "data sheet" properties, such as heat distortion temperature (HDT) and modulus–temperature behavior as elucidated through dynamic mechanical analysis (DMA).

The fundamental driving force behind the development and commercialization of polymer blends is the requirement to provide materials with a balance of properties not obtainable with available polymers alone, while avoiding the time and cost penalties associated with the invention of new molecules.

Most applications of polymer blends have been geared towards the replacement of traditional materials, like metals. To be viable in such applications, polymer blend systems must mimic many of the performance characteristics of metals, such as toughness, stiffness, and the ability to withstand high temperatures and other aggressive environments.

This chapter deals with the interaction of two of the aforementioned performance criteria: stiffness and high-temperature resistance. The apparent stiffness of a object is due to two contributions. One is the intrinsic material property characterized by the modulus. The second is the contribution of geometry to stiffness, which is characterized by the moment of inertia. However, when one refers to thermal performance, the meaning is far more ambiguous than that for stiffness. "High-temperature capability" can refer to a brief exposure to an elevated temperature, like that for infrared soldering of electronic connectors. This exposure can be on the order of seconds. It can also refer to longer exposures, such as for paint-bake ovens, common for automotive applications, which can last on the order of 0.5 to 1 hour. Still other applications require the capability for continuous use at elevated temperatures.

There are numerous ways in which one can characterize polymer blends in terms of their ability to withstand elevated temperatures. Among these methods include the use of properties that are governed in large measure by transition temperatures, such as heat distortion temperature, and those governed by oxidative or chemical stability, such as relative thermal index, continuous-use temperature, and so on. This chapter deals with the former type of property. The objective of the chapter is to provide a framework for tailoring the properties of polymer blends by manipulating the contributions of the blends' components with respect to the modulus and the transition temperatures.

II. HEAT DISTORTION TEMPERATURE

The HDT, which is often referred to as the deflection temperature under load (DTUL) [1], is the temperature at which an arbitrary deformation occurs in a molded part (or sheet) subject to an arbitrary loading condition. In this measurement, a bar with a rectangular cross section is tested as a beam simply supported at both ends, with the load placed at its center. The load provides maximum fiber stresses of 0.46 MPa (66 psi) or 1.82 MPa (264 psi). For adequate temperature control, the bar is submerged in a bath containing a heat-transfer medium (i.e., an oil bath). For typical 12.5 mm × 12.5 mm (0.5 × 0.5 inch) bars, the force is applied perpendicular to the direction of molding pressure.

The procedure consists of submerging a conditioned bar of the appropriate geometry in the bath. A load is then applied to the center of the bar. A dial gauge that measures the deflection of the bar under load is zeroed. Temperature is increased at 2°C/min, and the point at which the bar deflects 0.25 mm (0.01 inch) is recorded as the deflection temperature under load, or the heat distortion temperature.

III. DYNAMIC MECHANICAL ANALYSIS

Dynamic Mechanical Analysis (DMA) is a method that measures the stiffness and mechanical damping (i.e., the internal friction and thermal dissipation) of a cyclically deformed material as a function of temperature [2]. The combination of stiffness and damping properties is a reflection of the unique viscoelastic nature of polymers. One of the most common uses of DMA is to study molecular-level thermal transitions in polymeric materials. The most conspicuous of these transitions is the glass transition temperature T_g. Below this temperature, an amorphous polymer is a glass, and thermal energy is insufficient to cause rotation and translation of molecular segments. Above the transition temperature, micro-Brownian motion of the molecular segments can take place, and the resin is rubbery or can flow readily.

In the DMA test, small cyclic deformations are applied to the sample. Because of the dynamic nature of the test, material response is composed of two elements. The first, the storage modulus, or elastic modulus, denoted E', is the in-phase portion of the response. It represents the purely elastic component of the material's behavior— that is, the portion of the viscoelastic response that is like an elastic spring, or Hookean material, which stores energy as it is deformed and can return all of this energy as the imposed deformation is released. Likewise, the out-of-phase component of the material's behavior is called the loss modulus, denoted E''. It represents the energy that is lost to viscous dissipation or internal friction between the molecules— that is, the portion of the viscoelastic response that dissipates energy like a viscous fluid. The ratio of the loss modulus to the storage modulus is the loss tangent, or tan δ. Please refer to Chapter 10 for further discussion of DMA.

Figure 29.1 shows a typical example of data obtained from a DMA test for an amorphous resin system. A semilog plot is used, since modulus values vary over several orders of magnitude for the temperature range of interest. As the temperature is

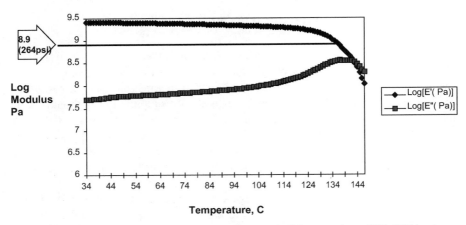

Figure 29.1   Dynamic mechanical spectra for a miscible amorphous PPE–PS blend.

increased, the storage modulus, E' gradually decreases, representing the softening of the resin. Simultaneously, viscous dissipation, which is represented by the loss modulus E'' increases until it reaches a maximum at the glass transition temperature—that is, until there is a peak in E''.

Some time ago, Takemori [3] reported on the relationship between the modulus and the HDT. From this perspective, it is easy to see how DMA can be used as a tool to measure the same thermal performance as HDT measures [4, 5]. For example, the equation for the center deflection Δ of a simply supported beam is given by

$$\Delta = FL^3/48EI, \tag{29.1}$$

where F is the load (i.e., 1.82 MPa), L is the distance between end supports, E is the modulus, and I is the beam's moment of inertia. This equation can be rewritten as

$$\Delta = \sigma_{max}L^2/6Ed, \tag{29.2}$$

where σ_{max} is the maximum outer-fiber stress and d is the width of the beam. Rewriting this equation, one can solve for the modulus E:

$$E = \sigma_{max}L^2/6\Delta d. \tag{29.3}$$

In Takemori's analysis, creep and thermal expansion effects were considered negligible, except in cases concerning very high temperatures and long times. For the case of the HDT test, one can replace the terms in Eq. 29.3 and solve for E as follows. The maximum stress is 1.82 MPa (i.e., for 264 psi), and the beam span is approximately 125 mm. The deflection is actually the the sum of the initial deflection and the deflection when HDT is registered, i.e., 0.25 mm. The initial deflection is typically 0.08 mm. Finally, the width, or thickness, of the bar is 12.5 mm. Therefore, E is approximately 0.75 GPa.

Since DMA is typically performed over a wide temperature range, it is common to represent the modulus data on a logarithmic scale. On a logarithmic scale (in units of Pa), the previous modulus value reduces to log $E' = 8.9$. Likewise, for the case of a 0.45-MPa load, this logarithmic scale value can be calculated to be 8.3.

These values give points at which HDT can be determined if modulus–temperature data are available. As discussed previously, DMA provides these data because E' and E'' are functions of temperature. E' is an ideal surrogate for E in Eq. 29.3, since over the temperature range of interest (up to approximately the glass transition temperature of the polymer), E' dominates E''. Furthermore, it is the glass transition temperature of an amorphous polymer that is responsible for the precipitous drop in modulus that is reflected in the HDT. It is not until well above T_g that E'' begins to dominate the viscoelastic behavior.

IV. AMORPHOUS BLENDS

A. Miscible Blends

Homopolymer blend pairs with complete thermodynamic miscibility over the entire composition range are quite rare. One exception is the system of poly(phenylene ether) with polystyrene (PPE–PS). Figure 29.1 shows a plot of the dynamic mechanical behavior of a commercial PPE–PS system, primarily made up of the PS component, wherein the temperature was increased at 2°C/min to mimic HDT conditions. If one draws a horizontal line at approximately 8.9 on the log E' (Pa) axis (corresponding to the high-load HDT), the intersection with the E' curve will correspond to the heat distortion temperature, read directly from the x-axis. In this case, the HDT is about 134°C, which corresponds to the value obtained from the ASTM HDT test [1]. As seen in Fig. 29.1, it is the glass transition temperature of the blend that determines the heat distortion temperature.

Figure 29.2 makes this concept easier to see via a schematic representation of the behavior seen in Fig. 29.1. Since a completely miscible blend system, such as a poly(phenylene ether)–polystyrene (PPE–PS) blend, can be formulated to have any glass transition temperature between that of PS (about 100°C) and that of PPE (about 210°C), it is expected that the HDT for this range of compositions would run from just below 100°C to just below 200°C. In fact, this attribute is one key discriminating factor that determines applications for commercial PPE–PS compositions. In Fig. 29.2, it is easily seen that increasing the T_g of the miscible blend would result in a higher HDT.

B. Fiber-Filled Blends

Figure 29.3 shows an example of essentially the same base polymer system as that in Fig. 29.1, but with glass fibers present in the composition as a reinforcement. The glass transition temperature of the blend is similar to that of the material in Fig. 29.1, slightly above 134°C. The key feature of this composition is the reinforcing effect

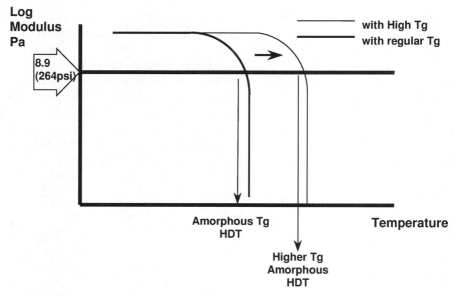

Figure 29.2 Modulus–temperature behavior for miscible amorphous blends.

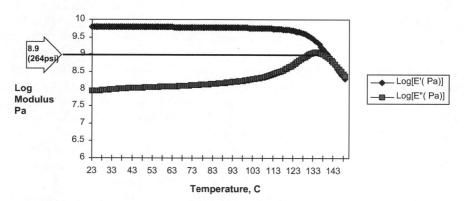

Figure 29.3 Dynamic mechanical spectra for a 30% glass-filled PPE–PS blend.

of the filler. Glass fibers are used to increase the storage modulus E' at the lower temperatures.

A schematic representation of the effect of fillers on the modulus–temperature behavior of miscible blends is shown in Fig. 29.4. Typically, unfilled polymers and blends are evaluated at the lower load (outer-fiber stress = 0.45 MPa, or 66 psi). Filled systems are evaluated at the higher load (outer-fiber stress = 1.8 MPa, or 264 psi). This is to allow for more differentiation among filled systems that have an inherently higher modulus relative to that of their unfilled counterparts. As Fig. 29.4 points out, higher load testing is more discriminating when one compares a filled

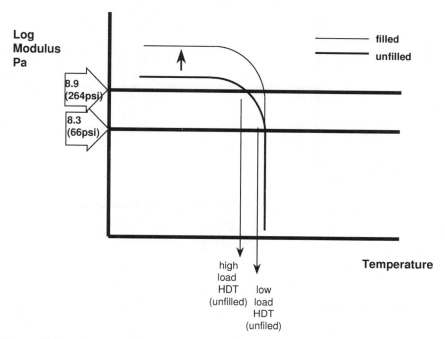

Figure 29.4 Modulus–temperature behavior for an $x\%$ glass-filled miscible amorphous blend.

system to the unfilled base system. Filler loading, filler type, resin modulus, and resin glass transition temperature all affect HDT. The major point that must be emphasized is that in increasing the HDT of a miscible blend system, there is far more advantage to be gained from significantly increasing the glass transition temperature of one or more components than from adding filler.

C. Partially Miscible Blends

Partially miscible blend pairs exhibit two distinct glass transition temperatures. However, the two T_g's may be shifted towards each other, with the shift being dictated by the amount of each phase that is dissolved in the other or by partitioning of oligomers or additives. With respect to the modulus and temperature, several possibilities exist.

For example, let us consider the case of polycarbonate–ABS (PC–ABS) blends. The SAN phase of the ABS resin has a T_g of about 110°C, while the PC has a T_g of about 145°C. In the blended formulation, the SAN T_g is shifted upward, while the PC T_g is depressed. Figure 29.5 is a representation of a PC–ABS blend formulated with PC (component 2) as the major component, such that the T_g of PC controls the HDT. Some of the reasons that PC and ABS are blended together are to provide exceptional low-temperature toughness (synergistic behavior), low melt viscosity for high shear molding of thin walls (an attribute of SAN), and high-temperature resistance (an attribute of PC). If the amount of PC in the formulation is not sufficiently high, the

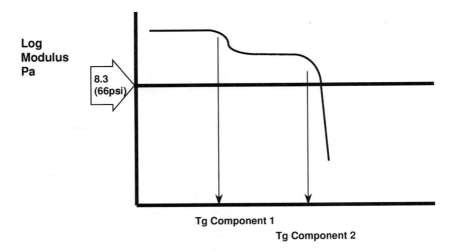

Figure 29.5 Modulus–temperature behavior for an immiscible or partially miscible amorphous blend.

PC will not act like a reinforcement, and the SAN T_g will determine the HDT of the blend. In other words, the HDT modulus threshold indicated by the horizontal line in Fig. 29.5 would intersect the modulus–temperature curve as the SAN T_g is crossed.

V. CRYSTALLINE–AMORPHOUS BLENDS

A. Fillers

Even though many semicrystalline polymers typically can have melting points in excess of 200°C, they are usually limited in their modulus–temperature capability, because they have relatively low glass transition temperatures. For example, nylon 66 has a melting point of approximately 260°C, but its glass transition temperature is about 200°C lower, at 65°C. Likewise, poly(butylene terephthalate) (PBT) has a melting point of about 220°C and a T_g of about 45°C. These values compare unfavorably with the glass transition temperatures of amorphous materials, such as PS ($T_g = 100°C$), SAN ($T_g = 110°C$), PC ($T_g = 145°C$), and PPE ($T_g = 210°C$). It is frequently the relatively low T_g of semicrystalline thermoplastics that limits their use when load-bearing capability and dimensional stability are required at higher temperatures.

It has already been mentioned that glass fibers act as reinforcements for polymers, enhancing the modulus. The same concept is readily applied to semicrystalline thermoplastics, as shown schematically in Fig. 29.6. On the addition of glass fibers to semicrystalline nylons and polyesters, the HDT is increased dramatically. However,

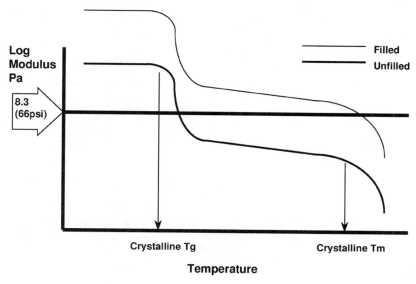

Figure 29.6   Effect of filler on the modulus–temperature behavior of a crystalline polymer.

glass fibers increase the specific gravity and, therefore, add weight and more cost per unit volume. They also cause a deterioration in surface appearance and increase the wear of injection molds and tooling.

To circumvent these issues, polymers are blended together in such a way that one of the components acts as an organic, melt-processable filler. The commercialization of nylon–PPE [6] and polyester–PC blends [7, 8] utilized this concept. Take the nylon–PPE case as an example; discussion of compatibility and compatibilization will be deferred to later in this chapter. If we consider the type of behavior depicted for unfilled nylon in Fig. 29.6 and combine that behavior with the concepts elucidated in Fig. 29.5 (wherein PC acts as a reinforcement for SAN), judicious formulation of the nylon–PPE ratio would result in a blend for which the T_g of PPE controls the heat distortion temperature. This situation is shown schematically in Fig. 29.7; actual DMA curves are shown in Fig. 29.8. In both of these figures, it is clearly seen that the PPE phase acts as a reinforcement up to the T_g of PPE. Formulations such as this one have been commercialized for external automotive components, which must be painted and then baked at elevated temperatures, and require Class A surface quality [9–14]. This concept is analogous to that for polyester–PC blends, for which the PC provides an enhanced modulus up to the T_g of PC.

Blends such as this can also incorporate fillers to boost the modulus incrementally, so that the nylon melting temperature controls the HDT. This approach allows less filler to be used, as compared to the amount of filler used for straight nylon–fiber compositions. A schematic representation of this case is provided in Fig. 29.9, with the experimental data for 0%, 10%, and 30% glass-filled PPE–nylon 66 provided in Fig. 29.10.

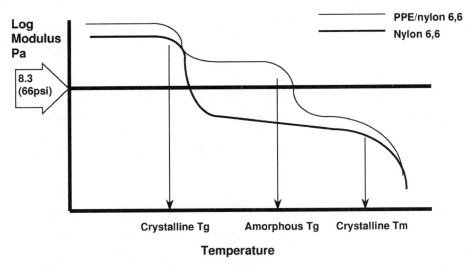

Figure 29.7 Modulus–temperature behavior for a crystalline–amorphous blend.

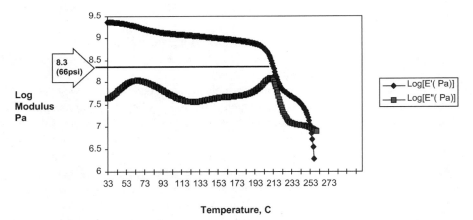

Figure 29.8 Dynamic mechanical spectra for a PPE–nylon 66 system.

B. Changing the Amorphous Phase

The concept of altering the glass transition temperature of either of the two components in a binary miscible blend to change the HDT can be applied to the more complicated semicrystalline–amorphous blends. That is, by blending the amorphous component with a third polymer with which it is mutually miscible, the T_g of the amorphous organic reinforcement phase can be altered, thus altering the HDT of the system. As is quite often the case, a miscible blend pair is not readily available for this purpose. However, there are frequently many available plasticizing schemes that the blend designer can utilize. By depressing the T_g, the HDT can be lowered. This

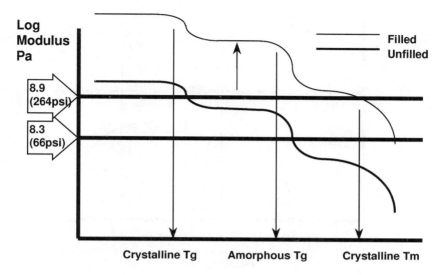

Figure 29.9 Effect of filler on modulus–temperature behavior for a crystalline–amorphous blend.

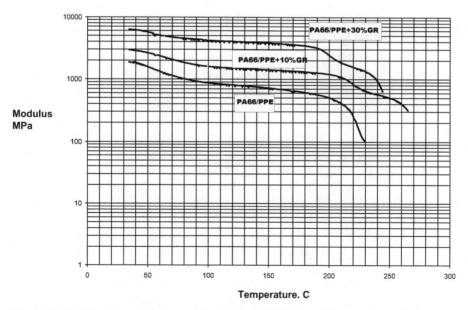

Figure 29.10 Modulus–temperature behavior for 0%, 10%, and 30% glass-filled PPE–nylon 66 blends.

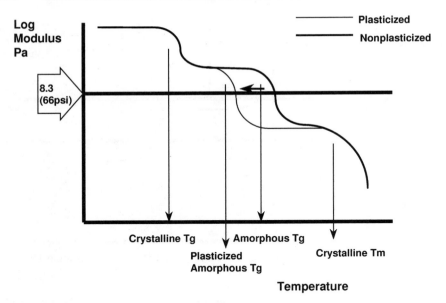

Figure 29.11   Effect of lower amorphous-phase T_g on the modulus–temperature behavior of a crystalline–amorphous blend.

effect is shown in Fig. 29.11. Likewise, antiplasticizers can be incorporated for the opposite effect.

C. Changing the Crystalline Matrix

The previous examples have focused on the manipulation of the amorphous polymer to alter HDT behavior. Changing the crystalline matrix provides additional degrees of freedom in design. For example, the crystalline phase may be "overengineered" for the demands of the end-use application. If heat distortion, which is controlled by the T_g of the amorphous phase, is the most critical parameter, then, all else being equal, the matrix phase should be chosen such that its T_m is at least as high as the T_g of the amorphous phase. A higher T_m does not afford benefits in HDT, but would require higher processing temperatures to fully melt the crystallites. A comparison of this sort is depicted in Fig. 29.12. Nylon 6 has a T_m just slightly above the T_g of PPE, so that the T_g of PPE and the T_m of nylon 6 merge to produce one precipitous drop in modulus. This concept can be applied to PET–PPE blends by replacing PET ($T_m = 255°C$) with PBT ($T_m = 220°C$) and to PET–PC blends, where in replacement of PET with PBT would provide a modulus plateau from above the T_g of PC up to the T_m of PBT.

Obviously, one can look at the aforementioned example from the opposite vantage point as well. For example, one can change the crystalline polymer to provide enhanced modulus–temperature behavior versus that of the "base case." Figure 29.13 is an example of such an approach, where in the PET matrix was replaced with

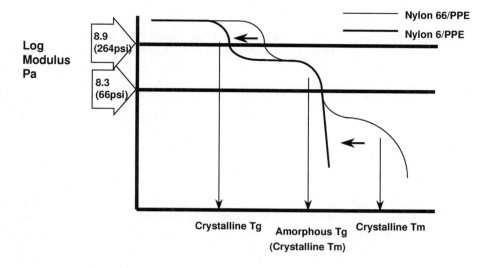

Figure 29.12 Effect of a lower T_g and lower T_m crystalline phase on the modulus–temperature behavior of a crystalline–amorphous blend.

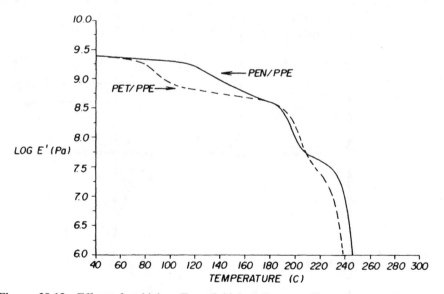

Figure 29.13 Effect of a higher T_g and higher T_m crystalline phase on the modulus–temperature behavior of a crystalline–amorphous blend.

poly(ethylene naphthalate), providing an enhanced modulus in the vicinity of the high-load HDT, without significantly affecting the low-load HDT behavior [15].

A blend of semicrystalline and amorphous polymers can be made in which the crystalline phase has a T_m that is below that of the amorphous material. Blends of polypropylene (PP) and PPE have been commercialized in Japan. In this case, the T_m of PP is approximately 165°C. Based on the physics of the aforementioned blend systems, the modulus–temperature behavior depicted in Fig. 29.14 is expected. In other words, the PPE phase provides little value above the T_m of PP and, in fact, would be expected to make the blend more difficult to compound and mold relative to a blend for which the T_g of the amorphous phase and the T_m of the crystalline phase T_m were more closely matched. In this case, it makes more sense to choose an amorphous polymer with a lower T_g and blend it with a third polymer or plasticize it.

An even more complicated case can be considered—namely, that of a ternary blend system consisting of two amorphous phases and one semicrystalline phase. An example of this type of system is a PBT–PC–PPE [16, 17]. The modulus–temperature behavior of this blend is seen in Fig. 29.15. For the particular formulation in this figure, the HDT is controlled by the T_g of PC. As previously discussed, the ratios of the components will dictate which phase controls the HDT. For the example in Fig. 29.15, there are several ways to increase the HDT. In one method, the T_g of the PC could be increased, as shown in the figure. In another method, one could alter the ratios of the components to make the T_g of PPE the controlling transition temperature. In a third scheme, a small amount of reinforcement could be added to shift the entire curve upwards, making the PPE phase control the HDT.

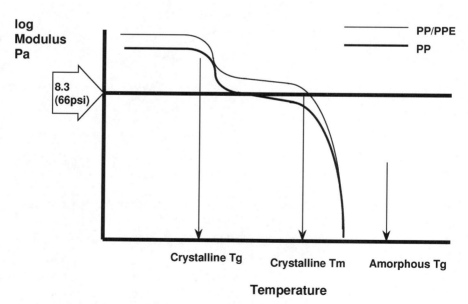

Figure 29.14   Modulus–temperature behavior of a crystalline–amorphous blend wherein the T_m of the crystalline phase is less than the T_g of the amorphous phase.

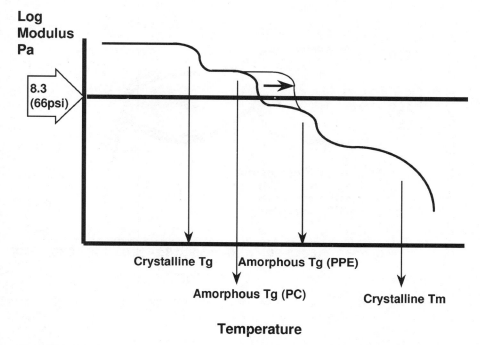

Figure 29.15 Modulus–temperature behavior for a crystalline–amorphous blend with two different amorphous phases.

D. Compatibilization

DMA has also been used to characterize the effect that compatibilization has on both the elastic and loss moduli of immiscible blends [18]. The specific subject of compatibilization is dealt with in great detail in Chapter 17, so only its relevance to modulus–temperature behavior will be discussed here.

Studies of compatibilization effects have focused on the behavior of the loss modulus, i.e., the shift in peak temperature and reduction in peak height that occur for the higher T_g [18]. An example of the effect of compatibilization on the DMA behavior of semicrystalline–amorphous blends is shown in Fig. 29.16. The effect of compatibilization on the DMA behavior of an amorphous–amorphous blend system is shown in Fig. 29.17. In each of these figures, data are presented as shear-storage and loss modululi (G' and G'', respectively). Noncompatibilized blends are indicated as having no copolymer, whereas compatibilized blends are labeled as "copolymer" to indicate that a compatibilizing copolymer is present.

It has been surmised that the attentuation in peak height is related to the chemical linking points between phases (i.e., the copolymer graft points) [18]. These links were thought to restrict the relaxation and energy dissipation at the higher T_g. The reason that little effect was seen at the lower T_g is that the presence of the glassy material of higher T_g in the interphase region provided constraints on molecular motion that dominated the constraints imposed by copolymer graft points. However,

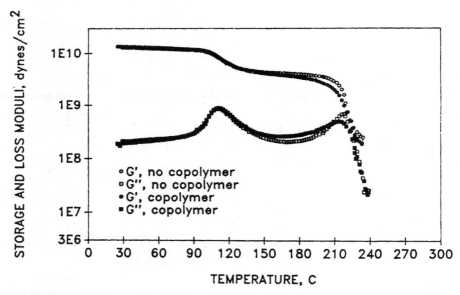

Figure 29.16 Effect of compatibilization on the dynamic mechanical spectra of a crystalline–amorphous blend of PPS and PEI.

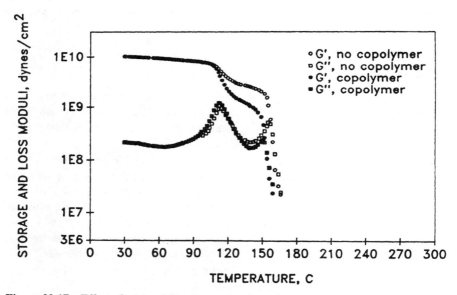

Figure 29.17 Effect of compatibilization on the dynamic mechanical spectra of an immiscible amorphous–amorphous blend.

in that work, little difference was seen in T_g values for the higher and lower T_g phases. (In those studies, rubber-phase transitions were not considered.)

It has further been noted that compatibilization increases the temperature dependence of the storage (elastic) modulus between the glass transition temperature T_g of the matrix and the discontinuous phases [19]. This effect is clearly seen in Figs. 29.16 and 29.17. Compatibilization was found to decrease the intensity of the loss peak at the higher glass transition temperature. This concept is especially important when one refers back to Fig. 29.8, wherein the modulus–temperature region of interest for HDT falls precisely in the vicinity where compatibilization effects are observed to be most significant.

To explain this behavior, it is beneficial to review a model for blend moduli. In the Nielsen model [20], a single adjustable parameter ϕ_{max} is used in conjunction with individual-phase moduli, Poisson's ratios, and geometrical parameters to predict the modulus of the blend, E. This model is given by

$$E = E_m(1 + AB\phi_d)/(1 - B\phi\phi_d), \qquad (29.4)$$

where

$$\phi = 1 + (1 - \phi_{max})\phi_d/\phi_{max}^2. \qquad (29.5)$$

In Eq. 29.4, A is a function of the dispersed-phase (filler) geometry and Poisson's ratio of the matrix and B is a function of the relative moduli of the maxtrix and the filler. The volume fraction of the filler phase is given by ϕ_d. The matrix modulus is E_m.

In this model, with all other parameters being constant, a higher ϕ_{max} gives a higher blend modulus. Only the phenomenological significance of this parameter will be considered here. This interaction parameter ϕ_{max} is proportional to $\{(R + \Delta R)/R\}^3$, where R is the radius of the dispersed phase (which is assumed to be spherical) and ΔR is the thickness of the immobilized matrix layer surrounding the dispersed phase. In the present context, we may consider this layer to be an interphase layer.

With the aforementioned considerations in mind, it was seen that the noncompatibilized systems have greater moduli than the compatibilized systems. If ϕ_{max} is considered to be the only adjustable parameter, then it would follow that a compatibilized system would have a smaller value of ϕ_{max} relative to that of the noncompatibilized system. For $1/\phi_{max}$ proportional to $\{(R + \Delta R)/R\}^3$, smaller particles would have a smaller ϕ_{max} value. Typically, compatibilized blends have dramatically reduced disperse-phase particle sizes. Therefore, the ratio of ϕ_{max} for compatibilized blends to ϕ_{max} of noncompatibilized systems would be enormous, assuming a constant ΔR.

Since ϕ_{max} is typically between 0.5 and 1.0, it is impossible to fit the experimental data to the model. The implication is that ΔR gets thicker as the particle gets smaller. Compatibilized systems are expected to have more interphase material with a composition gradient similar to that described by the model of Helfand [21]. (See Chapter 3.) Figure 29.18 is an idealized diagram of the phase morphology and interphase mixing for noncompatibilized and compatibilized blends. For detailed discus-

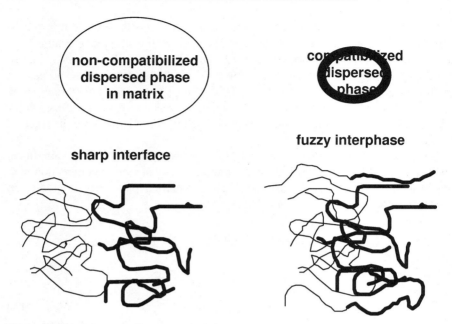

Figure 29.18   Schematic diagram of the interphase of noncompatibilized and compatibilized blends.

sions on aspects of polymer blend morphology, the reader is referred to Chapters 9, 15, and 16.

Returning to the modulus–temperature behavior, it can be argued that the increased temperature dependence of the modulus is a result of the thicker interphase region for the compatibilized blend. For the PPS–PEI system in Fig. 29.16, certain ratios of the two primary components of the blend can give rise to the behavior depicted in Fig. 29.19.

This modulus–temperature behavior appears to be a more specific example of the effects of morphological coarsening. Figure 29.20 shows similar storage-modulus behavior for a PC–SAN system as a function of morphology [22]. The trend is for finer morphology to show greater temperature dependence of the storage modulus.

It is widely accepted that compatibilization improves many of the physical properties of blends relative to those of noncompatibilized systems. However, increased dependence of the modulus on temperature can adversely affect the heat distortion behavior.

VI. HEAT DISTORTION TIME AND STRESS RELAXATION

The previous discussion has focused on one aspect of thermal behavior for these numerous applications, namely, the modulus as a function of temperature with a constant time scale, related to frequency. Another aspect, which is often overlooked

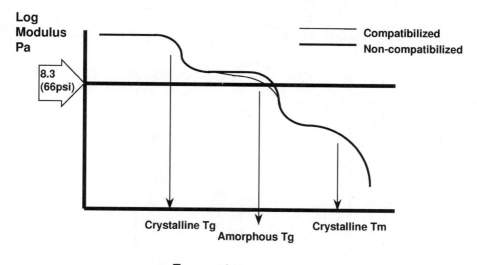

Figure 29.19 Modulus–temperature behavior showing the effect of compatibilization for a crystalline–amorphous blend.

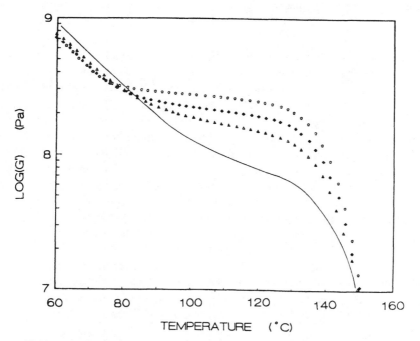

Figure 29.20 Modulus–temperature behavior as a function of morphology for a PC–SAN blend. Morphologies vary from the finest morphology (solid line) to the coarsest morphology (pentagons). Reproduced with permission of the Society of Plastics Engineers [22].

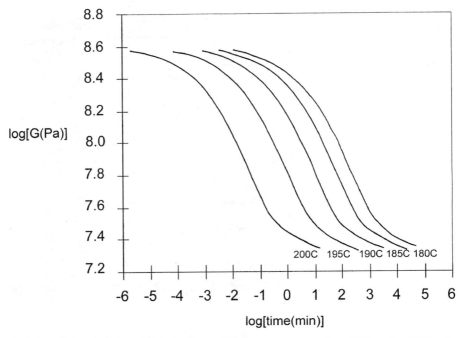

Figure 29.21 Modulus–time behavior at different temperatures for a PPE–nylon 66 blend.

from a thermal analysis point of view, is the modulus as a function of time at constant temperature. Data for these analyses are generated by stress relaxation experiments.

Figure 29.21 shows the modulus–time behavior for a blend of nylon 66 and PPE plotted as master curves for several reference temperatures. The range of temperatures on this plot represents a reasonable potential range for applications involving automotive paint-process ovens. It should be noted that although the use of time–temperature superposition is not strictly applicable to multiphase systems, plots of this sort are useful for comparing relative behavior. The drop in modulus that is observed in these plots is representative of the softening of the PPE phase. In this temperature range, the blend is sufficiently above the glass transitions of the nylon and rubber phases (the compostion also contains impact modifier) and is significantly below the melting transition of the nylon phase, so that these transitions have little effect on the curves.

One obvious feature of Fig. 29.21 is the dramatic difference in relaxation rates (softening rates) for 180°C and 200°C tests. At 200°C, the PPE phase softens almost instantly, which we would expect from inspection of modulus–temperature curves. However, for 180°C testing, 1000 minutes are required for the same degree of softening to occur, equivalent to 1 minute at 200°C. Figure 29.22 makes this effect clearer by replotting the data of Fig. 29.21 as isochronous curves (i.e., the modulus as a function of temperature for a constant time scale). A plot like that in Fig. 29.22 can be extremely useful, as is described next.

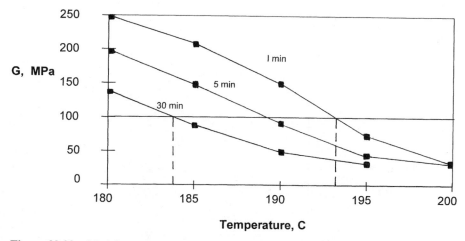

Figure 29.22   Modulus–temperature behavior for different time scales for a PPE–nylon 66 blend.

Let us assume that a criterion exists for thermal performance that is similar to HDT. However, whereas HDT is most concerned with temperature, this new criterion would deal with both temperature and time. As with HDT, such a test could be used to determine whether a material is suitable for a given application. Arbitrarily, let us say that if a material's shear modulus G softens to 100 MPa, then we conclude that the material it is too soft or flexible for the application. (This is exactly the concept behind HDT measurements.) But now, let us look at this process in the context of Fig. 29.22. As with the modulus–temperature plots we have examined, one can draw a line across the plot at our arbitrary softening criterion of $G = 100$ MPa. Again, when the material's modulus decreases to this level, the material fails the criterion for thermal performance. This line is included in Fig. 29.22. In a plot such as that in Fig. 29.22, one can now see how fast a material will "fail" under particular temperature conditions. For example, for a nylon–PPE blend to meet this criterion, only 1 minute is required for the blend to fail at a temperature of about 193°C. However, at 184°C, only 9°C less, this time is extended to 30 minutes. Knowledge of this time dependence cannot be obtained from HDT testing.

The previous example illustrates the following point: In the elevated temperatures of an environment such as an automotive paint-process oven, it is not only the oven temperature that must be taken into account, but also the time during which a plastic component is in the oven. Theoretically, short temperature spikes can be just as damaging to the part as long times, such as those experienced during production-line stoppages, at relatively milder temperatures. These considerations are important not only for exterior automotive applications, but also for any application in which the material is subject to elevated temperatures for any length of time.

Changes in material formulation alter the stress relaxation master curves in much the same way as they alter modulus–temperature curves. For example, a higher glass transition temperature of the dispersed phase would shift the curves to the right.

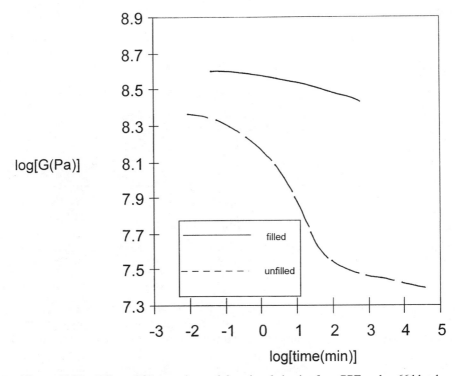

Figure 29.23   Effect of filler on the modulus–time behavior for a PPE–nylon 66 blend.

In any material, all else being equal, a higher glass transition temperature delays softening. For composite materials, the curves are shifted upwards and are flattened. An example of this behavior is shown in Fig. 29.23. In essence, all of the design parameters that one can use for improving the modulus–temperature performance of a blend, which have been discussed at some length in this chapter, can be applied to the design of blends with improved modulus–time performance.

VII. REFERENCES

1. ASTM D648, Standard Test Method for Deflection Temperature of Plastics Under Flexural Load, American Society for Testing Materials, West Coshocken, PA, 1997, p. 57.

2. T. Murayama, *Dynamic Mechanical Analysis of Polymeric Material*, Elsevier, New York, 1978.

3. M. Takemori, *SPE ANTEC*, **24**, 216 (1978).

4. J. J. Scobbo and P. L. Cometti, *SPE ANTEC*, **40**, 3326 (1994).

5. J. J. Scobbo, P. L. Cometti, and C. P. Strom, *SPE ANTEC*, **41**, 4024 (1995).

6. R. R. Gallucci, *SPE ANTEC*, **32**, 48 (1986).

7. S. Y. Hobbs, M. E. J. Dekkers, and V. H. Watkins, *J. Mat. Sci.*, **23**, 1219 (1988).

8. M. E. J. Dekkers, S. Y. Hobbs, and V. H. Watkins, *J. Mat. Sci.*, **23**, 1225 (1988).

9. C. Koevoets, D. Noordegraaf, and U. Hoffman, *Kunststoffe Plast. Europe*, p. 26 March 1996.

10. S. Y. Hobbs, M. E. J. Dekkers, and V. H. Watkins, *J. Mat. Sci.*, **24**, 2025 (1989).

11. S. Y. Hobbs and M. E. J. Dekkers, *J. Mat. Sci.*, **24**, 1316 (1989).

12. J. R. Campbell, S. Y. Hobbs, T. J. Shea, and V. H. Watkins, *Polym. Eng. Sci.*, **30**, 1056 (1990).

13. H. J. Sue and S. F. Yee, *J. Mat Sci.*, **26**, 3449 (1991).

14. H. J. Sue and A. F. Yee, *J. Mat Sci.*, **24**, 1447 (1989).

15. S. B. Brown, C. R. Hwang, S. T. Rice, J. J. Scobbo, and J. B. Yates, U.S. Patent 5,539,062, General Electric Co., 1996.

16. S. Y. Hobbs, M. E. J. Dekkers, and V. J. Watkins, *Polymer*, **29**, 1598 (1988).

17. S. Y. Hobbs, M. E. J. Dekkers, and V. J. Watkins, *Polymer*, **32**, 2150 (1992).

18. J. J. Scobbo, *Polym. Test.*, **10**, 279 (1991).

19. J. J. Scobbo, "Compabilization and Domain Size Effects in Multi-phase Polymer Blends as Observed by Dynamic Mechanical Analysis," Polymer Processing Society Conference, Akron, OH, April 1994.

20. L. E. Nielsen, *J. Appl. Phys.*, **41**, 4626 (1970).

21. E. Helfand, *Acc. Chem. Res.*, **8**, 295 (1975).

22. D. Quintens and G. Groeninckx, *Polym. Eng. Sci.*, **31**, 1207 (1991).

30 Barrier Materials by Blending

P. M. SUBRAMANIAN

S. P. M. Technologies
Hockessin, DE 19707

I. G. PLOTZKER

Central Research and Development
DuPont Company
Wilmington, Delaware 19880-0323

Polymer Blends, Volume 2: Performance. Edited by D. R. Paul and C. B. Bucknall.
ISBN 0-471-35280-2. © 2000 John Wiley & Sons, Inc.

I. INTRODUCTION

Prevention and control of the permeation of gases and liquids through plastic films, membranes, and other articles are required in several packaging and industrial applications, for purposes such as maintaining freshness and preventing degradation of food items, containing fuels and chemicals for storage, and preventing losses in pipes and tubes in the automotive, air conditioning, and other industries. In addition, many plastics are affected by exposure to solvents, which change their dimensions and other mechanical properties for engineering applications. For example, articles made from polyamides are significantly affected by moisture, and amorphous polymers, such as polycarbonate, are significantly affected by several solvents.

The general approach in formulating high-performance barriers is to use materials having low diffusivity and/or low solubility of the permeant. However, for good economic and processing reasons, these materials are often combined with other polymers that are of lower cost or that have other complementary properties. The combination of these polymers can be achieved in several ways. Thin coatings of the barrier polymer can be applied to the substrate from a solution, dispersion, or the melt. Surface coating with impermeable inorganic materials, such as aluminum or glass (SiO_2), is also practiced. Multilayer coextrusions with several layers of the barrier polymer are also practiced. However, while these techniques are highly effective for their respective needs, they involve expensive machinery and time.

Permeability control through polymer–polymer blending offers an alternative, unique, simple, and economical approach for several applications. The science and technology for creating such blends have been studied extensively.

Table 30.1 lists several of the commonly used polymers for polymer–polymer blending. Each of these polymers has exceptional properties in selected areas. However, they have high susceptibility to several fluids. Enhancement of their barrier properties and solvent resistance would extend their utility in several applications.

The barrier properties and solvent resistance of the polymers given in Table 30.1 can be improved by blending them with a polymer that has complementary permeation properties. Table 30.2 lists several polymers that act as good permeation barriers, giving significantly lower solubility and diffusion coefficients against several fluids. Poly(vinylidene chloride) and poly(vinyl alcohol) have limited thermal stability and are used primarily in the form of coatings from solutions and dispersions. Poly(vinyl alcohol) is an outstanding barrier to oxygen when dry, but has poor properties under high humidity.

Table 30.1   Polymer Classes and Fluids with High Permeabilities

Polymer	Permeant
polyethylene	hydrocarbons, oxygen
polypropylene	hydrocarbons, oxygen
polyamide (aliphatic)	moisture
poly(ethylene terephthalate)	oxygen, carbon dioxide

Table 30.2   Polymers with Good Barrier Properties and Corresponding Permeants

Polymer	Permeant
aliphatic polyamides	hydrocarbons
semiaromatic polyamides (6-I, MXD-6)	oxygen
polyethylene	water vapor
polypropylene	water vapor
poly(ethylene-co-vinyl alcohol)	oxygen
poly(ethylene-co-carbon monoxide)	oxygen, solvents
aromatic liquid crystalline polymers	oxygen, water vapor, and most solvents and vapors
poly(vinyldene chloride)	oxygen, water vapor
poly(vinyl alcohol)	oxygen (when dry), solvents

The permeability properties and solvent sensitivity of polymers depend upon their crystallinity and polarity, characterized by the solubility parameter. Polyolefins, such as high-density polyethylene (HDPE), are outstanding barriers to moisture, but are highly permeable to hydrocarbons. Aliphatic polyamides, such as nylon 6 and nylon 66, have outstanding resistance to hydrocarbons, but they are highly permeable to water vapor.

In general, the goal is to add small amounts of a polymer with high barrier properties (generally more expensive) to the selected matrix polymer (generally low cost). The primary goal in studies of the barrier properties of a polymer blend is to learn how to optimize the amount of the barrier polymer, so as to minimize permeability without significantly affecting the desirable properties of the host, or matrix, polymer. Understanding and controlling the phase morphology of such blends is essential for optimizing performance.

In Section II of this chapter, we consider the permeability of miscible polymer blends. We examine how permeability may increase or decrease, depending on the nature and degree of interaction between the polymers in the blend and between the permeating substance and the polymers. In Section III, we describe models that predict the effects of specific phase morphologies on the permeability of immiscible blends. Section IV includes specific examples of the manipulation of morphology to improve blend performance in practical applications. Finally, in Section V, we discuss the special challenges pertaining to the effective use of liquid crystalline polymers as barrier materials in polymer blends.

II. PERMEABILITY OF MISCIBLE BLENDS

A. Overview

While immiscible polymer blends are far more common, miscible blends offer possibilities for fine-tuning material properties through specific interactions between the components. Such blends could be produced, for example, with the intention of tailoring gas permeability and can perform better than anticipated from the properties of the individual components. For example, Lee et al. found that blends of certain alkylsulfonylmethyl-substituted oxyalkylene polymers are better barriers against oxygen permeation than is either polymer individually; in fact, the best such blends rival EVOH as barriers, without exhibiting that polymer's moisture sensitivity [1].

As another example, polyimide separation membranes show high selectivity and excellent high-temperature performance and chemical resistance, but are expensive and are seriously plasticized by carbon dioxide above about 15 atm (the "critical plasticization pressure"). Polysulfone is less heat and chemical resistant, but it has similar permeability to H_2, CO_2, O_2, and N_2; is not plasticized by carbon dioxide below 50 atm; and is less expensive. Kapantaidakis and colleagues found that a miscible, 50/50 polysulfone–polyimide blend was chemically and thermally more resistant than polysulfone and had a critical plasticization pressure of about 30 atm CO_2, double that of polyimide alone [2].

Another goal can be to retain certain desirable features of the component polymers by suitably functionalizing one component to achieve miscibility with the other. Percec and Melamud did this in blending poly(ethylene-co-maleic anhydride) with poly(acrylonitrile-methacrylate-butadiene) terpolymer to produce a material with the low moisture permeability of polyethylene and the high gas barrier of the nitrile polymer [3].

Conversely, permeation of small gas molecules can be used to probe the morphology and the nature of interactions between polymer components. For example, oxygen permeability was used to demonstrate that "miscible" blends (i.e., blends that exhibit one glass transition temperature T_g) of poly(n-butyl acrylate) and n-butyl acrylate–vinylidene chloride copolymers are inhomogeneous on a molecular scale; the activation energy for permeation matched that of the more permeable butyl acrylate component, and the permeability was higher for the blends than for the compositionally equivalent copolymers [4].

B. Modeling Transport through Miscible Blends

Much work has been done over the last two decades, particularly by Paul and his colleagues, to address such issues as how to predict miscibility, differences between block copolymers and miscible blends, and the way in which component interactions can increase or decrease permeability. In an early example [5], they found that carbon dioxide sorption and transport in miscible blends of polycarbonate (PC) with a copolyester based on 1,4-cyclohexanedimethanol and isophthalic and terephthalic acids can be described using a "dual mode" model that is useful in explaining sorption and transport of a gas in a glassy polymer [6]. Part of the population of gas molecules is considered essentially immobilized in unrelaxed microvoids (Langmuir sites), while the remaining gas obeys Henry's law. Thus,

$$C = k_D p + C'_H bp/(1 + bp) \tag{30.1}$$

and

$$P = k_D D_D + C'_H b D_H/(1 + bp_2), \tag{30.2}$$

where C is the concentration of gas in the polymer, k_D is the Henry's law parameter, p is the gas-phase pressure, C'_H is the Langmuir capacity parameter, b is the Langmuir affinity parameter, P is the permeability coefficient, p_2 is the upstream driving pressure, and D_D and D_H are the diffusion coefficients for Henry's law species and Langmuir species, respectively. For many polymers, the parameter C'_H is proportional to the difference between the polymer's glass transition temperature and the measurement temperature. The parameters k_D, C'_H, and b are obtained from sorption measurements of C vs. p. Diffusion coefficients were lower than expected from additivity on a logarithmic scale of the component polymers' diffusion coefficients (ln D vs. composition). This result is consistent with the reduction in volume that occurs on mixing. The authors also used the Flory–Huggins theory to find a value for the polymer–polymer interaction parameter; this value was negative, suggesting strong interaction between the two polymers.

In the dual mode model used by Paul and Koros [6], the permeating species' concentration gradient in each mode is the driving force for diffusion. Petropoulos [7] expressed the total gas flux in terms of the local chemical potential gradient in order to derive the following expression for permeability:

$$P = k_D D_{TD} + C'_H b D_{TH}[\ln(1 + bp)]/bp. \tag{30.3}$$

In this equation, the T subscript denotes that these variables are thermodynamic diffusion coefficients for the Langmuir and Henry's law populations.

These two models both imply that there are no "cross currents," i.e., that there is no exchange of molecules between Langmuir and Henry's law modes in the steady state. A third model developed by Barrer [8] and Fredrickson and Helfand [9] includes exchange fluxes, with the following expression for permeability:

$$P = k_D D_D + 2C'_H bA[\ln(1 + bp)]/bp + [C'_H b D_H/(1 + bp)]. \tag{30.4}$$

This equation reduces to the Petropoulos expression if the mobility factor A is much larger than D_H and to the Paul–Koros expression if A is much smaller than D_H and D_D.

Chin, Gaskins, and Durning [10] studied the steady-state permeation of carbon dioxide through miscible, glassy blends of polyarylate with poly(butylene terephthalate), using these dual-mode models to determine CO_2 permeability in the limits of high and low pressure. With increasing pressure, CO_2 permeability decreased nonlinearly in a manner that could be explained satisfactorily by both the Petropoulos and Paul–Koros models, while the data were insufficient to determine all of the parameters in the Barrer–Fredrickson and Helfand model to any degree of certainty.

In a study by Harris, Paul, and Barlow [11], blends of aliphatic polyesters with the polyhydroxy ether of bisphenol-A ("phenoxy") were characterized by using melting-point depression and sorption measurements to obtain polymer–polymer interaction parameters. Of interest is the deviation from linear additivity of the equilibrium sorption of small amounts of probe solvent (denoted by subscript 1) molecules in blends of polymer (represented by subscript 2) and polymer (subscript 3). By Flory–Huggins analysis, Harris, Paul, and Barlow obtained the expressions

$$\ln a_1 = \ln \phi_1 + (1 - \phi_1) + (1 - \phi_1)^2 \chi_{1b} \tag{30.5}$$

$$\chi_{1b} = \chi_{12} \phi_2' + \chi_{13} \phi_3' - \chi_{23} \phi_2' \phi_3', \tag{30.6}$$

where a_1 is the activity of solvent vapor in equilibrium with a polymer, ϕ_i is the volume fraction of component i in the mixture, ϕ_i' is the volume fraction of component i in the mixture on a solvent-free basis, χ_{ij} is the Flory–Huggins interaction parameter for the interaction between components i and j in the mixture, and χ_{1b} is the Flory–Huggins interaction parameter between solvent 1 (e.g., carbon tetrachloride in this case) and the polymer blend. The solvent–polymer interaction parameters were nearly independent of solvent concentration above about 1% uptake. χ_{23} is then found from sorption data for a series of blends ranging in composition from pure polymer 2 (phenoxy) to pure polymer 3 (polyester), using Eq. 30.6. The polymer–polymer interaction parameters found using sorption data were reasonably similar to those found from melting-point depression and analogue calorimetry, all associating polymer–polymer miscibility with an exothermic heat of mixing—in this case, probably from hydrogen bond formation between phenoxy hydroxyls and ester groups.

One of the first miscible systems studied in depth was poly(phenylene oxide)–polystyrene (PPO–PS), whose importance in commercial applications derives primarily from its toughness. Morel and Paul [12] found that sorption and transport of CO_2 could be described quantitatively using the dual-mode approach described previously. Blend permeabilities were lower than expected from simple additivity of the component polymers' permeabilities (arithmetic or logarithmic). This finding is consistent with the strong interaction between PPO and PS, which leads to densification on mixing and thus to impeded transport. The diffusion coefficient D_D increased monotonically with increasing PPO content, while D_H exhibited a minimum at about 70% PPO.

Paul derived mixing rules for gas permeation in miscible polymer blends using both activated-state theory and free-volume theory [13]. Empirically, it had been determined that the permeability coefficient P for blends and copolymers often follows what is referred to as the semilogarithmic additivity rule:

$$\ln P = \phi_1 \ln P_1 + \phi_2 \ln P_2. \tag{30.7}$$

Here, ϕ_i is the volume fraction of the ith component and P_i is the component's permeability coefficient.

In the free-volume approach, the permeability is correlated with the specific volume V of the polymer, minus the volume "occupied" by the polymer chains, V_o [14, 15, 16]. One such correlation, used successfully by Lee [14], is simply

$$P = Ae^{-B/(V-V_o)}, \tag{30.8}$$

where A and B are constants for a specific gas. If the free volume of the mixture, $(V - V_o)$, is assumed to be just the sum of the contributions from each component, weighted by its volume fraction, one can show that

$$\ln(P/A) = \left[\phi_1 / \ln(P_1/A) + \phi_2 / \ln(P_2/A)\right]^{-1}. \tag{30.9}$$

This equation predicts that, where interactions between the two components are weak, such that there is no volume change on mixing, the permeability will show a positive deviation from the semilogarithmic additivity predicted by Eq. 30.7. If $\ln A \gg \ln P$ or $\ln P_i$, the expression for $\ln P$ reduces to that in Eq. 30.7. However, there might be a volume change on mixing, in which case another term, ΔV_{mix}, which varies with composition, needs to be added to the expression for the free volume:

$$(V - V_o) = \phi_1(V - V_o)_1 + \phi_2(V - V_o)_2 + \Delta V_{\text{mix}}(\phi_1). \tag{30.10}$$

Consequently, the permeability of the mixture is predicted to increase if $\Delta V_{\text{mix}}(\phi_1)$ is positive and to decrease if it is negative.

In the activated-state approach, it is known that the diffusion coefficient depends on temperature in an Arrhenius fashion, such that

$$D = D_o e^{-E_D/RT}, \tag{30.11}$$

where E_D, the activation energy for diffusion, can be written as the sum of contributions from each component, plus a deviation term:

$$E_D = \phi_1 E_{D1} + \phi_2 E_{D2} + \Delta E_{12}. \tag{30.12}$$

The logarithm of the pre-exponential term D_o has been found to be linearly related to the activation energy for diffusion:

$$\ln D_o = aE_D + b. \tag{30.13}$$

From ternary solution theory, S, the solubility coefficient for the mixture, is given by

$$\ln S = \phi_1 \ln S_1 + \phi_2 \ln S_2 + (\mathbf{B}V_3/RT)\phi_1\phi_2, \qquad (30.14)$$

where $\mathbf{B}$ is an interaction parameter for mixing components 1 and 2 and V_3 is the molar volume of the penetrating molecule, molecule 3. Since $P = D \cdot S$, the previous relationships can be combined and rearranged to give

$$\ln P = \phi_1 \ln P_1 + \phi_2 \ln P_2 + (aRT - 1)(\Delta E_{12}/RT) + (\mathbf{B}V_3/RT)\phi_1\phi_2. \quad (30.15)$$

If components 1 and 2 do not interact, i.e., if ΔE_{12} and $\mathbf{B}V_3$ are both zero, Eq. 30.15 reduces to the expected simple form. Deviations can be either positive or negative, depending on the nature and magnitude of the interaction of the two components.

Paul also demonstrated that the additive free-volume approach works well for oxygen and carbon dioxide permeation through methacrylonitrile–styrene random copolymers, but not for miscible blends of polycarbonate with a copolyester or for miscible blends of poly(phenylene oxide) with polystyrene [13]. In both blend systems, negative deviation from the additive free-volume prediction is observed, even when the prediction corrected for volume contraction on mixing. Paul did not apply activated-state theory for this case, because not all necessary parameters were known.

Maeda and Paul [17] extended the earlier study of the PPO–PS system to include He and CH_4 in addition to CO_2. Besides further demonstrating the utility of the dual-mode model, they determined that the ideal separation factor for pairs of gases P_{gas_1} and P_{gas_2} did not vary monotonically with composition, but instead reached a maximum at about 50% PPO for He–CH_4 and at about 75% PPO for CO_2–CH_4. This effect was shown to be due primarily to the strong PPO–PS interactions, which led to a reduction in volume and inhibited gas-molecule mobility.

Recently, Conforti, Barbari, and Pozo de Fernandes [18] chose CO_2 sorption in the PPO–PS system as a means of testing the applicability of a glassy polymer lattice sorption model (GPLSM) to miscible blends, to determine a blend interaction parameter and, from that, an enthalpy of mixing. The system was represented as a lattice containing polymer segments and holes on which gas molecules are placed. The polymer lattice was compressible, but had a constant lattice-cell volume. The pure gas was also represented on a lattice containing gas molecules and holes.

These authors represented the free energy of mixing in terms of the number of lattice sites in the polymer–gas system, the number of sites (polymer segments plus holes) in the pure polymer, the number of sites in the pure gas, the fraction of pure gas sites occupied by gas molecules, the molecular lattice-site volume, the fraction of polymer segments that interact noncovalently (a function of the coordination number), a parameter analogous to the Flory χ parameter, and interaction energies (segment–segment, gas–gas, segment–gas). From this expression for the free energy of mixing, they developed an activity coefficient model. The polymer blend was treated as a single component in this model, and the blend's segment–segment interaction energy was expressed in terms of the segment–segment interaction energies for the homopolymers, the occupied volumes of each homopolymer, their weight fractions in the blend, and the specific volume of the blend.

The gas sorption and polymer dilation isotherm data were then fitted using this model to get the interaction energies needed to calculate the enthalpy of mixing. These authors' data for CO_2 sorption by PPO, PS, 75/25 PPO–PS and 50/50 PPO–PS agree well with the data obtained by Morel and Paul [12]. The calculated enthalpies of mixing are in reasonable agreement with those obtained using heats of solution or calorimetry with low-molecular-weight model compounds.

C. PMA and PMMA Blends

In studies of another miscible blend system of two amorphous polymers, poly(methyl acrylate) (PMA) and poly(epichlorohydrin) (PECH), Chiou, Barlow, and Paul [19] found the specific volume of the blend to follow the rules of linear additivity, implying minimal interaction between the component polymers. Linear additivity was also found for the logarithms of the permeability, solubility, and diffusion coefficients for helium, argon, nitrogen, and methane, similarly implying that the interaction terms in Eqs. 30.12, 30.14, and 30.15 are zero, within experimental error. The permeability coefficient of carbon dioxide increased with pressure, and the diffusion coefficient was shown to be an exponential function of concentration, such that $D = D_o \exp(\beta C)$, with the solubility coefficient being essentially independent of pressure. Thus, for this system and these five gases, both the excess activation energy of diffusion ΔE_{12} and the interaction energy density $\mathbf{B}$ are minimal.

Chiou and Paul applied both activated-state theory and free-volume theory to the transport of He, H_2, O_2, N_2, Ar, CH_4, and CO_2 in miscible blends of PMMA with another amorphous polymer, bisphenol chloral polycarbonate (BCPC) [20]. Specific volume varied linearly with composition, within experimental error, implying minimal interaction between the two polymers. Permeabilities showed a negative deviation from linearity in plots of $\ln P$ versus composition. In contrast, permeabilities calculated using free-volume theory predicted positive deviation from linearity. With activated-state theory (Eqs. 30.11 through 30.15), the interaction terms $(aRT - 1)\Delta E_{12}/RT$ and $\mathbf{B}V_3\phi_1\phi_2/RT$ were both negative, in accord with the observed negative deviation from linearity. Also, the extent of deviation increased with the size of the gas molecule (atom). Consequently, Chiou and Paul proposed that the ideal-gas separation factors P_{gas_1} and P_{gas_2} in a miscible blend can be greater than the separation factor for either parent polymer if gas_2 is much larger than gas_1, since the permeability of gas_2 will be lowered more in the blend than will the permeability of gas_1.

Another miscible blend system showing negative deviation from semilogarithmic additivity of gas transport properties, SAN (9% AN) with tetramethyl bisphenol A polycarbonate (TMPC), shows volume contraction on mixing, indicating strong polymer–polymer interaction [21]. The permeability and diffusion coefficients exhibit strongly negative deviation from semilogarithmic additivity, while the solubility coefficients deviate only slightly. As in the PMA–PECH case, deviations are greatest for the largest gas molecules, so ideal separation factors are greater for the blends.

Free-volume theory predicts a decrease in P, because of the volume contraction, but agrees with observations only qualitatively. In activated-state theory, the interaction term **B** (see Eqs. 30.14 and 30.15), while negative, is small, and the decrease in P is due primarily to a positive value for ΔE_{12}; that is, the activation energy for diffusion is higher in the blends than in the parent polymers.

Although free-volume theory predicts the occurrence of miscible blend systems that exhibit higher permeabilities than those predicted by Eq. 30.7, wherein interaction between the components is minimal, Chiou and Paul were the first to report such a case, for blends of PMMA with random styrene–acrylonitrile copolymers (SAN), containing 13.5 and 28 wt% acrylonitrile (AN) [22]. The basis of miscibility for such copolymer systems is explained in Chapter 3. Blend films were cast from solution in tetrahydrofuran, and transport data were obtained for He, H_2, O_2, N_2, Ar, CH_4, and CO_2. For all of these molecules, permeability and diffusion coefficients were higher than predicted by Eq. 30.7. Unlike with the PMMA–bisphenol chloral polycarbonate system [20], in which the polymer components interact strongly with each other, the extent of deviation from semilogarithmic additivity does not increase with the size of the gas molecule. Consequently, semilogarithmic plots of ideal separation factors P_{gas_1}/P_{gas_2} versus blend composition are linear; no composition is observed for which the separation factor is greater than for both parent polymers.

Water-vapor diffusion in the PMMA–SAN (19 and 30% AN) system was studied by Gsell [23]. He introduced into the blend a polystyrene (PS) that contained 5 to 20 mol% pendant hydroxyl groups and observed a large increase in specific volume on mixing, which did not occur in the PMMA–SAN system. The activation energy for diffusion decreased significantly, the diffusion coefficient increased, and equilibrium sorption levels were lower than predicted by the semilogarithmic additivity rule. These changes were attributed to inhibited hydroxyl-group self-association and less efficient packing, which increased the specific volume, while the hydrogen bonding of water molecules with the hydroxyls lowered E_D. Without the modified PS (30% AN), these deviations essentially disappeared, unlike in the Chiou and Paul observations discussed previously for other gases.

Li and colleagues [24] studied N_2, CH_4, and He transport in the PMMA–SAN (30% AN) system at temperatures of 35 to 140°C, particularly near T_g (96°C). N_2 and CH_4 permeability and diffusion coefficients were larger than expected from semilogarithmic additivity only below T_g, but this was not the case for He. Arrhenius plots of $\ln P$ and $\ln D$ versus $1/T$ were discontinuous at T_g for all three gases and all blend compositions studied. The authors attributed the discontinuity to a large value for $\Delta\alpha/\alpha_g$, the change in the thermal expansion coefficient α at T_g, relative to its value in the glassy state (α_g). Alternatively, this system might exhibit dual-mode sorption below T_g and just Henry's law behavior above T_g [25].

Li, Kwei, and Myerson [26] subsequently tried to study gas transport in the same polymer system in the vicinity of the lower critical solution temperature (LCST), hoping to use N_2 and CH_4 to probe concentration fluctuations. However, they were able to measure D only up to 175°C, barely above the LCST of 173°C. A plot of $\ln D$ versus $1/T$ for a 50/50 blend, the critical composition, revealed no obvious discontinuity, possibly because the diffusion coefficients are not very different at that temperature (for CH_4, $D_{SAN} = 2.7 \times D_{PMMA}$). By contrast, the CH_4 diffusion coefficients differ by about an order of magnitude for the PS–poly(vinyl methyl ether) system, and accordingly, the authors found a clear break in the plot of $\ln D$ versus $1/T$ at the LCST (123°C) and critical composition (20% PS) for this blend.

Sorption and transport in PVF_2–PMMA blends [27, 28] involves the added complication of the semicrystallinity of PVF_2 and thus the presence of a mixed amorphous phase and a crystalline PVF_2 phase. CO_2 plasticizes the blends, inducing added PVF_2 crystallization. The concentration C of the gas in the polymer at equilibrium with gas at pressure p is best described by modifying the dual-mode sorption model to incorporate the volume fraction α of amorphous material, under the assumption that none of the CO_2 will dissolve in the crystalline phase. As in the PMA–EPCH system [19], D depends exponentially on concentration, except here only mobile gas molecules are included, i.e., the portion obeying Henry's law.

The other gases tested (i.e., He, Ar, N_2, and CH_4) have much lower solubility in this system and did not plasticize it [28]. In fact, He sorption was too small to be measured reliably. Henry's law was followed for rubbery compositions (less than about 50%), while sorption isotherms were slightly concave for glassy, PMMA-rich blends, following a dual-mode sorption model. For Ar, a plot of k_D divided by the volume fraction α of amorphous material, versus the PVF_2 volume fraction, agreed reasonably well with a calculated curve that showed negative deviation from the tie line. The curve had been calculated using an independently determined interaction-energy density B of -3.85 cal/cm^3 and a molar volume for Ar of 57 cm^3/mol. Permeability coefficients for Ar, N_2, and CH_4 in the PMMA-rich blends were independent of pressure, while the time lag decreased with pressure, indicating that a total immobilization transport model is appropriate. Curves both of permeability coefficients and of diffusion coefficients versus composition showed breaks at 60% PVF_2, reductions that occurred because of significant crystallinity.

III. MODELING TRANSPORT THROUGH MULTIPHASE POLYMER BLENDS

The general principles of mass transport in polymeric materials have been summarized elsewhere [29]. Our intent here is to focus on the effects of various phase morphologies on the permeation of fluids through a film or sheet made from a polymer blend.

The presence of an impermeable dispersed phase increases the tortuosity of the path a molecule must traverse in permeating through a film. It is useful to discuss this concept in terms of a tortuosity factor τ, which is the effective path length divided by the actual thickness of the film. Maxwell derived an expression for τ in considering the conductivity of a system in which a conducting phase contains a volume fraction ϕ_d of spherical, nonconducting particles [30]:

$$\tau \cong 1 + \phi_d/2. \tag{30.16}$$

The tortuosity can be used to calculate the permeability of the composite, as shown by Barrer et al. [31] and Michaels and Bixler [32] in the expression

$$P_c/P_m = \phi_m/\tau, \tag{30.17}$$

where P_c is the permeability of the composite, P_m is the permeability of the matrix polymer, and ϕ_m is the volume fraction of matrix polymer.

Barrer has reviewed a wide range of models of diffusion and permeation in heterogeneous media wherein the filler phase is impermeable and represented by many different geometrical shapes and arrangements in the continuous phase [33]. Petropoulos [34] has compared various approaches to the permeability of two-component polymeric materials in which each component forms a well-defined phase and does not interact with the other or with the permeant. He also has identified the circumstances under which each model (for example, dilute dispersion of spheres; various packing geometries of spheres, cylinders, or square rods) works best.

For nonspherical particles, a term denoting the anisotropy of the filler has to be incorporated. For thin, flat, square platelets, Nielsen [35] has modified the toruosity equation as

$$\tau = 1 + (L/2W) \cdot \phi_d, \tag{30.18}$$

where L and W are the length and the width of the platelet, respectively, and ϕ_d is the volume fraction of the dispersed phase. The permeability P can be calculated as

$$P = P_m(1 - \phi_d)/[1 + (L/2W)\phi_d], \tag{30.19}$$

where P is the permeability of the composite and P_m is the permeability of the matrix.

Fricke [36] extended Maxwell's model to describe the conductivity of a two-phase system in which permeable ellipsoids are dispersed in a more permeable continuous matrix. According to this model, the permeability of a composite system consisting of a blend of two permeable materials in which the dispersed phase is distributed as ellipsoids having two equal-length axes perpendicular to the direction of transport can be expressed as

$$P = (P_m + P_d F)/(1 + F), \tag{30.20}$$

where

$$F = [\phi_2/1 - \phi_2][1/1 + (1 - M)(P_d/(P_m) - 1)], \tag{30.21}$$

$$M = \cos\phi/\sin^3\phi[\phi - 1/2\sin 2\phi], \tag{30.22}$$

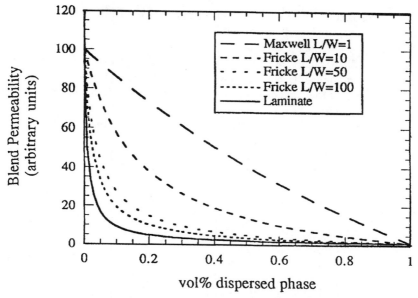

Figure 30.1 Permeability of a hypothetical blend vs. vol % of the dispersed phase. Matrix permeability $= 100$. Dispersed-phase permeability $= 1$ [37]. Reproduced with permission from Polymer Engineering and Science.

and

$$\cos \phi = W/L. \tag{30.23}$$

W is the dimension of the axis of the ellipsoid parallel to, and L the dimension perpendicular to, the direction of transport, and ϕ is in radians. When L/W approaches unity, this expression reduces to Maxwell's equation.

The permeability plotted for different L/W values against the volume fraction of the dispersed phase, ϕ_d is given in Fig. 30.1 [37]. At large values of L/W, substantial reductions in permeability can be seen at small volume fractions of the additive.

Kit, Schultz, and Gohil [37], in their studies of oriented films from PET–EVOH blends, found that the oxygen barrier properties were a function of the morphology of the EVOH phase and fit the model proposed by Fricke.

Similar observations have been made by Toy et al. [38] in studies of blends of poly(1-(trimethylsilyl)-1-propyne) (PTMSP) and poly(1-phenyl-1-propyne) (PPP). PTMSP is highly permeable to gases, whereas PPP, with its lower free volume, is much less permeable. In blends of these polymers, Toy et al. noted that as PPP increases from 3 to 20%, there is a dramatic reduction in permeability to carbon dioxide and nitrogen, sigmoidally with the concentration of PPP. Also the CO_2/N_2 selectivity increases dramatically. The PPP was found to have been dispersed as platelets, and the permeability could be explained using a modified Maxwell equation for platelets in a permeable media. This blend film was prepared by casting from a solution of the polymers and evaporation of the solvent.

Robeson [39] extended Maxwell's work by applying it to blends for which both components are permeable. While developed in the context of block copolymers of polysiloxanes and polysulfones, the analysis allows a straightforward comparison of the effect of morphology and of concentration on permeability. The composite morphologies considered and the permeabilities P_c derived by Robeson are given in the following equations, where the component polymers are denoted by the subscripts 1 and 2 and ϕ_i stands for the volume fraction of phase i:

Series laminate (layers normal to permeant flow):

$$P_c = P_1 P_2 / (\phi_1 P_2 + \phi_2 P_1). \tag{30.24}$$

Parallel laminate (layers parallel to permeant flow):

$$P_c = \phi_1 P_1 + \phi_2 P_2. \tag{30.25}$$

Continuous matrix phase m with dispersed, conducting, spherical filler, d:

$$P_c = P_m[P_d + 2P_m - 2\phi_d(P_m - P_d)]/[P_d + 2P_m + \phi_d(P_m - P_d)]. \tag{30.26}$$

In these equations, both polymers contribute to the continuous phase: Eq. 30.26 is used for each polymer, weighted by its fractional contribution, and the results for both polymers added.

This model allows us to consider the practical implications of attempts to increase the barrier performance of one material, polymer 1, by adding a small amount of a second, relatively impermeable material, polymer 2. Let $P_1 = 10$ (arbitrary units), and allow P_2 to have different values, from 1 to 0.001. We can then calculate what volume fraction of polymer 2 we need to reduce the permeability by one half, that is, to produce a blend with $P_c = 5$. (See Table 30.3.) We see clearly that the most efficient structure is for the barrier polymer to form layers normal to the direction of permeant flow, and the second most efficient structure is one in which polymer 2 is continuous.

Table 30.3   Barrier-Phase Volume Fraction Needed to Halve Permeability

Barrier-Polymer Permeability*	1	0.1	0.01	0.001
Morphology				
Series laminate	0.111	0.0101	0.001	0.0001
Parallel laminate	0.556	0.505	0.501	0.500
Continuous Pol.1	0.467	0.406	0.401	0.400
Continuous Pol.2	0.238	0.0291	0.00299	0.0003
Co-continuous**	0.347	0.127	0.0431	0.0140

*Arbitrary units, relative to the permeability of matrix polymer (Pol. 1), which is equal to 10.
**Equal contribution of Pol. 1 and Pol. 2 to continuous phase.

One limitation of Robeson's approach is that there is no way to predict the fractional contributions of polymers 1 and 2 to the two-phase continuous matrix; when Eq. 30.26 is used in such a case, the contribution of each polymer is adjusted arbitrarily. Kolařík and Geuskens [40] have developed a model for the permeability of binary polymer blends based on a two-parameter equivalent box model (EBM) and the concept of phase continuity. The EBM describes the contributions of polymers 1 and 2 to the permeability by letting a certain volume fraction from polymer 1 and another volume fraction from polymer 2 contribute as if they were coupled in parallel, i.e., like layers parallel to the flux. The remaining volume fractions of polymers 1 and 2 contribute as if they are coupled in series, i.e., normal to the flow. These two coupled systems are themselves then coupled in parallel.

The parallel-system volume fraction ϕ_p is the sum of the parallel coupled contributions from polymer 1 and polymer 2, that is, $\phi_{1p} + \phi_{2p}$. Similarly, the series-system volume fraction ϕ_s is $\phi_{1s} + \phi_{2s}$. We also know that

$$\phi_1 + \phi_2 = 1 = \phi_p + \phi_s. \tag{30.27}$$

Using Eqs. 30.24 and 30.25 for the permeability of the series system and of the parallel system, respectively, then gives us

$$P_p = (P_1\phi_{1p} + P_2\phi_{2p})/\phi_p \tag{30.28}$$

$$P_s = \phi_s/[(\phi_{1s}/P_1) + (\phi_{2s}/P_2)]. \tag{30.29}$$

Since the parallel and series systems are themselves coupled in parallel, the blend permeability P_b is $P_p\phi_p + P_s\phi_s$, or

$$P_b = (P_1\phi_{1p} + P_2\phi_{2p}) + \phi_s^2/[(\phi_{1s}/P_1) + (\phi_{2s}/P_2)]. \tag{30.30}$$

The challenge then is to determine the various volume fractions. The authors of Ref. 40 used percolation theory [41] to derive expressions for ϕ_{1p} and ϕ_{2p} in terms of their critical volumes, or percolation thresholds, ϕ_{1cr} and ϕ_{2cr}:

$$\phi_{1p} = [(\phi_1 - \phi_{1cr})/(1 - \phi_{1cr})]^{T1} \tag{30.31}$$

$$\phi_{2p} = [(\phi_2 - \phi_{2cr})/(1 - \phi_{2cr})]^{T2}. \tag{30.32}$$

$T1$ and $T2$ are the critical percolation exponents, with typical values near 1.8. Once ϕ_{1p} and ϕ_{2p} are determined, the other volume fractions can be calculated from their definitions. The authors then combined Eqs. 30.30 through 30.32 and adjusted ϕ_{1cr} and ϕ_{2cr} to fit existing data on permeability versus concentration for several permeant-blend systems: water through HDPE–LDPE, toluene through

HDPE–polyamide, methanol through polyethylene–silicone, and oxygen through PVC-chlorinated polyethylene. They obtained excellent agreement with experimental data throughout the concentration range.

IV. BARRIER MATERIALS THROUGH CONTROL OF BLEND MORPHOLOGY

The role of phase morphology in blends has been extensively studied, primarily in order to understand its influence on mechanical properties. Less understood has been the role of morphology in transport properties, such as permeation of gases and liquids through films and sheets and other articles made from polymer blends. The morphologies of these blends are influenced by the chemistries of the polymers involved and the processes used for preparation of the articles. In this section, we consider blends of different polymers prepared by melt processes, such as injection molding and extrusion.

Permeation through polymer blends generally is a function of the volume fraction of the barrier polymer in the blend. In blends wherein the dispersed polymer is distributed isotropically as uniform spherical particles, for good barrier properties, one has to resort to the addition of 30–60% of the "barrier polymer." The second polymer, ideally, needs to be used in small concentrations in order to minimize adverse influences on the properties of the primary polymer component. Understanding and optimizing the morphologies of the dispersed polymer phase is essential for such performance.

As previously mentioned, isotropic blends in which the matrix polymer contains the barrier polymer in the form of small, spherical particles show barrier enhancements as a linear function of their volume fraction. However, if the dispersed particles have the form of thin, flat platelets with large areas, then their effectiveness increases significantly. In such cases, the fluid pathways become more tortuous, thereby decreasing the permeation rate. Understanding platelet formation and preservation has been the subject of several investigations in the recent past. Processes that allow the creation of such platelet morphologies need to be combined with techniques capable of stabilizing them without relaxation into spheres. Parameters that influence such morphologies include the type of mixing involved, shear rates in the processing equipment, melt viscosities of the individual polymers, melting points, interfacial tension, and the use of compatibilizers.

The most common fluids for which permeation control is required are water, hydrocarbons, alcohols, ketones and ethers, and gases like oxygen and carbon dioxide. Depending upon the particular fluid, one chooses an appropriate barrier polymer. Consideration should be given to compatible melting points, viscosities, the interfacial properties, and the preferred processes for fabrication. A significant amount of work has been carried out in the areas for which the matrix polymer is a polyolefin and the dispersed phase is a polyamide or an oxygen barrier polymer, such as EVOH [42–44]. Studies have also been carried out towards barrier enhancement of poly(ethylene terephthalate) (PET) [45–47] through blending. For PET, the objective has been the improvement of oxygen and carbon dioxide barrier properties. As barrier polymers, poly(ethylene-co-vinyl alcohol) (EVOH), poly(ethylene naphthalate) (PEN), and some semiaromatic polyamides have been investigated [37, 48, 49].

A. Morphology Development in Blends

Morphology development is a complex process influenced by several parameters. (See Chapter 16.) Often, principles from studies on droplet deformation [50–53] can be applied to understand the morphology of the dispersed phase. Most of these studies involve Newtonian fluids, small deformations, and simple flow fields.

A critical parameter influencing the dispersed-phase morphology is the interfacial tension between the matrix and the additive polymer. Van Oene [54] has carried out rigorous calculations for interfacial tension in the flow of two polymers in a blend and predicted criteria for droplet and fiber formation and stratification into ribbons.

More recently, Levitt and Macosko [55, 56] have studied shearing of a polymer droplet (thin strand) against another polymer, with and without reaction. They studied a droplet of a polymer sandwiched between discs of another polymer, at high temperatures. The discs were rotated under low shear stresses, and the specimen was examined for deformation and area generation. The deformation of a polypropylene droplet in a matrix of polystyrene, as a function of the stress, is shown in Fig. 30.2 [55]. Levitt and Macosko have observed that when some reactivity is involved, the area generated under similar shear-stress conditions is significantly larger than the area when no reaction took place. Droplets of polystyrene containing anhydride groups and an amine-ended poly(methyl methacrylate), under similar shear stresses, created platelets with a significantly larger total area than those created when the same polymers were used without reactive functionality, as shown in Fig. 30.3 [56].

Sundararaj et al. [57] have found that in an extrusion process, as pellets of the polymer blend melt and advance down the extruder, each dispersing-polymer pellet is first stretched into sheets. As the mixing continues, the sheets develop into cylinders, which, on further stretching, break into spherical droplets.

These studies have contributed to a significant understanding of the mechanism of laminar blends with a polyamide or EVOH as a dispersed-barrier polymer and polyolefins as the matrix resins. Such blends are widely used for packaging, automotive fuel tanks, and other applications.

B. Role of Compatibilizers and Processing Conditions in the Development of Morphologies

Blends of highly immiscible polymers have a coarse morphology and are of limited value for making good films and containers. The dispersed-phase particles are often large and have weak interphase boundaries, creating "holes" for permeation. Furthermore, these systems have poor mechanical properties, e.g., low tear strength and impact strength. Compatibilizers enhance interfacial adhesion and mechanical properties. In addition, as Macosko has shown [56], compatibilizers enable creation of platelets with larger areas on shearing and thereby enhance the efficiency of these

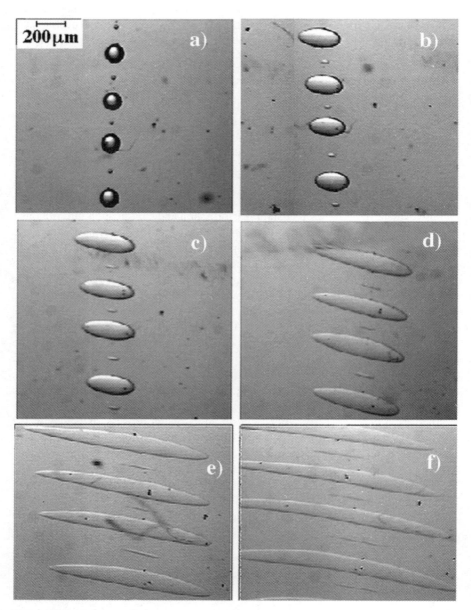

Figure 30.2   PP drop in PS matrix as a function of stress γ. Viscosity ratio $\eta_r = 2.4$; elasticity ratio $G_r' = 7$; interfacial tension $= 5$ mN/m. (a) $\gamma = 0$. The drop in diameter is about 130 μm. (b) $\gamma = 2$. c) $\gamma = 3$. An ellipse is formed. (d) $\gamma = 5$. (e) $\gamma = 9$. Still, the width of the sheet remains approximately equal to the diameter of the initial drop. (f) $\gamma = 13$. The sheet contracts due to the interfacial tension [55]. Reproduced with permission from Polymer Engineering and Science.

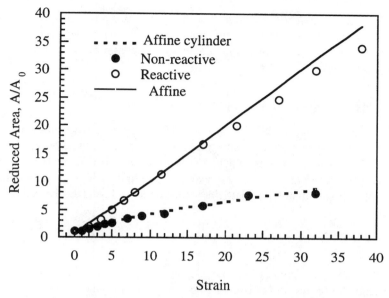

Figure 30.3   Reduced area vs. strain for the reactive and nonreactive PMMA in a PS–MA matrix [56]. Reproduced with permission from Prof. C. S. Macosko.

polymers as barriers. Yeh et al. [58–60] have examined different compatibilizers for blends of polyolefins with polyamides and studied their effects on barrier properties. They have also shown that the amount of mixing has a significant effect on the barrier properties in these blends.

Lohfink and Kamal [61] have done extensive work to understand the formation of platelets of ethylene–vinyl alcohol copolymer in a matrix of polypropylene (PP), in the presence of a maleated PP as a compatibilizer, and of nylon in polyethylene, in the presence of an anhydride-grafted polyethylene as a compatibilizer. They controlled the formation of laminar platelets by designing low-shear-mixing extruder screws and die systems.

In extrusion processes, the shear rates and the mixing can be easily controlled. However, injection-molding processes involve high shear rates and discontinuous melt flows. Development of laminar structures during injection molding is more difficult and needs to be rigorously controlled by adjusting several molding parameters, the material composition, and melt rheologies. Holsti-Miettinen et al. [62] have investigated the morphologies developed during injection molding of polypropylene–nylon 6 blends, in which the nylon is the barrier polymer. An anhydride-modified polypropylene was used as the compatibilizer. They found that for a given blend, oxygen permeation increased with increasing amounts of compatibilizer and also resulted in the creation of spherical morphologies. However, when the nylon was preblended with the compatibizer by melt compounding and the polypropylene added later as a "pellet blend" in the injection-molding machine, the article had permeation barrier properties that were comparable to those of a coextruded multilayer article from the same components. While the mechanism of this significant change is not clear, it could be attributed to the significant changes in the relative viscosities of the components, making conditions unfavorable for dispersion of the nylon into small spherical particles.

C. Lamellar Injection Molding and Extrusion Processes

A unique, multilayer structure with two or more components wherein the polymers are distributed as several co-continuous layers has been developed by Schrenk and others [63–66]. The unique feedblock, inserted into the extruder system, splits the polymer stream several times, making several co-continuous layers without dispersing the polymer stream into fine particles. The number of layers can be derived from the equation

$$M = K^{N(n-1)} + 1, \tag{30.33}$$

where M is the total number of layers; K is the layer-multiplication factor, n is the initial number of layers, and N is the number of layer-multiplication stages.

Co-continuous layers numbering from just a few to a few thousand can be made by using these feedblocks. Molded articles or extruded films with several barrier layers can be made by using small quantitites of the barrier polymer. These microlayer composites show barrier properties similar to those of conventional, coextruded, multilayer articles. Besides the barrier properties, these composites can be engineered to have good optical properties as well. A schematic of a typical feedblock set up for creation of the co-continuous layers, and the injection-molding process incorporating this feedblock, are given in Fig. 30.4. Optical photomicrographs illustrating layers produced by the lamellar injection molding (LIM) process are shown in Fig. 30.5. The oxygen barrier properties of an HDPE–EVOH adhesive system made by this process are shown in Fig. 30.6.

D. Polyolefin Blends

Enhancing the performance of polyolefins as permeability barriers by making laminar blends has been studied extensively and practiced commercially. Small amounts of a polyamide (e.g., 2–5%), distributed as a multitude of very thin platelets in a polyolefin matrix, improve hydrocarbon barrier properties by 100–200 times. Similarly, oxygen transmission rates through polyolefins have been reduced substantially by the incorporation of EVOH as thin laminar platelets [67–69].

By extruding structures from polypropylene and EVOH, Lohfink and Kamal [67] made laminar blends with good oxygen barrier properties, under controlled processing conditions of a low-mixing extruder system and die design. McCord [68] showed that the excellent oxygen barrier properties result from a combination of the polyolefin, compatibilizer, and the EVOH.

Lee and Kim [69] extended this work to films made by an extrusion blowing process from low-density polyethylene (LDPE) and EVOH. Various laminar mor-

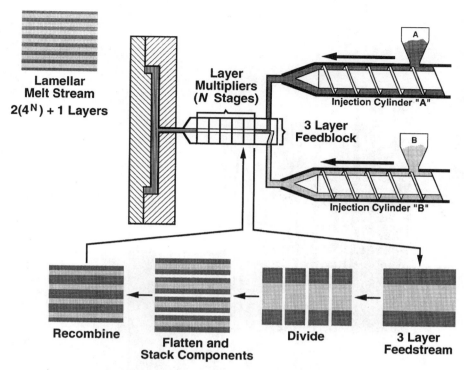

Figure 30.4 Schematic of a LIM process. Simultaneous injection through a feedblock and layer-multiplication stage are used to create a micron-scale lamellar-morphology blend [66]. Reproduced with permission from Hanser/Gardner Publications, Inc.

phologies were obtained. The main factors determining the morphology of the blends were the viscosity ratios, processing conditions, and compatibilizer levels. Optimized conditions gave the films a barrier enhancement of 740 times that of LDPE.

Mouzakis and Karger-Kocsis [70] have studied gasoline absorption (for 120 days at 20°C) in the HDPE-Selar®RB polymer–polymer composite (PPC) and calculated the sorption (S), diffusion (D), and permeability (P) coefficients from a plot of weight gain vs. the square root of time (in days). (See Table 30.4.) The diffusion coefficient was reduced by a factor of four with only 4% addition of Selar®RB 901. It did not change much as the Selar® concentration was increased further. The change in sorption was much less pronounced. The permeability coefficient showed a large drop with a small amount of Selar®RB. Further decrease was gradual.

E. Poly(ethylene terephthalate) Blends

1. *PET–EVOH Blends*   Poly(ethylene terephthalate) (PET) is the material of choice for a variety of films and containers for packaging applications such as soft drinks and juices. Several studies aimed at the enhancement of its oxygen and carbon dioxide barrier properties through blending have been carried out. Blends of PET

LIM Morphology - Effect of Number of Layers
16 Cavity Mold - Blue/White HIPS

33 Feedstream Layers

500 μm

129 Feedstream Layers

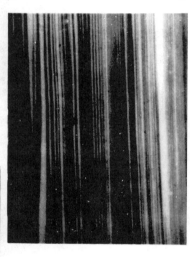

513 Feedstream Layers

Figure 30.5 Cross-sectional optical micrographs of LIM structures created from streams of blue- and white-pigmented polystyrene [66]. Reproduced with permission from Hanser/Gardner Publications, Inc.

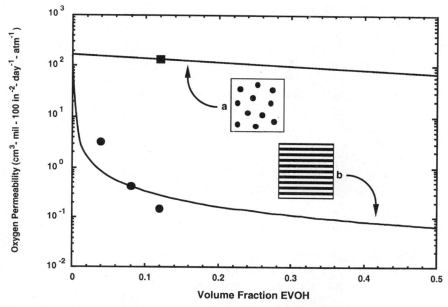

Figure 30.6 Oxygen permeability versus EVOH content for polyethylene–EVOH blends that exhibit (a) discontinuous morphology and (b) co-continuous lamellar morphology [66]. Reproduced with permission from Hanser/Gardner Publications, Inc.

Table 30.4   Permeability (P), Sorption (S), and Diffusion (D) Coefficients for Gasoline

PPC System	D $(10^{13} m^2 s^{-1})$	S (g/g)	P $(10^{14} m^2 g s^{-1} g^{-1})$
HDPE	19.0	0.077	14.63
HDPE+ 4% Selar® RB 901	5.62	0.065	3.62
HDPE+ 7% Selar® RB901	5.19	0.060	3.13
HDPE+ 14% Selar® RB901	5.93	0.051	3.05

with EVOH copolymers have been shown to have improved oxygen barrier properties when tested as oriented films. Schultz et al. [37, 71] have found that in oriented films, the morphology of the EVOH phase dictates the oxygen barrier properties of these blends to a large extent. The oxygen barrier improvement could be fitted to the models proposed by Fricke for a two-phase conducting system, wherein the EVOH particles are ellipsoids with large areas. The oxygen permeability of the PET–EVOH system was reduced by a factor of 4.2 on addition of 20% EVOH. Under similar conditions, blends of poly(ethylene naphthalate) (PEN) and EVOH showed oxygen barrier improvement by a factor of only 2.7 [37].

The development of phase morphologies induced by stretch orientation is a complex phenomenon influenced by several parameters: T_g, crystallinity, interfacial adhesion, stretching temperature, draw ratio, relative viscosities, and so on. In stretched specimens of PET–EVOH blends, the EVOH was deformed to high-aspect-ratio platelets. When such mechanical stretching was not involved (e.g., under extrusion blow-molding conditions), only spherical particles were observed by Shepherd et al. [48]. The complex nature of the parameters involved in the development of the geometry of the dispersed phase is shown by the fact under the aforementioned conditions, PEN and EVOH blends showed a lamellar structure for EVOH.

The water-vapor permeation of the oriented PET–EVOH films was not affected significantly, although the morphology of the EVOH phase was more platelet-like. This is not surprising, since the EVOH itself is water sensitive.

2. PET–Polyamide Blends   Aliphatic polyamides do not show good oxygen barrier properties. In addition, they are moisture sensitive. However, thermoplastic polyamides having aromatic content, such as the ones made from isophthalic acid or m-xylylene diamine (MXDA) monomers, have been found to have good oxygen barrier properties without the moisture sensitivity [72]. Blends of PET using such polyamides have been investigated. Blends of PET with MXD-6 nylon (Mitsubishi Gas Chemical Co.) have shown significant reduction of oxygen and carbon dioxide permeation and are currently being developed for use in containers and films.

F. EVOH Blends with Aromatic Polyamides

Ethylene–vinyl alcohol copolymers have excellent oxygen barrier properties. However, under conditions of high humidity, the barrier is reduced significantly, because hydrogen bonding of the water molecules with the hydroxyl groups of the EVOH polymer reduces the crystallinity of the copolymers significantly. Semiaromatic copolyamides (e.g., Selar®PA, an amorphous polyamide made from isophthalic acid, terephthalic acid, and hexamethylene diamine), unlike aliphatic polyamides, have been shown to have good oxygen barrier properties that, surprisingly, improve with exposure to moisture [72].

Chou and Lee [73] found that when Selar®PA ($\sim 30\%$) was blended with EVOH (32% ethylene), the blend retained most of its oxygen barrier performance, even when exposed to an environment with 80% relative humidity. The Selar® improved the processability of the blend as well. (See Section IV. H.)

G. Polyamide Blends with Polyethylene: Water-Vapor Barrier Properties

Aliphatic polyamides, such as nylon 66 and nylon 6, are highly permeable to water vapor; however, reduction of this moisture sensitivity is desired for many film applications. In addition, reducing the moisture sensitivity would improve the dimensional stability of parts molded from these polyamides for engineering applications. A potential route for such property improvement is to blend the polyamides with the highly moisture-impermeable high-density polyethylene (HDPE). Good blends of nylon and HDPE can be made by adding compatibilizers such as polyethylene-based ionomers or anhydride-grafted polyolefins [74].

Given in Table 30.5 are water-vapor permeabilities of nylon 66 blended with polyethylene. The nylon 66 has a number-average molecular weight of about 18,000. Marlex® 5202 is a high-density polyethylene made by Phillips Petroleum and has low moisture permeability. Surlyn®, a zinc-ion-neutralized ethylene–methacrylic acid copolymer, was added as a compatibilizer. As the polyolefin content increases, the water-vapor transmission rate (WVTR) decreases substantially.

H. Barrier Blends: Mechanical and Rheological Properties

So far, we have discussed the barrier properties and morphologies of blends. In addition to barrier properties, satisfactory mechanical properties are critical for the implementation of these technologies in practical applications. In general, barrier polymers and matrix polymers differ significantly in their polarities and crystallinities, which tends to worsen the mechanical properties of blends made from them. (See Table 30.6.)

The mechanical properties of a blend can be improved by adding a compatibilizing agent. However, since this addition facilitates dispersion of the barrier polymer, caution needs to be exercised in order to prevent too fine of a dispersion of the bar-

Table 30.5   Water-Vapor Transmission Rates of Nylon–Polyethylene Blends

Nylon 66	Surlyn® 9520[1]	Marlex® 5202[2]	WVTR[3]
100			28.3
70		30	5.8
70	10	20	10.6
70	10	30	4.3
60	20	20	3.3
55	20	25	2.7

[1] Zn ionomer of poly(ethylene methacrylic acid), Du Pont.
[2] HDPE, Philips Chemical.
[3] Units: g·mil/100 in^2·24 h at 38°C, 90% RH.

Table 30.6   Mechanical Properties of Polyethylene Blends with Nylon 66

Composition	a	b	c
Yield strength, MPa	26.8	—	34.4
Tensile strength, MPa	27.2	23.7	34.4
Elongation, %	300	—	100
Impact, Notched Izod, J/m	NB*	123	NB*

a = HDPE.
b = HDPE + 20% nylon 66.
c = HDPE + 20% nylon + anhydride-modified PE as a compatibilizer.
*No break.

rier polymer, which would decrease the platelet area and increase the permeation significantly.

Parts molded from blends often show weakness in the area of impact strength, as mentioned earlier. This factor is especially critical, even in compatibilized blends, in the area of the weld line, where the different molten streams of the polymer meet within the mold cavity. Under these conditions, some phase segregation and coalescence or agglomeration occur to form larger particles. These larger particles promote localized stress intensities on impact and contribute to failures, necessitating optimization of the mold design and processing conditions. With such modifications, one can, from blends of polyethylene and nylon, make large containers, such as gasoline tanks, that have excellent impact strength and are able to survive being dropped from a height of six meters, even at −40°C.

These additive polymers and compatibilizers with reactive groups alter the melt rheology of these blends, which can affect extrusion and molding processes significantly. Therefore, the viscosities of the matrix polymer, the barrier polymer, and the compatibilizer need to be considered and optimized for the final processes by which articles are to be manufactured.

Many barrier articles are made by thermoforming processes, wherein the sheet of the polymer is heated to softening and then shaped by applying pressure or a vacuum. The polymer melt should have adequate melt strength to stretch and conform to the mold. Often, the softened polymer sheet could be drawn to an extension ratio of 3 to 5, as in the thermoforming of multilayer sheets having a thin, sandwiched, barrier polymer layer. If the barrier polymer has poor elongation in the melt, it could lead to breakage and the formation of pinholes. This is particularly so in the case of the highly crystalline EVOH barrier layer sandwiched between layers of polyolefins. Chou and Lee [73] found that addition of small amounts of the amorphous, semiaromatic nylon Selar®PA to the EVOH significantly improved thermoformability.

The relationship between microstructure and impact properties in discontinuous-microlayer polymer–polymer composites (PPC) has been extensively investigated by Yuan, Karger-Kocsis, and others [70, 75–77]. They established the microstructure of HDPE–nylon specimens containing discontinuous layers of a modified polyamide (Selar®RB) by microscopy and subjected these specimens to instrumented impact at −40°C, in the Charpy and Izod modes. Failure modes of these microlayer composites were studied by fractography. It was found that the impact response of the PPC produced by extrusion blow molding depends strongly on its microstructure. High impact strength is linked to a regular (uniform) microstructure, especially when it is measured using a dart that strikes the planar wall of the molding in a direction normal to the plane of the Selar® layers. Major contributions to high impact strength include the thickness of the individual polyamide layers and their interlayer spacing. Local accumulations of the Selar® layers influence crack initiation and growth, through the thickness. Brittle fracture is favored when the thickness of the Selar® layer is greater than ∼ 10 μm. PPCs of regular buildup with thin layers of the Selar® show the same impact strength as that of HDPE. It was concluded that the in-plane impact response of this microlayer composite is superior to the out-of-plane one. They also found that a fine and regular microstructure contributes to a high impact resistance.

In instrumented tensile impact tests at room temperature and at −40°C, the tensile elongation of the PPC containing 7% Selar® exhibited lower elongation than did the HDPE itself (∼75% vs. 130%). After conditioning of these blends by exposing them to gasoline for three months, both became more extensible and had more toughness than in the dry condition. The polyethylene was plasticized and became more flexible. The PPC had a higher modulus and retained its stiffness to a greater extent.

The interesting aspect of the increased stiffness at small loadings of the stiffer, dispersed phase, when dispersed as flat laminae, is yet to be explored in detail. Such distribution of the dispersed phase influences the mechanical properties and could result in improved creep resistance, thermal properties, and so on.

I. Characterization of Morphology

The significant influence of morphology on the permeability barrier properties necessitates the identification of such morphologies using analytical methods. The phase morphology is most easily observed under an optical or scanning electron microscope on a specimen obtained by microtoming the article. Depending upon the polymer system, different dyes or vapor deposition may be used to distinguish the phases. Where polyamides make up one of the phases, one could use an organic dye specific to nylon. Light scattering and near-infrared (NIR) spectroscopy have also been used for fast characterizations, within a specified polymer blend system.

The microstructure can also be established indirectly, by using ultrasound techniques. In the case of a container blow molded from a blend of nylon in polyethylene, the specimen, in a water bath, is exposed to a pulse of a 10-MHz ultrasound. The echoes from the front and back surfaces of the specimen wall are recorded on an oscilloscope. These echoes are separated in time, by an amount proportional to the thickness of the specimen. The echoes from the discontinuities within the sample, such as layers or voids, appear in time, between the front and rear surfaces as a function of their location. These signals form a "fingerprint" pattern that depends on the morphology of the nylon particles and that can be correlated with the actual permeability obtained from weight-loss measurements [78, 79].

Another convenient technique is to examine the absorption of a solvent by a small section of the article and to correlate that absorption with the permeability and morphology established by control experiments.

V. THE SPECIAL CHALLENGES OF LIQUID CRYSTALLINE POLYMERS

Liquid crystalline polymers (LCPs) are known to be exceptional barrier materials, similar to polymers such as EVOH and poly(acrylonitrile) [80–87]. Their unique morphology results in both low diffusivity and exceptionally low solubility for gas molecules [80]. They are also expensive, primarily because of the high cost of the monomers they comprise. Thus, attempts to use the barrier properties of LCPs com-

Table 30.7   Permeability of LCP–PET Blown Films[a]

%LCP in PET	CO_2	O_2	N_2
0	5.15	0.96	0.23
2	3.51	0.67	0.12
10	2.96	0.54	0.14
30	1.55	0.25	0.09

[a]$(\times 10^{14}$ cm^3(STP)·cm/cm^2·s·Pa)

mercially have usually involved blending a small amount of an LCP in a conventional, inexpensive thermoplastic, such as PET. It is thus hoped that such blends have processing and mechanical characteristics similar to, or better than, the host thermoplastic.

However, the unique rheology of thermotropic LCPs makes it a challenge to use them as barrier materials in blends with conventional thermoplastics. Rheology and morphology depend in a complex manner on composition and applied shear, causing anomalous viscosity reductions, with the dispersed LCP present as droplets or fibrils [88], which are much less effective in increasing barrier properties than is a laminar or a platelet-containing structure.

Little has been published concerning the permeability of LCP blends. Thermoplastics used in these studies include polyesters, polyolefins, poly(ethersulfone) (PES), and thermoplastic polyimides.

A. LCP–Polyester Blends

One system that has been scrutinized is that of PET blended with the liquid crystalline copolymers of PET and p-hydroxybenzoic acid (HBA), originally developed at Eastman Kodak [89] and later sold by Unitika under the name Rodrun®. Two commercially available polymer grades of this system are LC3000, with a 60:40 HBA:PET molar ratio, and LC5000, with an 80:20 HBA:PET molar ratio. Despite the significant amount of PET copolymerized with HBA, such LCPs have only slight miscibility with PET, and the LCP phase tends to form fibrils and droplets [90]. Such morphology makes ineffective use of the LCP as a barrier material.

Motta and colleagues measured transport properties of PET–Rodrun® LC3000 LCP blown films containing 2, 10, and 30 wt% LCP [91]. The LCP was present almost entirely as spherical droplets at 2 wt% and as droplets and fibrils at 10 and 30 wt%. Permeability to a mixture of CO_2, O_2, and N_2 was measured. (See Table 30.7.) The films showed the same trend with each gas: After a sharp drop in permeability on addition of 2 wt% LCP, the permeability decreased more gradually, with ln P linear in wt% LCP. The initial sharp drop, which accounted for about half of the total decrease from 0 to 30 wt% LCP, may be due to increased PET crystallinity, with the LCP acting as a nucleating agent.

Modest decreases in oxygen and water-vapor permeability have been observed in films of PEN blended with HBA–PET LCP (62:38 mole ratio) [92]. The authors of

Table 30.8   Permeability of LCP–PEN Films

Wt% LCP	0	10	20
O$_2$ Permeability*			
As extruded	4.40	3.67	3.80
Biaxially stretched 4X by 4X	1.50	1.33	1.33
Water-vapor Transport Rate**			
As extruded	2.32	1.57	2.83
Biaxially stretched 4X by 4X	0.41	0.40	0.37

*Units: cm^3·mil/100 in^2·24 h·atm at 30°C, 68% RH, where one mil = 0.001 inch.
**Units: g·mil/100 in^2·24 h at 38°C, 90% RH.

that study did not discuss the superior barrier shown by as-extruded films containing 10% LCP as compared to those containing 20% LCP. (See Table 30.8). If this effect is real, it could be related to nucleation by the LCP, which has been shown in PET to be more effective at low LCP concentrations [93].

Bonis, Schuler, and Adur have filed patent applications concerning compatibilized LCP–polyester blends [94, 95]. They used a mixture of two compatibilizers that bonded to each other. One end was then available to react chemically with the polyester and the other to react with the LCP. One such "dual compatibilizer" is an ethylene–methyl acrylate–acrylic acid terpolymer with an ethylene–maleic anhydride copolymer. Another example is a copolyester elastomer with an ethylene–maleic anhydride copolymer. Bonis, Schuler, and Adur reported that such blends containing only 10% LCP processed like conventional PET, but exhibited at least a twofold improvement in both oxygen and water-vapor barrier behavior and 2 to 5.5 times better mechanical properties. Most work was done with Vectra® LCP A950, from Hoechst-Celanese, and Kodar® polyester A150 (a polymer of cyclo-hexanedimethanol, isophthalic acid, and terephthalic acid), from Eastman Chemical [96, 97, 98].

B. LCP–Polyolefin Blends

Because of their widespread use as packaging materials, polyolefins have also been blended with LCPs in attempts to improve stiffness and decrease permeability. Again, the LCP must be a minor component, due to cost and processing considerations.

Suokas and colleagues developed compatibilized polyolefin–LCP blends containing about 60 to 70 vol% LCP as the continuous phase [99–101]. The compatibilizer was a random terpolymer of ethylene, butylene acrylate, and glycidyl methacrylate, sold by Norsolor under the trade name Lotader AX 8660.

In one example, a polyethylene–LCP blend film exhibited a decrease in oxygen permeability from about 3000 to about 0.6 cm^3/m^2 · 24 h·atm as the LCP content was increased from 50 to 60 vol%. (See Fig. 4 in [99].) With the viscosity of the LCP and the polyolefin (PO) being nearly equal, a multilayer film in the form PO–blend–PO can be extruded. Though the adhesion between the outer PO layers and the blend is poor, this factor can be an advantage in separating the layers for recycling. LCPs studied include Vectra A950, Vectra B950 (a polyesteramide), and, cited as particularly suitable, Rodrun LC3000 and a copolyester of t-butylhydroquinone, terephthalic acid, HBA, and PET in the molar ratio of 25:25:35:15 [100–101].

C. LCP–Polyethersulfone Blends

James and colleagues examined the oxygen permeability of blends of poly(ether sulfone) (PES) (Victrex® PES 4100P) with an LCP, HBA–6-hydroxy-2-naphthoic acid (HNA) (molar ratio 73:27) [102]. The blends were extrusion compounded and then pressed into films. Data were analyzed in terms of Robeson's limiting series (laminar morphology, i.e., layers normal to permeant flow) and parallel (columnar morphology, i.e., layers parallel to flow) models of the permeability P_c of a two-component composite (See Eqs. 30.24 and 30.25) [39]:

$$\text{Series: } P_c = P_1 P_2/(\phi_1 P_2 + \phi_2 P_1)$$

$$\text{Parallel: } P_c = P_1\phi_2 + P_2\phi_1$$

Again, in these equations, P_i is the permeability of phase i and ϕ_i is the volume fraction of phase i.

The observed permeabilities fall between the parallel and series extremes, closer to the parallel case up to 20 wt% LCP and to the series case above about 40 wt% LCP, until the permeability is essentially indistinguishable from the series model at 70 wt% LCP.

Scanning electron photomicrographs of etched samples as well as weight-loss measurements indicate that the barrier improvement is caused by formation of a continuous LCP layer at the surface. This continuous layer forms while the LCP is still the minor component, probably because under these processing conditions, it is much less viscous than the PES. The LCP becomes the continuous phase through the bulk of the film at around 50–60 wt% LCP.

D. LCP–Thermoplastic Polyimide Blends

The use of LCPs in extruded forms has been limited by their highly anisotropic properties. Harvey and colleagues have developed processing techniques using counter-rotating dies that impose a controlled, continuously changing orientation throughout the material [103–106]. One application of this technology has been to produce extruded blends of LCPs with thermoplastic polyimides to give a planar or laminar morphology [107]. The LCPs used in such cases were Vectra A, from Hoechst-Celanese (HBA–HNA copolyester), and Xydar® LCP, from Amoco Performance Products (based on terephthalic acid, 4, 4′-biphenol, and HBA). The polyimides were LARC–TPI and Aurum®, from Mitsui-Toatsu Chemicals. The LCP concentration was about 5 to about 20%.

The planar morphology is made possible by high transverse shear produced by the rotating die, rotating rapidly enough to form layers instead of fibrils. Also, the LCP viscosity must be low relative to the thermoplastic phase. Laminar-flow conditions must be maintained in the die, and the film should be drawn equally in two orthogonal directions after exiting the die, while the blend is partly solidified. The LCP layers formed are about 0.1 to 1.5 µm thick and overlap one another. In one such polyimide–LCP blend, containing only 10 wt% Xydar LCP, permeability to oxygen decreased 8 times and permeability to water vapor over 16 times, as compared to the case for the polyimide.

VI. SUMMARY AND CONCLUSIONS

Polymer–polymer blends afford a unique, economic way for permeation barrier enhancement and control. In miscible blends, such control can be obtained through specific chemical interactions between the components, leading to a lower permeability of the blend than of either polymer individually.

In immiscible blends, the dispersed-phase morphology generally dictates barrier properties. Secondary effects, such as nucleation and retardation of the crystallization of the components, sometimes influence the barrier behavior. The size and the aspect ratios of the dispersed-phase particles depend upon their mechanism of formation—that is, whether they are formed by coalescence or agglomeration of smaller particles or obtained by deformation of the original polymer pellets. Models proposed by Robeson, Fricke, and others guide us in calculating the potential for achieving the barrier goals by blending. With proper choice of the component polymers and their morphologies, polymer–polymer blends have shown outstanding utility in permeability control for several applications in the packaging, automotive, and other industries.

In the area of engineering polymers, while not discussed in this chapter, polymer-polymer blends afford a novel way to influence solvent sensitivities and stress cracking behavior. Moisture sensitivity of polyamides is significantly reduced by the incorporation of polyolefins. Solvent susceptibilities of polycarbonates are significantly reduced by the incorporation of crystalline poly(butylene terephthalate). Through such modifications, polyamide blends and polycarbonate blends are extensively used for automotive, industrial, and electronic applications.

One of the aspects of immiscible polymer–polymer blends is that they tend to be opaque or white, because of the differences in the refractive indexes of their constituent polymers. (See Chapter 28). While this characteristic could be an asset in some areas (e.g., when the articles are pigmented) other techniques need to be explored when transparency is required.

The potential for using polymer–polymer blends to control barrier properties seems to be of increasing interest and research activity. New blends that act as good barriers to oxygen, carbon dioxide, and moisture would expand the utility of blends extensively. Blends of high-barrier polymers, such as LCPs, ethylene–carbon monoxide copolymers, syndiotactic polystyrene, and others, remain to be explored.

Further understanding the role of interfacial properties and polymer rheologies would enable the creation of designed polymer blend compositions to satisfy a multitude of permeability needs.

VII. REFERENCES

1. J.-C. Lee, M. H. Litt, and C. E. Rogers, *J. Polym. Sci: Part B: Polym. Phys.*, **36**, 75 (1998).

2. G. C. Kapantaidakis, S. P. Kaldis, X. S. Dabou, and G. P. Sakellaropoulos, *J. Membr. Sci.*, **110**, 239 (1996).

3. S. Percec and L. Melamud, *High Performance Polymers,* **1**, 73 (1989).

4. P. T. DeLassus, K. L. Wallace, and H. J. Townsend, *Polym. Prepr., Amer. Chem. Soc., Div. Polym. Chem.*, **26**, 116 (1985).

5. P. Masi, D. R. Paul, and J. W. Barlow, *J. Polym. Sci: Polym. Phys. Ed.*, **20**, 15 (1982).

6. D. R. Paul and W. J. Koros, *J. Polym. Sci: Polym. Phys. Ed.*, **14**, 675 (1976).

7. J. H. Petropoulos, *J. Polym. Sci., Part A-2*, **8**, 1797 (1970).

8. R. M. Barrer, *J. Membr. Sci.*, **18**, 25 (1984).

9. G. H. Fredrickson and E. Helfand, *Macromolecules*, **18**, 2201 (1985).

10. S. Chin, T. Gaskins, and C. J. Durning, *J. Polym. Sci.: Part B: Polym. Phys.*, **34**, 2689 (1996).

11. J. E. Harris, D. R. Paul, and J. W. Barlow, *Adv. Chem. Ser.*, **206**, 43 (1984).

12. G. Morel and D. R. Paul, *J. Membr. Sci.*, **10**, 273 (1982).

13. D. R. Paul, *J. Membr. Sci.*, **18**, 75 (1984).

14. W. M. Lee, *Polym. Eng. Sci.*, **20**, 65 (1980).

15. S. A. Stern and S. Troholaki, "Fundamentals of Gas Diffusion in Rubbery and Glassy Polymers," in *Barrier Polymers and Structures*, W. J. Koros (ed.), American Chemical Society, Washington, DC, 1990, Chapter 2.

16. J. Bicerano, *Prediction of Polymer Properties*, Marcel Dekker, New York, 1993.

17. Y. Maeda and D. R. Paul, *Polymer*, **26**, 2055 (1985).

18. R. M. Conforti, T. A. Barbari, and M. E. Pozo de Fernandes, *Macromolecules*, **29**, 6629 (1996).

19. J. S. Chiou, J. W. Barlow, and D. R. Paul, *J. Appl. Polym. Sci.*, **30**, 1173 (1985).

20. J. S. Chiou and D. R. Paul, *J. Appl. Polym. Sci.*, **33**, 2935 (1987).

21. J. S. Chiou and D. R. Paul, *J. Appl. Polym. Sci.*, **34**, 1503 (1987).

22. J. S. Chiou and D. R. Paul, *J. Appl. Polym. Sci.*, **34**, 1037 (1987).

23. T. C. Gsell, *Diss. Abstr. Int.*, **81**, 775-B (1990).

24. R. J. Li, W. P. Hsu, T. K. Kwei, and A. S. Myerson, *AIChE J.*, **39**, 1509 (1993).

25. S. A. Stern, U. M. Vakil, and G. R. Mauze, *J. Polym. Sci.: Polym. Phys. Ed.*, **27**, 405 (1989).

26. R. J. Li, T. K. Kwei, and A. S. Myerson, *AIChE J.*, **41**, 166 (1995).

27. J. S. Chiou and D. R. Paul, *J. Appl. Polym. Sci.*, **32**, 2897 (1986).

28. J. S. Chiou and D. R. Paul, *J. Appl. Polym. Sci.*, **32**, 4793 (1986).

29. H. B. Hopfenberg and D. R. Paul, "Transport Phenomena in Polymer Blends," in *Polymer Blends*, Vol. I, D. R. Paul and S. Newman (eds.), Academic Press, New York, 1978, Chapter 10.

30. J. C. Maxwell, *Electricity and Magnetism*, 3rd ed., Vol. 1, Dover, New York, 1891.

31. R. M. Barrer, J. A. Barrier, and M. G. Rogers, *J. Polym. Sci. Part A*, **1**, 2565 (1963).

32. A. S. Michaels and H. J. Bixler, *J. Polym. Sci.*, **50**, 413 (1961).

33. R. M. Barrer, "Diffusion and Permeation in Heterogeneous Media," in *Diffusion in Polymers*, J. Crank and G. S. Park (eds.), Academic Press, New York, 1968, Chapter 6, p. 165.

34. J. H. Petropoulos, *J. Polym. Sci.: Polym. Phys. Ed.*, **23**, 1309 (1985).

35. L. W. Nielsen, *J. Macromol. Sci.*, **A1**, 929 (1967).

36. H. Fricke, *Phys. Rev.*, **24**, 575 (1924).

37. K. M. Kit, J. M. Schultz, and R. Gohil, *Polym. Eng. Sci.*, **35**, 680 (1995).

38. L. G. Toy, B. D. Freeman, R. J. Spontak, A. Morisato, and I. Pinnau, *Macromolecules*, **30**, 4766 (1997).

39. L. M. Robeson, A. Noshay, M. Matzner, and C. N. Merriam, *Angew. Makromol. Chem.*, **29/30**, 47 (1973).

40. J. Kolařík and G. Geuskens, *Polym. Networks Blends*, **7**, 13 (1997).

41. P. G. De Gennes, *J. Phys. Lett. (Paris)*, **37**, L1 (1976).

42. P. M. Subramanian, "A Novel Thermoplastic Hydrocarbon Barrier System," *Conference Proceedings, Technical Association of Pulp and Paper Industry: Laminations and Coating Conference*, Boston, MA, September 24–26, 341 (1984).

43. P. M. Subramanian and V. Mehra, "Laminar Morphology in Polymer Blend—Structure and Properties," in *Society of Plastics Engineers (SPE) ANTEC Proceedings*, Boston MA, 301, April–May 1, 1986.

44. P. M. Subramanian, U.S. Patent 4,444,817 (1984).

45. P. M. Subramanian, *Polym. Eng. Sci.*, **25**, 483 (1985).

46. P. M. Subramanian and V. Mehra, *Polym. Eng. Sci.*, **27**, 663 (1987).

47. M. R. Kamal, H. Garmabi, S. Hoshabr, and L. Arghyris, *Polym. Eng. Sci.*, **35**, 41 (1995).

48. F. A. Shepherd, H. Gonzalez, and S. L. Hess, U.S. Patent 4, 835,214 (1989).

49. P. M. Subramanian, *Polym. Eng. Sci.*, **27**, 1574 (1987).

50. G. I. Taylor, *Proc. Roy. Soc. (London)*, **A138**, 41 (1932).

51. R. G. Cox, *J. Fluid Mech.*, **37**, 601 (1969).

52. A. Acrivos and T. S. Lo, *J. Fluid. Mech.*, **88**, 641 (1978).

53. H. B. Chin and C. D. Han, *J. Rheol.*, **23**, 557 (1979).

54. H. Van Oene, *J. Coll. Int. Sci.*, **40**, 448 (1972).

55. L. Levitt, C. Macosko, and S. D. Pearson, *Polym. Eng. Sci.*, **36**, 1647 (1996).

56. L. Levitt and C. Macosko, *Macromolecules* (private communication) (in press).

57. U. Sundararaj, Y. Dori, and C. Macosko, *Polymer*, **36**, 1957 (1995).

58. J.-T. Yeh, C.-C. Fan-Chiang, and M. F. Cho, *Polym. Bull.*, **35**, 371 (1995).

59. J.-T. Yeh, C.-C. Fan-Chiang, and S.-S. Yang, *J. Appl. Polym. Sci.*, **64**, 1531 (1997).

60. J.-T. Yeh and C.-C. Fan-Chiang, *J. Polym. Res.*, **3**, 211 (1996).

61. G. W. Lohfink and M. R. Kamal, *Polym. Eng. Sci.*, **33**, 1404 (1993).

62. R. M. Holsti-Miettinen, K. P. Perttila, J. V. Sippala, and M. T. Heino, *J. Appl. Polym. Sci.*, **58**, 1551 (1995).

63. W. J. Schrenk, U.S. Patent 3,884,606 (1975).

64. W. J. Schrenk, U.S. Patent 5,202,074 (1993).

65. W. J. Schrenk, *J. Plast. Film Sheeting*, **4**, 104 (1988).

66. M. A. Barger and W. J. Schrenk, "Lamellar Injection Molding Process for Multiphase Polymer Systems," in *Innovation in Polymer Processing: Molding*, J. F. Stevenson (ed.), Hanser Gardner Publications, Inc., Cincinnati, OH, 1996.

67. G. W. Lohfink and M. R. Kamal, *Polym. Eng. Sci.*, **33**, 1404 (1993).

68. E. F. McCord, U.S. Patent 4,971,864 (1990).

69. S. Y. Lee and S. C. Kim, *Polym. Eng. Sci.*, **37**, 463 (1997).

70. D. E. Mouzakis and J. Karger-Kocsis, *J. Appl. Polym. Sci.*, **68**, 561 (1998).

71. R. Gopalakrishnan, J. M. Schultz, and R. M. Gohil, *J. Appl. Polym. Sci.*, **56**, 1749 (1995).

72. T. D. Krizan, J. C. Coburn, and P. S. Blatz, "Structure of Amorphous Polyamides: Effect on Oxygen Permeation Properties," in *Barrier Polymers and Structures*, W. J. Koros (ed.), American Chemical Society, Washington, DC, 1990, Chapter 5.

73. R. Chou and I.-H. Lee, *J. Plast. Film Sheeting*, **13**, 74 (1997).

74. P. M. Subramanian, unpublished results.

75. J. Karger-Kocsis, E. Moos, and D. E. Mouzakis, *Plast., Rubber Compos. Process. Appl.*, **26**, 178 (1997).

76. Q. Yuan, K. Friedrich, and J. Karger-Kocsis, *Plast., Rubber Compos. Process. Appl.*, **22**, 29 (1994).

77. Q. Yuan, J. Karger-Kocsis, and K. Friedrich, *Adv. Comp. Lett.*, **2**, 135 (1993).

78. T. W. Harding (E. I. du Pont de Nemours), private communication.

79. R. L. Bell and V. Mehra, "Production of Plastic Fuel Tanks Using Laminar Barrier Technology," *SAE (Society of Automotive Engineers) Technical Paper #890442*, International Congress, February 27–March 3, 1989. Detroit, MI.

80. J. S. Chiou and D. R. Paul, *J. Polym. Sci.: Part. B: Polym. Phys.*, **25**, 1699 (1987).

81. D. H. Weinkauf and D. R. Paul, *J. Polym. Sci.: Part. B: Polym. Phys.*, **29**, 329 (1991).

82. D. H. Weinkauf and D. R. Paul, *J. Polym. Sci.: Part. B: Polym. Phys.*, **30**, 817 (1992).

83. D. H. Weinkauf and D. R. Paul, *J. Polym. Sci.: Part B: Polym. Phys.*, **30**, 837 (1992).

84. D. H. Weinkauf, H. D. Kim, and D. R. Paul, *Macromolecules*, **25**, 788 (1992).

85. Vectra® A950 Film Data Sheet, *Vectra Polymer Materials, Supplier Design Guide* (B 121 BR E9102/014), Hoechst AG 1991, in *Permeability and Other Film Properties of Plastics and Elastomers, Plastic Design Library*, Norwich, NY, 1995, 143.

86. Zenite™ 100A, 600A, 700A, and 800A *Preliminary Data Sheets*, DuPont Engineering Polymers, Wilmington, Delaware, 1996.

87. I. G. Plotzker, J. D. Citron, and M. G. Waggoner, U.S. Patent 5,688,895 (1997).

88. M. J. Rivera-Gastélum and N. J. Wagner, *J. Polym. Sci.: Part B: Polym. Phys.*, **34**, 2433 (1996).

89. W. J. Jackson, Jr., *Mol. Cryst. Liq. Cryst.*, **169**, 23 (1989).

90. W. Brostow, T. S. Dzremianowicz, J. Romanski, and W. Werber, *Polym. Eng. Sci.*, **28**, 785 (1988).

91. O. Motta, L. Di Maio, L. Incarnato, and D. Acierno, *Polymer*, **37**, 2373 (1996).

92. Anonymous, *Research Disclosure* **283**, 651, November, 1987.

93. C.-F. Ou and C.-C. Lin, *J. Appl. Polym. Sci.*, **56**, 1107 (1995).

94. L. J. Bonis, P. S. Schuler, and A. M. Adur, Foster-Miller, Inc., PCT Appl. WO 97/24403 (1997).

95. L. J. Bonis, P. S. Schuler, and A. M. Adur, Foster-Miller, Inc., PCT Appl. WO 96/00752 (1996).

96. A. M. Adur and L. J. Bonis,. "PET–LCP Compatibilized Alloys: A New Unique Development," *Future-Pak '94, the Eleventh International Schroeder Conference on Packaging Innovations*, November 16 and 17, 1994, Chicago, IL.

97. P. A. Toensmeier, "PET/LCP Alloy Offers Benefits in Barrier and Downgaging," in *Modern Plastics*, 25 (July 1995).

98. L. M. Sherman, "New Barrier Technologies Enhance Packaging Films and Bottles," in *Plastics Technology*, **41**, 17 (December 1995).

99. T. Heiskanen and E. Suokas, "A New Low Processing Temperature LCP with High Mechanical Properties and Suitable for Blending," in *Spec. Polyesters '95: Proceedings*, 49, 27–28 June, 1995, Brussels.

100. M. Härkönen, S. Kuusela, E. Laiho, M. Ora, E. Suokas, and R. Holsti-Miettinen, Neste Oy and Optatech Oy, PCT Appl. WO 95/23063 (1995).

101. E. Suokas and M. Härkönen, Neste Oy, PCT Appl. WO 95/23180 (1995).

102. S. G. James, A. M. Donald, I. S. Miles, and W. A. MacDonald, *High Perform. Polym.*, **4**, 3 (1992).

103. A. C. Harvey, R. W. Lusignea, D. M. Baars, D. Bretches, and R. B. Davis, U.S. Patent 4,973,442 (1990).

104. A. C. Harvey, R. W. Lusignea, D. M. Baars, D. Bretches, and R. B. Davis, U.S. Patent 4,939,235 (1990).

105. A. C. Harvey, R. W. Lusignea, J. L. Racich, D. M. Baars, D. D. Bretches, and R. B. Davis, U.S. Patent 4,963,428 (1990).

106. A. C. Harvey, R. W. Lusignea, and J. L. Racich, U.S. Patent 4,966,807 (1990).

107. L. S. Rubin, K. G. Blizard, R. R. Haghighat, and R. W. Lusignea, Foster-Miller, Inc., PCT Appl. WO 92/17513 (1992).

31 Reinforced Polymer Blends

JÓZSEF KARGER-KOCSIS

Institute for Composite Materials
University of Kaiserslautern
D-67663 Kaiserslautern, Germany

Polymer Blends, Volume 2: Performance. Edited by D. R. Paul and C. B. Bucknall.
ISBN 0-471-35280-2. © 2000 John Wiley & Sons, Inc.

I. INTRODUCTION

Thermoplastic composites with various fillers and reinforcements are well established for a variety of applications. It is, however, less well known that the matrix of a number of these composites is a polymer blend itself. Composites are defined as materials consisting of two or more distinct phases with an interface between them. This definition is generally used for materials containing reinforcements characterized by a high aspect ratio (i.e., the ratio of length to thickness), as is the case for fibers, platelets, and flakes. The incorporation of these materials into thermoplastic matrices results in improved, but possibly anisotropic, mechanical and thermal properties. On the other hand, fillers with a low aspect ratio are mostly used to reduce cost and may even sacrifice some mechanical properties (e.g., strength and toughness). Nevertheless, the distinction between reinforced and filled polymers is often arbitrary.

The roles of the matrix, reinforcement, and interface in composites are well defined. The matrix is responsible for transfering the load from the matrix to the reinforcement, for distributing the stress among the reinforcement elements, for protecting the reinforcement from environmental attack, and for positioning the reinforcing material. Meanwhile, the task of the reinforcement is to carry the load, due to its higher stiffness and strength compared with that of the matrix. The interface, (for two dimensions) or the interphase (for three dimensions) is a negligible or finite thin layer with its own properties, and its role is stress transfer from the matrix to the reinforcement [1].

The blended polymer, usually present as a fine dispersion in the matrix, can take on many different roles: flame retardant, processing aid, surface finish, impact modifier, adhesion promoter between the matrix and reinforcement, and so on. Nevertheless, in the majority of cases, blending is aimed at improving the toughness of the composites, or more exactly, to achieving the desired balance between stiffness and toughness. It is widely assumed that stiffness and strength characteristics are related to the reinforcement and that toughness is related to the matrix; however, this presumption generally does not hold. The toughness-enhancing mechanisms are quite complex, and both matrix deformation and fiber-related failure events are involved. It is appropriate to call attention to the large gap between the trade and patent literature and the scientific literature in terms of their attention to the field of reinforced polymer blends. Only limited information on this topic can be found in handbooks [2–9], in contrast to the abundance of patent and trade (e.g., product brochures, leaflets) literature on the subject.

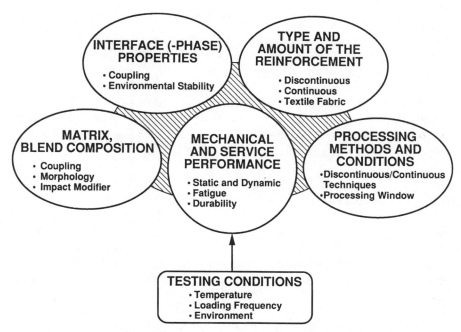

Figure 31.1 Factors influencing the mechanical performance of reinforced polymer blends.

Blending of the matrix polymer for thermoplastic composite applications is done when the matrix tends to undergo brittle fracture and/or exhibits notch sensitivity. Brittle fracture is favored, for example, by temperatures far below the glass transition temperature T_g, by low average molecular weight (MW), by deficient tie-molecule density, and by high crystallinity. Other goals of blending are to improve the heat distortion temperature (HDT) and to reduce some environmental effects (e.g., water uptake). Blending may also be the right tool for cost-reduction purposes. Last but not least, adding a second polymer may substantially increase the interfacial adhesion between fiber and matrix. By using such coupling agents, we can tailor the mechanical performance of the related composites. Based on the aforementioned aspects, blends are often composed of the following types of polymers: polycondensation polymers having a low mean MW and a tendency for hydrolytic decomposition, polymers with a high T_g, and polymers with low inherent adhesion to reinforcements. The trade literature seems to justify this assessment.

The mechanical performance of reinforced thermoplastic blends is affected by the following factors (cf. Fig. 31.1):

(a) blend composition and morphology
(b) type and amount of the reinforcement
(c) interface or interphase between matrix and reinforcement
(d) processing methods and conditions, and
(e) testing conditions.

There is a strong interrelation among items (a) to (d). For example, both the matrix morphology and reinforcement structuring may be highly dependent on the processing methods, as in the case of injection molding. On the other hand, the type and amount of the reinforcement dictate the selection of both suitable processing methods and conditions. Items (a) to (e) in Fig. 31.1 list some matrix-, reinforcement-, interface-processing- and testing-related factors. This chapter aims to give an overview of the manufacturing, structure, processing, and properties of thermoplastic composites containing various reinforcements in polymer blend matrices.

II. POLYMER MATRIX

The weakest site in a composite is either the matrix (which has low strength, low ductility, and low thermal resistance), the interface (which has poor adhesion between the matrix and the reinforcement, as well as incomplete fiber wetting), or both. In concert with Fig. 31.1, it should be noted that matrix modification may affect the interfacial properties (e.g., the use of polymeric couplants), and alterations in the interphase morphology may also have an impact on the matrix properties (e.g., transcrystallization in a unidirectional, continuous fiber-reinforced composite at a high volume fraction of fiber). According to the this scenario, improvements in the stiffness, strength, ductility (toughness), or thermal resistance of the polymeric matrix should yield beneficial properties for the related composites. Considering the fact that it is exactly these properties that are covered by 65% of the patents related to polymer blends and alloys (specifically, 38% for high-impact strength, 11% for tensile strength, 8% for rigidity and modulus, and 8% for HDT), one could state that blending should fuel the development of thermoplastic composites. This value (i.e., 65%) may be even higher when aspects of the processability and weld-line strength improvement (18%) are considered [10].

The matrix in reinforced polymer blends often consists of one of the following: polyamide (PA), poly(ethylene terephthalate) (PET), poly(butylene terephthalate) (PBT), poly(phenylene sulphide) (PPS), polycarbonate (PC), poly(phenylene oxide) (PPO), acrylonitrile–butadiene–styrene polymers (ABS), polyoxymethylene (POM), and polypropylene (PP). Note that many of these polymers are produced by polycondensation (and thus have a low molecular weight) and/or are known to be notch-sensitive materials (e.g., PA, PC, PBT), while others have poor cold impact resistance (e.g., PP homopolymers).

Compatibilization and the effects of morphology on the mechanical response of blends are beyond the scope of this chapter; for details on these topics; see Chapters 15, 16, 17, and 23. Here, it is sufficient to point out that blend morphology and the interfacial adhesion between the components of a blend are of crucial importance. Various nonreactive and reactive polymers may be used as compatibilizers. Reactive or functional modifiers that react with the functional groups of the matrix may also be produced *in situ*. The modifier is often referred to as "rigid" or "soft," depending on whether its stiffness is comparable or smaller (e.g., for rubbers) than that of the matrix polymer. Accordingly, blends are denoted as "rigid–rigid" or "rigid–soft" combinations. A very useful review of commercial blends, including trade names, is given by Utracki [10, 11].

The goal of blending in some cases is to improve the interfacial adhesion between the reinforcement and matrix. The wetting out of glass fibers (GF) of high polarity by apolar PP is very modest. This is reflected in very low values of the interfacial shear strength τ_{int} that are less than one sixth to one seventh of the yield (≈ 35 MPa) or shear strength of PP. Apart from suitable fiber sizing, the other opportunity is to use polymeric couplants (e.g., maleated PP or rubber) that migrate into the vicinity of the GF and improve the value of τ_{int} [12].

III. REINFORCEMENTS

Platy or fibrous materials of both natural and artificial origin may be used as reinforcements, provided that their aspect ratio is higher than about 5. Platelike (e.g., talc, clay, mica, glass flake) and fibrous (e.g., wollastonite, GF, carbon fiber (CF), aramid fiber (AF)) reinforcements are also often used in combination with nonreinforcing fillers (e.g., glass bead, glass microsphere, kaolin, chalk) in order to decrease the anisotropy in selected properties and to avoid some anisotropy-related problems (e.g., shrinkage, warpage). The main properties of reinforcements are summarized in handbooks (e.g., [13, 14]). Attention should be paid to the fact that fibers with a diameter in the range 5 to 20 μm are inherently stronger than the related bulk materials, because the probability of inherent material flaws reduces considerably with decreasing size.

The various reinforcing options are summarized in Fig. 31.2. The basic differences between them are related to the packing and thus to the reinforcement volume fraction ϕ_f. It is obvious that ϕ_f can be highly enhanced by aligning the fibers. This enhancement is normally associated, however, with an amplification of the anisotropy (i.e., direction dependence of some properties). On the other hand, the setup of the requested reinforcement achitecture is related to special manufacturing and processing techniques. Three-dimensional textile reinforcing structures can be produced, for example, by using commingled or intermingled yarns in which both the reinforcement and matrix-giving polymer are present in the form of fibers [1, 15]. The related textile assembly is a nonconsolidated preform in which the reinforcement is not wetted by the matrix. Apart from the packing (i.e., ϕ_f), the other main difference between the reinforcing options shown in Fig. 31.2 is whether the reinforcement is fully impregnated (as in the case of injection-moldable granules or impregnated unidirectional endless fiber-reinforced tapes) or not (as in the case of nonconsolidated textile preforms) [1]. As a consequence, the related processing and shaping techniques are different. For nonimpregnated textile preforms, a consolidation step should also be inserted in the production cycle [1, 15].

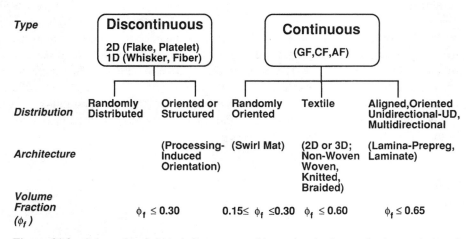

Figure 31.2 Schematic of the reinforcement architecturing in thermoplastic matrix-based composites. *Note*: The unidirectionally (UD) aligned discontinuous reinforcement is omitted, due to its practical irrelevance.

IV. INTERFACE/INTERPHASE

In order to fully exploit the potential of the matrix, τ_{int} should be matched with the shear strength of the matrix. Assuming a perfect bonding between the reinforcement and matrix, the critical aspect ratio of the reinforcement can be deduced from the force balance between the tensile strength of the fiber, σ_f and τ_{int}:

$$\left(\frac{1}{d}\right)_{\text{crit}} = \frac{\sigma_f}{2\tau_{int}}. \tag{31.1}$$

By adopting a τ_{int} value of 60 MPa (equal to the shear strength of the matrix) for GF–PA-66 blends [7] and letting $\sigma_f = 2400$ MPa, a critical aspect ratio of 20 can be deduced. According to Eq. 31.1 the critical aspect ratio is higher for fibers that are stronger than GF. Owing to incomplete wetting, poor bonding, or fiber-bundling effects, the critical aspect ratio is substantially higher in reality. Useful techniques for the determination of τ_{int} are tests on single-fiber microcomposites (e.g., fragmentation; microdroplet pulloff, pullout, pullthrough) [16–17] and on unidirectional fiber-reinforced composite laminates (e.g., indentation, short-beam shear, transverse tensile, etc.) [18]. A further rough estimation of the critical aspect ratio is based on fractographic inspection: The critical fiber length is approximately twice the maximum observed pullout length. This technique is mostly used for short-fiber-reinforced injection-molded composites [19]. The pros and cons of the various techniques [20] for assessment of τ_{int} are beyond the scope of this chapter.

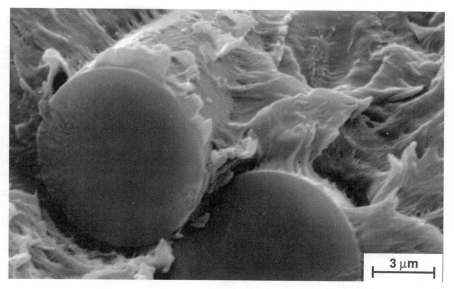

3 μm

Figure 31.3   Scanning electron microscope (SEM) picture of the fatigue fracture surface of a PBT reinforced by coupled GF (20 wt%, $\phi_f = 0.11$).

In order to improve the adhesion between matrix and fibers, the surface of the latter is treated accordingly, by sizing or by use of coupling agents [21]. Functionalized (reactive) silanes that are able to react with both the matrix and fiber are especially effective [7, 21]. Figure 31.3 shows the good bonding between PBT and GF. It should be emphasized here that the interfacial properties are very sensitive to temperature, moisture, and loading frequency. Therefore, the widely practiced fractographic inspection of bonding may be misleading. By selecting a suitable loading mode and temperature, both poor and good bonding can be shown for the same compound.

There is a further aspect that should be discussed here: the effect of interfacial morphology. In semicrystalline polymers and related blends, the reinforcement may act as a heterogeneous nucleant and produce transcrystallization [22–24]. An essential prerequisite of transcrystallization is the presence of active nuclei on the surface of the substrate (i.e., fillers, reinforcements) in high density. The closely spaced nuclei hinder the lateral extension of spherulites, which are then forced to grow in one direction, namely, perpendicular to the substrate's surface (i.e., oriented crystallization or oriented growth). Since the density of the nuclei on the substrate's surface is higher than in the bulk polymer, a columnar morphology appears. Transcrystallization can take place, however, only in a quiescent melt (e.g., consolidation by hot pressing), which is only seldom encountered in the processing of composites. Furthermore, GF does not induce transcrystallization unless specially treated. If the melt velocity differs from that of the reinforcement during processing, which is almost always the case, then a local shear field emerges in the vicinity of the fiber. This may generate row nuclei that initiate cylindritic growth [22, 25]. The presence of the cylindritic morphology has often been overlooked as well as regarded as transcrystalline, so the related results [26] should be treated with caution. The presence of a cylindritic superstructure does not yield any improvement in τ_{int}, as has been shown for the example of GF–PP blends [12, 27]. However, irrespective of the intense scientific interest behind this topic, transcrystallization may have only a small effect on the composites' performance (note that this effect is amplified by increasing ϕ_f), even in the case of semicrystalline polymer blends with CF and AF reinforcement [22].

The role of the interphase can well be demonstrated using the example of unidirectional continuous fiber-reinforced advanced composites. Assuming a two-dimensional interface and a quadratic packing of CFs of 7 μm diameter, the volume-related interface occupies 2300 and 4000 cm^2/cm^3 for $\phi_f = 0.4$ and 0.7, respectively. As mentioned in Section II, the blend composition e.g., may strongly affect the interphase behavior.

V. BLENDS WITH DISCONTINUOUS REINFORCEMENT

A. Manufacturing

Until recently, the market for thermoplastic molding composites was dominated by short-fiber-reinforced versions (SFRTs). SFRTs are produced by extrusion (melt) compounding of chopped fiber strands and thermoplastic resins. In order to minimize fiber breakage during compounding, the reinforcement is often introduced into the molten polymer by side feeding. This method also reduces the erosion of the extruder screw, compared with the traditional feeding via the hopper. The reinforcement aspect ratio in the melt-compounded granules is in the range of 30 to 60. As the stiffness and strength characteristics are increased by increasing the fiber length and providing denser packing (i.e., increasing ϕ_f), efforts were focused on these two items. Several approaches were and are still intensively explored, like *in situ* polymerization (restricted to poly(methyl methacrylate) and PA-6), powder impregnation, melt coating (analogous to cross-head wire coating), and melt impregnation. The breakthrough was achieved, however, through the thermoplastic pultrusion technique in the mid 1980s [28].

The related long (or pellet size) fiber-reinforced thermoplastics (LFRTs), in which the collimated endless fibers are completely wetted in a pressurized melt bath, are pultruded in rods of ca. a 3-mm diameter and chopped after cooling [28]. The impregnation of the spread and tensioned rovings may also be carried out using an array of melt-injection nozzles or an impregnation "wheel" of controlled melt permeability [29].

The fiber length for injection-moldable compounds is ca. 10 mm, while for compression molding the length is over 25 mm. Accordingly, the aspect ratios in these LFRTs are ≈ 1000 and ≥ 2500, respectively. LFRT compounds are marketed under the trade names Verton® (formerly ICI, now LNP), Celstran® and Compel® (formerly Höchst, now Ticona), Vari-Cut® (Polymer Composites, Inc.), Nepol® (Borealis), and Pryltex® (Appryl Composites). The matrices of Pryltex® and Nealid® (SFRTs of Borealis) compounds are blends consisting of PA and PP. It is obvious that a rather high ϕ_f value ($\phi_f \approx 0.35$) can be achieved using the unidirectional arrangement of the reinforcing fibers in LFRTs. In order to circumvent the various patented and expensive alternatives to LFRT production, attempts were also made to combine the fiber and matrix directly during processing (e.g., [30, 31]). The idea behind this development is very straightforward, since processing of the LFRT pellets is associated with severe fiber damage and thus also with property degradation.

B. Processing

The overwhelming majority of short- and long-fiber-reinforced thermoplastic compounds are processed by injection molding. During processing, the retention of the fiber length is of crucial importance. According to numerous works devoted to this issue, fiber breakage is due to the following factors: fiber–fiber interaction; converging flow and flow around sharp corners; spatial hindrance in the sprue, gate, or mold; and the presence of a partially molten phase with a solid phase. As a consequence, the target of process optimization and mold design is to avoid these effects [28, 32, 33].

Since fiber attrition is mostly encountered in the feeding zone of the reciprocating screw injection-molding machine, attempts have also been made to modify the screw and barrel configuration [31]. Fiber breakage is obviously more severe for LFRTs than for SFRTs. Nevertheless, the residual aspect ratio in molded parts of LFRTs is 10 to 30 times higher than that for SFRTs. Fiber-length retention also depends on the matrix properties. Long-glass-fiber (LGF) reinforced PA-66 moldings can gave a monomodal fiber-length distribution [34], whereas in an LGF–PP blend, a bimodal fiber-length distribution has been observed [35]. The presence of granule-sized GFs in LGF–PP molding indicates some cooperative motion (i.e., a bundling or bunching effect) of the GF during molding [36]. This bunching could well be observed on the fracture surface or on polished sections of the related composites [34, 35, 37]. Recall that the aspect ratio of the reinforcement is strongly reduced by fiber bundling. Increasing ϕ_f also causes more pronounced fiber fracture. The basic difference between short- and long-discontinuous-fiber reinforcement is that the latter becomes tangled, or "nested." As a consequence, the reinforcement remains in a skeletal structure after ashing [28].

Injection molding results in peculiar structuring (i.e., layering and orientation) of the reinforcing fibers, as shown in Fig. 31.4. The fibers are aligned in the mold-filling direction (MFD) in the surface (S), whereas the fibers adopt a transverse orientation to the MFD in the central (C) layer [2–4, 28, 34–37]. The explanation of this orientation pattern is given by the flow model of Rose [38] and Tadmor [39].

Depending on the shear and extensional-flow components, the fiber layering may show five or more odd-numbered layers instead of three with various fiber-orientation distributions in each layer [2]. The fiber-length distribution also changes across the thickness of the molded part: The mean fiber length increases from the surface toward the central layer [2, 34]. In case of semicrystalline matrices, a matrix-related skin–shear-core morphology [40] may be superimposed on the fiber layering. As the thickness of the skin, surface, core, and central layers are not identical, it is ap-

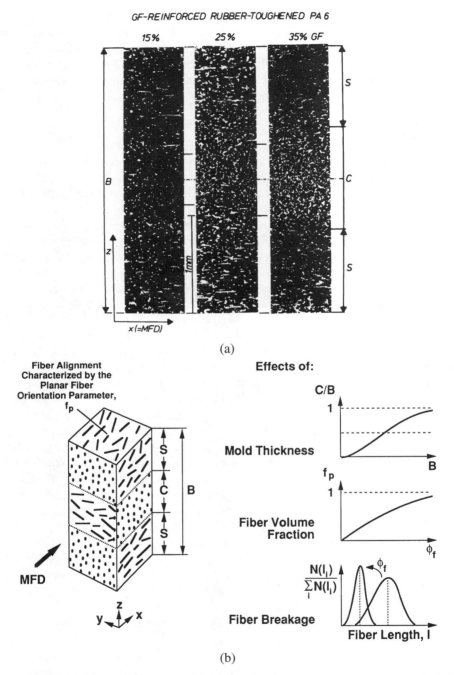

(a)

(b)

Figure 31.4   (a) Fiber layering as a function of ϕ_f for an SGF-reinforced rubber-toughened PA-6. *Note*: 15,25, and 35 wt% GF correspond to $\phi_f = 0.073, 0.129$, and 0.193, respectively. (b) Microstructural parameters of S(L)FRT compounds, shown schematically.

propriate to keep this differentiation in the terminology. Both the fiber layering and orientation (see Fig. 31.4) become more pronounced with increasing ϕ_f [2, 4, 34, 35, 37]. The fiber structuring is not affected by the fiber aspect ratio. On the other hand, the relative layer thickness (e.g., C/B, where B is the thickness of the molding—see Fig. 31.4) and the mean planar fiber orientation, described by a Hermans-type orientation factor [41, 42], may be substantially different between SFRTs and LFRTs at the same ϕ_f [2–4, 34, 37]. This is due to differences in the shear and elongational viscosities and thus in the flow-front profile [2, 36] between SFRT and LFRT melts. It is worth noting that the determination of the in-plane fiber orientation becomes complicated in LFRTs, owing to fiber bending [19, 34, 37].

This peculiar fiber structuring affects the mechanical and thermal performance of the composites and, thus, should be considered accordingly [2, 34, 37, 43]. This is the reason that considerable attention is focused on the prediction of the fiber structuring and structuring-caused effects (e.g., warpage, distortion, etc.) in molded parts [44–47]. Predictions work well for SFRTs, but are less persuasive for LFRT compounds. It is worth noting that directly compounded LFRT can also be used for extrusion blow-molding operations [48]. Hot pressing of LFRT compounds is discussed with respect to glass-mat-reinforced thermoplastics (GMT) next.

C. Properties and Their Prediction

1. *Mechanical Performance*   Stiffness, strength, and toughness all depend on the aspect ratio and fiber-orientation distribution. The best studied cases are when all fibers are oriented along—that is, longitudinal (L)—or transverse (T) to the loading direction or are present in a random distribution. Figure 31.5 shows how the normalized predicted stiffness [49], strength [50], and impact resistance [51] change with the aspect ratio for unidirectionally aligned discontinuous GF-reinforced PP at $\phi_f = 0.13$ [52]. This figure highlights the sensitivity of strength and impact responses to changes in the l/d ratio. A further lesson from Fig. 31.5 is that a compromise should be found with respect to the aspect ratio when balanced mechanical performance is targeted. When the fibers are misaligned in plane, the Young's modulus can be predicted by using the shear-lag theory developed by Cox [49] and improved by Krenchel [53] by considering the fiber orientation with the equation

$$E_c = \eta_0 \eta_1 \phi_f E_f + (1 - \phi_f) E_m, \tag{31.2}$$

where η_0 and η_1 are the fiber orientation and length efficiency factors, respectively, and E_c, E_f, and E_m are the Young's moduli of the composite, fiber, and matrix, respectively. This modified rule-of-mixture description works well for injection-molded single-gated dumbbells, which are usually used for the standardized determination of the tensile properties [54–56]. A widely used alternative approach is the Halpin–Tsai equation [57], which uses an analogy to laminated composites. Good overviews on this issue are available in the literature [43, 58–60].

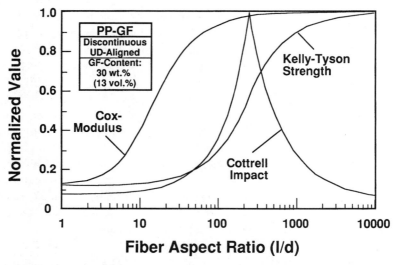

Figure 31.5 Normalized predicted stiffness (Cox–Modulus [49]), strength (Kelly–Tyson [50]), and impact toughness (Cottrell [51]) for discontinuous, aligned, GF-reinforced PP composites at 30 wt% ($\phi_f = 0.13$) reinforcement [52]. Reprinted from Composites A [52], copyright 1998, with permission from Elsevier Science.

The strength σ_c of the unidirectionally aligned discontinuous fiber-reinforced composites has been given by Kelly and Tyson [50, 61] as

$$\sigma_c = \sum_{l_i < l_{\text{crit}}} \frac{\tau_{\text{int}} l_i \phi_{f,i}}{d} + \sum_{l_j > l_{\text{crit}}} \sigma_f \phi_{f,j} \left(1 - \frac{l_{\text{crit}}}{2l_j}\right) + \sigma'_m (1 - \phi_f), \qquad (31.3)$$

where $\phi_{f,i} + \phi_{f,j} = \phi_f$; l_i and l_j represent the sub- and supercritical fiber-length populations, respectively, and σ'_m is the stress in the matrix at composite failure. Equation 31.3 can also be modified in order to consider the fiber orientation, as done in Eq. 31.2 by introducing the term η_0. This description works well for injection-molded composite dumbbells, including reinforced blends [54, 55].

According to the Cottrell model [51], the maximum impact toughness is reached when the length of all aligned fibers equals the critical length l_{crit}. This is in agreement with other predictions [61]. Nevertheless, the toughness of discontinuous fiber-reinforced composites can hardly be predicted, especially for composites with fiber structuring and polymer blend matrices. This is due to the fact that the fiber-related energy dissipation mechanisms (i.e., debonding, pullout, fracture; see Fig. 31.6) may trigger or hamper the matrix-related ones. Fig. 31.7 shows the interaction between fiber- and matrix-related failure mechanisms: Matrix crazing is terminated by fiber-pullout-induced shear yielding, the overall outcome of which is toughness improvement. On the other hand, the impact strength may also be reduced, since the ductility (i.e., the maximum strain value) is substantially reduced by the reinforcement [62]. Stress concentrations at the fiber ends, fiber crossings, or unwetted sites may also reduce the toughness, depending on whether or not their effect is compensated for by matrix deformation. If the stress is not redistributed by proper matrix deformation, resulting in an enlarged damage zone, the toughness may drop. In toughened reinforced systems, like GF-reinforced high-impact polystyrene (HIPS) [63], matrix failure becomes more and more hampered with increasing ϕ_f. By studying the effects of fiber ends by using fibers of various diameters at the same reinforcement loading, it was found that both optimum stiffness and toughness are related to a given fiber diameter and, thus, the aspect ratio [64].

This scenario becomes even more complicated because of interphase effects. Recall that l_{crit} decreases with increasing τ_{int} (see Eq. 31.1), which is often the case in reinforced polymer blends. Preferential wetting or encapsulation of the reinforcement by one of the blend components occurs frequently (see later). The fracture surface of rubber-toughened reinforced PAs showed, for example, sheathed pullout [6, 65]. This was explained by assuming that the shear strength of the matrix became smaller than τ_{int}, due to rubber modification [65]. On the other hand, Gaymans [6] attributed this sheathed pullout to the onset of cavitation. This interpretation could also be applied to the observed dependence of the loading frequency. However, some effects of preferential wetting cannot be excluded. It was reported that the adhesion between GFs and the semicrystalline matrix increases with the amorphous content of the matrix [27], which is definitely the case in rubber toughening. The modeling of toughness is a great challenge, which, in contrast to the modeling of stiffness and strength, has not at all been solved.

2. Microstructure-Related Fracture and Fatigue   Fiber orientation also influences toughness. Fiber structuring should be considered as a means of improving the toughness of injection-molded items. Toughness is often assessed by the methods of fracture mechanics, which are treated in Chapter 20 of this book. It was shown by Friedrich [66] that the fracture toughness, measured by the critical stress-intensity factor, of SFRTs can be estimated by the microstructural efficiency concept according to a simple empirical expression:

$$K_{C,c} = MK_{C,m} = (a + nR)K_{C,m}. \tag{31.4}$$

Here, $K_{C,c}$ and $K_{C,m}$ are the fracture toughness of the composite and matrix, respectively; M is the mirostructural efficiency parameter; a is the matrix stress condition factor; n is the energy-absorption ratio; and R is the reinforcing effectiveness parameter. Originally, R included the effects of fiber layering, orientation, and ϕ_f [66]. This was extended later to include the aspect ratio and aspect-ratio distribution (i.e., the effects of fiber ends [34]) and was generalized in the form [2, 34]:

$$R = \sum_i T_{\mathrm{rel},i} f_{p,\mathrm{eff},i} \phi_{f,i} \left(\frac{l}{d}\right)_{\mathrm{equ},i} \frac{(l/d)_{n,i}}{(l/d)_{m,i}}, \tag{31.5}$$

where $T_{\mathrm{rel},i}$ is the relative thickness of the ith layer normalized to the sample thickness (B; see Fig. 31.4), $f_{p,\mathrm{eff},i}$ is the effective orientation in the ith layer calculated using the function of planar orientation, f_p, vs. $f_{p,\text{eff}}$, introduced by Friedrich [66], $\phi_{f,i}$ is the fiber volume fraction in the ith layer; $(l/d)_{\text{equ},i}$ is the equivalent aspect ratio in the ith layer; and $(l/d)_{m,i}$ and $(l/d)_{n,i}$ are the mean mass- and number-average aspect ratios in the ith layer, respectively. Equation 31.5 considers the effects of fiber structuring (layering and orientation with respect to the crack path), molding-induced enrichment or depletion of the reinforcement, and the aspect ratio (affected by bunching and bending) and its distribution (the effects of fiber ends). The empirical nature of R is given by the fact that its constituent terms are interrelated (cf. Fig. 31.4b), but they are considered in Eq. 31.5 as independent parameters.

This empirical approach was adopted to describe the static and dynamic fracture behavior of a variety of short- and long-fiber-reinforced composites, including those with polymer-blend matrices [2–4, 34, 35, 37, 43, 66]. Furthermore, this microstructural efficiency concept has proved to be a valuable tool for describing the fatigue-crack propagation (FCP) behavior (see Chapter 27) of S(L)FRTs [2–4, 19, 35, 43, 67–69], as well as of rubber-toughened PA-6 [67] and compatibilized PA-66–PP matrices [69]. This concept was also recommended for design purposes [70]. Based on this microstructural efficiency approach even the FCP response of rubber-toughened PA-6 could be estimated [67], a value that cannot otherwise be experimentally determined, owing to a superimposed creep process.

D. Property Modification

1. *Polymeric Coupling Agents*   The interfacial shear strength τ_{int} between GF and PP is very low (between 5 and 8 MPa [12, 16, 27]), as the surface tension of PP is lower than that of GF, resulting in poor wetting. In addition, crystallization produces some "dewetting," due to shrinkage [12, 27]. Further, the GF surface itself offers no sites for mechanical anchoring. A promising way to improve τ_{int} is the use of polymeric coupling agents, also termed "adhesion promoters," which are mostly grafted polyolefins and elastomers. These agents include maleic-anhydride- or acrylic-acid-grafted PP (e.g., PP-g-MA and PP-g-AA, respectively), grafted ethylene–propylene rubbers (EPR-g-MA), functionalized styrene–ethylene-butylene–styrene block polymers (SEBS-g-MA), and the like. Such functionalized polymers are now commercially available from several producers (e.g., Uniroyal Chemical, Exxon, Shell, DuPont). By using these polymeric couplants, the τ_{int} between GF and PP can be increased up to 18–20 MPa [12, 27]. Polymeric couplants have also found application in platy reinforced thermoplastics [71].

Table 31.1 collates the basic mechanical properties for SGF–PP and LGF–PP systems with and without polymeric coupling. The data in Table 31.1 demonstrate that stiffness does not change with either the aspect ratio or the coupling [72]. By contrast, the strength values are strongly affected by them. The change in Izod impact resistance with aspect ratio is very pronounced; however, coupling has an adverse effect on the toughness, owing to restricted matrix deformation. Note that the Izod impact strength is five times higher for LGF–PP than for SGF–PP [72]. Surprisingly, the toughness of LGF–PP tends to increase with decreasing temperature (cf. Table 31.1). In another paper comparing the dynamic fracture response of SGF–PP

Table 31.1 Effects of Fiber Length^a and Polymeric Couplant^b on the Basic Mechanical Performance of SGF- and LGF-Reinforced PP^c

Property	Unit	Standard (ISO)	Short Glass Fiber (SGF–PP) Commercial	Long Glass Fiber (LGF–PP)		
				Without Coupling*d	With 5 wt% Coupling*d	Commercial
Tensile Strength	MPa	527	85	69	107	123
Tensile Modulus	GPa	527	9.3	8.6	8.6	9.1
Elongation at Break	%	527	1.7	1.4	1.8	2.1
Flexural Strength	MPa	178	131	131	173	199
Flexural Modulus	GPa	178	8.1	8.5	7.9	8.2
Deflection at Break	%	178	2.3	2.7	2.8	3.1
Izod Impact Energy	kJ·m^{-2}	180				
23°C			7	33	24	27
−30°C			6	41	29	37

^{a}diameter = 17 μm.

^{b}PP-g-MA; Polybond® 3002 of BP.

^{c}At the same fiber volume content: $\phi_f \approx 19$ vol% = 40 wt%.

^{d}Notes: The number-average fiber length of the compounded LGF–PP (*) was ca. 1 mm; the commercial SGF–PP and LGF–PP were products of ICI (Procom® GC 40H252 and Verton® MFX 7008, respectively) [72].

and LGF–PP, a similar large difference in the initiation fracture energy was observed between SGF and LGF reinforcement. Further, the fracture energy did not vary with temperature over a rather broad interval ($T = -40$ to $+60°C$). This has been attributed to the development of an enlarged damage zone controlled by both fiber structuring and matrix deformation (the latter was argued to be influenced also by adiabatic heating) [73].

Analysis of acoustic emission signals has proved to be a useful tool to distinguish between fiber-related failure mechanisms and to determine the failure sequence [74, 75]. It was demonstrated using acoustic emission that the superior impact resistance of one LGF–PP compound was due to a shift in the failure mechanisms from fiber fracture to fiber pullout [74].

By adding a third, elastomeric component to the binary system of PP-g-AA with mica, the toughness was increased or decreased, depending on whether the mica was encapsulated by the elastomer [76]. Encapsulation of glass beads (GBs) by a functionalized rubber (SEBS-g-MA) in the PP matrix increased the specific essential work of fracture, but strongly decreased the nonessential work constituent and thus the resistance to crack propagation [77].

It is noteworthy that the efficiency of the polymeric coupling agent is higher if it cocrystallizes with the matrix polymer [78]. The results achieved with grafted PPs in various PP composites seem to corroborate this conclusion.

2. *Compatibilizers* Graft or block copolymers are extensively used as compatibilizers in reinforced polymer blends. The driving force for blending of PP with PA is that PP is low priced and exhibits low density, low water absorption, and good resistance to solvents and chemicals. On the other hand, PA has a high HDT, excellent oil resistance, high toughness, and good paintability. By combining PP with PA, the deficient properties of both components can be eliminated, and the resulting blend has outstanding performance, especially for automotive applications [79]. How can we improve, however, the poor compatibilty between PP and PA? This can be achieved by incorporating various polymeric compatibilizers. Their effect in GF-reinforced blends depends strongly on the blend composition. If the matrix-forming polymer is PP, the PA as the minor component encapsulates the GF [80], so the effect of compatibilizers is marginal [79]. On the other hand, if the matrix-forming polymer is PA, PP domains and GF are separately dispersed. In that case, property improvement can be achieved only by adding compatibilizers [81, 82]. The encapsulation or preferential wetting by one of the blend components is controlled either thermodynamically or kinetically. This can be elucidated by a well-selected series of tests [80].

Obviously, the type of GF sizing (PP- or PA-compatible) determines the mechanical properties of PP–PA blends. The GF sizing must be matched to the matrix-forming polymer in order to achieve optimum performance [83–85].

3. *Impact Modifiers* The impact resistance of polymers is determined by the type and morphology of the dispersed-phase polymer. This determination can be done either in a chemical way (via copolymerization), a physical way (melt blending) or by their combination (reactive blending). Rubber-toughened PA-6 can also be produced by adding amine-terminated butadiene–acrylonitrile copolymer (ATBN) to the anionically polymerizing e-caprolactam. This *in situ* polymerization may be combined with a pultrusion process. During pultrusion, the continuous GF is wetted out by the prepolymer before the polymerization is completed. In this way, which is also suitable for the production of LGF-reinforced PA-6, the reinforcement is wetted out much better than by melt impregnation. In addition, it was found that the adhesion between GF and the matrix is enhanced in the presence of ATBN modifier (which becomes a dispersed phase by phase segregation) [86].

It was recognized early on that a good balance can be achieved between impact resistance and stiffness while maintaining cost effectiveness by using filled [87] and platelet-reinforced (e.g., mica, talc) rubber-toughened PPs. The demands of the automotive industry are met mostly by such compounds. The filler is either sheathed by the rubber or dispersed separately. The resulting morphology depends on the type of rubber chosen [88–90]. This choice affects both the stiffness and toughness reponse. Reinforcement encapsulation decreases the Young's modulus [91], but increases the toughness. So, improvement in toughness can be achieved at the cost of stiffness and vice versa [92], which is a well-known rule of thumb.

For reinforced PP-based blends (i.e., PP–rubber with or without functionalization), strength improvement and toughness deterioration were reported on applying functionalized modifiers (e.g., PP-g-MA, EPR-g-MA). This was explained by the local enrichment of the modifier in the vicinity of GF, which affected the pullout process [93]. Recall that the overall energy absorption may be reduced when pullout is not accompanied by an efficient stress redistribution via various matrix deformation processes (cf. Figs. 31.6 and 31.7).

The marketing success of fiber-reinforced rubber-toughened polymer blends is due to the beneficial combination of stiffness, strength, toughness, and environmental behavior that can be achieved by combining elastomer and fiber rather than by using fiber alone [59]. The ϕ_f value for optimum toughness is usually less for the toughened system than for the neat polymer. This difference is a clear hint of the competition between fiber- and matrix-related energy absorption. Figure 31.8 shows the Charpy impact energy and fracture toughness as functions of both ϕ_f and rubber content for PA-66 composites. The related data are available from the literature [4, 65, 94, 95]. The improvement in the fracture toughness depends also on the type and amount of the "soft" modifier.

The development of the damage zone in reinforced toughened systems can be studied using the crack-resistance-curve (K_R curve) analysis. The essential prerequisite of this approach, which works well for both brittle and ductile systems, is that crack growth can be identified and followed [95, 96]. For PAs, K_R reaches a maximum value as a function of both GF reinforcement and impact-modifier content [95].

Table 31.2 compares the basic mechanical and thermal properties of neat, rubber-toughened, GF-reinforced, and toughened GF-reinforced PA-66 resins. Based on Table 31.2, one can see how efficiently the notch sensitivity of the neat PA-66 is reduced by adding rubber and/or GF reinforcement. This table also demonstrates the stiffness and strength deterioration due to moisture uptake. Based on the decrease in the stiff-

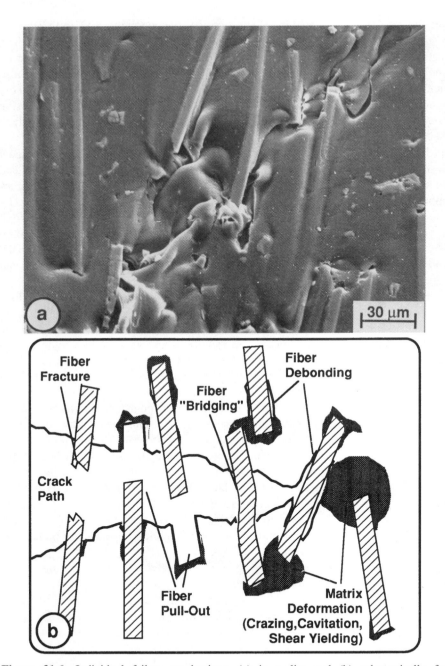

Figure 31.6 Individual failure mechanisms (a) in reality and (b) schematically for short-fiber-reinforced polymeric composites. *Note*: picture (a) was taken on the crack tip of a partially loaded compact tension (CT) specimen cut of an injection-molded LGF–PBT compound. (GF content is 20 wt%, $\phi_f = 0.11$, and $T = 60°$C.)

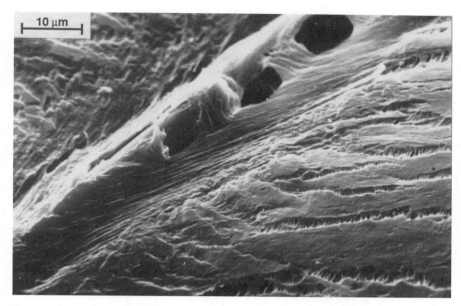

Figure 31.7   Fiber pullout in a rubber-toughened PA-6 composite. *Note*: This picture was taken on the surface of a CT specimen partially loaded at $T = 180°C$.

ness and strength parameters in Table 31.2, it is obvious that water attacks primarily the interface region. We have previously learned the effects of preferential wetting and encapsulation. Can we make use these effects to protect the interface between the matrix and GF more effectively against hydrolytic attack? The outcome of a recent study on the hygrothermal ageing of SGF-reinforced PBT compounds with various rubber modifiers showed that this goal is indeed realistic [97].

Incorporation of GF may also cause a deterioration in the toughness, as was shown for PC. The deterioration in toughness is amplified on increasing ϕ_f [66, 95]. This is due to a peculiar failure mode: The ductile PC is easily detached from the fibers, resulting in rodlike holes that coalescence suddenly via plastic deformation of the matrix. Therefore, PC is reinforced with specially sized GFs at low concentrations (ca. 10 wt%) and/or blended with other polymers, like PBT (e.g., Xenoy® of GE Plastics), PET, and ABS (e.g., Bayblend® of Bayer). These blends are rigid–rigid combinations, in contrast to the rigid–soft types of blends discussed earlier. Though the composition of these blends may vary throughout the whole range, the semicrystalline polymer is usually the larger component when high HDT values (see Chapter 28) must be guaranteed. The presence of the blended noncrystalline polymer may affect the crystallization kinetics of PET [98], PBT, or the PAs significantly. Apart from PC, other notch-sensitive polymers, such as PAs, are also blended with rigid polymers, such as PPO or ABS. The PA-66–PPO compound is marketed as Noryl® GTX by GE Plastics. The resulting stiffness and HDT values are usually superior to those of rubber-toughened variants. Figure 31.9 shows the fracture surface of a 30 wt% GF-reinforced PA-66–PPO blend Noryl® GTX of GE Plastics; composition

Table 31.2 Effects of Rubber Toughening^a, GF Reinforcement, and Relative Humidity^b on the Basic Mechanical and Thermal Properties of PA-66 Compounds^c

Property	Unit	Standard (ISO)	PA-66 Compound							
			(a) Neat		(b) RT		(c) Neat + GF		(d) RT + GF	
			dry	50% rh	dry	50% rh	dry	50% rh	dry	50% rh
Yield Strength	MPa	527	83	59	50	43	–	–	–	–
Tensile Strength	MPa	527	–	–	–	–	208	135	139	94
Tensile Modulus	GPa	527	3.0	1.5	2.0	1.2	10	7.5	8.8	6.8
Notched Charpy	kJ·m^{-2}	179								
23°C			6	16	85	110	14	16	25	26
−30°C			4	3	18	18	14	16	16	16
HDT	°C	75								
0.45 MPa			235		219		260		259	
1.82 MPa			80		66		254		245	

^{a} RT.
^{b} rh.
^{c} Data based on the following grades of DuPont: (a) Zytel® 101L, (b) Zytel® ST801 (contains 20 wt% modified EPDM rubber), (c) Zytel® 70G30HSL (contains 30 wt% GF in matrix (a)), and (d) Zytel® 80G33HS1L (contains 33 wt% GF in matrix (b)).

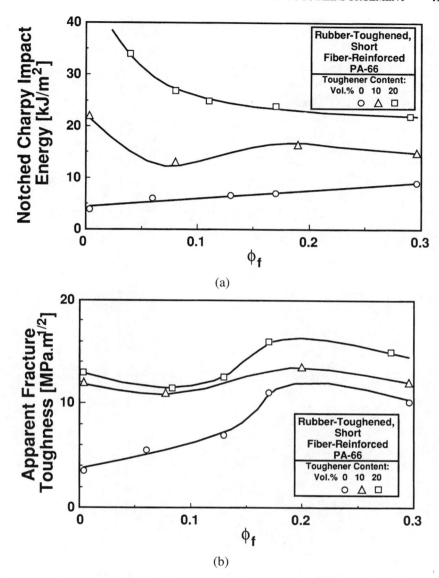

Figure 31.8 (a) Charpy impact energy and (b) fracture toughness as a function of rubber and glass fiber content (ϕ_f) for rubber-toughened PA-66, shown schematically [4, 65, 95]. *Note*: The apparent fracture toughness was computed by considering the maximum load registered.

$\approx 50/50$ wt%. It can clearly be seen that the dispersed-phase polymer (Fig. 31.9a) induces massive cavitation, followed by plastic deformation of the matrix ligaments between the dispersed particles (see Fig. 31.9b). These mechanisms are responsible for the outstanding toughness of this blend. Owing to the high ductility of such reinforced polymer blends, the J-integral method is preferred for toughness deter-

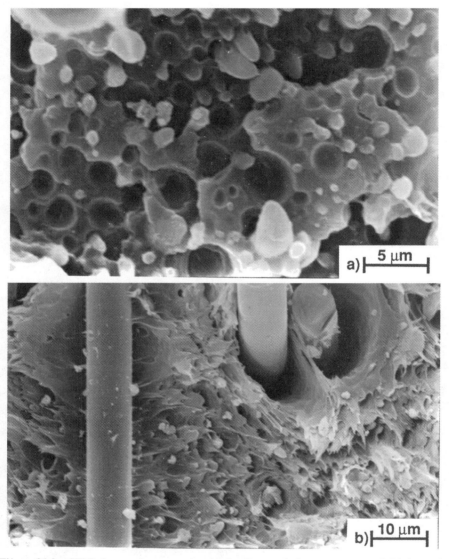

Figure 31.9   SEM pictures demonstrating (a) the dispersed-phase structure and (b) the cavitation-induced failure mechanism in a GF-reinforced PA-66–PPO blend (20 wt% GF; Noryl® GTX of GE Plastics).

mination. (See Chapter 20.) This approach was adopted for GF-reinforced PA-66–ABS [99], PA-66–styrene-acrylonitrile (SAN) [100], PA-66–PPO [101, 102], and PC–PBT [101, 102] blends. The J-integral terms related to fracture initiation and propagation did not follow the additivity law (rule of mixtures) either for the unreinforced or for the fiber-reinforced blends. The positive deviation (synergism) observed for the reinforced blends was attributed to an alteration in the failure mode (extensive shear deformation) and interfacial effects [99, 100]. Note that the aforementioned results were achieved at a constant ϕ_f value of 0.16. On the other hand, the critical value of the J-integral strongly decreases with increasing ϕ_f [2, 35, 56]. It can be predicted that an alternative approach to the J-integral, viz., the essential-work-of-fracture method (see Chapter 20), will be established for the toughness determination of reinforced polymers in the future. This prediction is based on the simplicity of the experimental procedure and data reduction of the latter method. Though this work-of-fracture approach is currently widely used for tough polymers, only preliminary tests have been done on reinforced polymer blends, namely on GF-reinforced PA-6.6–PP blends compatibilized by SEBS-g-MA [103]. It should be noted that the applicability and limitations of various nonlinear elastic fracture mechanical methods are still the topic of intensive dispute.

VI. BLENDS WITH CONTINUOUS REINFORCEMENT

Among composites with polymer-blend matrices, only the mat-reinforced systems have reached practical importance so far. Glass-mat-reinforced thermoplastic (GMT) sheets were first introduced in the early 1970s to produce large parts by hot pressing [104]. Though GMTs are available with various matrices, including PP, PET, PBT, PA, ABS, and even polymer blends (viz., PP–EPR, PC–PBT (Xenoy®-based GMT commercialized under the trade name Azloy® by Azdel BV, Holland) [104–106]), the market is controlled by GMT–PP. On the other hand, the recent achievements of thermoplastic composites with various textile reinforcements are very promising [1, 15]. These developments will definitely initiate interest on composites with polymer-blend matrices. Therefore, this section surveys some reinforcing options other than mats.

A. Manufacturing

1. *Composites with Mat Reinforcement*   The manufacture and formation of GMTs occur separately. The sheet production technologies use either melt impregnation (dry-laid techniques) or slurry deposition or papermaking processes (wet-laid techniques). The principles of these two variants are depicted schematically in Fig. 31.10. The basic advantage of melt impregnation is that wet-out and consolidation (compaction) of the composite occur in one step. By contrast, in the papermaking process, the "carrier material" (e.g., water, solvent) must be removed prior to sheet consolidation. The other difference is that the papermaking process can produce only discontinuous long-GF (length < 30 mm) reinforced composites. On the other hand, continuous swirl mat or advanced (containing UD plies as reinforcement) GMTs can be manufactured by the melt-impregnation route.

It is obvious that both manufacturing variants in Fig. 31.10 are suitable for the production of GMTs based on polymer blends. The blend constituents may be added in fiber or powder form in the papermaking process [106, 107]. Segregation due

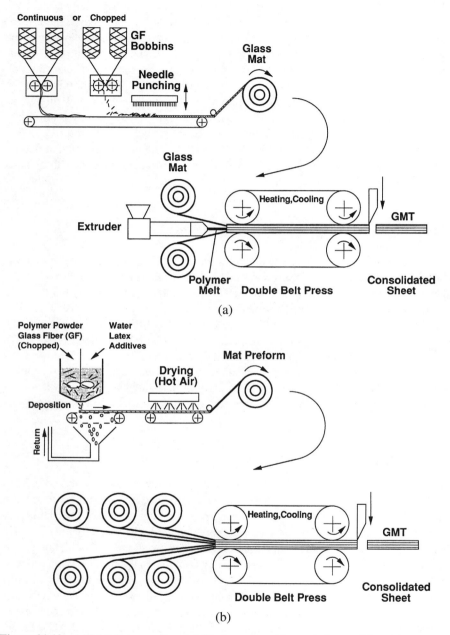

Figure 31.10 (a) Melt impregnation and (b) slurry deposition techniques of GMT sheet manufacturing, shown schematically.

to the density difference between GF and the polymer is circumvented by a proper consistency, like shaving foam, or by suitable additives (e.g., a latex binder).

Other options in GMT production differ mostly in the way the preform to be consolidated is produced: fluidized deposition of chopped GF strands and polymer powder [108], needle punching of GF or natural-fiber mats and PP nonwoven layers [109], needling of GF mats and polymer staple fibers, etc.

2. Composites with Other Reinforcements   Nonconsolidated preforms with two-dimensional or three-dimensional textile reinforcements can be produced via weaving, braiding, and knitting from commingled yarns [1, 15, 110] or by direct intermingling of the reinforcing and polymer fiber during the related textile operation. Commingled yarns composed of GF–PP, GF–PET (e.g., Twintex® of Vetrotex), and other blends are now available commercially. The GF–PP version usually contains a polymeric couplant, so that the related matrix can be treated as a blend. By using two or more polymer fibers or special fibers (e.g., bi- or multicomponent fibers) for intermingling with reinforcing ones, various textile architectures can be generated. After consolidation, the matrix in the related composites will be a polymer blend. Alternative approaches are the wet or dry coating of continuous reinforcing fiber tows by polymer powders. The powder-impregnated tow may be jacketed in another polymer as well [15, 43, 111]. The powder-coated tow, or the "towpreg," can also be partially compacted (e.g., hot compacted, calendered). Precompacted woven fabric-reinforced sheets are also available on the market (e.g., Twintex® of Vetrotex). Unidirectional fiber-reinforced fully wetted prepregs can be manufactured by exploiting the *in situ* anionic polymerization of ε-caprolactam and related compounds [86].

B. Processing

1. Composites with Mat Reinforcement   GMT-PPs with swirl-mat or long-GF reinforcements are nowadays almost exclusively processed by flow molding. The preheated blanks placed in the mold have a thickness of 3–4 mm, but are smaller in surface area than the final part. The fast compression of the preheated blanks, or their "packages," requires high pressures (ca. 10–30 MPa). On the other hand, the processing cycle is very fast (< 40 s) [106].

The recently developed direct-processing methods eliminate the step of producing consolidated GMT sheets. This step elimination is achieved, however, at the cost of the fiber length. Products of such on-line techniques are (randomly distributed) long-discontinuous-fiber (e.g., LGF—see section V.A) reinforced composites. Though, during pressing, the LGFs adopt a random orientation and by their "nesting" even a skeletal, matlike structure develops, they are not always considered as GMTs. The on-line techniques work according to the principle of extrusion melt compounding [106, 108, 112] or via melt impregnation of directly fed GF rovings [112, 113]. In the latter process, of Menzolit–Fibron (Germany), the fiber length can be exactly controlled, which is a great advantage over the melt-compounding processes. Owing to economic benefits, the direct or on-line processes present a great challenge for the conventional dry- or wet-laid GMT production lines. Polymer blends, especially rubber-toughened PPs, are often used as matrix materials in these direct processes in order to improve toughness.

High innovation potential characterizes the structural reaction injection-molding (S-RIM) process. This process is analogous to the resin transfer molding practiced with thermoset resins: A low-viscosity prepolymer mixture is pumped into the mold, which was previously loaded with the reinforcing structure. After wetting out, the prepolymer polymerizes *in situ* in the heated mold. This process is viable by the Nyrim® systems of DSM (Holland) and produces composite parts with PA block copolymer matrices [106, 114, 115].

2. Composites with Other Reinforcements   Nonconsolidated textile preforms can be compacted and shaped by different batchwise or continuous methods, e.g., hot pressing, thermoforming, pultrusion, roll forming, and filament winding [1, 43, 116–118]. The processing window is generally set to maximize wetting and minimize the void content. Some textile preforms, like two-dimensional knits, allow us to reach high deep-draw ratios, while others provide us with near-net-shape processing capability (e.g., pultrusion of a braided preform). These processing opportunities have not yet been explored for reinforced polymer blends.

C. Properties and Their Modification

1. Composites with Mat Reinforcement   Owing to the mat structure of GMTs, their properties exhibit a very great scatter that can reach ±30%. By location of acoustic emission events and infrared thermographic mapping it has been shown that the size of the damage zone may be as large as 30 mm [74, 112, 119, 120]. This large damage zone is due to a stress-relief and redistribution process controlled by both mat (e.g., needling-induced "mesh" structure and its resulting deformability), matrix (failure mode), and interface (τ_{int}) characteristics (cf. Fig. 31.11). It is therefore recommended that the deformability of the mat be matched with that of the matrix in order to upgrade toughness [119, 120]. This task can be easily accomplished using polymer blends.

The mechanical performance of continuous swirl-mat-reinforced GMT–PP blends is superior to that of LGF versions. The difference is less pronounced for stiffness and strength, but very pronounced with respect to toughness (e.g., transverse perforation impact). The reason behind this is highlighted by the scheme in Fig. 31.11. To obtain the required toughness, LGF–PP manufacturers (and also direct compounders) often use rubber-toughened PPs (copolymers and blends). Furthermore, in the proprietary formulations for flow-moldable GMT–PP blends, polymeric couplants (e.g., PP-g-MA) are also present. The same holds for natural-fiber (e.g., flax, sisal, jute) mat-reinforced composites [122].

The basic mechanical and thermal properties of GMTs with various matrices at the same values of ϕ_f are collated in Table 31.3. It is noteworthy that GMTs contain usually less than than 40 wt% GF reinforcement. This limit is imposed by wetting (void content) and packing (random distribution) problems; see also Fig. 31.2. The tensile modulus of GMT–PP blends can be approached by using the Cox–Krenchel

Discontinuous Long Fiber

Swirl Mat

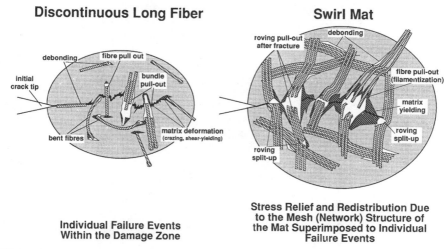

**Individual Failure Events
Within the Damage Zone**

**Stress Relief and Redistribution Due
to the Mesh (Network) Structure of
the Mat Superimposed to Individual
Failure Events**

Figure 31.11   Difference in the damage zone and related failure mechanisms between long discontinuous (LGF) and mat-reinforced PP (GMT–PP) composites. Reprinted from Composites Science and Technology [75], copyright 1998, with permission from Elsevier Science.

relationship (see Eq. 31.2 [49, 53]) [52], whereas for tensile strength, the Kelly–Tyson formula (see Eq. 31.3 [50, 61]) holds [123]. It has also been demonstrated that the strength data depend strongly on the GF sizing and thus on τ_{int} [123, 124]. Thomason et al. [124, 125] have also successfully estimated the directional dependence of linear thermal expansion coefficients of GMT–PP blends produced using the wet-laid technique.

2. *Composites with Other Reinforcements*   It has been shown with balanced and non-balanced woven textile fabric-reinforced rubber-toughened PP composites that the acoustic emission technique is a suitable tool to estimate the size of the damage zone and determine the failure sequence [126]. The authors, emphasizing always the similarities with mat-reinforced composites, found in another study that the ranking of the textile fabrics with respect to their reinforcing efficiency is as follows: woven fabric > knitted fabric > mat [127].

The effects of GF sizing and polymeric couplants (e.g., PP-g-MA) were studied in knitted GF fabric-reinforced PP composites produced from commingled yarn [128–130]. A strong dependence was found for both static and dynamic responses as a function of τ_{int}. (τ_{int} varied between 7 and 20 MPa.)

In the case of textile reinforcements, attention should be paid to the homogeneous distribution of the reinforcement. In a single-layer knit "reinforced" system, the reinforcement distribution is highly inhomogeneous and follows an all-nil (i.e., reinforcement-only matrix) pattern. This is the reason that a given number of textile fabric layers are required to get a homogenized stress-transfer unit [131]. Stacking of several knit layers prior to consolidation results, however, in a nonplanar reinforcing structure. Woven fabrics can be treated as more or less planar (i.e., more or less

Table 31.3 Effects of Matrix Polymer and GF-Mat Content on the Basic Mechanical and Thermal Properties of GMTs[a]

Composition Property	Unit	Standard (ISO)	Azdel® (Matrix:PP)		Azmet®-F (Matrix:PBT) Valox®	Azloy®-F (Matrix:PC/PBT) Xenoy®
			P100	P112		
Reinforcement Content	wt%		40	30	30	30
	vol%		≈13	≈19	≈20	≈17
Tensile Strength	MPa	527	95	85	115	120
Tensile Modulus	GPa	527	7.0	5.5	6.5	8.0
Flexural Strength	MPa	178	160	140	170	165
Flexural Modulus	GPa	178	5.5	4.5	6.5	6.0
Unnotched Charpy 23°C	kJ·m^{-2}	179	120	100	35	65
HDT 1.82 MPa	°C	75	165	165	220	180

[a]Note: Azdel® is produced by melt impregnation, whereas Azmet®-F and Azloy®-F are produced by the papermaking process of Azdel B.V. Valox® and Xenoy® are grades of GE Plastics.

two dimensional) reinforcements, which is not the case with two-dimensional knitted [127–131] and (of course) three-dimensional braided fabrics. This aspect should be considered during modeling studies as well.

VII. OUTLOOK AND FUTURE TRENDS

As the revolutionary metallocene-catalyzed synthesis allows the copolymerization of olefins with polar monomers [132], it can be predicted that the use of polymeric couplants in polyolefin composites will become less important. It can also be anticipated that polymer blends reinforced with various textile assemblies will become more important. Better thermoformability and reduced "read through" (i.e., surface appearance showing the reinforcement structure beneath) effects are the major arguments for using polymer blends. A further promising development is the production of composites with thermoset and thermoplastic blends. The unique feature of this method is that the thermoset resin encapsulates the reinforcement, and by curing-induced phase segregation the matrix becomes the thermoplastic resin [133]. This encapsulation strategy can also be followed for platy reinforced composites. The goal here is to develop inexpensive compounds with new or upgraded properties (e.g., barrier systems [134]). A vigorous development can be predicted also for the shape-memory thermoplastic composites. The shape memory in special polyurethanes and related blends is guaranteed by the T_g (which can be set accordingly) for which the modulus drop is in the range of two or three orders of magnitude. This modulus drop is responsible for the elastic memory. The inelastic memory is given by the segmental motion due to which even 400% deformation can be recovered [135]. A further challenging field of activity involves tailoring the morphology of polymer blends to produce reinforced polymer–polymer microlayer or microfibrillar composites. (See Chapters 30 and 33.) Last but not least: the concept of nanocomposites (distribution of organophilic silicates in nanometer scale) will be extended for polymer blends (especially for these containing polyolefins.)

VIII. ACKNOWLEDGEMENTS

This work was supported by the following grants: German Science Foundation (DFG: Ka 1202/7 and 1202/8) and European Commission (Brite: BRPR-CT96-0291; Inco-Copernicus: PL 964056; and TMR program: BRRT-CT97-5004).

IX. REFERENCES

1. J. Karger-Kocsis, in *Polymeric Materials Encyclopedia*, Vol. 2, J. C. Salamone (ed.), CRC Press, Boca Raton, FL, 1996, p. 1378.
2. J. Karger-Kocsis, in *Application of Fracture Mechanics to Composite Materials*, K. Friedrich (ed.), Elsevier Science, Amsterdam, 1989, p. 189.

3. K. Friedrich and J. Karger-Kocsis, in *Fractography and Failure Mechanisms of Polymers and Composites*, A. C. Roulin-Molonay (ed.), Elsevier Applied Science, Barking, United Kingdom, 1989, p. 437.

4. K. Friedrich and J. Karger-Kocsis, in *Solid State Behavior of Linear Polyesters and Polyamides*, J. M. Schultz and S. Fakirov (eds.), Prentice Hall, Englewood Cliffs, NJ, 1990, p. 249.

5. G. W. Becker and D. Braun (eds.), *Kunststoff Handbuch*, Hanser, München, 1992.

6. R. Gaymans, in *Rubber Toughened Engineering Plastics*, A. A. Collyer (ed.), Chapman and Hall, London, 1994, p. 210.

7. H. A. Scheetz, in *Nylon Plastics Handbook*, M. I. Kohan (ed.), Hanser, München, 1995, p. 387.

8. C. A. Cruz, in *Plastics Additives: An A–Z Reference*, G. Pritchard (ed.), Chapman and Hall, London, 1998, p. 386.

9. H. Domininghaus, *Die Kunststoffe und ihre Eigenschaften*, 5th Ed., Springer (VDI), Berlin, 1998.

10. L. A. Utracki, *Polymer Alloys and Blends*, Hanser, Munich, 1990, p. 10.

11. L. A. Utracki, *Commercial Polymer Blends*, Chapman and Hall, London, 1998.

12. F. Hoecker and J. Karger-Kocsis, *J. Adhesion*, **52**, 81 (1995).

13. H. S. Katz and J. V. Milewski, *Handbook of Fillers for Plastics*, Van Nostrand Reinhold, New York, 1987.

14. S. M. Lee (ed.), *International Encyclopedia of Composites*, VCH Publishers, New York, 1991.

15. K. Friedrich, in *Polypropylene: An A–Z Reference*, J. Karger-Kocsis (ed.), Kluwer Academic, Dordrecht, 1999, p. 81.

16. M. R. Piggott, in *Composite Applications*, T. L. Vigo and B. J. Kinzig (eds.), VCH Publishers, New York, 1992, p. 221.

17. J. J. Elmendorp and G. E. Schoolenberg, in *Polypropylene: Structure, Blends and Composites*, J. Karger-Kocsis (ed.), Vol. 3, Chapman and Hall, London, 1995, p. 228.

18. L. T. Drzal and P. J. Herrera-Franco, in *Engineered Materials Handbook*, Vol. 3: *Adhesives and Sealants*, ASM International, Materials Park, OH, 1991, p. 391.

19. J. Karger-Kocsis, K. Friedrich, and R. S. Bailey, *Adv. Comp. Mater.*, **1**, 103 (1991).

20. M. R. Piggott, *Comp. Sci. Technol.*, **57**, 965 (1997).

21. E. P. Plueddemann, *Silane Coupling Agent*, Plenum, New York, 1982.

22. J. Karger-Kocsis and J. Varga, in *Polypropylene: An A–Z Reference*, J. Karger-Kocsis (ed.), Kluwer Academic, Dordrecht, 1999, p. 348.

23. M. J. Folkes, in *Polypropylene: Structure, Blends and Composites*, Vol. 3, J. Karger-Kocsis (ed.), Chapman and Hall, London, 1995, p. 340.

24. H. Ishida and P. Bussi, in *Materials Science and Technology*, Vol. 13: *Structure and Properties of Composites*, T.-W.Chou (ed.), VCH, Weinheim, Germany, 1993, p. 339.

25. J. Varga and J. Karger-Kocsis, *J. Polym. Sci.: Part B: Phys.*, **34**, 657 (1996).

26. S. F. Xavier, in *Two-Phase Polymer Systems*, L. A. Utracki (ed.), Hanser, Munich, 1991, p. 347.

27. F. Hoecker and J. Karger-Kocsis, *Polym. Bull.*, **31**, 707 (1993).

28. J. M. Crosby, in *Thermoplastic Composite Materials*, L. A. Carlsson (ed.), Elsevier Science, Amsterdam, 1991, p. 139.

29. A. Lutz and T. Harmia, in *Polypropylene: An A–Z Reference*, J. Karger-Kocsis (ed.), Kluwer Academic, Dordrecht, 1999, p. 301.

30. F. Truckenmüller and H.-G. Fritz, *Polym. Eng. Sci.*, **31**, 1316 (1991).

31. T. Moriwaki, *Composites A*, **27A**, 379 (1996).

32. C. R. Gore, *Comp.Polym.*, **1**, 280 (1988).

33. R. Bailey and H.-G. Kraft, *Int. Polym. Proc.*, **2**, 94 (1987).

34. J. Karger-Kocsis and K. Friedrich, *Comp. Sci. Technol.*, **32**, 293 (1988).

35. J. Karger-Kocsis, in *International Encyclopedia of Composites*, Vol. 5, S. M. Lee (ed.), VCH Publishers, New York, 1991, p. 337.

36. A. G. Gibson, in *Polypropylene: Structure, Blends and Composites*, Vol. 3, J. Karger-Kocsis (ed.), Chapman and Hall, London, 1995, p. 71.

37. D. E. Spahr, K. Friedrich, J. M. Schultz, and R. S. Bailey, *J. Mater. Sci.*, **25**, 4427 (1990).

38. W. Rose, *Nature*, **191**, 242 (1961).

39. Z. Tadmor, *J. Appl. Polym. Sci.*, **18**, 1753 (1974).

40. M. Fujiyama, in *Polypropylene: An A–Z Reference*, J. Karger-Kocsis (ed.), Kluwer Academic, Dordrecht, 1999, p. 668.

41. P. H. Hermans, *Contribution to the Physics of Cellulose Fibers*, Elsevier, Amsterdam, 1946.

42. R. B. Pipes, R. L. McCullough, and D. G. Taggart, *Polym. Comp.*, **3**, 34 (1982).

43. A. Savadori and J. Schuster, in *Polymeric Materials Encyclopedia*, Vol. 11, J. C. Salamone (ed.), CRC Press, Boca Raton, FL, 1996, p. 8326.

44. F. Folgar and C. L. Tucker III, *J. Reinf. Plast. Comp.*, **3**, 98 (1984).

45. J.-F. Agassant, P. Avenas, J.-P. Sergent and P. J. Carreau, *Polymer Processing. Principles and Modeling*, Hanser, Munich, 1991.

46. T. Matsuoka, in *Polypropylene: An A–Z Reference*, J. Karger-Kocsis (ed.), Kluwer Academic, Dordrecht, 1999, p. 233 and p. 859.

47. M. Vincent and J. F. Agassant, in *Two-Phase Polymer Systems*, L. A. Utracki (ed.), Hanser, Munich, 1991, p. 277.

48. M. Thielen, *Kunststoffe*, **84**, 1406 (1994).

49. H. L. Cox, *Br. J. Appl. Phys.*, **3**, 72 (1952).

50. A. Kelly and W. R. Tyson, *J. Mech. Phys. Solids*, **13**, 329 (1965).

51. A. H. Cottrell, *Proc. Roy. Soc.*, **A282**, 2 (1964).

52. J. L. Thomason and M. A. Vlug, *Composites A*, **27A**, 477 (1996).

53. H. Krenchel, *Fibre Reinforcement*, Akademisk Forlag, Kopenhagen, 1964.

54. J. Denault, T. Vu-Khanh, and B. Foster, *Polym. Comp.*, **10**, 313 (1989).

55. T. Vu-Khanh, J. Denault, P. Habib, and A. Low, *Comp. Sci. Technol.*, **40**, 423 (1991).

56. K. J. Din and S. Hashemi, *J. Mater. Sci.*, **32**, 375 (1997).

57. J. C. Halpin and J. L. Kardos, *Polym. Eng. Sci.*, **16**, 344 (1976).

58. D. Hull, *An Introduction to Composite Materials*, Cambridge University Press, Cambridge, 1981.

59. M. G. Bader and A. R. Hill, in *Materials Science and Technology*, Vol. 13: *Structure and Properties of Composites*, T.-W. Chou (ed.), VCH, Weinheim, 1993, p. 291.

60. J. R. White and S. K. De, in *Short Fibre-Polymer Composites*, S. K. De and J. R. White (eds.), Woodhead Publishing Ltd., Cambridge, 1996, p. 1.

61. A. Kelly and N. H. MacMillan, *Strong Solids*, Clarendon Press, Oxford, 1986.

62. L. E. Nielsen and R. F. Landel, *Mechanical Properties of Polymers and Composites*, 2nd Ed., Marcel Dekker, New York, 1994, p. 488.

63. N.-S. Choi and K. Takahashi, *Coll. Polym. Sci.*, **270**, 659 (1992).

64. N. Sato, T. Kurauchi, S. Sato, and O. Kamigaito, *J. Comp. Mater.*, **22**, 850 (1988).

65. R. S. Bailey and M.G.Bader, in *Proceedings of the Vth International Conference of Composite Materials* (ICCM-V, San Diego, 29 July–1 August 1985, W. C. Harrigan, J. Strife, and A. K. Dhingra (eds.), The Metallurgical Society Inc., 1985, p. 947.

66. K. Friedrich, *Comp. Sci. Technol.*, **22**, 43 (1985).

67. J. Karger-Kocsis and K. Friedrich, *J. Thermoplast. Comp. Mater.*, **1**, 126 (1988).

68. J. Karger-Kocsis, *Composites*, **21**, 243 (1990).

69. T. Harmia, *Polym. Comp.*, **17**, 926 (1996).

70. T. Harmia and K. Friedrich, *Th. Appl. Fract. Mech.*, **26**, 47 (1997).

71. A. M. Ader, R. C. Constable, and J. A. Humenik, *J. Thermoplast. Comp. Mater.*, **1**, 196 (1988).

72. P. Peltonen, E. J. Pääkkönen, P. K. Järvelä, and P. Törmälä, *Plast. Rubber Comp., Proc. App.*, **23**, 111 (1995).

73. J. Karger-Kocsis, *Comp. Sci. Technol.*, **48**, 273 (1993).

74. T. Czigány and J. Karger-Kocsis, *Polym. Bull.*, **31**, 495 (1993).

75. J. Karger-Kocsis, T. Harmia, and T. Czigány, *Comp. Sci. Technol.*, **54**, 287 (1995).

76. W.-Y. Chiang, W.-D. Yang, and B. Pukánszky, *Polym. Eng. Sci.*, **32**, 641 (1992).

77. D. E. Mouzakis, F. Stricker, R. Mülhaupt, and J. Karger-Kocsis, *J. Mater. Sci.*, **33**, 2551 (1998).

78. J. F. Feller, A. Guyot, R. Spitz, B. Chabert, and J. F. Gerard, *Comp. Interface*, **3**, 121 (1995).

79. L. Setiawan, D. Schönherr, V. Schulze, and J. Siedentopf, *Gummi, Asbest, Kunststoffe*, **47**, 708 (1994).

80. D. Benderly, A. Siegman, and M. Narkis, *Polym. Comp.*, **17**, 86 (1996).

81. R. Cameron and R. C. Constable, *Polym. Polym. Comp.*, **1**, 45 (1993).

82. J.-Y. Wu, W.-C. Lee, W.-F. Kuo, H.-C. Kao, M.-S. Lee, and J.-L. Lin, *Adv. Polym. Technol.*, **14**, 47 (1995).

83. A. Perwuelz, C. Caze, and W. Piret, *J. Thermoplast. Comp. Mater.*, **6**, 176 (1993).

84. T. Harmia and K. Friedrich, *Plast. Rubber Comp., Proc. App.*, **23**, 63 (1995).

85. T. Harmia and K. Friedrich, *Comp. Sci. Technol.*, **53**, 423 (1995).

86. S. H. Lin, C. C. M. Ma, H. F. Meng, and L. H. Perng, *Mater. Sci. Res. Int.*, **2**, 87 (1996).

87. Gy. Marosi, Gy. Bertalan, P. Anna, and I. Rusznák, *J. Polym. Eng.*, **12**, 33 (1993).

88. J. E. Stamhuis, *Polym. Comp.*, **5**, 202 (1984).

89. J. E. Stamhuis, *Polym. Comp.*, **9**, 72 (1988).

90. D. L. Faulkner, *J. Appl. Polym. Sci.*, **36**, 467 (1988).

91. J. Jancar and A. T. Dibenedetto, *J. Mater. Sci.*, **29**, 4651 (1994).

92. Y. Long and R. A. Shanks, *J. Appl. Polym. Sci.*, **61**, 1877 (1996).

93. I. Kelnar, *Angew. Makromol. Chem.*, **189**, 207 (1991).

94. D. C. Leach and D. R. Moore, *Composites*, **16**, 113 (1985).

95. T. J. Pecorini and R. W. Hertzberg, *Polym. Comp.*, **15**, 174 (1994).

96. T. Vu-Khanh and J. Denault, *J. Mater. Sci.*, **29**, 5732 (1994).

97. T. Czigány, Z. A. Mohd Ishak, T. Heitz, and J. Karger-Kocsis, *Polym. Comp.*, **17**, 900 (1996).

98. V. E. Reinsch and L. Rebenfeld, *J. Appl. Polym. Sci.*, **59**, 1929 (1996).

99. S. V. Nair, S.-C. Wong, and L. A. Goettler, *J. Mater. Sci.*, **32**, 5335 (1997).

100. S. V. Nair, A. Subramaniam, and L. A. Goettler, *J. Mater. Sci.*, **32**, 5347 (1997).

101. N.-S. Choi, D. Oschmann, K. Takahashi, J. Karger-Kocsis, and K. Friedrich, *Trans. Japan. Soc. Mech. Eng.*, **61**, 62 (1995).

102. N. S. Choi, K. Takahashi, D. Oschmann, J. Karger-Kocsis, and K. Friedrich, *J. Mater. Sci.*, **33**, 2529 (1998).

103. S.-C. Wong and Y.-W. Mai, *Polym. Eng. Sci.*, **39**, 356 (1999).

104. D. M. Bigg, in *Polypropylene: Structure, Blends and Composites*, Vol. 3, J. Karger-Kocsis (ed.), Chapman and Hall, London, 1995, p. 263.

105. G. D. Tomkinson-Walles, *J. Thermoplast. Comp. Mater.*, **1**, 94 (1988).

106. J. Karger-Kocsis, in *Polymeric Materials Encyclopedia*, Vol. 4, J. C. Salamone (ed.), CRC Press, Boca Raton, FL, 1996, p. 2761.

107. P. L. Wallace, *Comp. Manufac.*, **1**, 109 (1990).

108. L. A. Berglund and M. L. Ericson, in *Polypropylene: Structure, Blends and Composites*, Vol. 3, J. Karger-Kocsis (ed.), Chapman and Hall, London, 1995, p. 202.

109. Zs. Fejes-Kozma and J. Karger-Kocsis, *J. Reinf. Plast. Comp.*, **13**, 822 (1994).

110. T.-W. Chou and F. K. Ko (eds.), *Textile Structural Composites*, Elsevier Applied Science, Amsterdam, 1989.

111. B. Z. Jang, in *Polypropylene: Structure, Blends and Composites*, Vol. 3, J. Karger-Kocsis (ed.), Chapman and Hall, London, 1995, p. 316.

112. J. Karger-Kocsis, in *Polypropylene: An A–Z Reference*, J. Karger-Kocsis (ed.), Kluwer Academic, Dordrecht, 1999, p. 284.

113. R. Liebold, *Kunststoffe*, **87**, 164 (1997).

114. C. W. Macosko, *RIM: Fundamentals of Reaction Injection Molding*, Hanser, Munich, 1989.

115. J. Karger-Kocsis, *J. Appl. Polym. Sci.*, **45**, 1595 (1992).

116. B. T. Åström, *Manufacturing of Polymer Composites*, Chapman and Hall, London, 1997.

117. S. T. Peters (ed.), *Handbook of Composites*, 2nd Ed., Chapman and Hall, London, 1998.

118. D. Bhattacharyya, in *Polypropylene: An A–Z Reference*, J. Karger-Kocsis (ed.), Kluwer Academic, Dordrecht, 1999, p. 721 and p. 841.

119. J. Karger-Kocsis and Zs. Fejes-Kozma, *J. Reinf. Plast. Comp.*, **13**, 768 (1994).

120. J. Karger-Kocsis, *Polym. Bull.*, **31**, 235 (1993).

121. J. Karger-Kocsis, in *Polypropylene: Structure, Blends and Composites*, Vol. 3, J. Karger-Kocsis (ed.), Chapman and Hall, London, 1995, p. 142.

122. K.-P. Mieck, in *Polypropylene: An A–Z Reference*, J. Karger-Kocsis (ed.), Kluwer Academic, Dordrecht, 1999, p. 527.

123. J. L. Thomason, M. A. Vlug, G. Schipper, and H. G. L. T. Krikor, *Composites A*, **27A**, 1075 (1996).

124. J. L. Thomason, in *Polypropylene: An A–Z Reference*, J. Karger-Kocsis (ed.), Kluwer Academic, Dordrecht, 1999, p. 407.

125. J. L. Thomason and W. M. Groenewoud, *Composites A*, **27A**, 555 (1996).

126. T. Czigány and J. Karger-Kocsis, *Polym. Polym. Comp.*, **1**, 329 (1993).

127. T. Czigány, M. Ostgathe, and J. Karger-Kocsis, *J. Reinf. Plast. Comp.*, **17**, 250 (1998).

128. J. Karger-Kocsis, E. Moos, and T. Czigány, *Adv. Comp. Lett.*, **6**, 31 (1997).

129. J. Karger-Kocsis and T. Czigány, *Appl. Comp. Mater.*, **4**, 209 (1997).

130. J. Karger-Kocsis and T. Czigány, *Composites A*, **29A**, 1319 (1998).

131. J. Karger-Kocsis, T. Czigány, and J. Mayer, *Plast. Rubber Comp., Proc. App.*, **25**, 109 (1996).

132. R. Mülhaupt, in *Polypropylene: An A–Z Reference*, J. Karger-Kocsis (ed.), Kluwer Academic, Dordrecht, 1999, p. 454.

133. R. W. Venderbosch, T. Peijs, H. E. H.Meijer, and P. J. Lemstra, *Composites A*, **27A**, 895 (1996).

134. C. D. Mueller, S. Nazarenko, T. Ebeling, T. L. Schuman, A. Hiltner, and E. Baer, *Polym. Eng. Sci.*, **37**, 355 (1997).

135. C. Liang, C. A. Rogers, and E. Malafeew, *J. Int. Mat. Syst. Struct.*, **8**, 380 (1997).

32 Liquid Crystalline Polymer Blends

DONALD G. BAIRD AND MICHAEL A. MCLEOD
Department of Chemical Engineering and the Center for Composite Materials
and Structures
Virginia Polytechnic Institute and State University
Blacksburg, Virginia 24061-0211

I. INTRODUCTION

Thermotropic liquid crystalline polymers (TLCPs) are typically melt-processable polyesters or polyesteramides that contain rigid aromatic units called mesogens in the main chain [1–4]. A listing of some TLCPs and their compositions is provided in Table 32.1. Because of their rigid-chain architecture, these polymers are rodlike, leading to nematic liquid crystalline order in the melt state. Also, the rigid-chain structure of TLCPs makes many of their properties superior to those of commodity thermoplastics. Specifically, in comparison to thermoplastics, TLCPs are known to have higher moduli, greater resistance to chemical attack, lower gas permeability, and lower coefficients of thermal expansion [5].

One of the attractions in using TLCPs is the exceptional mechanical properties they exhibit. Fiber spinning often yields fibers with tensile strengths over 1 GPa and tensile moduli on the order of 50 to 100 GPa [6–8]. Meanwhile, typical nylon and polyester yarns have strengths of around 1 GPa, with moduli ranging only from

Polymer Blends, Volume 2: Performance. Edited by D. R. Paul and C. B. Bucknall.
ISBN 0-471-35280-2. © 2000 John Wiley & Sons, Inc.

Table 32.1 Thermotropic Liquid Crystalline Polymers and Their Chemical Compositions

Name	Manufacturer	Chemical Composition[a]
Granular	Granmont	TA/PhHQ/StHQ
HX1000, HX4000, HX6000, and HX8000	DuPont	Based on HQ, TA, and other unspecified constituents
K161	Bayer	HBA/IA/TA/HQ/BP
Rodrun LC 3000	Unitika	PET/HBA (40/60 mol%)
Rodrun LC 5000	Unitika	PET/HBA (20/80 mol%)
SBH 1:1:2	Eniricerche	SA/BP/HBA (25/25/50 mol%)
SBHN 1:1:3:5	Eniricerche	SA/BP/HBA/HNA (10/10/30/50 mol%)
Vectra A900/950	Hoechst-Celanese	HBA/HNA (73/27 mol%)
Vectra B900/950	Hoechst-Celanese	HNA/TA/AP (58/21/21 mol%)
Xydar	Amoco	TA/BP/HBA

[a] AP: 4-aminophenol; BP: 4,4'-dihydroxybiphenol; HBA: 4-hydroxybenzoic acid; HNA: 2-hydroxy-6-naphthoic acid; HQ: hydroquinone; PET: poly(ethylene terephthalate); PhHQ: phenyl hydroquinone; SA: sebacic acid; StHQ: styryl hydroquinone; TA: terephthalic acid.
Sources: [3, 4, 6].

6 to 14 GPa [9]. This difference demonstrates that when TLCPs are oriented to a high degree through processes such as fiber spinning, properties exceeding those of commodity thermoplastics can be achieved. These oriented TLCPs are much stiffer than nylon or polyester yarns and offer strengths that are similar to or exceed those of many thermoplastics.

To take full advantage of the mechanical properties of TLCPs, it may be useful to use them as a reinforcing material, in a manner similar to that used for glass fiber. In work reported by Baird and coworkers [10], the tensile moduli of the TLCP phase in composite strands were found to range from 50 to 100 GPa. Also, the densities of TLCPs are typically around 1.4 g/cm^3, and the strengths of neat TLCP fibers have been reported to be in the range of 1 GPa as spun and over 2 GPa when annealed [7, 11]. These properties compare favorably with those of glass fiber, which has a tensile modulus of 69 to 83 GPa, a tensile strength that ranges from 1.72 to over 2.07 GPa, and a density of 2.52 to 2.61 g/cm^3 [12–15]. These values indicate that the properties of the TLCP fibers compete well with what glass fibers can provide, especially when compared on a weight basis, because TLCPs are significantly lighter than glass fibers.

The scope of this chapter is to discuss wholly thermoplastic composites in which TLCPs are the reinforcing phase. First, composites in which both the matrix and TLCP are processed as a molten blend and formed into a final part, called *in situ composites*, are discussed. In the course of covering *in situ* composites, issues of how fibrils are formed, the role of compatibilizers, and systems with partial miscibility will be examined. Second, the work that has been done with composites in which the TLCP fibrils are pregenerated is addressed. This chapter then finishes with conclusions and recommendations for realizing the full reinforcing potential of the TLCP fibrils in these composites.

Table 32.2   Tensile Moduli of *In Situ* Composites

Matrix Polymer[a]	Matrix Tensile Modulus (GPa)	TLCP Composition	Injection-Molded Plaques[b]	Spun Fibers and Strands
PP	1.3	20 wt% Vectra B950	MD: 2.55 GPa TD: 1.37 GPa [20]	13 GPa [10]
PP	1.3	30 wt% Vectra A950	MD: 2.96 GPa TD: 1.41 GPa [4]	19.0 GPa [21]
PET	2.2	20 wt% HX1000	MD: 6.87 GPa TD: 2.51 GPa [99]	12.2 GPa [99]
PP	1.8	30 wt% Vectra A950	2.94 GPa [18]	4.2 GPa [22]
PET	2.6	30 wt% Vectra A950	5.50 GPa [18]	8.0 GPa [22]
PPE/PS (70/30 wt%)	2.8	25 wt% Vectra B950	6 GPa [34]	16 GPa [34]
PPS	3.7	30 wt% Vectra A950	4.95 GPa [18]	6.0 GPa [22]
PS	3.2	25 wt% Vectra A950	3.7 GPa [34]	14 GPa [34]

[a]PET: Poly(ethylene terephthalate); PP: Polypropylene; PPE: poly-2,6-dimethyl-1,4-phenylene ether; PPS: Polyphenylene Sulfide.
[b]MD: Machine Direction; TD: Transverse Direction.
Sources are given in brackets within the tables.

II. COMPOSITES FROM DIRECTLY PROCESSING BLENDS

A wide variety of TLCP–matrix combinations have been used to produce *in situ* composites. For example, in one of the most comprehensive early studies, performed by Kiss [16], a thermotropic copolyester and a copolyesteramide were blended with several commodity thermoplastics, including polyethersulfone (PES), polyetherimide (PEI), polyarylate (PAR), poly(butylene terephthalate) (PBT), polycarbonate (PC), and poly(etheretherketone) (PEEK). It was found that regardless of the mixing and processing conditions used, TLCP domains 1 to 10 μm in diameter were formed. In addition to the thermoplastics used by Kiss, other research groups have blended TLCPs with matrix polymers, such as polypropylene (PP) [17–2], poly(ethylene terephthalate) (PET) [22–25], polystyrene (PS) [24, 26, 27], polyphenylene sulfide (PPS) [18, 19, 22, 28], and thermoplastic elastomers [29]. The general conclusion from this body of work has been that the TLCP phase is typically deformed into fibrils 1 to 10 μm in diameter that successfully reinforce the matrix, as the tensile moduli listed in Table 32.2 demonstrate.

For injection-molded *in situ* composites, it has been found that the level of reinforcement generally does not meet theoretical expectations. For example, O'Donnell and Baird [30] injection molded film-gated plaques composed of PP, maleated polypropylene (MAP), and Vectra B950 (63/7/30 wt%, respectively) and determined that they had a machine-direction tensile modulus of 4.94 GPa and a transverse-direction modulus of 1.71 GPa. These values were significantly greater than the modulus of 1.37 GPa for PP, but below the predicted level of stiffness. Applying the Halpin–Tsai equation [31], O'Donnell and Baird calculated that if the full reinforcing potential of the TLCP were realized, a planar isotropic sample would

Table 32.3   Tensile Moduli of Extruded Sheets of PET–Vectra A900

Material (wt%/wt%)	Draw Ratio[a,b]	Experimental Tensile Modulus (GPa)[a]	Theoretical Tensile Modulus (GPa)[c]
Neat PET (100/0)	5.0 (NC)	2.20 (0.18)	not applicable
PET–Vectra A900 (85/15)	6.2 (NC)	2.65 (0.28)	11.3
PET–Vectra A900 (82/18)	6.0 (NC)	2.95 (0.42)	13.1
PET–Vectra A900 (65/35)	4.8 (C)	1.99 (0.11)	23.6

[a] Standard deviations are given in parentheses.
[b] C: calendared, NC: not calendared.
[c] The vectra A900 modulus used in theoretical calculations is 65 GPa.
Source: [33].

have a modulus of 7.33 GPa. Note that this figure is based on all of the TLCP being deformed into fibrils with aspect ratios over 100 and tensile moduli of 75 GPa.

Likewise, sheet-extruded *in situ* composites also tend to have moduli that fall below theoretical predictions. The work of Sukhadia et al. [32, 33] on blends of PET and Vectra A900 helps illustrate this point. Composite sheets containing 15, 18, and 35 wt% Vectra A900 were made, as shown in Table 32.3. For each of these sheets, the tensile modulus was between 1.9 and 3 GPa, with the PET–Vectra A900 blend (65/35 wt%) having nearly the same modulus as that of neat PET. Furthermore, even in the cases wherein the modulus was improved, those values did not approach the levels of stiffness expected if all of the TLCP had been fully oriented. For example, by calculations using the Halpin–Tsai equation for a uniaxial composite containing 35 wt% Vectra A900, the sheet should have had a modulus of 23.6 GPa if the TLCP had been deformed into high-aspect-ratio fibrils. The reason cited for the inability to achieve higher mechanical properties was that it was not possible to apply high extensional strains. The result was that the TLCP phase was not fully deformed and oriented, so the full reinforcing ability of the TLCP was not realized.

However, in fiber spinning and strand extrusion, it has been determined that composites with mechanical properties that match theoretical expectations can be produced. Crevecoeur and Groeninckx [34] examined fibers spun from blends of poly-2,6-dimethyl-1,4-phenylene ether (PPE), polystyrene (PS), and Vectra B950. In these blends, the concentration ratio of PPE to PS was kept at 70/30 wt%, with loading levels of 1, 5, 10, 25, and 50 wt% Vectra B950. At all loading levels, it was found that the tensile moduli of the fibers matched the values predicted using the Halpin–Tsai equation. For instance, composite fibers of PPE–PS (70/30 wt%) blended with Vectra B950 (50/50 wt%) with a draw ratio of 40 possessed a tensile modulus of 32 GPa. This demonstrates that when the blend is melt spun into fibers, it is possible to realize the full reinforcing potential of the TLCP phase.

Ultimately, *in situ* composites have several significant limitations. First, the development of mechanical properties is affected by the processing method used. This factor is clearly demonstrated in Table 32.2, which shows that composites with the same composition have different properties, depending on whether they were injection molded or spun into fibers or strands. It should be recognized that theoretically predicted levels of reinforcement are attained only in spun fibers and strands. This is a severe drawback, because other processing methods, such as injection molding, are often used to form composite parts, making it more difficult for TLCPs to compete as reinforcing agents.

A second limitation of *in situ* composites is the anisotropy seen in injection-molded parts, which typically possess high machine-direction properties and low transverse-direction properties. Bafna and coworkers [35] demonstrated this factor in a study of PEI–HX1000 composite plaques, from which samples were cut and the machine- and transverse-direction flexural moduli were measured. At loadings of 50 wt% TLCP, over 4-to-1 anisotropy was found, with the machine-direction modulus being 12.92 GPa and the transverse-direction modulus being 3.20 GPa. The transverse-direction modulus was not significantly above that of neat PEI, which has a modulus of 3.00 GPa. Also, in the same study, it was established that *in situ* composites are more anisotropic than glass-filled thermoplastics at the same weight fraction of reinforcement.

The cause of this dependence of properties on the processing method used is the flow kinematics found in each process. In particular, it has been shown that extensional deformation is much more effective in deforming TLCP domains into reinforcing fibrils than is shear deformation [34, 36–41]. For injection-molded film-gated plaques, extensional deformation occurs only at the advancing front as the melt flows in the machine direction. The shear flow in the center of the material does not significantly deform the TLCP and does not contribute to property improvement. The result of the mold-filling flow kinematics is manifested in two ways. First, because the extensional deformation is predominantly in the machine direction, that is the only direction which is reinforced. Second, because the skin is produced through extensional deformation and the core is produced through shear deformation, a skin–core structural hierarchy is found in the plaques [42]. Meanwhile, although extruded sheets are subjected to extensional deformation, the levels of strain are not high enough to fully orient the TLCP. This means that for both of these processing methods, the mechanical properties of the composites fall below theoretical expectations, because not all of the TLCP is deformed into high-aspect-ratio fibrils with high levels of molecular orientation. In fiber spinning, it is possible to reach theoretically expected levels of stiffness, because all of the TLCP can be subjected to a high degree of extensional deformation.

A third limitation of *in situ* composites is that the TLCP and matrix must have overlapping melt-processing temperatures. This condition is a limitation because many TLCPs have to be taken to temperatures of over 300°C to be properly processed. For example, Vectra A950 must briefly be taken to temperatures of 320°C to 330°C to melt all high-temperature-melting crystallites [43, 44]. This factor is important because the presence of high-temperature-melting crystallites can prevent the mechanical properties of TLCPs from being maximized [45–47]. At these temperatures, commodity thermoplastics such as PET and PP would be severely degraded. Therefore, because of the need for overlapping melt processing temperatures, many TLCPs are limited to being processed with high-temperature engineering polymers.

To produce high-aspect-ratio TLCP fibrils so that the full reinforcing ability of the TLCP can be attained, it is necessary to understand droplet deformation and breakup in a two-phase system. For liquid–liquid blends, shear stresses due to viscous drag of the suspending fluid compete with the cohesive effect of surface tension. If viscous drag overcomes the effect of surface tension, the droplet will elongate into a cylindrical (threadlike) shape, eventually breaking up into small aligned droplets when the shearing and/or extensional stresses exceed the interfacial forces resisting this deformation [48]. What is critical in producing *in situ* composites is the ability to solidify the elongated TLCP fibrils before they break up into smaller spherical droplets. Therefore, the variables controlling the deformation of a droplet in a fluid need to be qualified. Also, the influence of the viscoelastic characteristics of polymer melts needs to be recognized.

The seminal work in the area of drop deformation and breakup was performed by G. I. Taylor [49–51], who examined two-phase systems using Newtonian liquids. In studying this subject, he found that drop deformation is controlled by the ratio of the viscosity of the dispersed phase, μ_d, to the viscosity of the liquid medium μ_m—that is, $p = \mu_d/\mu_m$—and the Weber, or capillary, number Ca. The Weber number is the ratio of the viscous stress in the fluid to the interfacial stress tending to prevent deformation. Specifically, $Ca = \mu_m \dot{\gamma} a/(\gamma)$, where $\dot{\gamma}$ is the shear rate, a is the initial droplet radius, and γ is the interfacial tension.

Taylor formulated two expressions for the limiting cases, that is, when either interfacial-tension effects or viscous effects dominate droplet deformation. Using the length L and breadth B of a droplet, Taylor defined a deformation parameter $D = (L - B)/(L + B)$. For the case in which interfacial-tension effects dominated viscous effects (i.e., $p = O(1)$ and $Ca \ll 1$, where $O(1)$ means "on the order of magnitude of 1"), the deformation parameter was

$$D = Ca\frac{(19p + 16)}{(16p + 16)},\tag{32.1}$$

and the orientation angle α of the droplet was 45°. Meanwhile, when viscous effects dominated interfacial-tension effects ($Ca = O(1)$, $p \gg 1$), the deformation parameter was

$$D = (5/4)p,\tag{32.2}$$

and $\alpha = 0°$. Also, it should be noted that based on Taylor's derivations, the critical Weber number needed for bursting equals one half the deformation parameter (i.e., $Ca_C = 0.5D$).

In examining the effects of deformation on droplet behavior, it has been found that droplets deform and break up over a greater range of viscosity ratios in extensional-flow fields than in shear-flow fields. Grace [52] established this finding in a comprehensive study that encompassed blends with viscosity ratios ranging from 10^{-6} to 10^3. In simple shear flows, the breakup of droplets appeared to occur more readily at $0.1 < p < 1.0$, with no breakup once $p > 3.5$. However, for extensional flows, breakup could be induced at any viscosity ratio. Also, for any given viscosity ratio, the critical Weber number was lower when the droplet was subjected to extensional deformation rather than shear deformation. This factor is important to consider when producing *in situ* composites, because the TLCP domains must be deformed and elongated into fibrils in order to reinforce the matrix polymer.

Studies on viscoelastic droplet behavior have indicated that there are some significant differences between viscoelastic fluids and Newtonian fluids [48, 53–61]. It appears that viscoelasticity can severely inhibit or even prevent thread breakup. From studies of viscoelastic droplets in Newtonian matrices, Elmmendorp and Maalcke [53] concluded that droplets with higher normal stresses appeared to be more stable. However, when Newtonian droplets were suspended in viscoelastic matrices, the normal stresses exhibited by the matrix tended to increase the deformation of the droplet. Similar results were reported by Tavgac [54], who compared the Deborah number for the viscoelastic fluid (that is, the ratio of the material's characteristic time to its characteristic flow time) to the critical Ca.

Another concern in producing *in situ* composites is the critical time to break up, that is, the time needed for a deformed, cylindrically shaped droplet to divide into a series undeformed, spherical droplets. This factor is important, because in order to develop a fibrillar morphology, the TLCP must be deformed into fibrils quickly and the fibrils must then be frozen into place. The work of Machiels and colleagues [62, 63] emphasized this point. By studying Vectra A900 fibrils encapsulated in Arnitel em630, a thermoplastic elastomer produced by DSM, it was shown that a fibril 9.4 µm in diameter held at 300°C divided into droplets in 80 seconds. For fibrils with smaller diameters, the breakup time was even less, with fibrils 1 µm in diameter breaking up into droplets after a few seconds at 300°C. Thus, in order to maximize the properties of *in situ* composites, the fibrils generated in the processing step must be cooled quickly so that they cannot relax back into droplets.

In addition to drop deformation and breakup, the effects of processing conditions on the development of mechanical properties are important. For injection-molded *in situ* composites, O'Donnell and Baird [30, 47, 64, 65] examined the factors of injection speed (i.e., fill time), mold temperature, melt temperature, and mold thickness. Of these variables, melt temperature, injection speed, and mold thickness were found to influence mechanical properties significantly. Varying the mold temperature from 20°C to 70°C had no discernible effect on the mechanical properties of the composites.

The influence of injection speed on mechanical properties is generally that faster speeds produce better properties in the machine direction, but also greater anisotropy [47, 64, 65]. This was demonstrated in a blend of PP, MAP, and Vectra B950 (63/7/30 wt%) [64, 65]. 1.5-mm-thick plaques showed the highest mechanical properties when injected in the fastest time of 1 second. The machine-direction flexural modulus and strength at this injection time were 5.10 GPa and 60.5 MPa, respectively, while the transverse-direction modulus and strength were 1.80 GPa and 35.1 MPa, respectively. Raising the injection time to 10.7 seconds produced plaques with a machine-direction modulus of 3.21 GPa and a strength of 52.1 MPa, an appreciable decline. Transverse-direction properties were not measured for this injection time, but for injection times up to 4.4 seconds, the transverse-direction moduli and strengths remained around 1.80 GPa and 35.1 MPa, respectively, matching what was found for the fastest injection time. Therefore, faster injection speeds improve machine-direction properties, but have no influence on reinforcing the transverse direction.

The effects of mold thickness on matrix–TLCP blends were found to be complex, because higher mechanical properties are not always obtained with thinner molds. Rather, an optimal thickness can exist. For example, O'Donnell and Baird [64] injection molded plaques 1.0 mm, 1.5 mm, and 2.3 mm thick composed of PP, MAP, and Vectra B950 (63/7/30 wt%). Comparing machine-direction results on plaques injection molded at a volumetric flow rate of about 14.5 cm^3/s, O'Donnell and Baird found that the 1.5-mm-thick plaque had the highest flexural modulus (5.10 GPa) and strength (60.5 MPa). This finding contrasts sharply with the properties of samples from the other two molds. Plaques from the 1.0-mm-thick mold had a modulus of 3.71 GPa and a strength of 54.2 MPa, while the 2.3-mm-thick plaques possessed values of 4.61 GPa and 54.5 MPa, respectively. However, it should be noted that when other TLCPs, viz., Vectra A950 and LC3000, were blended with PP, the expected trend of a thinner mold producing specimens with higher mechanical properties was observed [30, 47, 65].

Additionally, mold thickness is important in the production of *in situ* composites because it is a controlling factor in transferring heat from the sample. To illustrate this point, consider two plaques, 1.0 and 2.3 mm thick, respectively, that start with a temperature of 295°C and are in intimate thermal contact with a mold at 20°C. To reach a temperature of 200°C one third of the distance into the plaque, it takes 0.25 seconds for the 1.0-mm-thick sample and 1.4 seconds for the 2.3-mm-thick sample [47]. Exposing the TLCP fibrils to this thermal history could allow molecular or morphological relaxation. This would reduce the reinforcing ability of the TLCP, yielding a composite that would have mechanical properties lower than would be expected from theoretical calculations.

It was determined that, rather than attempting to correlate mechanical properties to mold thickness, the mechanical properties should be compared to the Graetz number [64, 65]. The Graetz number N_{GR} is a dimensionless group that relates the magnitude of the heat convected in the down-channel direction to the heat conducted in the transverse direction. It was found that for each plaque thickness, the flexural modulus began to drop dramatically for values of N_{GR} lower than 10. At these smaller values of N_{GR}, it was speculated that the Vectra B950 began to solidify, making the TLCP more difficult to deform into fibrils. Hence, it is believed that this dimensionless group is useful in relating TLCP solidification and deformation to the observed mechanical properties.

Processing temperature also has the potential to affect mechanical properties. O'Donnell and Baird [47] demonstrated that the starting melt temperature affected the rheological behavior of Vectra A950, by performing dynamic oscillatory cooling scans using an angular frequency of 1 radian/sec. When the Vectra A950 was cooled from 295°C, the complex viscosity was about a half an order of magnitude greater than when the Vectra A950 was cooled from 300°C. This finding is important, because it shows that the TLCP has to be stretched into high-aspect-ratio fibrils in order to maximize mechanical properties. For the Vectra A950 to obtain the desired morphology when processed at 295°C, the authors anticipated that either high volumetric-flow rates or high mold temperatures were needed, so that the TLCP would be deformed as the mold was filled.

When *in situ* composites are made through fiber spinning, one of the most important processing conditions to control is the amount of extensional deformation applied. The effect of draw ratio on the mechanical properties of fibers has been the subject of many studies [4, 10, 17, 27, 34, 66–71], which have generally shown that mechanical properties tend to rise with the draw ratio to some maximum value, above which they level off. For example, Kyotani and coworkers [66] spun fibers of a PET–Vectra A950 (80/20 wt%) blend with draw ratios of 10 to 120. At the low draw ratios, there was a steady increase in tensile modulus. However, once the draw ratios exceeded 80, the tensile modulus was consistently about 12 GPa. Note that for PET–Vectra A950 (80/20 wt%) fibers, theoretical calculations predict that the modulus should be 14.3 GPa; thus, the high-draw-ratio fibers had tensile moduli approaching theoretical expectations.

The effect of the die's length-to-diameter ratio on TLCP fibrillation has been the subject of several studies [72–76]. Blizard and Baird [72] showed that with a L/D of 7.82, fibrils were present in all samples extruded using shear rates from 45.7 to 457 s^{-1}, while with a L/D of 21.4, droplets were formed. It was suggested that the difference in morphology was due to relaxation of the fibrils as they passed through the capillary. Therefore, for fiber spinning or strand extrusion, it is preferable to use a die with a low L/D, so that fibrils formed at the converging entrance are retained in the final composite.

Mixing history can influence the mechanical properties of extruded strands and fibers. Sukhadia et al. [32, 77, 78] and Baird and Sabol [79] proved this condition by contrasting composites produced with one extruder with those obtained from a patented dual-extruder system. Consistently, strands of a PET–Vectra A900 (70/30 wt%) blend had higher mechanical properties when processed with the dual-

Table 32.4   Comparing the Properties of PET–Vectra A900 (70/30 Wt%) Composite Strands Produced Using a Single Extruder, Versus the Dual-Etruder Processing Method

Draw Ratio of Dual-Extruder Strands	Tensile Modulus (GPa) of Dual-Extruder Strands[a]	Draw Ratio of Single-Extruder Strands	Tensile Modulus (GPa) of Single-Extruder Strands[a]
2.36	5.45 (0.58)	4.55	3.98 (0.23)
3.25	6.97 (0.34)	7.10	7.08 (0.82)
3.80	8.93 (0.39)	13.0	8.05 (0.09)
39.0	13.31 (0.37)	20.0	8.49 (0.54)
43.2	17.21 (0.13)	49.0	13.39 (0.45)
49.7	18.99 (0.17)		

[a] Standard deviations are given in parentheses.
Source: [33].

extruder system, as shown in Table 32.4. This improvement in properties was attributed to the difference in morphology. For the strands made using the dual-extruder system, no skin–core morphology existed, with fibrils being present from the skin to the center of the strand. Also, the fibrils had essentially infinite aspect ratios, because the TLCP was present in the melt as axially continuous streams, not as droplets. A more thorough description of the dual-extruder system and its advantages is given later in this chapter.

Compatibilization has been noted to have dramatic effects on morphology, with the TLCP being dispersed more effectively, creating smaller fibrils, yielding smoother surfaces, and allowing less fiber pullout [20, 30, 80–84]. A more fibrillar structure has been observed with an increase in TLCP dispersion, thereby improving the ability of the TLCP to reinforce the matrix [30]. For example, O'Donnell and Baird [30] found that tensile bars of a PP–Vectra A950 (70/30 wt%) blend had a tensile modulus of 4.2 GPa without compatibilizer, but improved to 4.8 GPa when 20 wt% maleic-anhydride-grafted polypropylene (MAP) was added to the PP as a compatibilizer. However, when the *in situ* composite had 50 wt% MAP, the stiffness dropped to 3.6 GPa. Morphological examinations revealed that the TLCP domains were smaller than those in the other PP–Vectra A950 blends, leading the authors to conclude that the addition of too much MAP reduced the interfacial surface tension, thereby hindering fibril formation.

Compatibilization has also been reported to increase the tensile strength of *in situ* composites, because of improved adhesion between the two polymer phases [20, 30, 47, 65, 80–87]. O'Donnell [30, 65] showed this by studying injection-molded plaques of a PP–Vectra B950 (70/30 wt%) blend, with MAP content varying from 0 to 30 wt%. From contact angle measurements, the work of adhesion for PP–Vectra B950 blends was found to be 58.0 mN/m, while adding 10 wt% MAP to the PP increased the work of adhesion to 62.8 mN/m. This increase in adhesion was reflected in the tensile properties. The PP–Vectra B950 blend without MAP had a strength of 23.1 MPa in the machine direction and 11.8 MPa in the transverse direction. Meanwhile, the PP(MAP)–Vectra B950 (49/21/30 wt%) blend had strengths of 40.3 MPa and 17.8 MPa in the machine and transverse directions, respectively.

However, even with the creation of a finer morphology and increased mechanical properties, those properties usually do not reach theoretical expectations. For instance, Datta and Baird [84] injection molded an *in situ* composite composed of a PP–MAP–Vectra A950 (45/5/50 wt%) blend, which was found to have a machine-direction tensile modulus of 5.205 GPa and a transverse-direction modulus of 1.567 GPa. However, a random planar composite would have a modulus of 12.0 GPa if all of the Vectra A950 were being used to stiffen the matrix. This study demonstrates that compatibilizers can serve to improve the properties of injection-molded *in situ* composites, but do not lead to complete utilization of the TLCP phase.

Sabol and coworkers [70, 71] examined the effects of compatibilization in strand extrusion by spinning two sets of composite strands, one composed of a PP–MAP–Vectra B950 (63/7/30 wt%) blend and the second composed of a PP–Vectra B950 (70/30 wt%) blend. The addition of MAP did not have a significant effect on tensile strength, but consistently improved the modulus of the composite strands. For

Table 32.5   The Effect of PEsI on the Mechanical Properties of PEI–Vectra B950 Composite Strands

Strand Composition PEI/TLCP/PEsI (wt%/wt%/wt%)	Tensile Modulus (GPa)	Tensile Strength (MPa)	Percent Elongation at Break
75/25/0	13	350	4.6
74.25/25/0.75	14	400	5.2
73.5/25/1.5	16	520	6.3
72.75/25/2.25	12	300	4.8
71.25/25/3.75	8.5	150	3.2
67.5/25/7.5	7	125	3.4

Source: [89].

example, at a draw ratio of 8.2, adding compatibilizer resulted in an increase in the modulus from 9.8 GPa to 11.26 GPa. It was also noted that the presence of MAP led to higher draw ratios, because it helped stabilize the drawing process. Examining the two sets of strands, no discernible difference in morphology was revealed. From this observation, the authors speculated that the MAP may have caused the improvement in properties by affecting the adhesion between TLCP and PP, not by changing the compatibility of the materials.

The effect of compatibilizer concentration on the mechanical properties of composite strands has been investigated by Seo and colleagues [88, 89]. In their work, strands containing 25 wt% Vectra B950 were made using PEI as the matrix and a specially synthesized poly(ester imide) (PEsI) as the compatibilizer. Keeping the concentration of TLCP constant, the content was increased from 0 wt% to 7.5 wt%, with the best mechanical properties being obtained from strands containing 73.5 wt% PEI, 1.5 wt% PEsI, and 25 wt% Vectra B950. These strands had a tensile strength of 500 MPa, a modulus of 16 GPa, and an elongation at break of 6.3%. These values were significantly greater than the properties obtained from the uncompatibilized composite strands or from the strands that contained a higher concentration of PEsI, as presented in Table 32.5. From the morphology of the strands and the mechanical properties, Seo and colleagues concluded that at 1.5 wt% PEsI, enough compatibilizer was present to improve the adhesion at the surface without preventing fibril formation and inducing coalescence of the TLCP particles [89]. Based on Fourier-transform Raman spectra of compatibilized and uncompatibilized blends, they inferred that the improvement in adhesion between the TLCP and matrix was due to a chemical reaction between the PEsI and Vectra B950, resulting in the formation of a block or graft copolymer at the surface of the TLCP fibrils [88].

In a few cases, partial miscibility has been reported for TLCP–thermoplastic blends [35, 73, 90–93]. Bretas and Baird [90] investigated binary blends of PEEK and HX4000 and PEI and HX4000, as well as ternary blends of PEI, PEEK, and HX4000. Dynamic mechanical thermal analysis showed that all of the blends were at least partially miscible, with PEEK–HX4000 *in situ* composites being completely miscible for loadings of up to 50 wt% HX4000. Testing the mechanical properties of the *in situ* composites, Bretas and Baird found that some compositions had bett-

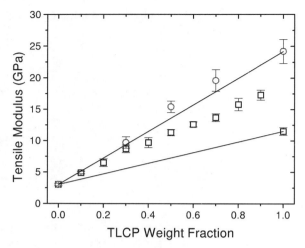

Figure 32.1 Tensile modulus versus TLCP weight fraction. □: PEI–HX1000 composites. ○: PEI–HX4000 composites. Lines are rule-of-mixtures predictions.

ter properties than any of the neat injection-molded materials. For a composite of PEI, PEEK, and HX4000 (10/10/80 wt%), the tensile modulus was 19.47 GPa, 29% higher than that of neat HX4000, 369% higher than that of PEEK, and 433% higher than that of PEI. It was also found that annealing at 200°C tended to cause a drop in strength for all composites containing greater than 10 wt% HX4000. It was speculated that this behavior was due to crystallization of the PEEK, causing phase separation in the composites. In addition to the changes in strength, annealing at 200°C for three days caused the fibrils to recoil, as determined in micrographs showing the changes in morphology.

Baird and coworkers [92] determined that the mechanical properties of some partially miscible PEI–TLCP systems exhibited positive deviations from the rule of mixtures for tensile and flexural moduli, relative to the values for injection-molded specimens of the neat materials, shown in Fig. 32.1. Particularly impressive were the PEI–HX1000 composites, with a PEI–HX1000 (50/50 wt%) blend having the same modulus as neat HX1000, within the error of the results. Although a cause for this behavior was not determined, some reasons were suggested. First, because of the partial miscibility, the interfacial adhesion between the Ultem–TLCP solid-solution phase and the free-TLCP phase (fibrils) could have been increased. This factor might have produced a higher reinforcing effect and thereby yielded moduli higher than predicted by the rule of mixtures. Second, the partial miscibility could have led to lower interfacial surface tension, producing finer fibers with higher aspect ratios. Also, the characteristics of the TLCP and the final blend morphology might have contributed to the positive deviations.

Because the PEI–TLCP composites exhibited such high tensile moduli in the machine direction, the TLCP reinforcement was found to be competitive with glass-fiber reinforcement. For example, PEI with 30 wt% glass fiber was shown to have a machine-direction modulus of 9.2 GPa, while loading PEI with 30 wt% HX4000 yielded a modulus of 9.8 GPa [92]. When the difference in density between the two composites was considered, the *in situ* composite showed superior mechanical properties. The specific modulus of the PEI–glass (70/30 wt%) blend was 6.1 GPa·cm^3/g, while that of the PEI–HX4000 (70/30 wt%) blend was 7.4 GPa·cm^3/g, confirming the potential of TLCPs to be a reinforcing agent.

De Souza and Baird [93] examined injection-molded *in situ* composites composed of PEI–HX1000 blends and PEI–Vectra A900 blends in order to determine the effects that partial miscibility has on mechanical properties. The TLCPs were selected for two reasons: (1) They had overlapping melt-processing temperatures with PEI and (2) HX1000 was partially miscible with PEI, unlike Vectra A900. By using contact-angle measurements, it was found that the PEI–HX1000 blends had an interfacial tension of 7.20 mN/m, versus 20.80 mN/m for the PEI–Vectra A900 blends. Because of the lower interfacial tension, the PEI–HX1000 composites would be expected to have finer, higher aspect-ratio fibrils, which was subsequently confirmed using microscopy. Furthermore, because of this difference in morphologies, the PEI–HX1000 blends had greater moduli than the PEI–Vectra A900 composites throughout the range of compositions investigated. For example, at a loading of 30 wt% TLCP, the HX1000-reinforced composites had flexural and tensile moduli of 9 GPa, while the composites containing Vectra A900 had flexural and tensile moduli of 5.5 GPa. This difference between the moduli existed despite the fact that neat injection-molded plaques of both TLCPs had roughly the same tensile moduli (10 GPa) and flexural moduli (14 GPa).

The effect of partial miscibility was also observed in the transverse-direction tensile strengths of film-gated injection-molded plaques [93]. Neat HX1000 had a transverse-direction strength of only 28.90 MPa, versus 45.70 MPa for Vectra A900. However, despite this difference, PEI–HX1000 composites had higher strengths than PEI–Vectra A900 composites at TLCP loadings of 10, 20, and 30 wt%. To illustrate this point, consider the fact that the PEI–HX1000 (80/20 wt%) composite had a strength of 57.10 MPa, while the PEI–Vectra A900 (80/20 wt%) composite had a strength of 45.00 MPa. This result suggests that the adhesion was improved because the PEI–HX1000 system was partially miscible.

III. COMPOSITES IN WHICH TLCP FIBRILS ARE PREGENERATED

In order to augment the limited numbers of TLCP–matrix combinations that are available, it is desirable to combine polymers with nonoverlapping processing temperatures. To do this, Sukhadia and coworkers [32, 33, 78] developed a novel processing method that they called the dual-extruder processing scheme, illustrated in Fig. 32.2. In this process, the matrix and TLCP are plasticated in separate extruders. The TLCP is then cooled and introduced into the matrix as continuous streams, using a phase-distribution system at the tee connecting the two extruders. Immediately past the phase-distribution system, the blend is passed through a mixing head containing static mixers to further divide the TLCP into smaller continuous streams.

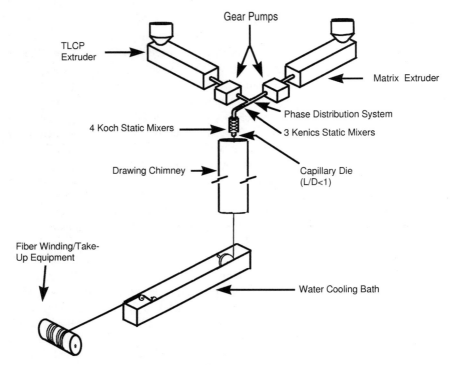

Figure 32.2   Diagram of the dual-extruder processing scheme.

The composite melt leaving the mixing head is then extruded through a capillary die ($L/D < 1$) and drawn to orient the TLCP.

One of the key points in this process is that the TLCP can be heated to a high temperature and then supercooled below its melting temperature while still remaining fluid. This factor is important for three reasons. First, it allows the TLCP to be exposed to temperatures high enough to melt any residual crystallites, such as those that Lin and Winter [43, 44] reported for Vectra A900. Second, because the TLCP can be cooled before being introduced into the matrix, it can be combined with the matrix without problems with degradation. This makes it possible to combine high-temperature-melting TLCPs with commodity thermoplastics such as PP and PET. Third, through independent control of the thermal history, it is often possible to vary the deformability of the TLCP phase. Specifically, it has been noted that the deformability of TLCPs is strongly influenced by previous thermal history, with exposure to higher temperatures often allowing the melt to be cooled to lower temperatures before solidifying [47, 94]. Thus, even in cases wherein processing temperatures do overlap, the TLCP may be less deformable than desired. The dual-extruder process offers a means of keeping the TLCP phase deformable so that it can be stretched into high-aspect-ratio fibrils.

In addition to the benefits gained by plasticating the two polymers in separate extruders, the dual-extruder process has another significant advantage. By using a single-screw extruder, droplet deformation is relied upon to create the TLCP fibrils. This is a potential problem, because if the droplets are not elongated to aspect ratios of at least 100, the fibrils will not behave like continuous reinforcement, and the highest possible stiffnesses will not be achieved [95]. However, this problem does not exist in the dual-extruder process, because continuous streams of TLCP are introduced into the matrix. This axial continuity is maintained in the final composite strands, so that the strands contain TLCP fibrils with essentially infinite aspect ratios dispersed throughout the matrix polymer.

Using the dual-extruder process, several workers have been able to spin composite strands with excellent properties [6, 10, 32, 70, 71, 78, 79, 96–99]. Robertson et al. [10, 96] demonstrated how well matrix polymers can be reinforced using this process by spinning PP–Vectra B950 (50/50 wt%) blends. Fibers produced with this process had tensile moduli of 44 GPa and strengths around 400 MPa. What makes these results striking is that the modulus of the TLCP phase was calculated from the rule of mixtures to be roughly 100 GPa. This value is much greater than the value of 75 GPa obtained when neat Vectra B950 is spun, showing that the dual-extruder process creates a synergistic reinforcing effect. One explanation given by Robinson et al. was that the dual extruder system oriented the TLCP more effectively. Specifically, Yee et al. [100] extrapolated a modulus of 110 GPa for Vectra B950, assuming complete molecular orientation. Hence, the PP may have served as an insulator, allowing the Vectra B950 fibrils to be drawn further and oriented prior to solidification more than is possible when neat Vectra B950 is spun.

Another study which demonstrated that strands produced using the dual-extruder system have exceptional moduli is that of Krishnaswamy and Baird [98]. In their investigation, strands of a blend of nylon 11 and HX8000 were spun at loadings of 13, 22, and 35 wt% HX8000. For neat HX8000, it has been shown that a maximum tensile modulus of approximately 47 GPa is reached at draw ratios over 50 [98, 99]. However, the properties of the composite strands were greater than predicted based on this modulus. By using the rule of mixtures, the modulus of the TLCP phase was calculated to be as high as 70 to 80 GPa. In accounting for this difference in properties, it was concluded that the TLCP fibrils in the composite strands had higher degrees of molecular orientation than did the neat TLCP fibers. Based on these experimental results, it is clear that the dual-extruder process allows consistently high levels of reinforcement to be attained, with the moduli of the composite strands often exceeding theoretical expectations.

Starting with strands that have pregenerated TLCP fibrils, several processing options are available. First, the strands can be compression molded by heating to a temperature below the melting point of the TLCP but above the melting point of the matrix [27, 70, 71, 98, 99, 101–103]. Second, they can be woven into prepregs and sent through various postprocessing steps, such as compression molding and thermoforming [10, 96, 98, 104, 105]. Third, they can be chopped and processed through an injection molder, using temperatures that are lower than the processing temperatures for the neat TLCP [6, 70, 99, 106, 107, 108]. Also, limited research has been undertaken in which melts containing pregenerated TLCP fibrils have been used in processes such as rapid prototyping [97], sheet extrusion [70, 109, 110], and extrusion blow molding [6, 111]. The aim in each of these processing techniques is to process the TLCP fibril-filled matrices in a manner that damages the TLCP reinforcement as little as possible.

Sabol et al. [70] compression molded drawn strands of PP–HX1000 (71/29 wt%) blends at 190°C to determine if the strand properties would be retained when consolidated. For both uniaxially and randomly aligned compression-molded specimens, it was found that the HX1000 fibrils generally did not lose their reinforcing ability. For example, strands with a draw ratio of 26.8 possessed a tensile modulus of 12.55 (± 1.60) GPa. (Standard deviations given in parentheses.) When consolidated into uniaxial composites, the modulus was the same within experimental error, at 13.55 (± 2.75) GPa. Furthermore, when the fibers were aligned randomly, it was possible to obtain a modulus as high as 4.55 (± 0.45) GPa. This value is over 80% of the theoretically predicted modulus for a planar isotropic composite, demonstrating that most of the stiffness in the composite strands was retained.

Robertson et al. [10, 96, 104, 105] performed a thorough study of woven preforms composed of PP and Vectra B950, examining the effects of the TLCP loading level and the fabric-layer distribution on mechanical properties, as well as on the formability of these materials. It was found that as the TLCP loading level was increased, the modulus of the composites greatly exceeded that of neat PP. Specifically, at a loading level of 9.84 wt% Vectra B950, the tensile modulus of the consolidated preforms was 3.87 GPa, versus 1.01 GPa for neat PP. Meanwhile, when the loading level was 31.5 wt%, the consolidated preforms were found to have a modulus of 10.3 GPa, a tenfold increase over that of PP. Note that this sample was predicted to have a modulus of 12.0 GPa based on composite theory, so the final composite retained over 85% of the strand modulus.

It was also determined that the placement of preform layers could significantly affect flexural properties. At a loading of roughly 9.5 wt% Vectra B950, it was found that the flexural modulus could be increased from 3.22 GPa to 5.23 GPa by changing the distribution of preform layers from even spacing through the thickness of the composite to selective placing of the layers near the composite surfaces. The reason for this behavior was that the reinforcing TLCP fibrils were located where the greatest tension and compression occurred during flexural loading.

To evaluate the formability of the PP–Vectra B950 ($\sim$80/20 wt%) composites produced from consolidated woven preforms, the tensile elongation at break was determined as a function of temperature. It was found that for an orthotropic composite with an even distribution of four fabric layers across the composite cross section, it was possible to reach 15% elongation when the composite was heated to 250°C. The authors noted that this degree of extension is not possible for composites reinforced with conventional fibers, such as fiberglass and carbon. Having established that the composites could be stretched, Robertson et al. performed tests to prove that the composites could be thermoformed. Thus, TLCP-reinforced preforms can be used to produce composite sheets without a significant loss in properties; these composite sheets can then be thermoformed into wholly thermoplastic composite parts.

Although good mechanical properties can be achieved by using woven performs, the thermal history used in the consolidating step can affect their mechanical prop-

Table 32.6   The Effect of Processing Conditions on the Mechanical Properties of Consolidated Nylon 11–HX8000 (65/35 Wt%) Preforms

Processing Conditions	Tensile Modulus (GPa)[b]	Flexural Modulus (GPa)[b]
Neat Nylon 11[a]	0.98 (0.07)	0.89 (0.08)
205°C for 10 min	3.63 (0.20)	4.16 (0.45)
195°C for 10 min	4.15 (0.88)	5.36 (0.27)
185°C for 5 min, 195°C for 5 min	5.18 (0.33)	5.67 (0.38)

[a] Nylon 11 properties are from injection molded specimens.
[b] Standard deviations are provided in parentheses.
Source: [98].

erties, as shown by Krishnaswamy and Baird [98]. In this study, woven preforms of blends of nylon 11 and HX8000 (65/35 wt%) were consolidated using three sets of processing conditions, as indicated in Table 32.6. It was found that increasing the consolidation temperature from 195°C to 205°C caused the flexural modulus to decline from 5.36 GPa to 4.16 GPa. Also, all of the consolidated preforms had moduli significantly below the theoretically predicted value of 9.9 GPa. To serve as a comparison to the woven preforms, uniaxial composites were also compression molded. It was determined that after compression molding at 195°C for 10 minutes, the tensile modulus was 13.4 GPa, only 67% of the original fiber modulus. Hence, it appeared for this system that exposing the HX8000 fibrils to these temperatures inherently caused a drop in stiffness. It was speculated that the reason for the strong dependence on consolidation temperature was molecular relaxation within the HX8000 fibrils. In particular, dynamic mechanical experiments on neat HX8000 fibers showed that heating the fibers above 135°C caused an irreversible drop in the storage modulus, which is consistent with orientation relaxations within the TLCP fiber.

Pregenerated TLCP fibrils have also been used to produce injection-molded composites. Heino et al. [106] performed the first study of this kind by using PP–Vectra A950 (80/20 wt%) composite strands. In this study, the effect of processing temperature on flexural and impact properties was investigated by producing three sets of composites. The first set was injection molded using barrel temperatures of 180°C to 200°C; the second set used temperatures of 230°C to 250°C, and the third set used temperatures of at least 280°C to produce *in situ* composites. The pregenerated microcomposites produced at 230°C to 250°C had poorer properties than those made at the lower temperatures. This factor was attributed to fibrils deforming and agglomerating into clusters, as confirmed by optical microscopy. The best flexural properties for the pregenerated microcomposites were a modulus of 1.77 GPa and a strength of 45 MPa. These values were greater than those of neat PP and comparable to the *in situ* composite values of 1.97 GPa and 42 MPa. The authors also noted that the pregenerated microcomposite had a much greater Charpy impact strength than that of the *in situ* composite (i.e., 39 kJ/m^2 versus 22 kJ/m^2). Other work examining the effect of processing temperature has shown similar results, with the highest properties being obtained at the lowest processing temperatures [6, 70, 99, 107].

One question addressed in the work of Handlos and Baird [6, 111] was the effect of strand draw ratio on the properties of injection-molded pregenerated microcomposites. Using PP–HX6000 (60/40 wt%) strands spun to draw ratios of 4, 13.5, and 25, the strands were chopped, dry blended with additional PP to lower the TLCP concentration to 30 wt%, and injection molded using a maximum barrel-temperature setting of 190°C. From the mechanical properties of these composites, it was shown that increasing the draw ratio yielded improved tensile strengths and moduli. For example, using strands with a draw ratio of 4 produced composites with a machine-direction tensile modulus of 3.20 GPa and a tensile strength of 35.0 MPa. Meanwhile, when strands with a draw ratio of 25 were used, the plaques had a modulus of 3.98 GPa and a strength of 38.5 MPa. Similar trends were also observed with injection-molded PP–Vectra A950 pregenerated microcomposites.

McLeod and Baird [99] examined whether diluting the composite strands with a low-viscosity matrix polymer would produce improved mechanical properties. This was done by injection molding two sets of composites containing 20 wt% HX1000 fibrils, using a film-gated plaque mold. The first set was made from PT 7067–HX1000 (80/20 wt%) composite strands, PT 7067 being a blow-molding grade of PET manufactured by DuPont. The second set was made from chopped PT 7067–HX1000 (50/50 wt%) strands, which were diluted to a loading of 20 wt% HX1000 by dry blending the chopped strands with Rynite, an experimental, low-viscosity grade of PET made by DuPont. The PT 7067–HX1000–Rynite (20/20/60 wt%) pregenerated microcomposites had a modulus of 4.586 GPa, versus just 3.106 GPa for the PT 7067–HX1000 (80/20 wt%) composites. Similar results were seen for the flexural moduli as well. Hence, the results demonstrated that better mechanical properties were obtained by diluting with a low-viscosity polymer, possibly because the low-viscosity polymer helped reduce the damage that the TLCP fibrils sustained in the injection-molding process.

In making pregenerated microcomposites using PET as the matrix polymer, it was determined that two advantages were obtained by diluting with Celenex 1600A, an injection-molding grade of poly(butylene terephthalate) manufactured by Hoechst-Celanese, rather than Rynite, the injection-molding grade of PET [99]. Specifically, because Rynite had a melting temperature of 257°C, versus 228°C for Celenex 1600A, it was possible to lower some of the injection-molding-zone temperatures, thereby minimizing the temperatures to which the HX1000 fibrils were exposed. Also, it was found that by diluting with PBT, a homogeneous, thoroughly mixed melt was formed earlier in the barrel of the injection molder. This meant that the PBT was able to wet the HX1000 fibrils earlier in the injection-molding screw, leading to better mechanical properties. For example, composites composed of PT 7067, HX1000, and Rynite (30/30/40 wt%) had a machine-direction flexural strength of 79.2 MPa, while those with a composition of PT 7067, HX1000 and Celenex 1600A had a flexural strength of 95.7 MPa. Similar increases in tensile strength, percent elongation, and tensile toughness were also observed, while tensile and flexural moduli remained roughly constant. Of particular interest is that Celenex 1600A had lower tensile and flexural strengths than Rynite (45.8 MPa vs. 49.4 MPa and 60.4 MPa vs. 69.0 MPa, respectively). This indicates that the improvements in strength were due to better re-

Table 32.7 Comparing the Mechanical Properties of Pregenerated Microcomposites with Glass-Reinforced PET at the Same Weight Fraction of the Reinforcement

	Tensile Properties		Flexural Properties	
	Machine Direction[b]	Transverse Direction[b]	Machine Direction[b]	Transverse Direction[b]
Pregenerated Microcomposite: HX1000/PT 7067/Celenex 1600A (30/30/40 wt%)[a]				
Modulus (GPa)	5.85 (0.70)	3.39 (0.60)	6.88 (0.89)	4.13 (0.72)
Strength (MPa)	72.0 (5.3)	42.6 (3.8)	101.0 (3.2)	91.8 (5.5)
% Elongation	1.41 (0.22)	1.43 (0.21)		
Toughness (MPa)	0.52 (0.10)	0.32 (0.06)		
Glass-Fiber-Filled PET: Rynite 530 (30 wt% glass)				
Modulus (GPa)	9.21 (0.64)	4.71 (0.55)	10.05 (0.35)	5.00 (0.33)
Strength (MPa)	117.8 (5.1)	59.1 (4.3)	205.4 (4.2)	98.2 (9.2)
% Elongation	1.72 (0.09)	1.37 (0.18)		
Toughness (MPa)	1.46 (0.21)	0.84 (0.17)		

[a] Strand draw ratio of 108.5 (19.6).
[b] Standard deviations are given in parentheses.
Source: [99].

inforcement from the HX1000 fibrils, which were probably damaged less when PBT was used as the diluting polymer.

Because pregenerated microcomposites have the potential to compete with glass-fiber-reinforced thermoplastics, it is important to compare the mechanical properties of these two systems. When this comparison has been made, it has generally been found that the glass-filled thermoplastics have better properties [6, 99, 110]. As shown in Table 32.7, at a loading of 30 wt% HX1000, a machine-direction tensile modulus of 5.8 GPa was obtained, versus 9.2 GPa when 30 wt% glass fiber was used. Additionally, in areas such as strength, percent elongation, and tensile toughness, the glass-filled systems tend to outperform pregenerated microcomposites.

Still, the potential to be competitive with glass-fiber reinforcement exists. Theoretically, a planar isotropic PET–HX1000 (70/30 wt%) composite should have a modulus of 7.26 GPa, versus 6.80 GPa for a planar isotropic sample of PET reinforced with 30 wt% glass fiber. This shows that if the losses in the modulus can be minimized, the injection-molded pregenerated microcomposites should be at least as stiff as glass-filled polymers. Also, it has been determined that because of the smaller size of the TLCP fibrils, the pregenerated microcomposites have smoother surfaces than do glass-filled thermoplastics [99]. Heat distortion properties of pregenerated microcomposites are competitive as well, as evidenced by dynamic mechanical thermal analysis of PET–HX1000-based pregenerated microcomposites, wherein the storage modulus closely matched that of glass-filled PET at temperatures up to the glass transition temperature of HX1000. Thus, if the mechanical properties of the composite strands can be retained, the potential exists to produce lightweight, wholly thermoplastic composites that are as stiff as glass-filled thermoplastics, have high heat distortion temperatures, and possess smoother surfaces.

IV. CONCLUSIONS AND RECOMMENDATIONS

There are many advantages inherent in using TLCPs as reinforcing agents [112]. First, they are melt processable with thermoplastics, so the reinforcement can be fully wetted as the composite is formed, eliminating the compounding step typically needed for other forms of reinforcement. Also, because both phases are thermoplastics, there are more possibilities for recycling. Second, TLCPs are much lighter than other forms of reinforcement, usually having densities of 1.3 to 1.4 g/cm^3 (versus 2.59 g/cm^3 for E-glass). Third, the reinforcing phase tends to be very small, with fibril diameters of roughly 1 µm, versus the 10- to 13-µm diameters typical of E-glass reinforcement. Fourth, the tensile moduli and strengths of TLCPs are competitive with other forms of reinforcement. Therefore, TLCPs possess a selection of useful characteristics that make them interesting alternatives to more conventional reinforcing materials.

However, despite these advantages, limitations exist that must be overcome to realize the full reinforcing potential of these materials. The processing limitations include the need for overlapping melt-processing temperatures in order to plasticate both components in a single extruder, the anisotropy present in injection-molded parts, and the incomplete use of the TLCP as a reinforcing material. In addition to the processing issues associated with TLCPs, cost is another limitation [113, 114]. Current costs for TLCPs are in the range of $26 to $48/kg [115], versus $1.65/kg for glass fiber [116]. However, it is anticipated that as the use of TLCPs become more prevalent and as cheaper monomers are developed, the costs will fall.

As shown by the work presented in this chapter, considerable attention has been focused on overcoming the processing limitations encountered with TLCP–thermoplastic blends. Specifically, the dual-extrusion process has been shown to offer some unique features that help reach the full reinforcing potential of the TLCPs. It allows TLCPs to be combined with matrix polymers that do not have overlapping processing temperatures, by utilizing the supercooling behavior of the TLCP. Also, because continuous streams of TLCP are introduced into the matrix, spun strands and fibers have fibrils of essentially infinite aspect ratios. Through this process, it is possible to consistently produce strands that have tensile properties that meet or exceed theoretical expectations. As further work is done with these composite strands, it is anticipated that it will be possible to use a variety of processing techniques to produce composites that retain the TLCP fibril properties and eliminate the problems of anisotropy found in injection-molded parts.

If the full reinforcing potential of these TLCPs can be achieved, they will be an attractive alternative to conventional glass-fiber-filled materials. Specifically, their light weight and smooth surfaces would be useful for areas such as the automotive industry, in which a premium exists on producing lightweight, paintable parts. For example, woven preforms have the potential to be thermoformed into panels and facias. Also, as the properties of injection-molded pregenerated microcomposites improve, the ability to form complex parts will develop. It is believed that as progress is made to take advantage of the light weight, surface smoothness, formability, and recycling potential that these novel composites possess, additional applications for them will be found.

V. REFERENCES

1. C. K. Ober, J. I. Jin, and R. W. Lenz, *Adv. Polym. Sci.*, **59**, 103 (1984).

2. P. Magagnini, "Molecular Design of Thermotropic. Main-Chain Liquid Crystalline Polymers," in *Thermotropic Liquid Crystalline Polymer Blends*, F. P. La Mantia (ed.), Technomic Publishing Company, Lancaster, PA, 1993, Ch. 1.

3. C. S. Brown and P. T. Alder, "Blends Using Thermotropic Liguid Crystalline Polymer," in *Polymer Blends and Alloys*, M. J. Folkes and P. S. Hope (eds.), Chapman and Hall, London, 1993, Ch. 8.

4. F. P. La Mantia, A. Roggero, U. Pedretti, and P. L. Magagnini, "Fibers of Blends with Liquid Crystalline Polymers: Spinnability and Mechanical Properties," in *Liquid-Crystalline Polymer Systems: Technological Advances*, A. I. Isayev, T. Kyu, and S. Z. D. Cheng (eds.), ACS Books, Washington, DC, 1996, Ch. 8.

5. A. Roggero, "Applications of Thermotropic Liquid Crystal Polymers and Thermotropic Liquid Crystal Polymer Blends," in *Thermotropic Liquid Crystalline Polymer Blends*, F. P. La Mantia (ed.), Technomic Publishing Company, Lancaster, PA, 1993, Ch. 6.

6. A. A. Handlos, "Microcomposites Based on Polypropylene and Thermotropic Liquid Crystalline Polymers," Ph.D. Dissertation, Virginia Polytechnic Institute, Blacksburg, VA, 1994.

7. H. N. Yoon, L. F. Charbonneau, and G. W. Calundann, *Adv. Mat.*, **4**, 206 (1992).

8. G. W. Calundann and M. Jaffe, "Thermotropic Liquid Crystalline Polymers," in *Proceedings of the Robert A. Welch Foundation Conferences on Chemical Research*, **26**, 247 (1983), Ch. 7.

9. M. C. Muir and R. S. Porter, *Mol. Cryst. Liq. Cryst.*, **169**, 91 (1989).

10. C. G. Robertson, J. P. de Souza, and D. G. Baird, "Development of In Situ Reinforced Polypropylene Fibers for Use in Woven Preforms," in *Liquid-Crystalline Polymer Systems: Technological Advances*, A. I. Isayev, T. Kyu, and S. Z. D. Cheng (eds.), ACS Books, Washington, DC, 1996, Ch. 6.

11. G. D. Figuly, "Liquid Crystalline Polymers Thermotropic Polyesters," in *Polymeric Materials Encyclopedia*, J. C. Salamone (ed.), CRC Press, Boca Raton, FL, 1996.

12. S. Kessler, "Properties of Glass Fiber Reinforcement," in *Plastics Additives and Modifiers Handbook*, J. Edenbaum (ed.), Van Nostrand Reinhold, New York, 1992, Ch. 48.

13. P. K. Mallick, *Fiber-Reinforced Composites*, Marcel Dekker, New York, 1988.

14. PPG Industries, Inc., in *Modern Plastics Encyclopedia '92*, McGraw-Hill, New York, 1992.

15. W. R. Graner, "Glass Fiber Reinforced Composites," in *Handbook of Composites*, Van Nostrand Reinhold, New York, 1982.

16. G. Kiss, *Polym. Eng. Sci.*, **29**, 410 (1987).

17. F. P. La Mantia, A. Valenza, and P. L. Magagnini, *J. Appl. Polym. Sci.*, **44**, 1257 (1992).

18. J. Seppälä, M. Heino, and C. Kapanen, *J. Appl. Polym. Sci.*, **44**, 1051 (1992).

19. M. T. Heino and J. V. Seppälä, *J. Appl. Polym. Sci.*, **44**, 2185 (1992).

20. A. Datta, H. H. Chen, and D. G. Baird, *Polymer*, **34**, 759 (1993).

21. A. M. Sukhadia, A. Datta, and D. G. Baird, *SPE ANTEC Tech. Papers*, **37**, 1008 (1991).

22. M. T. Heino and J. V. Seppälä, *J. Appl. Polym. Sci.*, **44**, 2185 (1992).

23. M. Amano and K. Nakagawa, *Polymer*, **28**, 263 (1987).

24. P. Zhuang, T. Kyu, and J. L. White, *Polym. Eng. Sci.*, **28**, 1095 (1988).

25. C. U. Ko, G. L. Wilkes, and C. P. Wong, *J. Appl. Polym. Sci.*, **37**, 3063 (1989).

26. R. A. Weiss, W. Huh, and L. Nicolais, *Polym. Eng. Sci.*, **27**, 684 (1987).

27. B. R. Bassett and A. F. Yee, *Polym. Compos.*, **11**, 10 (1990).

28. G. O. Shonaike, H. Hamada, S. Yamaguchi, M. Nakamichi, and Z. Maekawa, *J. Appl. Polym. Sci.*, **54**, 881 (1994).

29. H. Verhoogt, C. R. J. Willems, J. Van Dam, and A. Posthuma de Boer, *Polym. Eng. Sci.*, **34**, 453 (1994).

30. H. J. O'Donnell and D. G. Baird, *Polymer*, **36**, 3113 (1995).

31. J. C. Halpin and J. L. Kardos, *Polym. Eng. Sci.*, **16**, 344 (1976).

32. A. M. Sukhadia, A. Datta, and D. G. Baird, *Int. Polym. Proc.*, **7**, 218 (1992).

33. A. M. Sukhadia, "The In Situ Generation of Liquid Crystalline Polymer Reinforcements in Thermoplastics," Ph.D. Dissertation, Virginia Polytechnic Institute, Blacksburg, VA, 1991.

34. G. Crevecoeur and G. Groeninckx, *Polym. Eng. Sci.*, **30**, 532 (1990).

35. S. S. Bafna, J. P. de Souza, T. Sun, and D. G. Baird, *Polym. Eng. Sci.*, **33**, 808 (1993).

36. S. Kenig, *Polym. Eng. Sci.*, **29**, 1136 (1989).

37. G. G. Viola, D. G. Baird, and G. L. Wilkes, *Polym. Eng. Sci.*, **25**, 888 (1985).

38. D. E. Turek and G. P. Simon, *Polymer*, **34**, 2750 (1993).

39. K. G. Blizard and D. G. Baird, *Polym. Eng. Sci.*, **27**, 653 (1987).

40. D. G. Baird and R. Ramanathan, "The In Situ Generation of Liquid Crystalline Reinforcements in Engineerings Thermoplastics," in *Contemporary Topics in Polymer Science*, Vol. 6, B. M. Culbertson (ed.), Plenum Press, New York, 1989.

41. K. Fujiwara, M. Takahashi, and T. Asuda, *Int. Polym. Proc.*, **6**, 232 (1991).

42. M. S. Silverstein, A. Hiltner, and E. Baer, *J. Appl. Polym. Sci.*, **43**, 157 (1991).

43. Y. G. Lin and H. H. Winter, *Macromolecules*, **24**, 2877 (1991).

44. Y. G. Lin and H. H. Winter, *Macromolecules*, **21**, 2439 (1988).

45. H. Muramatsu and W. R. Krigbaum, *J. Polym. Sci.: Part B: Polym. Phys.*, **25**, 2303 (1987).

46. W. R. Krigbaum, C. K. Liu, and D.-K. Yang, *J. Polym. Sci.: Part B: Polym. Phys.*, **26**, 1711 (1988).

47. H. J. O'Donnell and D. G. Baird, *Polym. Eng. Sci.*, **36**, 963 (1996).

48. S. Wu, *Polym. Eng. Sci.*, **27**, 335 (1987).

49. G. I. Taylor, *Proc. Royal Soc.*, **A146**, 501 (1934).

50. G. I. Taylor, *Proc. Royal Soc.*, **A138**, 41 (1932).

51. G. I. Taylor, *Proc. Royal Soc.*, **A226**, 289 (1954).

52. H. P. Grace, *Chem. Eng. Comm.*, **14**, 225 (1982).

53. J. J. Elmendorp and R. J. Maalcke, *Polym. Eng. Sci.*, **25**, 1041 (1985).

54. T. Tavgac, "Deformation and Breaks Up of Drops," Ph.D. Dissertation, University of Houston, TX, 1972.

55. W. J. Milliken and L. G. Leal, *J. Non-Newt. Fl. Mech.*, **40**, 335 (1991).

56. R. W. Flumerfelt, *Ind. Chem. Eng. Fund.*, **11**, 312 (1972).

57. H. B. Chin and C. D. Han, *J. Rheol.*, **23**, 557 (1979).

58. H. B. Chin and C. D. Han, *J. Rheol.*, **24**, 1 (1980).

59. S. Torza, R. G. Cox, and S. Mason, *J. Coll. Inter. Sci.*, **38**, 395 (1972).

60. P. G. Ghodgaonkar and U. Sundararaj, *Polym. Eng. Sci.*, **36**, 1656 (1996).

61. L. Levitt and C. Macosko, *Polym. Eng. Sci.*, **36**, 1647 (1996).

62. A. G. C. Machiels, "Stability of Blends of Liquid Crystalline Polymer Blends," Ph.D. Dissertation, Delft University of Technology, Delft, The Netherlands, 1998.

63. A. G. C. Machiels, K. F. J. Denys, J. Van Dam, and A. Posthuma De Boer, *Polym. Eng. Sci.*, **36**, 2451 (1996).

64. H. J. O'Donnell and D. G. Baird, *Int. Polym. Proc.*, **11**, 257 (1996).

65. H. J. O'Donnell, "In Situ Composites of Compatibilized Polypropylene/LCP Blends," Ph.D. Dissertation, Virginia Polytechnic Institute, Blacksburg, VA, 1993.

66. M. Kyotani, A. Kaito, and K. Nakayama, *Polymer*, **33**, 459 (1992).

67. C. Carfagna, L. Nicolais, E. Amendola, and C. Carfagna, Jr., *SPE ANTEC Tech. Papers*, **37**, 1004 (1991).

68. C. Carfagna, E. Amendola, L. Nicolais, D. Acierno, O. Francescangeli, B. Yang, and F. Rustichelli, *J. Appl. Polym. Sci.*, **43**, 839 (1991).

69. F. P. La Mantia, F. Cangialosi, U. Pedretti, and A. Roggero, *Eur. Polym. J.*, **29**, 671 (1993).

70. E. A. Sabol, A. A. Handlos, and D. G. Baird, *Polym. Comp.*, **16**, 330 (1995).

71. E. A. Sabol, "The Development of Dual Extrusion Blending Process and Composites Based on Polypropylene and TLCPs," M.S. Thesis, Virginia Polytechnic Institute, Blacksburg, VA, 1994.

72. K. G. Blizard and D. G. Baird, *Polym. Eng. Sci.*, **27**, 653 (1987).

73. N. Chapleau, P. J. Carreau, C. Peleteiro, P.-A. Lavoie, and T. M. Malik, *Polym. Eng. Sci.*, **32**, 1876 (1992).

74. A. Kohli, N. Chung, and R. A. Weiss, *Polym. Eng. Sci.*, **29**, 573 (1989).

75. A. I. Isayev and M. J. Modic, *Polym. Comp.*, **8**, 158 (1987).

76. A. K. Mithal, A. Tayebi, and C. H. Lin, *Polym. Eng. Sci.*, **31**, 1533 (1991).

77. A. M. Sukhadia, A. Datta, and D. G. Baird, *SPE ANTEC Tech. Papers*, **37**, 1008 (1991).

78. D. G. Baird and A. M. Sukhadia, U.S. Patent 5,225,488 (1993).

79. E. A. Sabol and D. G. Baird, *Int. Polym. Proc.*, **10**, 124 (1995).

80. A. Datta, J. P. de Souza, A. M. Sukhadia, and D. G. Baird, *SPE ANTEC Tech. Papers*, **37**, 913 (1991).

81. H. J. O'Donnell, A. Datta, and D. G. Baird, *SPE ANTEC Tech. Papers*, **38**, 2248 (1992).

82. S. S. Bafna, T. Sun, and D. G. Baird, *Polymer*, **34**, 708 (1993).

83. R. M. Holsti-Miettinen, M. T. Heino, and J. V. Seppälä, *J. Appl. Polym. Sci.*, **27**, 573 (1995).

84. A. Datta and D. G. Baird, *Polymer*, **36**, 505 (1995).

85. M. T. Heino, in *Polymer Technology Publication Series No. 14*, National Technical Information Service, Springfield, VA, 1993.

86. M. T. Heino and J. V. Seppälä, *J. Appl. Polym. Sci.*, **48**, 1677 (1993).

87. H. J. O'Donnell, A. Datta, and D. G. Baird, *SPE ANTEC Tech. Papers*, **39**, 2248 (1993).

88. Y. Seo, S. M Hong, S. S. Hwang, T. S. Park, K. U. Kim, S. Lee, and J. Lee, *Polymer*, **36**, 515 (1995).

89. Y. Seo, S. M. Hong, S. S. Hwang, T. S. Park, K. U. Kim, S. Lee, and J. Lee, *Polymer*, **36**, 525 (1995).

90. Y. G. Lin, H. W. Lee, H. H. Winter, S. Dashevsky, and K. S. Kim, *Polymer*, **34**, 4703 (1993).

91. R. E. S. Bretas and D. G. Baird, *Polymer*, **33**, 5233 (1992).

92. D. G. Baird, S. S. Bafna, J. P. de Souza, and T. Sun, *Polym. Comp.*, **14**, 214 (1993).

93. J. P. de Souza and D. G. Baird, *Polymer*, **37**, 1985 (1996).

94. D. Done and D. G. Baird, *Polym. Eng. Sci.*, **30**, 989 (1990).

95. L. A. Utracki and T. Vu-Khanh, in *Multicomponent Polymer Systems*, I. S. Miles (ed.), John Wiley and Sons, New York, 1992.

96. C. Robertson, "Processing of Composite Fibers Based on Polypropylene and a TLCP," M.S. Thesis, Virginia Polytechnic Institute, Blacksburg, VA, 1995.

97. R. Gray, "Composites for Rapid Prototyping Applications," M.S. Thesis, Virginia Polytechnic Institute, Blacksburg, VA, 1997.

98. R. K. Krishanaswamy and D. G. Baird, *Polym. Comp.*, **18**, 526 (1997).

99. M. A. McLeod, "Injection Molding of Pregenerated Microcomposites," Ph.D. Dissertation, Virginia Polytechnic Institute, Blacksburg, VA, 1997.

100. Q. Lin, J. Jho, and A. F. Yee, *Polym. Eng. Sci.*, **33**, 789 (1993).

101. R. J. Amundsen and A. F. Yee, *Proceedings of the American Society for Composites: 5th Technical Conference: Composite Materials in Transition*, Technomic Publishing Company, Inc., 1990.

102. D. Dutta, R. A. Weiss, and K. Kristal, *Polym. Comp.*, **13**, 394 (1992).

103. A. A. Handlos and D. G. Baird, *SPE ANTEC Tech. Papers*, **39**, 1170 (1993).

104. C. G. Robertson, J. P. de Souza, and D. G. Baird, *SPE ANTEC Tech. Papers*, **41(II)**, 1649 (1995).

105. C. G. Robertson, J. P. de Souza, and D. G. Baird, "Composites Based on Fabric Prepregs Generated from In Situ Reinforced Thermoplastic Fibers," in *PMSE Polymeric Materials Science and Engineering (Proceedings of the ACS): Division of Polymeric Materials: Science and Engineering*, **72**, 605 (1995).

106. M. T. Heino, T. P. Vainio, and J. V. Seppälä, *Polym. & Polym. Comp.*, **1**, 439 (1993).

107. A. A. Handlos and D. G. Baird, *Int. Polym. Proc.*, **11**, 1 (1996).

108. M. A. McLeod and D. G. Baird, *SPE ANTEC Tech. Papers*, **44**, 2618–2622 (1998).

109. A. A. Handlos, E. A. Sabol, and D. G. Baird, *SPE ANTEC Tech. Papers*, **40**, 1594 (1994).

110. A. A. Handlos and D. G. Baird, *Polym. Compos.*, **17**, 73 (1996).

111. A. A. Handlos and D. G. Baird, *Polym. Eng. Sci.*, **36**, 378 (1996).

112. A. A. Handlos and D. G. Baird, *J. Macro. Sci.: Rev. Macro. Chem. and Phys.*, **C35**, 183 (1995).

113. A. Roggero, "Applications of Thermotropic Liquid Crystal Polymers and Thermotropic Liquid Crystal Polymer Blends," in *Thermotropic Liquid Crystal Polymer Blends*, F. P. La Mantia (ed.), Technomic, Lancaster, PA, 1993.

114. D. J. Williams, *Adv. Polym. Tech.*, **10**, 173 (1991).

115. L. M. Sherman and J. De Gaspari, *Plast. Tech.*, **46**, 71 (1996).

116. Anonymous, in *Modern Plastics Encyclopedia '92*, McGraw-Hill, New York, 1992.

33 From Polymer Blends to Microfibrillar Reinforced Composites

S. FAKIROV AND M. EVSTATIEV

Laboratory for Structure and Properties of Polymers
University of Sofia
1126 Sofia, Bulgaria

K. FRIEDRICH

Institute for Composite Materials, Ltd.
University of Kaiserslautern
67663 Kaiserslautern, Germany

I. INTRODUCTION

A. Definition and Classification of Composite Materials

Both polymer blends and composites are of prime commercial interest as polymeric materials; polymer blends alone currently make up over 30% of the polymer market [1].

Polymer Blends, Volume 2: Performance. Edited by D. R. Paul and C. B. Bucknall.
ISBN 0-471-35280-2.   © 2000 John Wiley & Sons, Inc.

Composites are defined as materials consisting of two or more distinct components with recognizable interfaces or interphases. This definition is generally restricted in practice to materials containing fibrous or other reinforcements, e.g., platelets or flakes, with different length and cross-section dimensions described by the aspect ratio, that are embedded in a continuous rigid matrix. Incorporating these reinforcements normally improves the mechanical performance of the matrix material. The effect on mechanical performance is the basic difference between composites and regular filled systems, in which fillers are used primarily to reduce cost, but often simultaneously degrade rather than improve the mechanical-property profile. Otherwise, the distinction between reinforced and filled systems is sometimes unclear [2].

Most natural materials derive their excellent properties from a combination of two or more components that can readily be distinguished when examined under the light or electron microscope. Thus, for example, many tissues in the human body, which have high strength combined with enormous flexibility, are made up of stiff fibers, such as collagen, embedded in a matrix of lower stiffness. The fibers are aligned in such a way that maximum stiffness is provided in the direction of high loads, but the fibers are also able to slide relative to each other so that the tissue is very flexible. Similarly, a microscopic examination of both wood and bamboo reveals a pronounced fibrillar structure, which is very apparent in bamboo when it is fractured. It is not surprising that bamboo has been called "nature's fiber glass" [3].

A peculiarity of polymeric composites that makes them very attractive for many applications is that their structure and, thus, their mechanical-property profile can be tailored for defined requirements and service conditions.

Composite materials have been classified in many ways, depending on the various ideas or concepts that need to be identified (e.g., classification according to the type of matrix, geometry of the reinforcement, etc.). For the purposes of this chapter, only classifications regarding the size of the reinforcing elements are considered. In this regard, three basic groups can be distinguished:

(1) Continuous-fiber composites (e.g., composites with a thermosetting or thermoplastic matrix),
(2) short-fiber composites (often having a thermoplastic matrix), and
(3) molecular composites (e.g., liquid crystalline polymers in a thermoplastic matrix).

An important characteristic feature of composites in the third group is that they are also prepared via blending. However, blending of two polymers usually results in immiscibility, so that often a third component—that is, a compatibilizer, such as a block copolymer—must be added to obtain a more uniform phase distribution. A similar problem exists in the case of fiber-reinforced composites, for which good adhesion between the fiber and the surrounding matrix can be reached only through the use of coupling agents coated on the fibers prior to impregnation with the polymeric matrix.

A new type of polymer composite, satisfying to a great extent the outlined peculiarities of polymer blends and composites and having reinforcing elements with sizes between those of the fiber-reinforced and molecular composites, has recently been developed [4–7]. Unlike the classical macrocomposites (e.g., fiber-reinforced macrocomposites) and the molecular composites (with single rodlike macromolecules as reinforcing elements), this new group of polymer composites is reinforced with polymer fibrils or, more frequently, bundles of them, and is called *microfibrillar-reinforced composites* (*MFCs*) [4].

A fibril can be defined as a structural entity with materials properties that are biased predominantly along a linear dimension or symmetry axis [8]. Similarly to natural materials, such as cellulosic structures and collageneous composites, man-made polymeric materials, such as ultrahighly drawn polymer solids, liquid crystalline polymers, and hard elastic materials, reveal outstanding mechanical properties, since their basic organizational units are microfibrillar in nature. Before describing the preparation of MFCs, let us comment briefly on the various reinforcement concepts applicable to polymeric materials.

B. Self-Reinforced Polymers and Molecular and Fibrillar Composites

The concept of self-reinforcement—that is, reinforcing a polymer with its own morphological entities, such as fibrils, crystallites, or fibers—is an important method for improving the mechanical characteristics of many polymers [9, 10]. The reinforcing elements grow in the amorphous matrix during the crystallization process. By selection of appropriate processing conditions, the reinforcing structures can be produced during injection molding or extrusion, for example [11–13]. A peculiarity of these systems is that they contain only one constituent, in contrast to the more common types of composites.

With the introduction of liquid crystalline polymers (LCP), new perspectives for polymer reinforcement were offered. It was believed that single, rigid, rodlike molecules could play the role of reinforcing elements; that is, they represented the typical case of molecular composites. This expectation appeared to be very attractive, since in this way, the aspect ratio—i.e., the length-to-diameter ratio—could be drastically increased, which, in turn, is very favorable for composite properties.

After development of the concept of molecular composites [14], this approach was followed by many researchers [15–19]. But despite the impressive mechanical properties of these LCP-reinforced thermoplastic polymers, which are comparable with those of conventional discontinuous-fiber-reinforced thermoplastic composites, it was found that they are not real molecular composites. For reasons predicted by Flory in 1978 [20], namely, that rigid and flexible molecules are thermodynamically immiscible, the phase-separated fibrous domains, which mainly contain the rigid-rod polymer, have diameters of up to 1000 nm or even more, as compared to the typical 3 nm for real molecular composites; that is, these domains form nanocomposites rather than molecular composites. While in the case of molecular composites, lyotropic LCPs were used as reinforcing elements, in the subsequent development of what were later named *in situ composites* [21], flexible polymers were blended with thermotropic LCPs. During their processing, e.g., by extrusion, the LCP is oriented into fine fibrils with high aspect ratios.

It is important to note once again that the types of composites described previously were supposed to be molecular composites, but actually were not. In order to obtain composites with a real molecular reinforcement, one has to use more sophisticated approaches, during both the synthesis (e.g., "hairy rods" in that the rigid and flexible molecules are chemically bonded) and preparation (e.g., LB technique) [22] of the composite, than those applied to the conventional LCP polymers and to their blending to produce composites via solution or melt. (For more details on LCP composites, see Chapter 32.)

In sum, it must be stressed that processing from (dilute) solution of a lyotropic LCP is a complicated process. In addition, the application of the self-reinforcing approach on a commercial scale requires a special set of processing conditions, which makes this technique rather expensive. At the same time, the development of real molecular composites still remains at the laboratory stage.

The recently developed new type of polymer–polymer composites, the MFC, satisfies to a great extent the basic requirements for polymer blends and composites; that is, it suppresses the incompatibility and improves the adhesion between fibers and matrix. Furthermore, unlike the classical macrocomposites (e.g., glass-fiber-reinforced composites) [23–25] or the LCPs and molecular composites [14–19, 22], this group of polymer composites is reinforced by microfibrils of *flexible* macromolecules. In contrast to the *in situ* composites, for which the microfibrils of LCP are also produced during processing, in the case of MFC a completely isotropic matrix is created during processing via special thermal-treatment conditions [4–7].

II. PREPARATION OF MFC

Macrocomposites and their "molecular" analogues are prepared in the same way, i.e., by blending the matrix material with the reinforcing material, usually in the melt. On the other hand, this approach is not applicable to MFCs, since the microfibrils are not available as a separate material. Instead, they are created by a process called MFC manufacturing. This process starts by blending immiscible polymeric partners. Not only must the partners be immiscible, but also there must be a difference between their melting points T_m. The essential stages of MFC preparation are as follows: (i) blending, (ii) extrusion, (iii) drawing (with good orientation of all components), and (iv) annealing at constant strain above the T_m of the component that melts at a lower temperature, but below the T_m of the component that melts at a higher temperature. During the drawing step, also called the fibrillization step, the blend's components are oriented and fibrils are created. In the subsequent annealing process, when melting of the component with the lower melting occurs, called the isotropization step, it is necessary to ensure that the oriented fibrillar structure of the component with the higher melting temperature is preserved. The preparation steps for MFCs are presented schematically in Fig. 33.1. It is important to note here that MFCs are based on polymer blends, but they should not be considered as "drawn blends," since the isotropization step results in the formation of an isotropic matrix

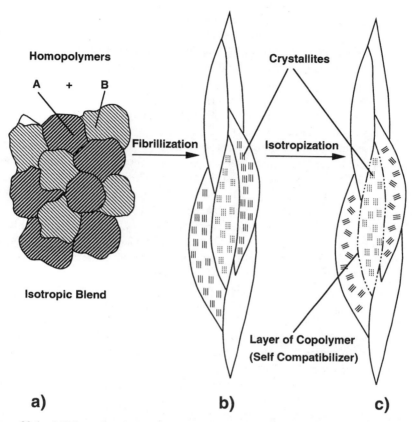

Homopolymers

A + B

Crystallites

Fibrillization

Isotropization

Isotropic Blend

**Layer of Copolymer
(Self Compatibilizer)**

a)   **b)**   **c)**

Figure 33.1 MFC can be obtained (a) from incompatible polymer blends (b) by extrusion and orientation (the fibrillization step) (c) followed by thermal treatment at a temperature between the melting points of the two components at constant strain (the isitropization step).

reinforced by fibrils of the component with the higher melting temperature; that is, one deals in the end with a typical composite material.

In some cases, it is advisable to perform annealing of the drawn blend (with fixed ends) before the isotropization step. The purpose of such a treatment is to improve the crystalline structure of the fibrils and thus, if necessary, increase their melting temperature. This effect in turn, enhances the processing conditions, since it prevents the microfibrils from melting.

III. DO MFC REALLY EXIST?

In order to answer the question posed in the tittle of this section, one has to look for evidence proving that MFCs have the two basic components of a composite material: the isotropized matrix and the reinforcing elements. In addition, one expects a synergistic effect to appear, in some of the properties, most often the mechanical ones, as compared to the properties of the neat starting materials [3]. In other words, one has to verify the preparation stages of MFCs, as illustrated schematically in Fig. 33.1.

To what extent are the two essential steps in the preparation of MFC—that is, the fibrillization and the isotropization—effective? The answer can be found in Fig. 33.2, which shows wide-angle X-ray scattering (WAXS) patterns for a blend of poly(ethylene terephthalate) (PET) with a melting temperature $T_m = 260°C$ and poly(butylene terephthalate) (PBT) with $T_m = 225°C$.

One can see that after blending, extrusion, and cold drawing, both the orientation and the crystallinity are very poor. (See Fig. 33.2a.) They improve drastically after annealing at 220°C with fixed ends. This pattern (see Fig. 33.2b) is typical for a highly oriented semicrystalline polymer, demonstrating the presence of fibrils of the two components PET and PBT, in accordance with the scheme in Fig. 33.1b.

The pattern in Fig. 33.2c also shows high crystallinity, but now for a highly oriented phase (PET) in an isotropic matrix (PBT). Occasionally, these two polymers have almost identical WAXS patterns, so that the aforementioned conclusions could be made only from their melting temperatures; that is, PBT is at 240°C in its molten state, at which the isotropization takes place, and crystallizes during subsequent cooling. Additional support for this conclusion can be found from observations of other systems, such as PET and polyamide 6 (PA-6) [4]. The observation of a crystalline phase in an isotropic state (shown in Fig. 33.2c for PBT and in [4] for PA-6 as circular reflexions) indicates that one really deals with an isotropic matrix, as a result of the final processing step. (See Fig. 33.1c.)

The best evidence for the existence of microfibrils that reinforce the isotropic matrix comes from scanning electron microscopy (SEM) studies of an MFC based on a blend of PET and PA-6 (1:1 by wt), as displayed in Fig. 33.3. The micrographs show the surfaces of the samples after selective extraction of PA-6 with formic acid.

The holes in Fig. 33.3a, previously occupied by PA-6 particles in the isotropic blend (before cold drawing), reflect the schematic idea of Fig. 33.1a. Rather perfect, separated PET fibrils can be seen after drawing and removal of the PA-6 matrix. (See Fig. 33.3b.) The same fibrils are no longer loose, but form aggregates after long periods (i.e., 25 hours) of annealing at 240°C. (See Fig. 33.3c.) The reasons for this will be discussed in the next section.

Finally, the mechanical properties (i.e., the tensile strength σ and tensile modulus E) for different blends of PET, PBT, PA-6, and polyamide-66 (PA-66) of various compositions, tested after the essential steps of MFC preparation, are summarized graphically in Fig. 33.4.

It is clear that both σ and E increase drastically after drawing of the as-extruded blends, which have properties typical of nondrawn (isotropic) PA-6; furthermore, they approach (as is the case for E; see Fig. 33.4b) or even surpass (as is the case for σ, Fig. 33.4a) the values of the reinforcing component (i.e., PET or PA-66) in the drawn state. As expected, the subsequent isotropization step (annealing at 240°C for 4 hours) results in a slight (for E) or strong (for σ) reduction of the values, but the important point at this stage is that one deals with a composite material (with properties that clearly are still better than those of the neat matrix) and not with a

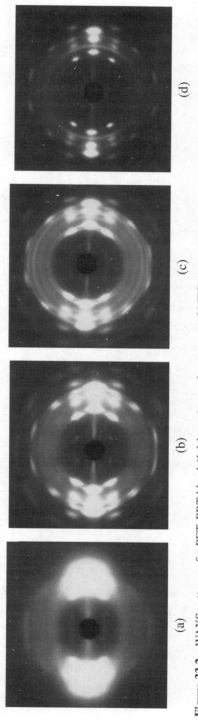

Figure 33.2 WAXS patterns of a PET–PBT blend (1:1 by wt) at various stages of MFC preparation: (a) as drawn, (b) after annealing at 220°C for 5 hours, and (c) after annealing at 240°C for 5 hours. (d) PET–PA-6 (1:1 by wt) blend after drawing and annealing at 240°C fo 25 hours.

(a) (b) (c) (d)

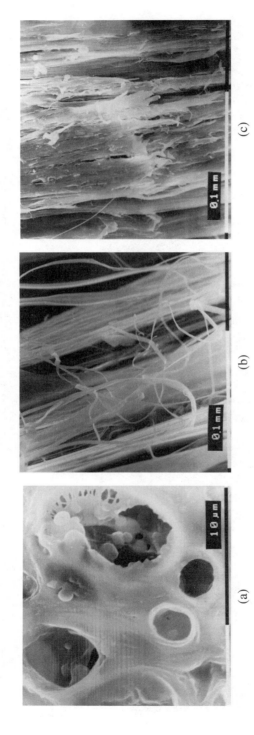

(a)   (b)   (c)

Figure 33.3 SEM micrographs of a PET–PA-6 (1:1 by wt) sample surface after selective extraction of PA-6 taken at various steps of MFC preparation: (a) after blending, (b) after drawing and annealing at 220°C for 5 hours, (c) after drawing and annealing at 240°C for 25 hours.

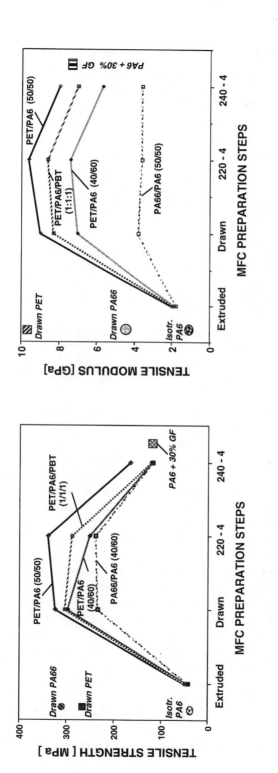

Figure 33.4 (a) Tensile strength and (b) tensile modulus for various blends and compositions (by wt) at different steps of MFC preparation. For comparison, the values of the neat components and of 30% glass-fiber-reinforced PA-6 are given. "220-4" and "240-4" mean annealing after drawing for 4 hours at 220 or 240°C, respectively.

drawn blend. In addition, the mechanical properties at this stage are about the same as, or slightly higher than, depending on the MFC composition, those measured for commercial short-glass-fiber (GF) reinforced PA-6 (PA-6 +30% GF in Fig. 33.4). Further, it should be noted that the values of the MFCs are 30–40% higher than the rule-of-mixtures values calculated from the properties of the individual components (isotropic PA-6 and drawn PET, for example [4]); that is, a synergistic effect is observed in the mechanical properties.

IV. CHEMICAL INTERACTIONS IN MFC AND THEIR SUBSEQUENCES

Over four decades, polymer scientists and engineers used the terms *miscibility* and *compatibility* loosely and/or interchangeably. For absolute clarity and understanding, we define them here. *Thermodynamic miscibility* describes polymer blends that are completely miscible and homogeneous down to the molecular level and that do not show any phase separation at all. By contrast, *practical compatibility* describes polymer blends which have properties that are useful in commercial practice. It should be emphasized that most of the commercially useful blends have practical compatibility, even though they do not have thermodynamic miscibility; in fact, they usually form multiphase morphologies that produce a synergistic advantage in the balance of properties, an effect that is not available from any single polymer [27].

Considering compatibility, one further has to note that a polymer blend, when thermodynamically immiscible, separates into two or more phases, thus resulting in poor practical properties. In such cases, it is generally assumed that either the sizes of the dispersed domains are not optimal or the immiscibility of the two phases produces weak interfaces, which in turn lead to easy failure of the blend under stress. As mentioned previously, the most popular way of trying to solve these problems is the addition of a third ingredient, a *compatibilizer*. In rare cases, the compatibilizer actually produces complete thermodynamic miscibility. In most cases, however, it simply acts as a surfactant to reduce the domain size or as an adhesive to strengthen the interface between the two immiscible polymer phases. Success is judged by improvements in various practical properties as a result of practical compatibility [27]. (For more on reactive compatibilization, see also Chapter 17.)

These introductory notes on compatibility are important for a better understanding of one of the main advantages of MFCs when they are prepared from condensation polymers. In such cases, in addition to isotropization during short (i.e., several hours) thermal treatment, chemical reactions (i.e., additional condensation and transreactions) between condensation polymers in the melt [28] as well as in the solid state [29] can take place at the interfaces, schematically shown as follows:

Additional condensation:

$$\cdots COOH + H_2N\cdots \quad \underset{\longleftarrow}{\longrightarrow} \quad \cdots CONH\cdots \qquad (33.1)$$

Transreaction:

$$
\begin{array}{ccc}
\cdots CO \quad\quad O\cdots & \longrightarrow & \cdots COO\cdots NHCO\cdots \\
\;\;\;|\;\; + \;\;| & \longleftarrow & + \\
\cdots NH \quad\quad OC\cdots & & \cdots NHOC\cdots COO\cdots
\end{array}
\tag{33.2}
$$

homopolymers   copolymer

These reactions result in the formation of a copolymeric layer on the interface. (See Fig. 33.1c.) This layer plays the role of a compatibilizer— that is, one deals with a self-compatibilization effect—in that there is no need to introduce to the blend an extra synthesized copolymer of the blend components according to the usual approach [30, 31].

Compatibilization can be effective only in the initial stages of chemical interaction. During further thermal treatment, the reaction goes on and involves finally all of the isotropic (molten) component and the amorphous portion of the fibrillized one in the formation of block copolymers, thus transforming the homopolymeric matrix into a copolymeric one:

$$
(A)_n + (B)_m \longrightarrow \cdots (A)_x(B)_y(A)_z \cdots .
\tag{33.3}
$$

In the case for which $(A)_n$ and $(B)_m$ are crystallizable homopolymers, the block copolymers also crystallize. As the chemical interactions progress, the block copolymers may, however, randomize and convert themselves into random copolymers:

$$
\cdots AABBBBBBBBBBBAAAAAAB \cdots \longrightarrow \cdots ABABBABABAA \cdots .
\tag{33.4}
$$

An important result of the randomization shown in Eq. 33.4, is the loss of the ability of the matrix to crystallize, which can influence the entire behavior of the MFCs. All of these transformations of the matrix, which result from chemical interactions in MFCs made from condensation polymers, are presented schematically in Fig. 33.5. (More about chemical interactions in blends of condensation polymers can be found in Chapter 14.)

At this point, the following question arises: What kinds of proof exist for the occurrence of the described chemical changes and of physical transformations in the matrix?

Figure 33.2d shows the WAXS of MFC based on PET and PA-6 after prolonged treatment (25 hours) at a relatively high temperature (240°C), at which PA-6 is in a molten state. One can see that after cooling, the PA-6 did not crystallize, in contrast to its behavior after shorter annealing periods of 4–5 h. (See Fig. 33.2b and c for PBT.) The same conclusion can be drawn from the DSC measurements carried out on the same PET–PA-6 system, as displayed in Fig. 33.6a [32]; that is, the melting peak of PA-6 is missing after this level of thermal treatment.

Figure 33.6b also summarizes dynamic mechanical thermal analysis (DMTA) curves, taken after the various steps of MFC preparation from PET and PA-6 [33].

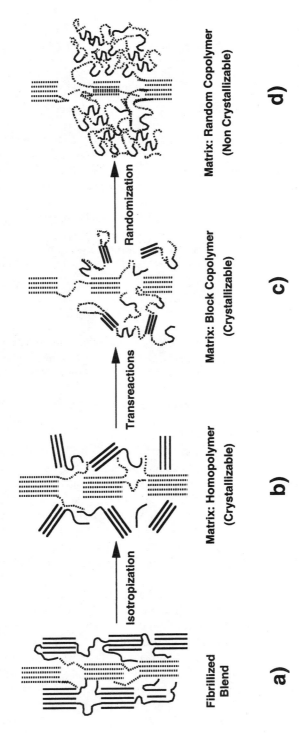

Figure 33.5 Schematic presentation of the changes in the matrix of MFC, in the case of condensation polymers, as a result of chemical interactions.

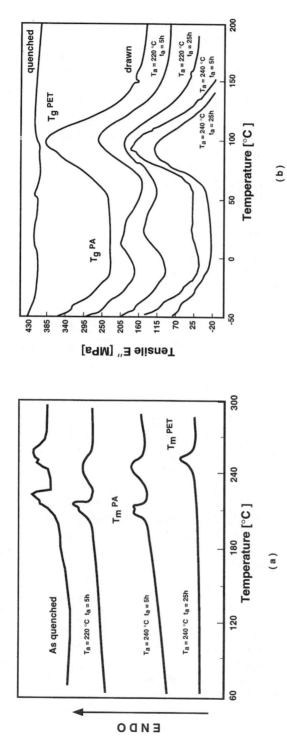

Figure 33.6 (a) DSC curves taken in second heating mode and (b) DMTA curves of a PET–PA-6 blend (1:1 by wt); treatment conditions are given for the respective curves.

These curves of the loss modulus E'' against temperature indicate that after the final treatment (25 hours at 240°C), PA-6 disappears as an amorphous phase, too, since its glass transition temperature is no longer observed. One might assume that PA-6 has been totally removed, e.g., by extraction from the system, but this is not the case; instead, it has reacted with PET to form a copolymer of random type. (See Eq. 33.4.) In such a case, PA-6 can neither crystallize nor form amorphous domains large enough to display their own T_g. That is, the matrix of the MFC in this case really is noncrystallizable. (See Fig. 33.5d.) In addition, and more importantly, the matrix is chemically bonded to the fibrils, the latter forming aggregates (see Fig. 33.3c) instead of loose, separated fibrils, as shown in Fig. 33.3b. This conclusion is supported by IR measurements after selective extraction [26], selective dying [7], and solubility tests [26]. All of these methods indicate an involvement of PA-6 in a copolymer with PET. Further details about the melting-, crystallization-, and miscibility-induced sequential reordering in condensation copolymers can be found in [34].

V. PROCESSING OF MFC VIA COMPRESSION OR INJECTION MOLDING

Compression molding is a widely used technique for processing continuous-fiber-reinforced thermoplastics. Since in the case of PET–PA-6 blends a very good consolidation between the partners is expected, compression-molding experiments with MFCs were also started recently [35]. In addition, a commercial injection-molding machine has been used to process MFCs into more complex geometries [36].

A cold-drawn PET–PA-6 blend (1:1 by wt) shaped as continuous rods 1–1.5 mm in diameter, that had not yet passed through the isotropization step was cut into 50-mm-long strings. These strings were placed into a mold (50 × 50 mm), which was then heated up to 240°C. At this temperature, the sample was kept for 10 minutes under a pressure of 8 MPa, followed by cooling to room temperature at a rate of about 15°C/minute. For reasons of comparison, plates of the same thickness (2 mm) were prepared under the same processing conditions from undrawn bristles of the PET–PA-6 blend, as well as from regular granules of neat PA-6 and of PA-6 reinforced by short glass fibers (30% by wt). The plates contained three layers of bristles placed either in parallel (0, 0, 0) or cross-plied (0, 90, 0) positions.

Results of static tensile tests to measure the strength σ and the modulus E are presented in Fig. 33.7. When measurements were carried out in the 0° direction, σ for the MFC with parallel alignment (0, 0, 0) was almost twice as high as that of the cross-plied (0, 90, 0) laminates or the glass-fiber reinforced PA-6 (see Fig. 33.7a), whereas the E values of the three samples were almost equal (see Fig. 33.7b). At the same time, σ increased in comparison to that of the neat PA-6 and that of the nondrawn PET–PA-6 blend by a factor of almost 3.

The injection-molding experiments were carried out using pellets of extruded and drawn PET–PA-6 blend in various ratios. It is very important to keep the processing temperature around 240°C in order to keep PET from melting, because otherwise its fibrillar morphology will disappear. The results of mechanical tests, including

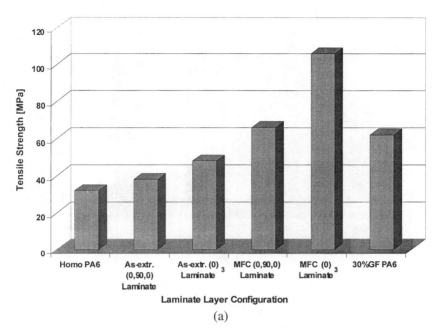

(a)

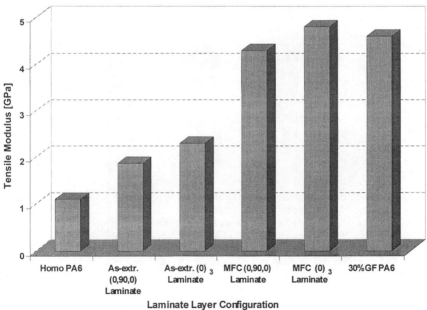

(b)

Figure 33.7   (a) Tensile strength and (b) tensile modulus of compressing-molded plates of neat PA-6, of an as extruded PET–PA-6 (1:1 by wt) blend and their MFCs with parallel (0, 0, 0) or cross-plied (0, 90, 0) bristles, as well as of 30% glass-fiber-reinforced PA-6.

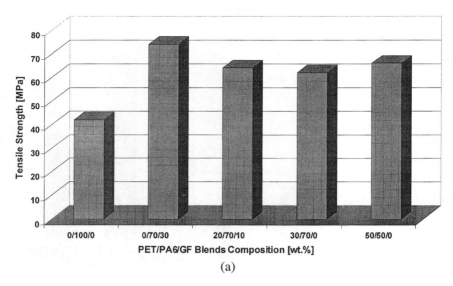

(a)

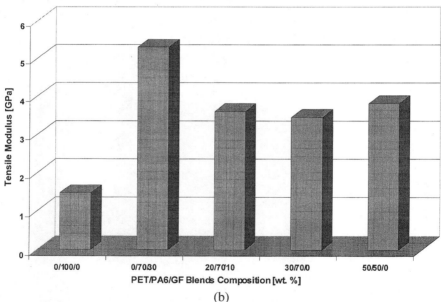

(b)

Figure 33.8 (a) Tensile strength and (b) tensile modulus of injection-molded MFC of a PET–PA-6 blend with various compositions, also containing short glass fibers, as indicated in the hystograms.

samples with short glass fibers, are presented in Fig. 33.8. One can see that the tensile strength σ of the final MFC product is higher by 50% than σ for the neat PA-6 and only 15% lower than σ for a 30% glass-fiber-reinforced PA-6. (See Fig. 33.8a.) At

(a)

(b)

(c)

Figure 33.9 SEM microphotographs of the fracture surface (in liquid nitrogen) of a PET–PA-6 (1:1 by wt) blend after (a) blending and extrusion, (b) compression molding at 240°C for 10 minutes and 8 MPa pressure on the extruded and drawn blend, and (c) injection molding of the extruded and drawn blend.

the same time, E for the final MFC product is more than twice as high as that of the neat PA-6 and 35% lower than the E for glass-fiber-reinforced PA-6. (See Fig. 33.8b.) The differences in the values of E in favor of glass-fiber-filled composites are related to the properties of the reinforcing material (glass and PET).

The significant differences in the mechanical behavior become understandable if one takes into account the structural peculiarities of the blends presented in Fig. 33.9.

While the undrawn blend is characterized by a rather uniform distribution of the components, which are almost of spherical shape (see Fig. 33.9a), the final MFC products, regardless of the processing technique, are distinguished by perfect fibrils of PET (mainly compression-molded plates) dispersed in the isotropic PA-6 matrix; see Fig. 33.9b and c, respectively.

VI. CONCLUSIONS AND OUTLOOK

As mentioned at the beginning of this chapter, polymer blends and composites are of prime commercial importance as polymeric materials. Scientific and commercial progress in this area is driven by the idea that new molecules are not always required to meet needs for new materials and that blending can be implemented more rapidly and economically than the development of new chemical compositions. For this reason, the aforementioned approach for the preparation of MFC from polymer blends seems to be a step in this direction. It is important to emphasize that, although, for the purposes of this chapter, MFCs made from condensation polymers have been discussed, the method described is applicable to any combination of partner polymers, provided that they meet the basic requirements for MFC preparation—that is, the partners have to be immiscible and they should have different melting or softening (in case of amorphous polymers) temperatures. In addition, when condensation polymers are used to make an MFC, other advantages arise that are related to the inherent ability of the condensation polymers to undergo additional chemical interactions.

The applicability of the MFC approach to polyolefins, as well as to blends of polyolefins with condensation polymers, was recently demonstrated [37, 38]. MFCs from polypropylene (PP) and polyethylene (PE) [37] and MFCs from blends of PET and PE, PET and PP, PA-66 and PP, and PA-66 and PE [38] showed the same improvement in mechanical properties in comparison to the respective isotropic matrix, and the quality of their mechanical properties approached that of the properties of the same matrix reinforced with short glass fibers, as demonstrated previously for the PET–PA-6 system (see Fig. 33.4). The effect of the partners' ratio in the MFC approach was also studied [37–39].

In addition to the synergistic effect in mechanical properties, MFC offer further important advantages: reinforcement of the polymer by conventional (flexible) thermoplastics, no mineral additives, reduction in weight, easy processing, no need for extra compatibilizers, control of the crystallization ability of the matrix and its solubility, complete recycling, and repetition of processing. The possibility of applying the MFC approach for recycling purposes is schematically presented in Fig. 33.10 for the best studied system: PET–PA-6 blends.

Another important advantage of MFCs arises from the known fact that by increasing the aspect ratio, either by increasing the length or by decreasing the diameter of the fibers, the Young's modulus of the composite can be improved by up to two orders of magnitude. In this respect, the properties of MFCs are approaching these of the molecular composites for which ultimate reinforcement is achieved by single, extended rigid-rod polymer molecules, or at least by bundles of them.

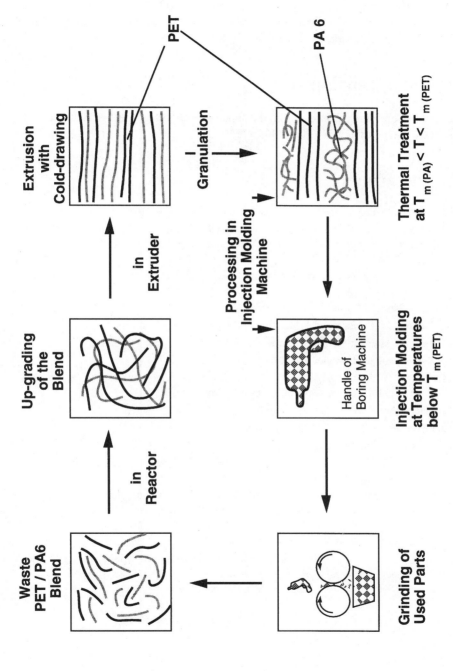

Figure 33.10 Schematic presentation of material upgrading, fibrilization, and recycling.

473

Finally, the difference in the size of the reinforcing elements, and particularly in the method of preparation of MFCs in contrast to that of macrocomposites, makes the MFCs similar to the natural composite materials in which the fibrils and the matrix grow simultaneously with interpenetration by formation of chemical bonds. Thus, MFCs mimic such natural materials.

VII. ACKNOWLEDGEMENTS

The authors gratefully acknowledge the financial support of the Deutsche Forsch-ungsgemeinschaft, Germany (DFG-FR 675/21-1).

VIII. REFERENCES

1. L. A. Utracki, *Polymer Alloys and Blends*, Hanser, München, 1989.

2. J. Karger-Kocsis, "Composites (Structure, Properties, and Manufacturing," in *Polymeric Materials Encyclopedia*, Vol. 2, J. C. Salamone (ed.), CRC Press, Inc., Boca Raton, Florida, USA, 1996, p. 1378.

3. D. Hull, *An Introduction to Composite Materials*, Cambridge University Press, Cambridge, 1987.

4. M. Evstatiev and S. Fakirov, *Polymer*, **33**, 877 (1992).

5. S. Fakirov, M. Estatiev, and J. M. Schultz, *Polymer*, **34**, 4669 (1993).

6. S. Fakirov, M. Estatiev, and S. Petrovich, *Macromolecules*, **26**, 5219 (1993).

7. S. Fakirov and M. Evstatiev, *Adv. Mat.*, **6**, 395 (1994).

8. D. W. Ihm, A. Hiltner, and E. Baer, "Microfiber Systems, A Review," in *High Performance Polymers*, E. Baer and A. Moet (eds.), Hanser, München, 1991.

9. J. Petermann, *Umschau*, **20**, 589 (1983).

10. J. Petermann, *J. Mater. Sci.*, **22**, 1120 (1987).

11. G. W. Ehrenstein, *Angew. Makromol. Chem.*, **175**, 187 (1990).

12. J. Song and G. W. Ehrenstein, "Friction and Wear of Self-Reinforced Thermoplastics," in *Advances in Composites Tribology*, K. Friedrich (ed.), Elsevier, Amsterdam, 1993.

13. J. Song, M. Prox, A. Weber, and G. W. Ehrenstein, "Self-Reinforcement of Polypropylene," in *Polypropylene, Structure, Blends, Composites*, J. Karger-Kocsis (ed.), Chapman and Hall, London, 1995, p. 273.

14. M. Takayanagi, *Pure & Appl. Chem.*, **55**, 819 (1983).

15. J. K. Stille, A. Parker, and J. Tsang, "Molecular Composites from Rod and Flexible Polyquinones," in *Contemporary Topics in Polymer Science*, Vol. 6 of *Multiphase Macromolecular Systems*, B. M. Culberton (ed.), Plenum, New York, 1989.

16. W. Brostow, *Kunststoffe*, **78**, 411 (1988).

17. O. Rötting and G. Hinrichsen, *J. Therm. Comp. Mater.*, **8**, 252 (1995).

18. G. Crevecoeur and G. Groeninckx, *Polym. Eng. Sci.*, **30**, 532 (1990).

19. R. Ding and A. I. Isayev, *J. Therm. Comp. Mater.*, **8**, 208 (1995).

20. P. J. Flory and A. Abe, *Macromolecules*, **11**, 1119 (1978).

21. G. Kiss, *Polym. Eng. Sci.*, **27**, 410 (1987).

22. M. Seifert, C. Fakirov, and G. Wegner, *Adv. Mater.*, **7**, 52 (1995).

23. G. W. Ehrenstein and R. K. Bayer, *Z. Plast.*, **32**, 1 (1981).

24. G. Menges and P. Geisbusch, *Coll. & Polym. Sci.*, **260**, 73 (1982).

25. W. Janzen and G. W. Ehrenstein, *Kunststoffe*, **81**, 231 (1991).

26. M. Evstatiev, N. Nikolov, and S. Fakirov, *Polymer*, **37**, 4455 (1996).

27. R. D. Deanin and C.-H. Chuang, "Polyolefin Polyblends," in *Handbook of Polyolefins: Synthesis and Properties*, C. Vasile and R. B. Seymour (eds.), Marcel Dekker Inc., New York, 1993.

28. P. J. Flory, *Principles of Polymer Chemistry*, Cornell University Press, Ithaca, NY, 1953.

29. S. Fakirov, "Solid State Reactions in Linear Polycondensates," in *Solid State Behavior in Linear Polyesters and Polyamides*, J. M. Schultz and S. Fakirov (eds.), Prentice Hall, Englewood Cliffs, New Jersey, 1990, p. 1.

30. R. Holsti-Miettinen, J. Seppalae, and O. T. Ikkala, *Polym. Eng. Sci.*, **32**, 868 (1992).

31. A. R. Padwa, *Polym. Eng. Sci.*, **32**, 1703 (1992).

32. S. Fakirov, M. Evstatiev, and K. Friedrich, "Interfacial Interactions in Microfibrillar Composites from Condensation Polymers," in *High Technology Composites in Modern Applications*, S. A. Paipetis and A. G. Youtsos (eds.), University of Patras, Patras, Greece, 1995, p. 12.

33. T. Serhatkoulu, I. Bahar, B. Erman, S. Fakirov, M. Evstatiev, and D. Sapundjieva, *Polymer*, **36**, 2371 (1995).

34. S. Fakirov and Z. Denchev, "Sequential Reordering in Condensation Polymers," in *Transreactions in Condensation Polymers*, S. Fakirov (ed.), Wiley-VCH, Weinheim, Germany, 1999, p. 319.

35. M. Evstatiev, S. Fakirov, and K. Friedrich, *J. Appl. Polym. Sci.*, submitted for publication.

36. M. Evstatiev, C. Fakirov, S. Fakirov, and K. Friedrich, NATO-ASI on "Structure Development in Processing for Polymer Property Enhencement," Caminha, Portugal, 17–28 May, 1999.

37. M. Evstatiev, S. Fakirov, and K. Friedrich, "Microfibrillar Reinforced Composites—Another Approach to Polymer Blends Processing," in *Structure Development During Polymer Processing*, A. M. Cunha and S. Fakirov (eds.), Kluwer Academic Publisher, Dordrecht, Boston, London, 2000.

38. M. Evstatiev, O. Samokovliski, S. Fakirov, and K. Friedrich, unpublished data.

39. M. Evstatiev, S. Fakirov, and K. Friedrich, *Appl. Comp. Mater.*, **2** , 93 (1995).

40. M. Evstatiev, S. Fakirov, M. Schnoor, and K. Friedrich, *Kunststoffberater*, **6**, 23 (1996).

34 Elastomer Blends

SUDHIN DATTA

Exxon Chemical Co.
5200 Bayway Drive
Baytown, TX 77520

Polymer Blends, Volume 2: Performance. Edited by D. R. Paul and C. B. Bucknall.
ISBN 0-471-35280-2. © 2000 John Wiley & Sons, Inc.

I. INTRODUCTION

Two technologically important processes have been developed to improve the properties of existing elastomers. These processes involve changes in either intermolecular or intramolecular composition. Simple, but effective, synthetic intramolecular strategies in copolymer elastomers include changing the composition or the microstructure of the intramolecular distribution of monomers. An example of this type of strategy is the change from random styrene–butadiene copolymers to block copolymers. These intramolecular procedures are limited by the availability of versatile polymerization processes to control the microstructure. The alternative strategy, which does not require the development of novel synthetic procedures, is blending different elastomers. Blends are commercially useful, because the synthesis of an entirely different elastomer is not required. This chapter describes some of the distinctive features of blends of elastomers.

An area of distinction between elastomer blends and elastomer–thermoplastic blends is the greater prevalence of miscible blends of elastomers. These miscible blends can show important physical and chemical differences as compared to a uniform copolymer of the same average composition and molecular weight. Blends of ethylene–propylene copolymers of different composition can result in an elastomer comprising a semicrystalline component of higher ethylene content and an amorphous component of lower ethylene content. These blends combine the higher tensile strength of the semicrystalline polymers and the favorable low-temperature properties of amorphous polymers. Similar compositional differences in styrene–butadiene polymers lead to miscible blends of different styrene contents that have a spectrum of molecular relaxation characteristics, due to the presence of high-styrene-content (high T_g) and low-styrene-content (low T_g) components. Chemical differences in miscible blends of ethylene–propylene and styrene–butadiene copolymers can also arise from differences in the distribution and the type of vulcanization sites on the elastomer. The uneven distribution of diene, which is the site for vulcanization in blends of ethylene–propylene–diene elastomers (EPDM), can lead to the formation of two distinct, intermingled vulcanization networks. Similar effects can be obtained in styrene–butadiene polymers by changing the insertion characteristics of butadiene from 1,4 to 1,2 in the components of the miscible blend. In the future, the development of miscible blends will encompass changes in the properties of the elastomer, due to the incorporation of closely related, but different, monomers (e.g., octene for propylene in ethylene copolymers) as components of miscible polymer blends. These existing and postulated changes in the monomer identity lead to a greater diversity in properties of the elastomer blends than is possible by simple changes in the ratio of the constituent monomers in a single, compositionally uniform elastomer. Miscible blends of elastomers are commonly used, though they have very rarely been recognized as such. Analysis of such blends, particularly after vulcanization, is difficult. The current analytical techniques are only slightly more capable than the classical techniques of selective precipitation of the components of an unvulcanized elastomer blend from solution [1].

Theoretically, blends of chemically dissimilar (that is, immiscible) elastomers can attain a wider variation in properties than blends of miscible, and thus chemically similar, elastomers. Combinations of immiscible elastomers lead to changes in properties, due to either intrinsic differences in the constituents or differences in the reinforcement and vulcanization of the constituents. An example of the former is the potential for improved heat ageing and resistance to solvents in elastomers. Figure 34.1 locates some of the more common elastomers with respect to these properties. This assignment of locations is approximate, since the properties are dependent both on the composition of the elastomers (i.e., comonomer content for SBR and EPR, degree of halogenation for chlorinated elastomers, random versus blocky distribution) as well as on the exact compound (i.e., filler, plasticizer, curative level) used for the evaluation. For a family of elastomers under comparable analytical conditions, the location stems from the chemical microstructure of the polymer chains. Blends of elastomers combining the extremes of these two properties can generate an elastomer with acceptable levels of resistance to these environmental factors. Changing the properties of elastomers by uneven distribution of fillers and vulcanization is, however, the more common use of blends of immiscible elastomers [2]. The engineering properties of elastomers (i.e., tensile strength, hysteresis, shifts in tan δ loss peaks) in vulcanized compounds depend not only on the elastomer itself, but also on the amount and identity of the fillers and plasticizers, as well as on the extent of cure. In an immiscible blend, the amount of these additives in any phase can be modulated by changes in the viscosity and chemical identity of the elastomer, the surface chem-

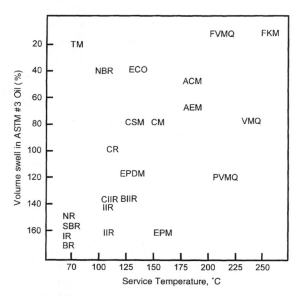

Figure 34.1   Elastomer distribution by resistance to aromatic solvents (ordinate) and temperature (abcissae).

istry of the filler, the chemical nature of the plasticizer, the sequence of addition of the components, and the details of the mixing procedure. A large body of experimental procedures (*vide infra*) has been developed to attain a thermodynamically metastable interphase distribution of additives in blends. On vulcanization, this distribution is rendered immobile, leading to desirable engineering of the blend. Both these differences and the greater flexibility of experimental design have generated a large body of commercially useful but proprietary procedures for immiscible elastomer blends.

Two notable reviews of elastomer blends exist. The first, by Hess, Verd, and Vegvari [3], reviews the applications, analysis, and the properties of the immiscible elastomer blends. The second, by Roland [4], has a discussion on the physics of mixing immiscible polymer blends and a more recent account of the analytical methods. Other reviews, such as those by Corish [5] and McDonel [6], deal with specific aspects of elastomer blends. These reviews are focused on immiscible blends of elastomers. In this chapter, we complement these previous works with information on miscible blends. However, this chapter does not deal with elastomeric heterophase blends (such as block copolymers [7] and thermoplastic vulcanizates [8]), which are not blends of individual elastomers.

II. MISCIBLE ELASTOMER BLENDS

A. Thermodynamics

The extension of thermodynamics to a blend of elastomers has been discussed by Roland [4]. Miscible blends are most commonly formed from elastomers with similar three-dimensional [9] solubility parameters. An example of this trend is blends from copolymer elastomers (e.g., ethylene–propylene and styrene–butadiene copolymers) from component polymers of different composition, microstructure, and molecular weight. Miscible blends of elastomers are also formed with a preferential stabilizing interaction between the components, such as hydrogen bonding and dipole–dipole interaction. While these synthetic methods are useful, their applicability is limited to fairly low-molecular-weight elastomers, since these specific interactions lead to increased viscosity for the elastomer blends. Further, the use of these elastomer blends requires the synthesis of new elastomers with specific types and levels of chemical functionality and reactivity. When the forces between the components of a polymer blend are mostly entirely dispersive, miscibility is achieved only in neat polymers with a very close match to Hansen's three-dimensional solubility parameter [9]. Simple solubility parameter calculations are reliable only in comparisons across a single family of elastomers (e.g., styrenics). Extension of the single-parameter procedures to elastomers of different families (e.g., polyolefins and acrylates) leads to exaggerated and erroneous conclusions about the miscibility of elastomers.

Miscible blends of elastomers differ from corresponding blends of thermoplastics in three important areas. First, the need for elastic properties as well as acceptable vulcanization characteristics require elastomers to be high-molecular-weight polymers with a limited polydispersity. This requirement reduces the miscibility of dissimilar elastomers by interdiffusion of the low-molecular-weight components of the blends. Second, elastomers are plasticized in the conventional compounding with process oils and reinforced with particulate fillers. Process oils are low-molecular-weight solvents for elastomers. The presence of plasticizers leads to a higher free volume for the blend components. This leads to an increase in the available conformations for the elastomer molecules. The addition of process oils also limits the formation of unlike $i–j$ contacts [4], which destabilize the blend. Process oils leads to an increase in the entropy and a decrease in the enthalpy for the formation of the miscible blend. The extent of this stabilizing influence is small, and the formation of miscible blends by plasticization has only been inferred in blends of closely related elastomers. The use of a blend of plasticizers, differing in solubility parameters, as a common solvent for a blend of dissimilar elastomers does not usually lead to a single plasticized phase. The thermodynamically stable system consists of two phases, with each phase containing a single polymer and a plasticizer containing predominantly a single component closest to the polymer in terms of the solubility parameter (coacervation). Third, elastomers are designed not to have either melting or glass transitions within the useful temperature range ($-35°C$ to $+120°C$). Changes in the miscibility of elastomers due to these effects is usually absent.

One area of comparatively little study is the effect of fillers on the miscibility of elastomers. We expect that chemisorption of elastomers on the surface of small-particulate fillers (e.g., carbon black) reduces the number of possible conformations available to a polymer chain and stabilizes unlike $i–j$ contacts between dissimilar polymer chains at the surface of the filler. Both effects act to aid miscibility of dissimilar elastomers, although quantitative data to support this speculation are not available. In sum, the thermodynamic predictions of the miscibility of elastomers are somewhat relaxed in compounds containg process-oil plasticizers and particulate fillers. In spite of this, truly miscible blends of dissimilar elastomers are most prevalent only in copolymers of different composition and in some selected blends of different polymers.

B. Kinetics

The miscibility of many elastomer mixtures depends on the reduction of the frequency, and strength of the unlike $i-j$ contacts. The uncoiling of a macromolecule in a shear field should presumably increase the number of these unlike contacts. Additionally, phase separation of polymers from solution is known to occur as a result of orientational crystallization [10a–c] during shear-induced flow. Both of these processes can lead to phase separation in miscible blends of elastomers. Strain-induced phase separation in a graft copolymer has been inferred from the resulting physical properties of the elastomer [11], although this has not been confirmed by morphological evidence. The reverse effect occurs for blends with specific interactions between the components of the mixture when an increase of unlike contacts could promote miscibility [12].

The formation of miscible rubber blends slows the rate of crystallization [13a, b] when one of the components is crystallizable. This phenomenon accounts for data which show lower heats of fusion that correlate to the extent of phase homogeneity [14] in elastomer blends. For example, while pure IR attained about a 30% crystallinity, the IR in blends, even after extensive annealing, was significantly less crystalline. A series of miscible blends of IR and 1,2-BR were crystallized, but no melting endotherm was detected for samples containing greater than 25% 1,2-BR. Additionally, the melting behavior of a polymer can be changed in a miscible blend. The stability of the liquid state by formation of a miscible blend reduces the relative thermal stability of the crystalline state and lowers the equilibrium melting point [15a, b]. This depression in the melting point is small for a miscible blend with only dispersive interactions between the components. In the blends of IR and 1,2-BR, no depression was observed. A small depression in the melting point of a semicrystalline EP polymer in a miscible blend with an amorphous EP polymer, in the composition range of 5% to 40% semicrystalline polymer, has been observed. The extent of the depression is 7°C for the blends containing the least amount of semicrystalline polymer [16].

C. Analysis

1. *Glass Transition* The principal effect of miscibility in blends of dissimilar elastomers is alteration of the glass transition temperature. The formation of miscible blends is similar to the formation of copolymers from the monomeric constituents of the blend. Since miscible blends should have negligible changes in the conformation of the polymer chains, the entanglement density of miscible blends should be a compositionally weighted average of the entanglement density of the pure components. This has been verified in blends of IR and 1,2-BR. Even in blends with weak specific interactions, there are only small effects on chain dimensions [17a–c]. Semiempirical correlations of rubbery plateau moduli with molecular characteristics [18a–e] neglect details of the chain structure and are of limited use. The prediction of entanglement density and plateau moduli in miscible blends is thus difficult.

2. Magnetic Resonance Imaging   Nuclear magnetic resonance has been applied to the study of homogeneity in miscible polymer blends and has been reviewed by Cheng [19a]. Roland [19b] has reviewed the different NMR techniques used for evaluation of elastomer blends. When the components of a blend have different T_g's, proton NMR can be used to assess the phase structure of the blend, by taking advantage of the rapid decrease of proton–proton coupling with nuclear separation [19c]. At a temperature between the glass transition temperatures of the components, observation of only a single, broad line width is evidence of spatial homogeneity on a scale of about one nanometer. This finding is confirmed only if the onset of mobility transpires at the same temperature for both components of a miscible blend [19d].

For blends containing elastomers of almost identical T_g's, slightly different techniques are required. Proton MAS NMR is applied to blends in which one of the components is almost completely deuterated [20]. Strong dipolar interactions between the residual protons occur only if components are intimately mixed. Another technique is crosspolarization MAS^{13}C NMR [21]. The transfer of spin polarization from protons to the ^{13}C atoms of the deuterated component can occur if these carbons are in proximity (nanometer) within one of the protons.

3. Crystallinity   Changes in polymer crystallinity have also been employed to study the homogeneity of elastomer blends [22]. Morris [22a] studied the rate of crystallization of cis-1,4-BR in blends with SBR. At any given blend composition, the BR crystallization rate diminished in the direction of greater blend homogeneity. Sircar and Lamond [22c] also studied the changes in BR crystallinity in blends with NR, IR, EPM, CIIR, NBR, and CR. The nature of the blend component had the greater effect, since the more compatible blends (smaller domains) had the greater loss in BR crystallinity.

4. Interdiffusion   Interdiffusion between a pair of polymers is a demonstration of their thermodynamic miscibility. The slow rates of diffusion of elastomers require a probe that is sensitive to the composition and structure at the interface between the polymers, since the bulk of the polymer may remain compositionally pure. Unless the components are of exceptionally high molecular weight or are highly branched, the time scales for significant interdiffusion are accessible to experiments. The adhesion between contacted rubber sheets parallels the extent of any interdiffusion of the polymer chains [23a]. When interdiffusion takes place, the adhesion is limited by the cohesive strength of the materials. If the contacted sheets are made up of immiscible rubbers, no interdiffusion occurs. Natural rubber (NR) and 1,2-polybutadiene (1,2-BR) are miscible even at high molecular weights [23b, c]. As a result, when the

NR is brought into contact with 1,2-BR, the initially separated polymers interdiffuse spontaneously. When some form of scattering contrast exists between the materials, interdiffusion enhances the scattering intensity (either X-ray or neutron) measured from the plied sheets, although the interdiffusion must be extensive for measurable scattering intensities to be observed. Infrared [23d, e] radiotracer diffusion [23f] and forward-recoil spectrometry (FRES) [23g] have been used to characterize concentration profiles of interdiffusing species directly. Light, neutron, and X-ray scattering [23h–j], as well as fluorescence spectroscopy, can be also be used.

5. Mechanical Properties   Miscible blends should have greater mechanical integrity than a multiphase structure. Miscible rubber blends that react chemically have a densification and a higher cohesive-energy density. These factors may provide improved mechanical properties, but they have been observed only below the T_g of the blend [24a]. The change in mechnical properties with phase morphology is indistinct, since there is no discontinuity in the mechanical properties of homopolymer blends near the critical solution temperature [24b, c].

D. Compositional Gradient Copolymers

A significant development [25] in the last decade is the use of miscible blends of compositionally different EPDMs. The blends are designed to balance viscoelastic properties, such as adhesion, and the rate of extrusion with physical properties, such as tensile strength, by changes in the relaxation characteristics. This balance is achieved by the components having different average molecular weight, through differences in the crystallinity (due to extended ethylene sequences), or by both. An example of the components of these blends is shown in Table 34.1. The blends of component A with one of the B polymers are miscible and are made by mixing hexane solutions of the elastomers. Blending by melt mixing the components is frequently unsuccessful, since it requires a molecular dispersion of components differing in viscosity. Figures 34.2 and 34.3 show the effect on tensile strength of the

Table 34.1   Ethylene–Propylene Blend Components Differing in Molecular Weight and Crystallinity[a]

Sample	Composition of C2, wt%	Viscosity, ML(1 + 4) 125°C
A	60	41
B_1	74	72
B_2	76	247
B_3	78	1900
B_4	68	189
B_5	84	291

[a] Viscosity determined according to ASTM D1646. C_2 = ethylene content, ML(1 + 4)125°C = mooney viscosity.

compounded but unvulcanized blends from the components in Table 34.1. The blends in Fig. 34.2 are different in terms of molecular-weight distribution, while in Fig. 34.3, they are different in terms of composition and crystallinity. Dispersion in the composition and the molecular weight of the miscible blends leads to improvements in tensile strength. This effect is expected, since both the introduction of crystallinity and the addition of the higher molecular-weight fraction slow down the viscous relaxation, leading to a higher tensile strength at ambient conditions. This dispersion in

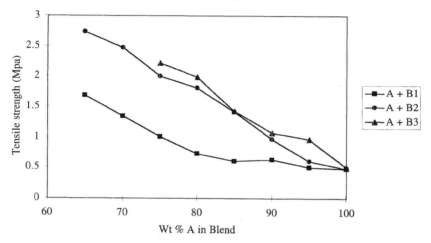

Figure 34.2 Variation in tensile strengh of unvulcanized, compounded blends of ethylene–propylene copolymer due to differences in molecular-weight distribution.

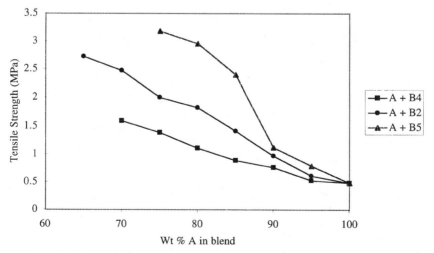

Figure 34.3 Variation in tensile strength of unvulcanized, compounded blends of ethylene–propylene copolymer due to differences in composition and crystallinity distribution.

molecular weight and crystallinity is also apparent in other viscoelastic properties, such as peel adhesion. Figures 34.4 and 34.5 show self-adhesion, measured by the force needed for the failure of a spliced portion, in blends identical to these shown in Figs. 34.2 and 34.3, respectively. Adhesion increases with increasing molecular weight and compositional dispersity of the blend. This relationship is shown by the comparative data for $B1$ and $B2$ in Fig. 34.4 and $B4$ and $B2$ in Fig. 34.5. Further increases in dispersion in molecular weight ($B3$ in Fig. 34.4) or composition ($B5$ in Fig. 34.5) are detrimental. Adhesion is promoted by intermingling of chains at the interface for long contact times and at low shear rates and by a resistance to

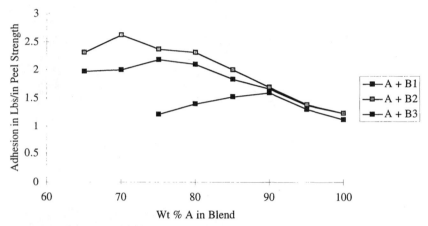

Figure 34.4 Variation in peel adhesion of unvulcanized, compounded blends of ethylene–propylene copolymer due to differences in molecular-weight distribution.

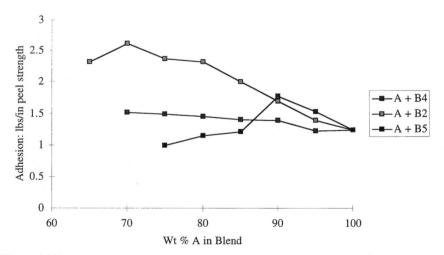

Figure 34.5 Variation in peel adhesion of unvulcanized, compounded blends of ethylene–propylene copolymer due to differences in composition and crystallinity distribution.

disentanglement during failure over a much shorter time and at higher shear rates. Small increases in the cystallinity and molecular-weight dispersion by blending promote adhesion by slowing down the latter without having a substantial effect on the former. Further increases in either factor severely retard the intermingling of chains necessary for the self-adhesion.

Nonuniform vulcanization networks in miscible blends of elastomers have a strong effect on tensile strength and elongation. These networks have an intermolecular distribution in cross-link density and are composed of different concentrations of cross-linkable sites in the components of the blend. Differences in the level of the enchained diene (5-ethylidene-2-norbornene) for EPDM copolymers [25, 26] or differences in the level of the vulcanizable chain-end unsaturation for siloxane polymers define these blends [27]. Nonuniform networks are most common in blends with a distribution of molecular weights greater than the most probable molecular weight, where the cross-link density is lower for the components with the higher molecular weights. Tensile elongation of these vulcanized blends leads to an increase in both the elongation and the tensile strength at high elongation, as compared to vulcanizates of similar viscosity having a uniform network. In particular, nonuniform networks display a nonlinear increase in the tensile modulus at high elongation. This nonlinearity is due to a nonaffine deformation of the network at the high elongation, which continually reallocates the stress during elongation to the lightly cross-linked component of the blend, which is most able to accommodate the strain. The large contribution from stress-induced crystallization (as, for example, in natural rubber) is unimportant, since, the effect persists at elevated temperatures.

Blends of polymer A and various amounts of polymer B as shown in Table 34.2, were blended in hexane solution, compounded, and vulcanized. Both polymers were amorphous, and the A polymers differed in terms of molecular weight and contained approximately 3% of vulcanizable diene (ENB). The B polymer was much lower in the amount of diene and had 0.9% ENB. The tensile strengths of the blends derived from all of the A polymers with varying amounts of the B polymers are shown in Fig. 34.6. In all cases where small amounts ($< 25\%$) of B were included in the blends, the tensile strength was higher for the blend than for the parent polymer A,

Table 34.2 Ethylene–Propylene Blend Components Differing in Cross-Link Density[a]

Sample	Composition of C2, wt%	Composition of ENB, wt%	Viscosity, ML(1 + 4) 125°C
A_1	57.0	3.2	20
A_2	60.2	2.9	32
A_3	60.3	2.8	41
A_4	59.4	2.6	51
A_5	60.5	3.2	67
B	64	0.9	2100

[a]Viscosity determined according to ASTM D1646.

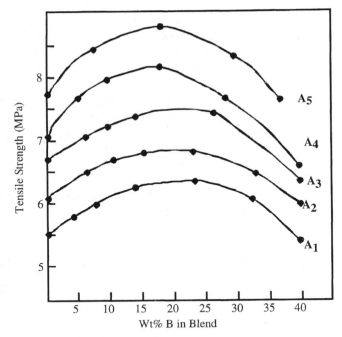

Figure 34.6   Tensile strength of the blends in Table 34.2 that differ in cross-link density.

even though the extent of vulcanization was lower in the blend than in the blend component A, due to reduction in the total amount of diene.

Similar results have been shown for miscible blends of SBR [28a, c] and NBR [28b] containing different styrene and acrylonitrile contents. In these experiments, it was difficult to determine the exact degree of inhomogeniety in molecular weight and composition for all of the components in the blend. In addition to differences in molecular weight and composition, blends of these two elastomers have different T_g's and solubility parameters, due to changes in the styrene content for SBR and acrylonitrile content for NBR. As a result, there has been some careful work to lessen or eliminate the effects of differences in all of these parameters in miscible blends in these systems [28d–g].

E. Distinct Elastomers

1. *IR–BR Blends*   Blends of 1,4-IR with 1,2-BR are unique, since they are misci-ble, chemically distinct, high-molecular-weight homopolymers [29a–e], even though there are no dipole or specific interactions, between the components [29c]. In blends containing less than 15% 1,4 isomers in 1,2-BR, no phase separation with IR has been observed. Infrared analysis of miscible mixtures confirms a sim-ple van der Waals interaction [29a] in these blends. As the concentration of 1,4 units in the BR increases, there is a decrease in miscibility with IR. This suggests equal polarizability between the monomer units of the 1,2-BR and the 1,4 IR.

The interaction parameter measured for blends of IR with BR of varying 1,2 and 1,4 geometry indicates that the exchange enthalpy between IR and BR becomes more endothermic as the concentration of 1,4 units increases [29b, d]. Density measurements for IR blended with 1,4-BR and 1,2-BR indicate additive volumes for blends with 1,2 BR, while there is a negative excess volume for blends with 1,4 BR. This indicates structural differences between 1,4-BR and IR. Differences in structure are apparent from the thermal expansion coefficients of the components. The difference in the thermal expansion coefficients of IR and BR is greatly diminished as the 1,2 content of the latter increases. Blends of syndiotactic 1,2-BR with IR have been found to be phase separated at molecular weights for which the corresponding atactic 1,2-BR blend with IR would be miscible.

2. Epoxidized PI–CPE Blends   The intramolecular epoxidation of 25 mol% of PI leads to miscible blends with CPE containing 25 wt% chlorine. At higher levels of epoxidation (e.g., 50 mol%), phase-separated blends were obtained, unless the chlorine content of the polyethylene was increased [30a, b]. The origin of the miscibility is the specific interactions involving the oxirane of the epoxy PI ring with the chlorine.

F. Reactive Elastomers

The formation of miscible blends of elastomers by mutual chemical reaction of the blend components has been explored most successfully by Coleman et al. [31a–d]. Chemical reaction in a miscible blend provides the excess negative free energy needed to compensate for the unlike $i-j$ contacts between dissimilar elastomers. Miscible blends were formed by stabilizing weak hydrogen bonding between a 2,3-dimethyl butadiene-4-vinyl phenol copolymer and carbonyl groups of polymers such as EVA and poly(alkyl methacrylates). Phase diagrams for miscible and immiscible blends were obtained for a number of EVAs of different vinyl acetate contents and alkyl methacrylates of varying alkyl-chain lengths. The extent of the hydrogen-bonding stabilization was modulated by the concentration of the phenol residues in the unsaturated copolymer. The formation of the miscible blends was inferred from the changes in the infrared spectrum of the carbonyl group, as well as from optical observation of miscibility. The choice of the stabilizing chemical bond between the elastomers is important. A stable, localized chemical bond leads to the loss of viscoelastic properties characteristic of an unvulcanized elastomer. Weak hydrogen bonding is fluxional and allows flow in response to stress in the elastomer blend. Steric crowding of the phenol by introducing alkyl groups in the 2 and 6 positions in the aromatic ring leads to greater miscibility with EVA and alkyl methacrylates. This effect is believed to be due principally to increased intermolecular hydrogen bonding between the dissimilar elastomers in the hindered system, at the expense of sterically hindered intraphase hydrogen bonding in the butadiene–phenol copolymer.

III. IMMISCIBLE ELASTOMER BLENDS

A. Formation

Immiscible blends can be made from any two dissimilar elastomers. They can be made by *in situ* copolymerization, blending polymer solutions or suspensions, and mechanical mixing. These methods have been reviewed by Corish [5]. In contrast to blends of other polymers, elastomers can be either preblended or phase mixed with each other [32]. Preblended systems are those in which the compounding ingredients (e.g., fillers, plasticizers, and curatives) are added simultaneously to a mixture of the immiscible polymers. In phase-mixed blends, ingredients are added separately to each of the individual polymers, in separate compounding operations. The compounded elastomers are then blended together. Phase-mixed blends provide a greater certainty of the initial interphase location of fillers, plasticizers, and curatives than do the preblends.

Preblending is the one of the most commonly used procedures for mixing elastomer blends. The blend components are usually mixed for a short time prior to addition of the compounding ingredients. Latex [33a–d] and solution mixing [33e] have both been employed for the preparation of elastomer preblends. However, the advantages in terms of uniformity of blend for these methods compared to the advantages to blending bulk polymers, are small. In phase mixing, separate compounds of the individual elastomers are mixed, in order to locate specific types and amounts of curing agents [34a–c] or fillers [34d–i] in the separate phases. In initial studies on phase blending, NR and BR masterbatches at different carbon-black loadings were mixed with *cis*-1,4-BR to produce tire-tread compounds with variations in carbon-black phase distributions. Extensive studies on phase-blended NR, BR, and SBR were carried out by Hess et al. [34e, h, i]. It is important to point out that phase blending provides the certainty only of the *initial* location of the polymer additives, since fillers, curatives, and process oils migrate across phase boundaries during subsequent operations.

B. Kinetics of Blend Morphology

The kinetics of the formation of blend phase morphology are dominated by two competitive factors. First, the viscoelastic flow of elastomers, the presence of plasticizers, the long resting times (days) between mixing and cross-linking, and the low-shear processing steps (e.g., calendering) are ideal for the development of the equilibrium morphology. However, the sluggish diffusion of high-molecular-weight elastomers, the presence of particulate fillers, and the ultimate cross-linking generate persistent nonequilibrium morphologies. Tokita [35a] and Avergoupolous et al. [35b] have studied the mixing of EPDM with the diene rubbers—NR or BR—as a function of relative viscosity. Smaller domain sizes of the dispersed phase were achieved with more intensive mixing and a diminishing concentration of the dispersed phase. Extensive mixing studies on EPDM–BR blends of widely varying rheological properties indicate that the best dispersion was obtained for elastomers with similar viscosities under the conditions of mixing.

The detailed morphology of elastomer blends depends on (a) the mixing procedure, (b) the rheology of the components of the blend, and (c) the interfacial energy. The usual elastomer blend is a dispersion of one component in a matrix of the other. As with other polymer blends, the elastomer of lower viscosity tends to be the continuous phase [36a, b]. Co-continuous blend morphology is observed only for elastomers with similar viscosities. The viscoelastic forces developed during the formation of the blend from two rheologically dissimilar elastomers are principally responsible for the morphology. Interfacial tension due to chemical differences between the elastomers are less important [36c]. Similar results have been reported by Hess et al. [36d] for carbon-black-filled elastomers. In studies with phase-mixed NR–BR and SBR–BR blends, the dispersion of the BR phase in the NR or SBR matrix was poorer when the BR phase was much higher in viscosity and slowly improved when the viscosity of BR was lowered [34c].

C. Analysis

1. *Microscopy* Phase-contrast light microscopy has been applied extensively to analyses of unfilled binary elastomer combinations. This method is based on differences in the refractive indexes of the polymers and has been reviewed by Kruse [37a]. Callan et al. [37b–d] have shown the versatility of the method for a wide range of binary blends containing NR, SBR, BR, CR, NBR, EPDM, IIR, and CIIR. They also employed the first image-analysis measurements to classify the compatibility of different blends by using phase-contrast images. The results of these experiments are shown in Table 34.3, which lists the measured areas of the dispersed phase in more than 50 combinations of Banbury-mixed 75/25 binary blends containing eight different elastomers. IIR–CIIR and SBR–BR blends are excluded, since the contrast was low. These blend proportions were used to assure that there would be a dispersed phase, rather than a co-continuous mixture. It can be seen that NBR produced the greatest heterogeneity in all blends except those with CR. The phase-contrast method is limited by low resolution, occasional poor contrast with certain blends, distortion of the polymer phases from swelling, and, most importantly, inapplica-

Table 34.3   Average Areas, in μm^2, of the Dispersed Phase in 75/25 Pure Elastomer Blends

Disperse phase 25%	Matrix (75%)							
	NR	CR	BR	SBR	NBR	EPDM	IIR	CIIR
NR	–	45	1.5	1.2	300	1.5	2.0	3.2
CR	35	–	4.0	2.5	1.5	25	20	15
BR	0.7	4.5	–	–	15	2.2	2.1	2.5
SBR	0.5	2.7	–	–	20	2.1	12	10
NBR	400	1.3	17	30	–	250	100	225
EPDM	3.5	75	2.8	2.6	225	–	2.0	1.5
IIR	3.0	15	3.0	4.2	75	1.0		
CIIR	2.2	25	2.3	2.5	85	1.2		

bility to carbon-black-filled systems. Carbon black absorbs light and eliminates the phase-contrast mechanism. However, this method is applicable if the dispersed phase does not retain any of the filler, due to migration to the continuous phase.

Transmission electron microscopy (TEM) (see Chapter 8) is applicable to both filled and unfilled elastomer blends. However, for most elastomer combinations, there is no contrast between the polymer phases in a transmission electron microscope. If the polymers differ significantly in unsaturation, osmium tetroxide (OsO_4) or ruthenium tetroxide (RuO_4) staining is the best method for obtaining contrast. The metal oxides selectively oxidize the unsaturation and the location of the metal atoms in the phase, with greater amounts of unsaturation providing the electron-density contrast for TEM. Andrews [38a] and Ban et al. [38b] employed osmium tetroxide staining to study the morphology of NR–EPDM blends. Contrast for TEM analysis of elastomer blends of diene rubbers is achieved through the sulfur-hardening method developed by Roninger [39a] and utilized by Smith and Andries [39b]. In this method, small rubber specimens are immersed in a molten mixture of sulfur, an accelerator, and zinc stearate. There is selective absorption of the zinc salts into the SBR phase, which renders the SBR phase darker than the BR phase in a transmission electron microscopy. The technique of pyrolytic etching was developed by Hess and Chirico [34i] and also used by Shehuplesov et al. [40] to study blends of NR with BR and SBR. This etching procedure utilizes differences in the thermal degradation peaks for these polymers. An extension of the Roninger method, developed by Kresge [41], uses a solvent-extraction procedure for preparing specimens of blends for SEM analysis. With uncured blends of SBR and EPM, the SBR phase was cross-linked with sulfur monochloride, and the un-cross-linked EPM phase was then extracted with n-heptane [41].

Scanning electron microscopy (SEM) involves simpler specimen preparations than does TEM. Both OsO_4 and RuO_4 staining techniques work with SEM and can be applied to both bulk specimens and films. Contrast is achieved by imaging with backscattered electrons, wherein emission increases in the direction of the elements with a higher atomic number. Therefore, the OsO_4-stained phase in a blend appears brighter with SEM than does the unstained polymer phase; the opposite is the case for TEM.

Atomic-force microscopy is an extension of scanning tunneling microscope (STM) and has the potential for atomic resolution. The atomic-force microscope [42a–d] acts as a miniature surface profilometer and provides topographical images. Lateral-force microscopy (LFM) is based on the frictional resistance of the probe as it scans across the sample. The technique of force modulation creates images that are based on variations in the elastic modulus at the surface. An alternative, tapping-mode image of the probe is based on the damping of the tip of the probe as it oscillates while tracking across the surface. This method of tracking produces higher resolution, because there is no lateral-force interaction of the tip of the probe with the polymer surface. The potential advantages of these techniques include higher resolution, simplicity of specimen preparation, and greater versatility in varying the mechanisms for achieving image contrast. Digital-image analysis has expedited particle-size analysis from micrographs [42e–h] at resolution limits of a few angstroms.

2. *Glass Transition Temperature*   The most common method of estimating the degree of homogeneity in elastomer blends is the measurement of the temperature of the transition from rubber to glass (see Chapter 10) [43]. For elastomers, this transition usually occurs at subambient temperatures. However, measurements of the glass transition temperature do not provide any information on blend morphology. The observation of distinct transitions, corresponding to the respective components of the blend, indicates the existence of a multiphase structure. A potential source of error is that vulcanization tends to raise the T_g of the blend, due to restricted motion of the chains; this increase in T_g might be interpreted as an indication of miscibility. Usually, calorimetry or dynamic mechanical measurements are used to measure T_g. Nuclear magnetic resonance and dielectric response are less commonly used. Note that the T_g methods for measuring T_g are misleading when the elastomers are nearly miscible. When the domain size is small, the thermal or mechanical response in the determination of T_g is not sensitive to the heterogeneous structure. For example, when the domains of a blend of NR and polypentenamer were 5 to 10 nm in diameter, dynamic mechanical measurements revealed only a single glass transition, with a temperature between those of the pure components [44a]. There is also evidence suggesting that when the interfacial energy for a pair of polymers (e.g., NR and 1,4-BR [44b] or BR and SBR [44c–e]) is small the intermediate interfacial region developed during vulcanization can cause disappearance of the expected distinct glass transitions.

3. *Magnetic Resonance Imaging*   NMR imaging of solids has been used to characterize the phase sizes with a spatial resolution of less than 50 µm in immiscible mixtures containing polybutadiene. Butera et al. [44f] have reported improvement over the ^{13}C-NMR method by using magic-angle spinning (MAS) in polymer blends containing a deuterated and a protonated polymer. More recently, Koenig et al. [44g–j] used NMR for the determination of vulcanization efficiencies in elastomer blends.

4. *Light, X-Ray, and Neutron Scattering*   The angular dependence of elastic scattering from X-ray, light, or neutrons provides morphological information, since scattering results from heterogeneities in the structure of the irradiated material. (See Chapters 12 and 28.) In unfilled, multiphase polymer blends, the angular distribution of the scattering reflects the size and spatial distribution of the phases, both of which can be calculated using published methods (see Chapter 11) [45a,b]. Light-scattering techniques [45c, d] have been applied to the study of polymer blend solutions, but they have limitations in the analysis of solid systems, due to difficulties with multiple scattering and sample transparency. For elastomer blends, much of the work on light scattering has focused on block copolymers [45e, f]. Phase separation in elastomer blends is sometimes judged from the appearance of enhanced light scattering (i.e., cloud-point measurements). Measurements of small-angle scattering by X-rays (SAXS) and by neutrons (SANS) (see Chapters 11 and 12) have both been applied to the study of the homogeneity of multicomponent polymer systems. Glatter and Kratky [46] have discussed SAXS and its analysis of heterophase polymer systems.

Roland and Bohm [47a] used both SAXS and SANS in conjunction with TEM to assess the size of butadiene domains (5% by weight) in a CR matrix. These three different methods indicated comparable size distributions, with the results of SAXS being somewhat closer than the results of SANS to the results of TEM. Roland et al. used different masterbatches containing deuterated and protonated butadiene, respectively. These masterbatches were each blended on a two-roll mill to study the coalescence of the polybutadiene phase. Coalescence, measured by SANS, was minimized by low-temperature mixing or by significantly increasing the molecular weight of the dispersed or continuous phase to create a greater disparity in viscosity. The substitution of deuterium for hydrogen in a polymer chain provides a contrast mechanism for neutron scattering without altering the chemistry of the system. (See Chapter 12.) Cooper and Miller [47b] used SANS to examine chain conformation in bulk polymer samples. Some preliminary SANS work on IR blended with deuterated 1,2-BR has been reported [47c, d]. All scattering techniques are affected by the presence of fillers, which lead to more extensive scattering than does the phase morphology. This relationship is a severe restriction on the wide applicability of these techniques to most elastomers.

D. Interphase Distribution of Filler, Curatives, and Plasticizers

At ambient and processing temperatures, elastomers are viscous fluids with persistent transport phenomena. In immiscible blends, these transport phenomena lead to changes in the size and shape of the elastomer phases and migration of the fillers, plasticizers, and curatives from one phase to another. These changes are accelerated by processing and plasticization, but are retarded by the ultimate vulcanization. Retention of the favorable properties of a metastable blend, which is often attained only at a select interphase morphology and filler–plasticizer distribution, thus requires careful control of both the processing and the vulcanization procedures.

1. *Curative and Plasticizer Migration in Elastomer Blends*   During mixing, curatives are initially located within the continuous phase [48a]. Since the curatives dissolve in the elastomer, curative migration across phase boundaries can occur [48b–e]. Owing to the higher solubility of sulfur in elastomers containing diene or styrene groups and the greater affinity of many accelerators for polar rubbers, large differences in cross-link density of the different phases result on vulcanization. Further, an increased rate of vulcanization in the diene- or styrene-containing elastomers can cause depletion of the curatives in this component, leading to even greater curative migration (by Le Châteliers principle). For most elastomer blends, these effects are in concert, and the result is a large cure imbalance [48f] between the phases. Curative migration results not only in different cross-link densities in the distinct phases, but also in a difference in conentration of cross links between the bulk and the phase interface. When a monomeric component (a curative or an accelerator) soluble in both polymers is present in a blend, the interface may be richer in this component than are the bulk phases. This results from the dilution in energetically unfavorable interactions between dissimilar polymer chains at the interface by virtue of the presence of

Table 34.4   Curative Diffusion Coefficients in Elastomer Blends

Curative	From	To	$D \times 10^7$ cm^2/s
Accelerator (TDDC)	IIR	BR	12.66
		EPDM	1.09
		CR	1.08
		SBR	0.58
		NR	0.70
Sulfur	IIR	SBR	4.73
		SBR & 50 PHR N700 CB	17.2
		NR	2.82

the low-molecular-weight plasticizers and curatives [48g]. In addition, the boundary layers of the two elastomers, adjacent to the interface, can be different than the bulk, since the greatest migration of curatives and plasticizers occurs in proximity to this region. The transfer of curatives to one phase at the expense of the other creates a morphology wherein the boundary layers separate a curative-depleted, slightly cross-linked elastomer near a curative-enriched, tightly cross-linked elastomer. Obviously, these problems are particularly significant when the elastomer components are dissimilar. Phase blending of the rubbers can improve the blends cross-link distribution, [48a] although such blends tend to prevulcanize.

Amerongen [49a] reviewed curative migration in heterophasic elastomer blends based on optical and radiochemical analyses. A later, more detailed work by Gardiner [49b–d] used optical analysis to study curative diffusion across the boundaries of elastomer blends consisting of binary combination of polymers of CIIR, IIR, EPDM, CR, SBR, BR, and NR. Gardiner measured a diffusion gradient for the concentration change as a function of distance and time. His measurements for the diffusion of accelerator and sulfur from IIR to other elastomers are listed in Table 34.4. The optical analyses were confirmed by extraction analyses of the phase-blended elastomers with solvent to determine the amount of un-cross-linked material. The lower functionality elastomer was always the extractable polymer, irrespective of the initial location of the curatives. An important finding was that curative migration is related to diffusion during curing and not to transfer during mixing.

2. Cure Compatibility   Adequate properties in a vulcanized rubber blend depends on the covulcanization across the phases. Covulcanization is the formation of a single network structure including cross-linked macromolecules of both polymers. The degree of vulcanization is at similar levels in both phases, with cross-linking across the phase interfaces. The formation of the network depends on the presence of curatives, fillers, and plasticizers in each of the phases. In the absence of covulcanization, either one of the phases or the interface can be substantially unvulcanized. Shershnev [50a] and Rehner and Wei [50b] have summarized the experimental requirements for covulcanization of the components of elastomer blends as follows:

1. Phase blending followed by very short, high-temperature cure cycles [50c];

2. Vulcanization agents chemically bound to the elastomer [50d–f];
3. Accelerators similar with solubility parameters similar to those of less reactive elastomers [50g–j];
4. Insoluble vulcanizing systems that cannot migrate;
5. Nonpolar vulcanizing agents that distribute uniformly and have similar reactivities towards different elastomers (e.g., peroxides and reactive resins).

Gardiner [49b–d], Woods, and Davidson [51] have improved the covulcanization of EPDM–NR and IIR–NR blends by using an insoluble lead-based curative for the less unsaturated elastomer. The insoluble curative retards the diffusion of curatives. The improved covulcanization of these blends generates properties of the blend as a compositionally weighted average of the two individual elastomers. The blends with soluble curatives from the same elastomers were much inferior to the blends with insoluble curatives. Coran [50f] covulcanized EPDM–NR blends by modifying the EPDM with maleic anhydride. The modified EPDM can be cross-linked with zinc oxide to form an ionic cross-link network. This type of cross-linking is neither migratory nor competitive with the sulfur system for NR. The superior mechanical properties of this blend, compared with the blend with pure EPDM, indicate the occurrence of covulcanization. Maleic anhydride modification has also been applied to IIR and EPDM by Suma et al. [52] for blends with NR. The efficacy of this procedure was determined by solvent extraction.

Bauer and Dudley [53] improved the covulcanization of CIIR–BR blends by utilizing a combination of different accelerators in a sulfur cure system. The curatives were added separately to the individual polymers prior to blending, and improvements were noted through changes in the damping coefficients. The migration of curatives was investigated by Bhowmick and De [50c] in binary blends of NR, BR, and SBR. Cure rates decreased in that order, and the migration of curatives was the greatest for NR–SBR blends.

The state of cure of the phases in a blend can be determined from changes in the magnitude of the damping peaks [54a] and from freezing-point-depression measurements on swollen networks. Honiball and McGill [54b] have used the freezing-point depression of solvent-swollen NR–BR blends to determine the cross-link densities of the individual polymer phases. The freezing-point depression of a solvent in a swollen vulcanizate is dependent on the volume fraction of solvent in the elastomer phase. The equilibrium swelling is restricted by the cross-link density. This restriction enables the freezing point of a solvent in a swollen vulcanizate to be correlated with the volume fraction of rubber. The Flory–Rehner equation relates the volume fraction to the cross-link density. The swelling measurements provide a measure of the average cross-link density of the blend, while the freezing-point depression characterizes the degree of cross-linking in the more lightly cross-linked phase. In 50/50 NR–BR blends, the cross-link density of BR varied from about two to four times higher than that of NR [54c].

Zapp [55] developed a differential swelling procedure to determine the extent of interfacial bonding in blends of CIIR with BR and SBR. If interfacial bonds are formed, then the lightly swollen dispersed phase in a blend restricts the swelling of the continuous phase, so that it falls below the rule of mixtures. If there is no swelling restriction, due to low interfacial bonding, the observed values for the blends are near the additive line. Interfacial bonding in the CIIR–SBR and CIIR–BR blends was demonstrated.

Cross-linking by ionizing radiation avoids complications with curative distribution, although uniform cross-link densities are not necessarily achieved. Generally, the extent of radiation cross-linking in a given component is not significantly affected by the nature of the other component, at least for blends of the typical hydrocarbon rubbers [56].

3. *Interphase Filler Transfer*   The distribution of fillers (e.g., carbon black and silica) and various processing aids in heterogeneous elastomer blends can be nonuniform. Interphase transfer of fillers has been observed in blends of both diene and saturated elastomers. This migration is due to a greater solvation between a filler and one of the polymers in the melt-mixed blends. Hess, Scott, and Callan [57a] showed significant transfer of carbon black from NR to BR and silica from BR to NR in solution and latex blends. Subsequent studies [57b] indicated that heat history and viscosity of the master batch, as well as polymer unsaturation, had a major influence on the amount of transfer. There was no indication of significant transfer to BR or SBR from typical carbon black–NR Banbury melt mixes. This finding was confirmed by Sircar and Lamond [57c] for binary blends of SBR, NR, and BR. However, Cotton and Murphy [57d] have reported contradictory results for similar blends.

Extensive carbon-black transfer from an elastomer with low unsaturation (e.g., IIR and EPDM) to elastomers with high unsaturation (e.g., diene) occurs regardless of the mixing procedure used for the master batch [57a]. This is seen in blends of EPDM or IIR with NR or SBR. During mechanical mixing, chemisorption of the unsaturated polymer onto the carbon black occurs, thus preventing any subsequent transfer of the carbon black. In all instances, the carbon black was located almost entirely in the NR phase when it was added to the preblended elastomers. Chemisorption is retarded and has not been observed in solution-blending experiments.

Transfer and preferential retention of carbon black in one phase does not occur in blends of saturated hydrocarbons. Similar interphase transfer of oils, resins, and other compounding ingredients, due to different affinities for the components of a blend, has been seen [58a–c]. The transfer can be retarded if the carbon black is partially graphitized by heating in an inert atmosphere to 1500°C. Such carbon blacks behave as inert fillers for polymer interaction. In the phase mixes for 50/50 EPDM–NR blends, some carbon black did appear to migrate from EPDM to NR. Partially graphitized carbon black transferred from either master batch into the other polymer, thereby resulting in a more uniform phase distribution. Similar results were obtained for blends of IIR with NR and in blends of CIIR and NR [57c], wherein the normal carbon black was almost completely transferred into the NR phase. The transfer of partially graphitized carbon black was only slightly less complete. SBR has a greater affinity for carbon black than does NR. In tertiary blends of EPDM with a mixture of NR and SBR, there was extensive transfer of carbon black, either active or graphitized, to the unsaturated phase. Transfer of carbon black from EPDM to NBR can be

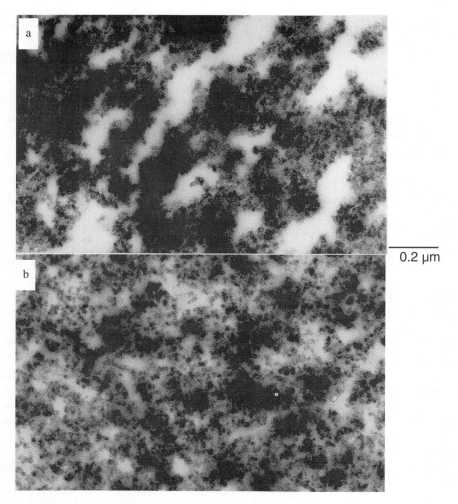

0.2 µm

Figure 34.7 Micrographs of an N234 carbon-black distribution in 70:30 blend of NBR and EPR. (a) The EPR is a copolymer of 42 mol% ethylene. (b) The EPR has 43 mol% ethylene and 0.9 mol% primary-amine functionality.

reversed if the EPDM is functionalized with a primary amine [58d]. Amines are tenaciously chemisorbed to the surface of carbon black, and the functionalization allows the EPDM phase to retain significant amounts of filler. The carbon-black distribution is shown in Fig. 34.7a for a blend of EPDM with NBR and may be compared with Fig. 34.7b for the blend of the amine-functionalized EPDM with NBR.

Hess and et al. [57a, b] used the differential swelling technique (*vide infra*) for a ranking of the relative affinity of carbon black to various elastomers. The data was obtained for 50/50 binary blends, and the carbon black was added to about 25 binary combinations of elastomer. The retention of carbon black in each phase was

Table 34.5 Effect of Type of Carbon Black on its Distribution in 50:50 Natural-Rubber–SBR Blends Containing 45 Phr Black

Carbon-Black Type	Surface Area (CTAB) m^2/g	Bound Rubber, %	Carbon-Black Loading, phr	
			NR	SBR
N560	43	17.0	41.2	48.8
N347	87	28.0	33.9	56.1
N339	95	30.2	31.1	58.9
N234	120	30.4	30.0	60.0
Vulcan (10H)	136	33.1	27.2	62.8
N326	84	22.5	31.9	68.1
N330	81	25.0	34.0	66.0

estimated from TEM micrographs of the blend. The ranking of affinity to carbon black for different elastomers was found to be in the order SBR > BR, CR, NBR > NR > EPDM > CIIR, IIR. In a complementary experiment, Cotton and Murphy [57d] have determined the distribution of carbon black for seven different carbon blacks in preblended SBR–BR and SBR–NR blends. The data are shown in Table 34.5. The carbon blacks differ in terms of surface structure and size: They ranged in surface area (CTAB) from 43 to 136 m^2/g. For all cases of SBR–NR blends, the carbon black was preferentially located in the SBR phase. The extent of the transfer strongly correlated with surface area, i.e., from 54% carbon black in the SBR phase at the lowest surface area to 70% for the highest. The transfer in SBR–BR blends [32e] was dependent on the molecular weights of the polymeric components, with the higher molecular-weight polymer being preferentially adsorbed by the carbon black [57f, 59]. For components whose molecular weights were equal, a slight preference by carbon black for the SBR phase was apparent. There were no significant variations in phase distribution as a function of the surface area of carbon black.

E. Analysis of Interphase Transfer

1. Microscopy Electron microscopy is a common technique for determining the filler distribution in heterogeneous elastomer blends. For TEM and SEM analyses, however poorly defined interphase contrast is a major problem. The solution to this is extensive staining to improve interphase contrast. Shong Lin [60a] has used scanning auger microscopy (SAM) for the analysis of elastomers. Finally, Dias et al. [60b] have used time-of-flight secondary mass spectrometry (TOF-SIMS) to determine simultaneously the morphology and the curative diffusion of BIMS–diene elastomer blends.

2. Differential Swelling The method of differential swelling of thin sections of blends of EPDM and IIR was first utilized by Callan, Topcik, and Ford [61a]. In blends with EPDM, the IIR phase absorbs more solvent; therefore, the IIR domains are thinner and appear lighter in the transmission electron microscope. Contrast is further improved by the degradation and removal of the IIR by electron bombardment during microscopy. Hess, Marsh, and Eckert [61b] applied the same method of differential swelling to the analysis of carbon-black distribution at low filler loadings. The method was used principally for IR–BR blends. A complication in this procedure was found by Marsh et al. [61c], who showed that the swelling is affected by the presence of substantial amounts of carbon black and that the technique is of limited utility for practical compounds that are highly loaded with carbon black. Wang et al. [61d] have improved the technique of differential swelling by swelling separately with two solvents, one for each of the phases. Hess and et al, [57a, b] also improved the technique of differential swelling by polymerizing a mixture of imbibed butyl methacrylate and methyl methacrylate with peroxides inside the thin specimens of the blend. This method allows for easier sectioning of the blend.

3. Staining   Staining with volatile reactive metal oxides, namely, OsO_4 and RuO_4, is the preferred method for achieving interphase contrast for TEM analyses. It is applicable to blends of elastomers with different degrees of unsaturation, such as NR–EPDM blends. In these blends, the carbon black is located predominantly in the unsaturated phase. Lee and Singleton [62] used the procedure of imbibing zinc and sulfur into the unsaturated phase in order to provide contrast to study carbon-black distribution in preblended and phase-blended SBR–BR mixtures.

4. Differential Pyrolysis   Differential pyrolysis is applicable to blends containing polymers with significantly different thermal degradation temperatures. It has been used to analyze the carbon-black distribution in NR–SBR and NR–BR blends [63a, b], since the NR phase is significantly less thermally stable and is removed in pyrolysis. Experiments with phase-blended polymers show that carbon black does not transfer during mixing from one highly unsaturated elastomer to another.

5. GC Analysis of Bound Rubber   Bound rubber is an elastomer that is insoluble in solvents, due to chemisorption onto the carbon black during mixing. It is extracted by swelling the unvulcanized polymer in a solvent for an extended period of time. Any soluble lower molecular-weight polymer that is not bound to the carbon black is removed. Analysis of this solution for elastomer content and composition reflects the distribution of carbon black in an elastomer blend. This method was first used by Callan, Hess, and Scott [57a] for a number of elastomer blends to study the transfer of carbon black to the unsaturated elastomers in blends of IIR with NR and SBR.

6. Mechanical Damping   The value of tan δ at the T_g of an elastometer is lower for a filled elastomer than for the pure elastomer [64a, b]. This difference is due to an increase in the dynamic elastic modulus of the filled compound for the higher temperature side of the T_g peak. The effect is governed by filler concentration and loading. Maiti, De, and Bhowmick [64c] have used this lowering of tan δ at T_g to estimate the distribution of filler in an immiscible elastomer blend. They studied the distribution of carbon black and silica in 50/50 blends of NR with epoxidized NR (ENR). A much greater change in tan δ was shown for ENR compared with

NR, thus indicating a higher proportion of the filler, either carbon black or silica, in ENR. At similar levels of filler, silica had a higher concentration than that of carbon black in ENR, indicating interaction between the epoxide groups of ENR and the silanol groups on the silica. On the other hand, carbon black had a greater affinity for NR in comparison to silica. The preferential location of silica in the ENR occurred irrespective of the polymer in which it was initially located. In studies of phase-blended elastomers, the final filler distribution was 70% of the carbon black in the ENR and 30% in the NR.

F. Compatibilization

Compatibilization of highly incompatible elastomers has been used only to a limited extent [65a–d]. Compatibilization is the addition of a minor amount of an interfacial agent and serves only to stabilize the extended surface of the dispersed phase in a finely dispersed morphology. The amount and composition of the interfacial agents are not designed to affect the bulk properties of either of the phases. Properties of elastomer blends are determined by intensive properties such as cohesive energy, cross-link densities, and chemisorption of fillers of the components; these properties are unaffected by the addition of compatibilizers. However, in binary blends of elastomers with large differences in solubility parameters, such as blends of polyolefin elastomers with polar elastomers, properties are dominated by the large domain size and the lack of interfacial adhesion. These elastomer blends are significantly improved by the addition of a compatibilizer.

Setua and White [66a, b] used chlorinated polyethylene (CM) as a compatibilizer to improve the homogeneity of binary and ternary blends of CR, NBR, and EPR. NBR–EPM and CR–EPDM blends homogenize more rapidly when small amounts of CM are added. The presence of the compatibilizer leads to reductions in both the time needed for mixing, observed by flow visualization, and the size of the domain of the dispersed phase, observed by SEM. Arjunan et al. [66c] used an ethylene acrylic acid copolymer and an EPR-g-acrylate as a compatibilizer for blends of EPDM and CR. The addition of the compatibilizer lead to a reduction in the phase size of the dispersed EPDM phase, as well as to an increase in the tensile tear strength of the blend. These changes was correlated to increased adhesion at the phase interface for the compatibilized blends. Surfactants with a oligomeric hydrocarbon end and a phenol, which is capable of hydrogen bonding to NBR, have been used as a compatibilizer in blends with EPR [66d, e]. TOR is also an effective compatibilizer for blends of EPDM and NBR [66f].

Intensive properties of the components of a blend that dominate vulcanizate properties of the elastomer are improved if the compatibilizer is the predominant fraction of the elastomer blend. Davison et al. [67a] described the formation of compatibilized blends of poly(alkyl acrylates) and preformed EPDM-g-acrylate, which, on vulcanization, are resistant to solvents. The acrylate-grafted EPDMs were copolymers of methyl, ethyl, or n-butyl acrylate. Each of these vulcanized blends has an excellent tensile strength and modulus, similar to that of a single elastomer. The solvent resistance increased with decreasing bulk of the alkyl group on the acrylate. A similar approach by Paddock [67b] led to the formation of useful compatibilized blends of EPDM and SBR, while Shedd [67c] used graft polymers of EPDM and acrylonitrile in intimate blends of NBR and EPDM. Datta et al. [58d] were able to generate the graft polymer *in situ* during compounding. A primary-amine-copolymerized EPDM blended with NBR to form a graft polymer by the amidine reaction of the amine with the nitrile. The grafting reaction was catalyzed by the presence of the Lewis acid phosphite plasticizer for the NBR phase. The compatibilizer promoted the formation of very small dispersed-phase domains. Figure 34.8a and b show micrographs of the dispersion of the EPDM and amine-functionalized EPDM, respectively, in the NBR matrix. In this case, the previously cited micrographs (see Fig. 34.7) also show the retention of carbon-black filler in the bulk of the EPDM phase. Vulcanization of the amine-functional EPDM and NBR blends with nonpolar peroxides that are expected to distribute equally in both phases led to blends with excellent solvent and temperature resistance.

G. Properties of Immiscible Blends

While true miscibility may not be required for elastomer properties, adhesion between the immiscible phases is required. Immiscible polymer blends that fulfill this criterion provide a significant opportunity to change the rheological, tensile, and wear properties of elastomer blends, compared with miscible blends.

1. *Processing*   Blends are often used to improve the processability of rubbers. This improvement may consist of either lowering the viscosity or producing a material less prone to melt fracture during flow. Secondary elastic effects, such as die swell, can also be affected by blending. Avgeropoulos et al. [68a] showed that the low-viscosity phase in a binary blend tends to become continuous. This effect is accelerated under shear as the morphology of the blend responds to the applied stresses. In the vicinity of a wall, the shear rates tend to be highest, and the component with the lower viscosity accumulates at the surface [68b]. Incorporation of only a few percent of EPDM into a fluoroelastomer or of PDMS into an SBR [68c] was found to reduce steady-state viscosities. This is because the component of lower viscosity resides at the interface. Hess, Swor, and Vegvari [68e,f] studied the properties of SBR–BR and NR–BR blends as a function of the composition (i.e., wt% BTR, wt% carbon black) and the type of carbon black. They correlated viscosity and die swell with the compositional parameters and concluded that the effects were due to an accumulation of filler in the SBR phase and a rise in viscosity of the BR phase on addition carbon black.

Much of the knowledge in this area is either derived from practical experience or anecdotal. Theoretical predictions of the viscosity of elastomer blends [68f, g] are of limited use, since the morphology of the inhomogeneous phase of a elastomer blend changes easily in response to applied stress and the nonuniform distribution of fillers and plasticizers in the phases responds to flow. These structural changes in elastomer blends under shear lead to anomalous rheological properties that are quite different from the expected average properties of pure components.

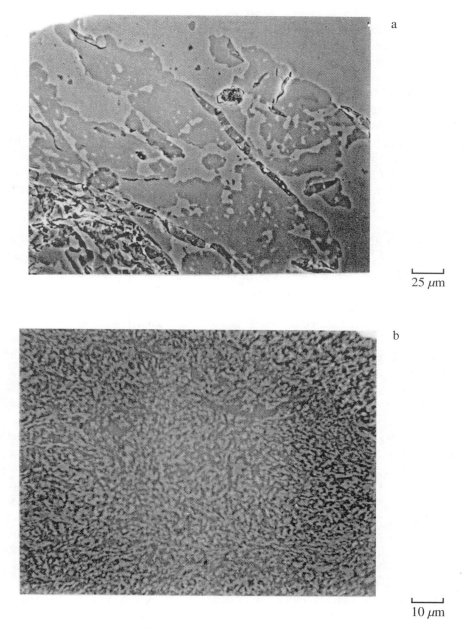

Figure 34.8   Micrographs of a dispersion of 70:30 blend of NBR and EP elastomer. (a) The EP elastomer is a copolymer of 42 mol% ethylene. (b) The EP elastomer has 43 mol% ethylene and 0.9 mol% primary-amine functionality.

2. *Modulus* In a heterogeneous blend, the details of the morphology generally do not exert much influence on the stress–strain tensile response. However, contrary to the expectation that the continuous phase would have more influence, the stress–strain response of unfilled EPDM–BR blends was found to be unaffected by a change in the BR domains from continuous to dispersed [69a]. Further, dependence of the modulus on domain size of the dispersed phase was not seen in blends of NR with either SBR or BR [69b] and in EPDM mixtures with BR [69c].

Carbon-black distribution has a profound effect on the modulus. Meinecke and Taftaf [69d] have shown that an increase in the nonuniformity of the distribution of this filler results in a lower modulus of the blend. The transfer of a portion of the carbon black from one phase to another lowers the modulus more than the increase in modulus of the phase with the higher concentration of carbon black. This effect is related to the nonlinear dependence of the modulus of rubber on carbon-black loading [69e, f].

3. *Tack and Adhesion* Adhesion is a surface phenomenon that is due to entangling of chains at the surface. Adhesion of different elastomers before (tack) and after vulcanization (cocure) can often be obtained only through blending the dissimilar elastomers with a single component. However, the composition of the surface and, thus, the characteristics of adhesion can be altered without using a high concentration of a particular elastomer in the blend. In addition, for filled elastomers, improvements in adhesion depend on the composition of the blend, the filler distribution, and the method of preparation. Roland and Bohm [23a,c] have related the adhesion of elastomers to their miscibility. Adhesion was reported for blends of NR with 1,2-BR (a miscible pair), but not for immiscible blends of NR with cis-1,4-BR.

Blends with components that differ in viscosity tend to have the rubber of lower viscosity concentrated at the surface during processing. Baranwal [70a] has obtained improved tack by the addition of NR to synthetic elastomers such that NR had the lower relative viscosity and was concentrated at the surface. A similar demonstration of the influence of relative viscosities on the distribution of components in a blend was given by Lee [70b]. The green strength of blends of NR and synthetic elastomers was improved by controlling the carbon-black distribution to create a NR-rich minor phase in the blend with a lower carbon-black content. This phase presumably the phase of lower viscosity in the blend. Nedden and coworkers [70c] have reported significantly improved values of green strength for NR compounds by using blends with polyoctanomer (TOR). Blends containing as little as 10% TOR, which acts as an interfacial agent, produced a 50% or more increase in green strength.

The most common procedure for increasing the tack and cocure of elastomer blends is to have a single elastomer as the predominant phase in each of the blends. Morrisey [70d] showed that the tack of dissimilar elastomers blended with NR improves monotonically with the amount of NR in the blend. Increasing the proportion of a single elastomer in both blends enables the ability of the plied surfaces to fuse together. When BR was blended with a BR containing grafted isopropylazodicarboxylate groups, a polymer that exhibits very high autoadhesion, high tack in the blend stocks was obtained only when the copolymer rubber constituted the continuous phase [70e].

This use of a single elastomer as a component of a blend is not limited to NR. In blends with chlorobutyl rubber, polychloroprene produces a larger increase in tack than does NR [71a]. An example of the use of blending to improve cocure adhesion is when highly unsaturated rubbers are bonded to elastomers with low unsaturation. Good adhesion between blends of IR or BR with chlorinated butyl rubber and blends of EPDM with butyl rubber was obtained only when the level of chlorinated butyl exceeded 75% [71b]. A reduction in the level of chlorinated butyl rubber and an increase in polyisoprene, on the other hand, gave superior adhesion to SBR. The adhesion of epichlorohydrin rubber to unsaturated rubbers was accomplished by blending the epichlorohydrin with 25–50 phr of polychloroprene [71a].

4. Hysteresis   A principal use of elastomer blends is in sidewalls of automotive tires. The reduction of hysteresis losses ("rolling resistance") is a principal target of design. Lower hysteresis in a single elastomer requires reduced carbon-black loading or increased cross-link density. These changes adversely affect other aspects of performance. Blending of elastomers provides a lower hysteresis with few of these adverse effects. The hysteresis of a filled elastomer containing zones of different carbon-black concentrations is lower than that of a uniformly filled elastomer. Creating blends of dissimilar elastomers, with differing abilities to retain carbon black, provides an easy experimental process to obtain this nonuniform carbon-black distribution. When the filler-depleted phase is the continuous phase, a low blend hysteresis may result [34i]. The hysteresis reduction accompanying a nonuniform distribution of carbon black is due to the nonlinear relationship between hysteresis and carbon-black loading. The effect of phase morphology on hysteresis properties appears to be less significant than that filler distribution.

Hess and Chirico [34i] studied the effects of interphase distribution and the aggregate-size distribution of the carbon blacks on the hysteresis of different 50/50 blends containing NR, SBR, and BR. In NR–SBR blends, higher loading (i.e., 75%) of carbon black in the NR phase reduced heat buildup and increased resilience. For NR–BR blends, heat buildup indicated a similar pattern of lower hysteresis loss with reduced carbon black in the BR phase. The blends with the largest difference in E'' between the polymeric components (i.e., NR–SBR and NR–BR) showed the greatest benefit with regard to hysteresis, with the carbon black in the polymer with the lowest loss modulus (i.e., NR).

5. Failure   Greater resistance to failure can result from the blending of elastomers, which can even lead to a performance of the blend that exceeds the performance of either pure component. An important aspect of the structure of a rubber blend is the nature of the interphase bonding. The mechanical integrity of an interphase cross-linked morphology usually leads to superior performance. In blends of SBR and chlorobutyl rubber, for example, an increase in fatigue life was obtained by the introduction of interphase cross-linking [71b]. Similarly, interfacial coupling improves the tensile strength of EPDM–silicone-rubber blends [73a].

The most detailed work in this area is concerned with blends containing NR and is directed toward the components of an automotive tire. Hess and Chirico [34i], Krakowski and Tinker [72a], and Lee [72b] have shown that for NR–SBR and NR–BR blends, maximizing the tear resistance requires the polymer of highest strength to be the continuous phase. This is requirement dictated mostly by the composition and the relative viscosities of the blend. A further improvement is achieved through the preferential location of carbon black in the continuous phase. For 50/50 NR–BR blends, the NR phase remains co-continuous, even with a high carbon-black content. Therefore, higher tear strength was obtained with a high carbon-black loading in the NR phase. With the nonuniform distribution of carbon black, greater tear resistance is found for 1,4-BR–NR and SBR–NR blends when the reinforcing filler is deposited principally in the continuous phase, which is the synthetic elastomer [72c]. This observation is consistent with the concept that the strength of the continuous phase increases the strength of the blend. Lee [72d] has also found that a higher carbon-black loading in the major (continuous) phase improves crack resistance. He stated that elongated domains for the dispersed phase, if oriented, would lead to anisotropic resistance to cracking. However, this simple model is sometimes invalid for blends with a co-continuous NR phase. Greater strength is obtained when the reinforcing filler is in the other component, since the ability of NR to strain crystallize provides a measure of self-reinforcement absent from other noncrystallizing rubbers.

There is improved abrasion resistance associated with a preferential carbon-black–BR phase distribution in blends of NR and BR and of SBR and BR. The first abrasion studies on the effects of carbon-black phase distribution in NR–BR blends were conducted by Hess, Scott, and Callan [34e], followed by Krakowski and Tinker [72a]. Resistance to tread wear was found to increase progressively with increasing carbon-black content in the BR phase, which was determined from TEM analyses. Tse et al. [72e] have shown in blends of dispersed BIMS in a BR matrix that failure due to fatigue can be retarded if the mean distance between the cross-links of the BIMS is less than 60 nm.

The incidence of cracking due to ozone attack has been investigated for NR–EPDM blends [73b–f]. Andrews [73g] showed that small zones of EPR in an EPR–NR blend provide a barrier that inhibits ozone-caused crack growth. Ambelang and coworkers [73h] found that it is important for EPM to have a small domain size in EPDM–SBR blends. Matthew [73b] has shown that carbon black improves the ozone resistance of NR–EPDM blends. An improvement was obtained in the blends with a balanced carbon-black phase obtained by phase inversion. This result is due to (a) better reinforcement of the EPDM phase and (b) the fact that carbon black in the EPDM expands the volume of that phase. In BIMS–BR blends, the ozone failure can be retarded by reducing the size of the BIMS dispersion [72e].

6. Transport The transport properties of polymer blends are of interest both in terms of the practical applications of blends and in terms of providing insight into the morphology of the blend. Measurement of the effect of blend composition on permeability in various rubbers has been described [73i]. The permeability of elastomer blends depends on the concentration of the continuous phase and the morphology of

Table 34.6   Elastomer Blends in Automotive Tires[a]

Component	Passenger Tires	Commercial-Vehicle Tires
Tread	SBR–BR	NR–BR or SBR–BR
Belt	NR	NR
Carcass	NR–SBR–BR	NR–BR
Sidewall	NR–BR or NR–SBR	NR–BR
Liner	NR–SBR–IIR	NR–IIR

[a] From [3].

the dispersed phase. Extended dispersed-phase structures, particularly when lying in a stacked or lamellar configuration, can lead to a reduced permeability, because of the more tortuous path that must be taken by penetrants [73j].

H. Applications

1. *Unsaturated Elastomer Blends*   The most commonly used blends of unsaturated elastomers are those used in various sections of automotive tires. Table 34.6 lists the important component of tires and the typical blends used for them. Much of the literature on elastomer blends reflects this important application. However, it is outside the scope of this chapter to discuss each of the applications of unsaturated elastomers. Instead, we have outlined most of the important principles used in the generation of such blends.

2. *Saturated and Unsaturated Elastomer Blends*   The use of blends of polyolefin elastomers, such as IIR and EPDM, as substantial components in blends of unsaturated elastomers is a rapidly developing area. Mouri et al. [74a] have compared the properties of EPDM–NR and BIMS–NR blends as sidewall components. In many of the applications, the saturated elastomer is considered to be a polymeric antioxidant for the diene rubber. It is believed that the higher molecular-weight polyolefins are better in these applications, due to limited interdiffusion and a more stable morphology. Some of the benefits in terms of the tensile properties and abrasion resistance of the blends may be due to the interdiffusion of high molecular weight chains of dissimilar elastomers across the phase interface. Significant advances have been made in modifying the structure of polyolefin elastomers to increase their compatibility with unsaturated elastomers. Tse et al. [72e] have shown that uncompatibilized blends of saturated elastomers and unsaturated elastomers are possible if the former contains substantial amounts (> 12%) of styrene residues. This topic is expected to be an important area of development in the future, with the advent of new synthesis procedures for polyolefins.

IV. CONCLUSION

The formulation and use of elastomer blends is technologically demanding. Miscible blends are widely used, but not usually recognized, since analytical separation of the vulcanized elastomer is experimentally impossible. Immiscible blends, just like all other polymer blends, require excellent phase dispersion and interfacial adhesion. In addition, they require control of filler distribution and cross-link density in each component. This requirement is due to the need for mechical integrity in vulcanized elastomers. The current design criteria of North-American automotive tires require treads to last for 80,000 miles with less than 0.4 inch of wear. Vulcanized roofing membranes require 35 years of outdoor exposure with minimal change in elongation and tensile strength. The technical complexity of the analysis and use of elastomer blends has leads to nondisclosure for many of the formulations and uses. In spite of the difficulties of analysis and the gaps in understanding, the use of blends containing elastomers continues to be an active and increasingly popular area. A part of the impetus for the great interest in those blends is the availability of directed synthesis of many of the older elastomers. The new synthetic tools that make this possed include new catalysts ("single sited") and process designs for manufacture of ethylene-based polyolefin elastomers, group-transfer polymerization for acrylates, and anionic polymerization for diene elastomers. The availability of elastomers with a narrow compositional and molecular-weight distribution in these syntheses makes the utility of blending more apparent and useful. In addition, a significant body of effort is currently directed towards the synthesis of elastomeric blends containing a small amount of semicrystalline thermoplastic. This is a rapidly increasing area of application of elastomers, and one aspect of it is covered in the next chapter.

V. APPENDIX. ACRONYMS FOR COMMON ELASTOMERS

ABR	acrylate-butadiene rubber
ACM	copolymer of ethylacrylate and a comonomer (acrylic rubber)
ANM	ethylacrylate-acrylonitrile copolymer (acrylate rubber)
BIMS	Brominated isobutylene paramethyl styrene rubber
BR	butadiene rubber (polybutadiene)
BIIR	bromobutyl rubber
CIIR	chlorobutyl rubber
CFM	polychlorotritluoroethylene (fluoro rubber)
CM	chloropolyethylene (previous designation CPE)
CO	epichlorohydrin homopolymer rubber (polychloromethyloxiran)
CR	chloroprene rubber (polychloroprene)
CSM	chlorosulfonylpolyethylene
EAM	ethylene–ethyl acrylate copolymer (e.g., Vamac)
ECO	copolymer of ethylene oxide (oxiran) and chloromethyloxiran
ENM or H-NBR	proposed code for hydrogenated NBR
ENR	epoxidized NR
EPDM	ethylene–propylene–diene terpolymer
EPM	ethylene–propylene copolymer
EVM	ethylene–vinylacetate copolymer (previous code: EVA or EVAC)
FMQ	methyl silicone rubber with fluoro groups (previous designation FSI)
FPM / FPM	rubber having fluoro and fluoroalkyl or fluoroalkoxy substituent group

IIR	isobutylene-isoprene rubber (butyl rubber)
IR	isoprene rubber (synthetic)
MQ (PVMQ)	methyl silicone rubber (with vinyl and phenyl end groups)
NBR	acrylonitrile-butadiene rubber (nitrile rubber)
NR	isoprene rubber (natural rubber)
PUR	generic code for urethane rubbers
Q	generic code for silicone rubbers
SBR	styrene-butadiene rubber
TM	polysulfide rubbers
TOR	*trans*-polyoctenamer
VMQ	methyl silicone rubber with vinyl groups

VI. REFERENCES

1. R. German, R. Hank, and G. Vaughan, *Rubber Chem. Technol.*, **40**, 569 (1967).

2. The sampling of polymer blends in *Polymeric Compatibilizers*, by S. Datta and D. J. Lohse (Hanser-Gardner, 1996), indicates that > 93% of polymer blends are combinations of thermoplastics and elastomers. However, automotive tires (4000 kt/yr in 1995), made from a blend of elastomers, represents the largest application for any immiscible polymer blend.

3. W. M. Hess, C. R. Herd, and P. C. Vergari, *Rubber Chem. Technol.*, **66**, 329 (1993).

4. C. M. Roland, *Rubber Chem. Technol.*, **62**, 456 (1989).

5. P. J. Corish, in *Science and Technology of Rubber*, F. R. Eirich (ed.), Academic Press, New York, 1978, Ch. 12.

6. E. T. McDonel, K. C. Baranwal, and J. C. Andries, in *Polymer Blends*, Vol. II, D. R. Paul and S. Newman (eds.), Academic Press, New York, 1978, Ch. 19.

7. J. R. Wolfe, Jr., in *Thermoplastic Elastomers*, N. R. Legge, G. Holden, and H. E. Schroeder (eds.), Hanser, New York, 1987, Ch. 6.

8. (a) A. Y. Coran and R. Patel, *Rubber Chem. Technol.*, **63**, 141 (1980).
(b) S. Abdou-Sabet and S. Datta, in *Polymer Blends*, D. R. Paul and C. Bucknall (eds.), Wiley, New York, 1999, Ch. 35.

9. (a) C. M. Hansen, *The Three Dimensional Solubility Parameter and Solvent Diffusion Coefficients*, Danish Technical Press, Copenhagen, 1967.
(b) C. M. Hansen and A. Beerbower, "Solubility Parameters," in *Encyclopedia of Chemical Technology*, Supplement Volume, 2nd Ed., Wiley, New York, 1971.

10. (a) A. Keller, *J. Polym. Sci., Polym. Symp.*, **68**, 395 (1977).
(b) A. Y. Malkin, S. G. Kulichikhin, and A. E. Chalykh, *Polymer*, **22**, 1373 (1981).
(c) G. Ver Strate and W. Philippoff, *J. Polym. Sci., Polym. Lett. Ed.*, **12**, 267 (1974).

11. G. R. Hamed, C.-H. Shieh, and D. N. Schulz, *Polym. Bull.*, **9**, 525 (1983).

12. H. Takahashi, T. Matsuoka, T. Ohta, K. Fukumori, T. Kurauchi, and O. Kamigaito, *J. Appl. Polym. Sci.*, **36**, 1821 (1988).

13. (a) J. P. Runt and L. M. Martynowicz, *Adv. Chem. Ser.*, **211**, 111 (1985).
(b) H. D. Keith and F. J. Padden, *J. Appl. Phys.*, **36**, 1286 (1964).

14. A. Ghijsels, *Rubber Chem. Technol.*, **60**, 278 (1977).

15. (a) T. Nishi and T. T. Wang, *Macromolecules* **8**, 909 (1975).
    (b) B. Rim and J. P. Runt, *Macromolecules*, **17**, 1520 (1984).

16. S. Datta, unpublished results.

17. (a) M. Shibayama, H. Yang, and R. S. Stein, *Macromolecules*, **18**, 2179 (1985).
    (b) S. Wu, *J. Polym. Sci., Polym. Phys. Ed.*, **26**, 2511 (1987).
    (c) S. Wu, *Polymer*, **28**, 1144 (1987).

18. (a) T. G. Fox and V. R. Allen, *J. Chem. Phys.*, **41**, 344 (1964).
    (b) D. W. van Krevelen, *Properties of Polymers*, 2nd Ed., Elsevier, Amsterdam, 1976.
    (c) R. F. Boyer and R. L. Miller, *Rubber Chem. Technol.*, **61**, 718 (1978).
    (d) P. Privalko, *Macromolecules*, **13**, 370 (1980).
    (e) S. M. Aharoni, *Macromolecules*, **16**, 1722 (1983).

19. (a) H. N. Cheng, in *Polymer Analysis and Characterization IV*, H. G. Barth and J. Janca (eds.), John Wiley and Sons, New York, 1992, p. 21.
    (b) C. M. Roland, *Rubber Chem. Technol.*, **62**, 456 (1989).
    (c) B. Albert, R. Jerome, P. Teyssie, G. Smyth, N. G. Boyle, and V. J. McBrierty, *Macromolecules*, **18**, 388 (1985).
    (d) C. M. Roland and C. A. Trask, *Polym. Mater., Sci. Eng.*, **60**, 832 (1989).

20. (a) D. L. Vander Hart, W. F. Manders, R. S. Stein, and W. Herman, *Macromolecules*, **20**, 1726 (1987).

21. (a) J. Schaefer, M. D. Sefcik, E. O. Stejskal, and R. A. McKay, *Macromolecules*, **14**, 188 (1981).
    (b) T. R. Steger, J. Schaefer, E. O. Stejskal, R. A. McKay, and M. D. Sefcik, *Annals NY Acad. Sci.*, 371 (1981).

22. (a) M. C. Morris, *Rubber Chem. Technol.*, **40**, 341 (1967).
    (b) A. Ghijsels and H. J. M. A. Mieras, *J. Inst. Rubber Ind.*, **6**, 259 (1972).
    (c) A. K. Sircar and T. G. Lamond, *Rubber Chem. Technol.*, **46**, 178 (1973).

23. (a) C. M. Roland and G. G. A. Bohm, *Macromolecules*, **18**, 1310 (1985).
    (b) C. M. Roland, *Rubber Chem. Technol.*, **61**, 866 (1988).
    (c) C. M. Roland, *Macromolecules*, **20**, 2557 (1987).
    (d) J. Klein, *Prilos. Mag.*, **43**, 771 (1981).
    (e) J. Klein, D. Fletcher, and L. J. Fetters, *Nature (London)*, **304**, 526 (1983).
    (f) Y. Kumugai, K. Watanabe, T. Miyasaki, and J. Hata, *J. Chem. Eng. Jpn.*, **12**, 1 (1979).
    (g) T. Gilmore, R. Falabella, and R. L. Lawrence, *Macromolecules*, **13**, 880 (1980).
    (h) T. Hashimoto, Y. Tsukahara, and H. Kawai, *Macromolecules*, **14**, 708 (1981).
    (i) J. E. Anderson and J.-H. Jou, *Macromolecules*, **20**, 1544 (1987).
    (j) G. C. Summerfield and R. Ullman, *Macromolecules*, **20**, 401 (1987).

24. (a) L. W. Kleiner, F. E. Karasz, and W. J. MacKnight, *Polym. Eng. Sci.*, **19**, 519 (1979).
    (b) E. V. Gouinlock and R. S. Porter, *Polym. Eng. Sci.*, **17**, 535 (1977).
    (c) F. S. Bates, H. E. Bair, and M. A. Hartney, *Macromolecules*, **17**, 1987 (1984).

25. (a) S. Datta and E. N. Kresge, U.S. Patent 4,722,971 (1988).
    (b) S. Datta, L. Kaufman, and P. Ravishankar, U.S. Patent 5 571,868 (1996).

26. F. Morrar, L. L. Ban, W. G. Funk, E. N. Kresge, H.C. Wang, S. Datta, and R. C. Keller, U.S. Patent 5,428,099 (1996).

27. (a) Z. M. Zhang and J. E. Mark, *Polym. Sci., Polym. Phys. Ed.*, **20**, 473 (1982).
    (b) J. E. Mark and A. L. Andrady, *Rubber Chem. Technol.*, **54**, 366 (1981).

28. (a) D. I. Livingston and R. L. Rongone, paper No. 22, *Proc. 5th, Int. Rubber Conf.*, Brighton, England (1967).

    (b) G. M. Bartenev and G. S. Kongarov, *Rubber Chem. Technol.*, **36**, 668 (1983).

    (c) P. A. Marsh, A. Voet, L. D. Price, and T. J. Mullens, *Rubber Chem. Technol.*, **41**, 344 (1968).

    (d) R. Couchman, *Macromolecules*, **20**, 1712 (1987).

    (e) T. K. Kwei, E. M. Pearce, J. R. Pennacchia, and M. Charton, *Macromolecules*, **20**, 1174 (1987).

    (f) M.-J. Brekner, H. A. Schneider, and H.-J. Cantow, *Polymer*, **29**, 78 (1988).

    (g) M. Aubin and R. E. Prud'homme, *Macromolecules*, **21**, 2945 (1988).

29. (a) C. M. Roland, *J. Polym. Sci., Polym. Phys. Ed.*, **26**, 839 (1988).

    (b) C. A. Trask and C. M. Roland, *Polym. Comm.*, **29**, 332 (1988).

    (c) C. M. Roland, *Macromolecules*, **20**, 2557 (1987).

    (d) C. M. Roland and C. A. Trask, *Rubber Chem. Technol.*, **62**, 456 (1989).

    (e) C. A. Trask and C. M. Roland, *Macromolecules*, **22**, 256 (1988).

30. (a) I. R. Gelling, *NR Technol.*, **18**, 21 (1987).

    (b) A. G. Margaritis, J. K. Kallitsis, and N. K. Kalfoglou, *Polymer*, **28**, 2122 (1987).

31. (a) M. M. Coleman, G. J. Pehlert, and P. C. Painter, *Macromolecules*, **29**, 6820 (1996).

    (b) G. J. Pehlert, P. C. Painter, B. Veytsman, and M. M. Coleman, *Macromolecules*, **30**, 3671 (1997).

    (c) M. M. Coleman, G. J. Pehlert, X. Yang, J. Stallman, and P. C. Painter, *Polymer*, **37**, 4753 (1996).

    (d) G. J. Pehlert, X. Yang, P. C. Painter, and M. M. Coleman, *Polymer*, **37**, 4763 (1996).

32. P. J. Corish and B. D. W. Powell, *Rubber Chem. Technol.*, **47**, 481 (1974).

33. (a) B. S. Gesner, in *Encyclopedia of Polymer Science and Technology*, Vol. 10, N. M. Bikales (ed.), Wiley, New York, 1969, p. 697.

    (b) D. C. Blackley and R. C. Charnock, *J. Inst. Rubber Ind.*, **7**, 60 (1973).

    (c) D. C. Blackley and R. C. Charnock, *J. Inst. Rubber Ind.*, **7**, 113 (1973).

    (d) Japan Synthetic Rubber, Ltd., British Patent 1,046,215 (1966).

    (e) D. l. Livingstone and R. L. Rongone, *Proc. Int. Rubber Conf.*, Brighton, England, 1968, p. 337.

34. (a) C. E. Scott, J. E. Callan, and W. M. Hess, *J. Rubber Res. Inst. Malays.*, **22**, 242 (1969).

    (b) J. B. Gardiner, *Rubber Chem. Technol.*, **42**, 1058 (1969).

    (c) R. F. Bauer and E. A. Dudley, *Rubber Chem. Technol.*, **60**, 35 (1977).

    (d) R. W. Smith and J. C. Andries, *Rubber Chem. Technol.*, **47**, 64 (1974).

    (e) J. E. Callan, W. M. Hess, and C. E. Scott, *Rubber Chem. Technol.*, **44**, 814 (1971).

    (f) P. A. Marsh, A. Voet, and L. D. Price, *Rubber Chem. Technol.*, **40**, 359 (1967).

    (g) A. Marsh, A. Voet, L. D. Price, and T. J. Mullens, *Rubber Chem. Technol.*, **41**, 344 (1968).

    (h) W. M. Hess, R. A. Swor, and P. C. Vegvari, *Kautsch. Gummi Kunstst.*, **38**, 1114 (1985).

    (i) W. M. Hess and V. E. Chirico, *Rubber Chem. Technol.*, **50**, 301 (1977).

35. (a) N. Tokita, *Rubber Chem. Technol.*, **60**, 292 (1977).

    (b) G. N. Avgeropoulos, R. C. Weissert, P. H. Biddison, and G. G. A. Bohm, *Rubber Chem. Technol.*, **49**, 93 (1976).

36. (a) M. Takenaka, T. Izumitani, and T. Hashimoto, *Macromolecules*, **20**, 2257 (1987).

    (b) H. Yang, M. Shibayama, and R. S. Stein, *Macromolecules*, **19**, 1667 (1986).

    (c) J. Kumaki and T. Hashimoto, *Macromolecules*, **19**, 763 (1986).

    (d) W. M. Hess, R. A. Swor, and P. C. Vegvari, *Kautsch. Gummi Kunstst.*, **38**, 1114 (1985).

37. (a) J. E. Kruse, *Rubber Chem. Technol.*, **46**, 653 (1973).
    (b) J. E. Callan, W. M. Hess, and C. E. Scott, *Rubber Chem. Technol.*, **44**, 814 (1971).
    (c) C. E. Scott, J. E. Callan, and W. M. Hess, *J. Rubber Res. Inst. Malays.*, **22**, 242 (1969).
    (d) J. E. Callan, B. Topcik, and F. P. Ford, *Rubber World*, **161**, 60 (1965).

38. (a) E. H. Andrews, *J. Polym. Sci.*, **10**, 47 (1966).
    (b) L. L. Ban, M. J. Doyle, and G. R. Smith, *Rubber Chem. Technol.*, **69**, 176 (1986).

39. (a) F. H. Roninger, Jr., *Ind. Eng. Chem., Anal. Ed.*, **6**, 251 (1933).
    (b) R. W. Smith and J. C. Andries, *Rubber Chem. Technol.*, **47**, 64 (1979).

40. V. G. Shehuplesov, S. V. Obekhov, and V. N. Kuleznev, *Polym. Sci., U.S.S.R. (Engl. Transl.)*, **23**, 1325 (1981).

41. E. Kresge, *J. Appl. Polym. Sci., Appl. Polym. Symp.*, **39**, 37 (1984).

42. (a) E. W. Stroup, A. Pungor, V. Hilady, and J. D. Andrade, *Polym. Prepr. ACS., Div. Polym. Chem.*, **34(2)**, 86 (1993).
    (b) W. Stocker, B. Bickmarm, S. N. Magonov, and H. J. Cantow, *Ultramicroscopy*, **42**, 1141 (1992).
    (c) Y. H. Tsao, S. X. Yang, and D. F. Evans, *Langmuir*, **8**, 1188 (1992).
    (d) E. Hamada and R. Kaneko, *Ultramicroscopy*, **42**, 184 (1992).
    (e) J. E. Sax and J. M. Ottino, *Polymer*, **26**, 1073 (1985).
    (f) T. Nishi, T. Hayashi, and H. Tanaka, *Makromol. Chem., Macromol. Symp.*, **16**, 91 (1988).
    (g) J. Kruse, *Rubber Chem. Technol.*, **46**, 653 (1973).
    (h) D. Vesely and D. S. Finch, *Makromol. Chem., Macromol. Symp.*, **16**, 329 (1988).

43. (a) R. Buchdahl and L. E. Nielsen, *J. Polym. Sci.*, **15**, 1 (1955).
    (b) M. C. Morris, *Rubber Chem. Technol.*, **40**, 341 (1967).
    (c) H. K. de Decker and D. J. Sabatine, *Rubber Age*, **99(4)**, 73 (1967).
    (d) L. Bohn, *Rubber Chem. Technol.*, **41**, 495 (1968).
    (e) W. Scheele, *Kautsch. Gummi Kunstst.*, **24**, 387 (1971).
    (f) A. R. Ramos and R. E. Cohen, *Polym. Eng. Sci.*, **17**, 639 (1977).
    (g) K. A. Mazich, M. A. Samus, P. C. Kilgoar, Jr., and H. K. Plummer, *Rubber Chem. Technol.*, **59**, 623 (1986).

44. (a) H. G. Braun and G. Rehage, *Angew. Makcromol. Chem.*, **131**, 107 (1985).
    (b) R. F. Bauer and E. A. Dudley, *Rubber Chem. Technol.*, **60**, 35 (1977).
    (c) T. Inoue, F. Shomura, T. Ougizawa, and K. Miyasaka, *Rubber Chem. Technol.*, **68**, 873 (1985).
    (d) Y. Takagi, T. Ougizawa, and T. Inoue, *Polymer*, **28**, 103 (1987).
    (e) S. Lipatov, T. S. Chramova, L. M. Sergeeva, and L. V. Karabanova, *J. Polym. Sci., Polym. Chem. Ed.*, **16**, 427 (1977).
    (f) R. J. Butera, J. B. Lando, and B. Simic-Glavaski, *Macromolecules*, **20**, 1724 (1987).
    (g) M. R. Krejsa and J. L. Koenig, *Rubber Chem. Technol.*, **64**, 635 (1991).
    (h) M. R. Krejsa and J. L. Koenig, *Rubber Chem. Technol.*, **65**, 956 (1992).
    (i) S. R. Smith and J. L. Koenig, *Macromolecules*, **24**, 3496 (1991).
    (j) S. N. Sarkar and R. A. Komoroski, *Macromolecules*, **25**, 1420 (1992).

45. (a) O. Glatter and O. Kratky, *Small-Angle X-Ray Scattering*, Academic Press, New York, 1982.
    (b) J. S. Higgins and R. S. Stein, *J. Appl. Crystal.*, **11**, 346 (1978).
    (c) J. S. Higgins and R. S. Stein, *J. Appl. Crystal.*, **11**, 346 (1978).
    (d) B. J. Berne and R. Pecora, *Dynamic Light Scattering*, Wiley, New York, 1976.

(e) M. Moritani, T. Inoue, M. Motegi, H. Kawai, and K. Kato, in *Colloidal and Morphological Behavior of Block and Graft Copolymers*, G. E. Molau (ed.), Plenum Press, New York, 1971.

(f) E. R. Pico and M. C. Williams, *Polym. Eng. Sci.*, **17**, 573 (1977).

46. (a) O. Glatter and O. Kratky, *Small Angle X-ray Scattering*, Academic Press, New York, 1982.

47. (a) C. M. Roland and G. G. A. Bohm, *J. Polym. Sci., Polym. Phys. Ed.*, **22**, 79 (1984).

(b) S. L. Cooper and J. L. Miller, *Rubber Chem. Technol.*, **68**, 899 (1985).

(c) C. A. Trask and C. M. Roland, *Macromolecules*, **22**, 256 (1989).

(d) C. M. Roland and G. G. A. Bohm, *J. Polym. Sci., Polym. Phys. Ed.*, **22**, 79 (1984).

48. (a) J. L. Leblanc, *Plast. Rubber Proc. Appl.*, **2**, 361 (1982).

(b) G. J. Van Amerongen, *Rubber Chem. Technol.*, **37**, 1065 (1964).

(c) V. A. Shershnev, *Rubber Chem. Technol.*, **66**, 537 (1982).

(d) M. G. Huson, W. J. McGill, and R. D. Wiggett, *Plast. Rubber Proc. Appl.*, **6**, 319 (1985).

(e) R. F. Bauer and A. H. Crossland, *Rubber Chem. Technol.*, **61**, 585 (1988).

(f) A. K. Bhowmick and S. K. De, *Rubber Chem. Technol.*, **63**, 960 (1980).

(g) R. L. Zapp, *Rubber Chem. Technol.*, **46**, 251 (1973).

49. (a) G. J. Amerongen, *Rubber Chem. Technol.*, **37**, 1065 (1964).

(b) J. R. Gardiner, *Rubber Chem. Technol.*, **41**, 1312 (1968).

(c) J. R. Gardiner, *Rubber Chem. Technol.*, **42**, 1058 (1969).

(d) J. R. Gardiner, *Rubber Chem. Technol.*, **43**, 370 (1970).

50. (a) V. A. Shershnev, *Rubber Chem. Technol.*, **66**, 537 (1982).

(b) J. Rehner, Jr. and P. E. Wei, *Rubber Chem. Technol.*, **42**, 985 (1969).

(c) A. K. Bhowmick and S. K. De, *Rubber Chem. Technol.*, **63**, 960 (1980).

(d) K. C. Baranwal and P. N. Son, *Rubber Chem. Technol.*, **47**, 88 (1974).

(e) K. Hashimoto, M. Miura, S. Takagi, and H. Okamoto, *Int. Polym. Sci. Technol.*, **3**, 84 (1976).

(f) A. Y. Coran, *Rubber Chem. Technol.*, **61**, 281 (1979).

(g) R. W. Amidon and R. A. Gencarelli, U.S. Patent 3,674,824 (1972).

(h) R. P. Mastromatteo, J. M. Mitchell, and T. J. Brett, *Rubber Chem. Technol.*, **44**, 1065 (1971).

(i) Sumitomo Chemical Company, British Patent 1,325,064 (1973).

(j) M. E. Woods and T. R. Mass, U.S. Patent 3,830,881 (1974).

51. M. E. Woods and J. A. Davidson, *Rubber Chem. Technol.*, **49**, 112 (1976).

52. N. Suma, R. Joseph, and D. J. Francis, *Kautsch. Gummi Kunstst.*, **43**, 1095 (1990).

53. R. F. Bauer and E. A. Dudley, *Rubber Chem. Technol.*, **60**, 35 (1977).

54. (a) M. G. Huson, W. J. McGill, and P. J. Swart, *J. Polym. Sci., Polym. Lett. Ed.*, **22**, 143 (1984).

(b) D. Honiball and W. J. McGill, *J. Polym. Sci., Polym. Phys.*, **26**, 1529 (1988).

(c) L. D. Loan, *Rubber Chem. Technol.*, **40**, 149 (1967).

55. R. L. Zapp, *Rubber Chem. Technol.*, **46**, 251 (1973).

56. G. G. A. Bohm and J. O. Tveekrem, *Rubber Chem. Technol.*, **66**, 575 (1982).

57. (a) J. E. Callan, W. M. Hess, and C. E. Scott, *Rubber Chem. Technol.*, **44**, 814 (1971).

(b) J. E. Callan, W. M. Hess, and C. E. Scott, *Rev Gen. Caoutch. Plast.*, **48**, 155 (1971).

(c) A. K. Sircar and T. J. Lammond, *Rubber Chem. Technol.*, **46**, 178 (1973).

(d) G. Cotton and L. J. Murphy, *Rubber Chem. Technol.*, **61**, 609 (1988).

58. (a) J. E. Lewis, M. L. Deviney, and L. E. Whittington, *Rubber Chem. Technol.*, **42**, 892 (1969).

(b) B. G. Corman, M. L. Deviney, and L. E. Whittington, *Rubber Chem. Technol.*, **43**, 1349 (1970).

(c) M. L. Deviney and L. E. Whittington, *Rubber Chem. Technol.*, **44**, 87 (1971).

(d) S. Datta, R. C. Keller, and F. T. Morrar, American Chemical Society, PMSE Preprints 76, 95 (1997).

59. B. L. Lee, *Polym. Eng. Sci.*, **21**, 294 (1981).

60. (a) S. Shong Lin, *Rubber Chem. Technol.*, **68**, 885 (1985).

(b) A. J. Dias and A. Galuska, *Rubber Chem. Technol.*, **69**, 615 (1996).

61. (a) J. E. Callan, B. Topcik, and E. P. Ford, *Rubber World*, **151**, 60 (1951).

(b) W. M. Hess, P. A. Marsh, and F. J. Eckert, Paper presented at a meeting of the Rubber Division, American Chemical Society, Miami Beach, Florida, 1965.

(c) P. A. Marsh, A. Voet, and L. D. Price, *Rubber Chem. Technol.*, **40**, 359 (1967).

(d) Y. F. Wang and H. C. Wang, *Rubber Chem. Technol.*, **70**, 63 (1997).

62. B. L. Lee and C. Singleton, *J. Appl. Polym. Sci.*, **24**, 2169 (1979).

63. (a) W. M. Hess, R. A. Swor, and P. C. Vegvari, *Kautsch. Gummi Kunstst.*, **38**, 1114 (1985).

(b) W. M. Hess and V. E. Chirico, *Rubber Chem. Technol.*, **50**, 301 (1977).

64. (a) A. Medalia, *Rubber Chem. Technol.*, **51**, 437 (1978).

(b) W. P. Fletcher and A. N. Gent, *Br. J. Appl. Phys.*, **8**, 1984 (1957).

(c) S. Maiti, S. K. De, and A. K. Bhowmick, *Rubber Chem. Technol.*, **65**, 293 (1992).

65. (a) M. E. Woods and T. R. Mass, *Adv. Chem. Ser.*, **142**, 386 (1975).

(b) E. T. McDonel, K. C. Baranwal, and J. C. Andries, in *Polymer Blends*, D. R. Paul and S. Newman (eds.), Academic Press, New York, 1978, p. 263.

(c) L. Leibler, *Makromol. Chem. Macromol Symp.*, **16**, 1 (1982).

(d) J. Noolandi and K. M. Hong, *Macromolecules*, **16**, 482 (1982).

66. (a) D. K. Setua and J. L. White, *Kautsch. Gummi Kunstst.*, **44**, 821 (1991).

(b) D. K. Setua and J. L. White, *Poly. Eng. & Sci.*, **31**, 1742 (1991).

(c) P. A. Arjunan, R. B. Kusznir, and A. Dekmezian, *Rubber World*, **21** (1997).

(d) R. H. Schuster, R. Schmidt, and G. Pampus, *Kautsch. Gummi Kuntst.*, **42**, 582 (1989).

(e) M. S. Sillars, *Kautsch. Gummi Kuntst.*, **43**, 412 (1990).

(f) Y. Mori, Y. Kawamura, H. Matsumoto, and Y. Takemura, U.S. Patent 4,849,478 (1989).

67. (a) J. A. Davison, W. Nudenberg, and Y. Rim, U.S. Patent 4,316,971 (1982).

(b) C. F. Paddock, U.S. Patent 3,758,435 (1973).

(c) C. H. Shedd and A. L. Stone, U.S. Patent 4,374,952 (1983).

68. (a) G. N. Avgeropoulos, R. C. Weissert, P. H. Biddison, and G. G. A. Bohm, *Rubber Chem. Technol.*, **49**, 93 (1976).

(b) C. M. Roland and M. Nguyen, *J. Appl. Polym. Sci.*, **36**, 141 (1988).

(c) A. C. Pipkin and R. I. Tanner, *Ann. Rev. Fluid Mech.*, **9**, 13 (1977).

(d) W. M. Hess, P. C. Vegvari, and R. A. Swor, *Rubber Chem. Technol.*, **68**, 350 (1985).

(e) W. M. Hess, R. A. Swor, and P. C. Vegvari, *Kautsch. Gummi Kunstst.*, **38**, 1114 (1985).

(f) R. F. Heitmiller, R. Z. Maar, and H. H. Zabusky, *J. Appl. Polym. Sci.*, **8**, 873 (1964).

(g) S. Uemura and M. Takayanagi, *J. Appl. Polym. Sci.*, **10**, 113 (1966).

69. (a) H. Yang, M. Shibayama, and R. S. Stein, *Macromolecules*, **19**, 1667 (1986).

(b) M. H. Walters and D. N. Keyte, *Trans. Inst. Rubber Ind.*, **38**, 40 (1962).

(c) F. C. Weissert and G. N. Avgeropoulos, *J. Elastomers Plast.*, **9**, 102 (1977).

(d) E. A. Meinecke and M. I. Taftaf, *Rubber Chem. Technol.*, **37**, 1190 (1988).

(e) A. I. Medalia, *Rubber Chem. Technol.*, **61**, 437 (1978).

(f) E. A. Meinecke and M. I. Taftaf, *Rubber Chem. Technol.*, **61**, 534 (1988).

70. (a) K. C. Baranwal, *Rubber Age*, **102(2)**, 52 (1970).

(b) B. L. Lee, U.S. Patent 4,455,399 (1984).

(c) K. Z. Nedden, K. M. Diedrich, and G. Huhn, *Rubber Plast. News*, **19(26)** 14 (1990).

(d) R. T. Morrisey, *Rubber Chem. Technol.*, **44**, 1029 (1971).

(e) C. M. Roland, G. G. A. Bohm, and P. Sadhukhan, *J. Appl. Polym. Sci.*, **30**, 2021 (1985).

71. (a) H. J. Barager, Paper No.78, presented at a meeting of the Rubber Division, American Chemical Society, Houston, Texas, Fall 1983; abstract in *Rubber Chem. Technol.*, **67**, 407 (1984).

(b) N. S. Maksimova and A. G. Shvarts, *Int. Polym. Sci. Tech.*, **11(11)**, T/20 (1984).

72. (a) F. J. Krakowski and A. J. Tinker, *Elastomerics*, **122(6)**, 34 and **122(7)**, 24 (1990).

(b) B. L. Lee, *Polym. Eng. Sci.*, **26**, 729 (1985).

(c) V. L. Folt and R. W. Smith, *Rubber Chem. Technol.*, **46**, 1193 (1973).

(d) B. L. Lee, *Adv. Chem. Ser.*, **206**, 185 (1984).

(e) M. F. Tse, K. O. McElrath, H.-C. Wang, Y. F. Wang, and A. L. Tisler, Paper 24, 153rd Rubber Division Meeting, Americal Chemical Society, Indianapolis, Indiana, May 5–8, 1998.

73. (a) J. M. Mitchell, *Rubber Plast. News*, June 3, 1985, p. 18.

(b) N. M. Matthew, *J. Polym. Sci, Polym. Lett. Ed.*, **22**, 135 (1984).

(c) W. von Hellens, Rubber Division, American Chemical Society, Detroit, Michigan, Fall 1989.

(d) F. C. Cesare, *Rubber World*, **201(3)**, 14 (1989).

(e) W. Hong, *Rubber Plast. News*, **19(11)**, 14 (1989).

(f) W. von Hellens, D. C. Edwards, and Z. J. Lobos, *Rubber Plast. News*, **20(6)**, 61 (1990).

(g) E. H. Andrews, *J. Polym. Sci.*, **10**, 47 (1966).

(h) L. Ambelang, F. H. Wilson, Jr., L. E. Porter, and D. L. Turk, *Rubber Chem. Technol.*, **42**, 1186 (1969).

(i) J. Barrier, *Rubber Chem. Technol.*, **28**, 814 (1956).

(j) J. Kinning, E. L. Thomas, and J. M. Ottino, *Macromolecules*, **20**, 1129 (1987).

74. H. Mouri and Y. Tonosaki, Paper 65, 152rd Rubber Division Meeting, American Chemical Society, Cleveland, OH, October 21–24, 1997.

35 Thermoplastic Vulcanizates

SABET ABDOU-SABET

Advanced Elastomer Systems L. P.
388 S. Main Street
Akron, OH 44311

SUDHIN DATTA

Exxon Chemical Co.
5200 Bayway Drive
Baytown, TX 77520

Polymer Blends, Volume 2: Performance. Edited by D. R. Paul and C. B. Bucknall.
ISBN 0-471-35280-2.   © 2000 John Wiley & Sons, Inc.

I. INTRODUCTION

Thermoplastic vulcanizates (TPVs) have become very useful as thermoplastic elastomers since their introduction in 1981. These materials have the processing characteristics of a thermoplastic and the functional performance of a conventional thermoset rubber, hence the name "thermoplastic vulcanizates." TPVs are a special class of thermoplastic elastomers (TPEs) generated from synergistic interaction of an elastomer–thermoplastic polymer blend to give properties demonstrably better than those of a simple blend. This interaction is best illustrated through dynamic vulcanization, wherein the elastomer is preferentially vulcanized under dynamic shear to generate fine, cured rubber particles in a thermoplastic polymer matrix. Elastomeric products are then fabricated for many end-use applications using fast thermoplastic processing.

The dynamic vulcanization process was first discovered by Gessler et al. [1] in their attempt to improve the impact properties of PP through the partial vulcanization of halo butyl rubber with zinc oxide. The first commercial application, however, was based on patents by Fischer [2], wherein dynamic partial vulcanization of EPDM in a polypropylene matrix was accomplished by limiting the amount of peroxide to maintain the thermoplastic processability of the blend.

Significant improvement in the properties of these blends was achieved by Coran et al. [3] by fully vulcanizing the rubber phase under dynamic shear, without affecting the thermoplasticity of the blend. This discovery was further advanced by Abdou-Sabet et al. [4] through the use of preferred curatives to achieve improvement in elastomeric properties and flow characteristics, which have aided the successful commercialization of dynamic vulcanization technology (e.g., Santoprene® thermoplastic rubber). The successful commercialization of these proprietary products have led to significant interest and the proliferation of patents and publications, which have exceeded 500 since 1980.

One of the advantages of this technology, versus elastomeric block copolymers, is that new products are generated from blends of existing polymers using processes that require low capital. This contrasts with the conventional, capital-intensive processes for new materials; these processes are also associated with environmental concerns and requirements for high-volume polymerization units. Other advantages of the TPV technology over block copolymers as a source for thermoplastic elastomers are in the areas of upper service temperature, resistance to hydrocarbon fluids, and resistance to compression setting [5]. Dynamic vulcanization can be applied to all possible combinations of elastomer and plastic polymer blends; however, only a limited number of these combinations have found useful applications. Many of these dynamically vulcanized blends are reviewed here. Generally, these TPVs possess significantly improved properties over simple blends. Some of these improvements are

- stable phase morphology and consistent processing,
- higher ultimate tensile strength,
- improved upper service temperature performance,
- improved fatigue resistance, and
- greater resistance to attack and swell by fluids.

The improvements in properties are generally obtained when (a) the surface energies of the elastomers and plastic are matched, (b) the molecular length of the entanglement of the elastomer is low, and (c) the thermoplastic polymer has some crystallinity. In blends for which there is a large difference between the solubility parameters (i.e., the surface energy of the elastomer and of the plastic), a compatibilizer can be used to improve the mixing of the blend and the generation of useful TPV materials from immiscible blends. The presence of the compatibilizer allows the formation of very small rubber particles in the matrix, thus generating useful products from a mixture of two thermodynamically incompatible polymer pairs.

Dynamic vulcanization for the generation of thermoplastic elastomers involves the melt mixing under shear of an elastomer with a rigid thermoplastic and the subsequent vulcanization of the elastomer phase under continuous mixing. The temperature must be above the melting point of the thermoplastic and sufficiently high to activate the vulcanization system. It is preferable to use a vulcanization system that will not attack the plastic phase. The vulcanization system generates the same crosslinks or three-dimensional polymer structure as in static vulcanization. In dynamic vulcanization, however, these structures are generated in small rubber particles that are dispersed in the thermoplastic polymer matrix as microgels, thus generating at least a two-phase morphology in which the plastic phase is the continuous phase. The size of these particles or microgels play a significant role in the improvement of mechanical properties. To obtain optimum properties, the cross-linked rubber particles should be less than 2 μm in diameter. Figure 35.1 illustrates commercial grades of EPDM–PP TPVs wherein the average particle size is between 1 and 2 μm. Figure 35.2 shows the variation of ultimate tensile strength and elongation with rubber particle size [6]. This figure depict a composite stress–strain curve constructed from different cured blends of rubber and plastic wherein the statically cured rubber has been ground and sieved into the different particle sizes, which were then blended in PP. The best properties were obtained from compositions generated by dynamic

10KV   2.00KX   5.00 μ

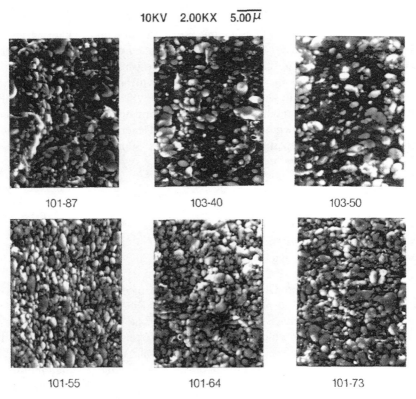

Figure 35.1   Morphology of commercial EPDM–PP TPVs.

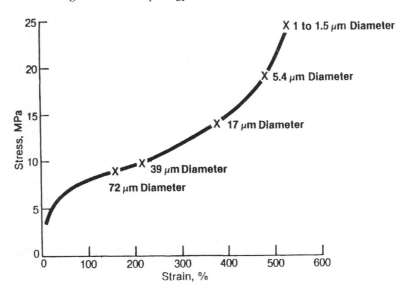

Figure 35.2   Effect of vulcanized-rubber particle size on the mechanical properties of EPDM–PP TPVs.

vulcanization, where the smallest average particle size was obtained. Currently, there are no other commercial means of producing such small sizes of rubber particles.

II. CRITICAL FEATURES OF THERMOPLASTIC VULCANIZATES

A. Compatibility

There is a wide variety of commercially available rubbers and plastics that can be considered for blending. However, relatively few of them have been of technological importance, since most polymers are incompatible with one another, at least in a thermodynamic sense [7]; that is, the polymers are immiscible. As a rule, the important materials are compositions of the more nearly thermodynamically compatible polymers that are capable of forming a fine heterophase morphology (e.g., hydrocarbon rubbers and plastics). This principle is best illustrated with blends of ethylene–propylene (diene monomer) rubbers and isotactic polypropylene. Lohse [8] described these blends as unstable, because the polymers are not miscible. Under molten and static conditions, the rubber phase coalesces and agglomerates. (See Fig. 35.3.) Grossly thermodynamically incompatible polymers do not melt mix, and they appear inhomogenous, even to the naked eye. In such cases, a compatibilizer is needed to make the blend useful. Techniques that improve the properties and hence the usefulness of such mixtures are designated "technological compatibilizations" [9]. Such techniques generally do not cause blends to become compatible in the thermodynamic sense, though they do drive the blend to form a fine, dispersed-phase morphology.

An example of the large differences in surface-energy or solubility parameters occurs between acrylonitrile–butadiene copolymer rubber (NBR) and polypropylene. The application of a compatibilizer and dynamic vulcanization to these materials generated a commercially proprietary composition called Geolast® thermoplastic rubber [10].

Phase Growth After 30 min 180°C

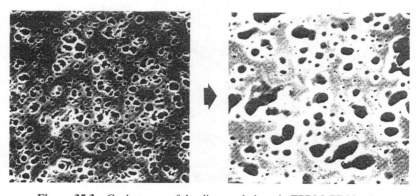

Figure 35.3   Coalescence of the dispersed phase in EPDM–PP blends.

On the other hand, miscible, or thermodynamically compatible polymers, generate a single-phase morphology, which is not conducive to generating useful TPVs. Blends of single-phase morphology can at best partially vulcanized, [11]. The basic fundamentals of dynamic vulcanization can be illustrated for polyolefin blends, which are described as technologically compatible. These blends are best represented by EPDM–polypropylene polymer blends.

B. Degree of Cure

Coran et al. [3] demonstrated the beneficial effects of complete vulcanization over partial dynamic vulcanization [2]. The effect of the cross-link density on stress–strain and swelling characteristics in hydrocarbon oil can be seen in Fig. 35.4. The cross-link densities were determined on the basis of measurements of solvent swelling, using the Flory–Rehner equation [12], by press curing the EPDM alone under conditions similar to dynamic vulcanization. The effect of PP on the cross-link density during dynamic vulcanization were very little, if anything.

The complete vulcanization [13] of the rubber phase leads to significant improvement of mechanical properties of the blend, both at room temperature and, more importantly, at 100°C. (See Fig. 35.4.) Significant improvement can also be seen for resistance to oil. Elastomeric recovery, as measured by tension set or compression set, shows similar dramatic improvement. The degree of cure can also be determined from the amount of un-cross-linked rubber. This measurement is normally done by a series of extractions to isolate quantitatively the cross-linked and un-cross-linked

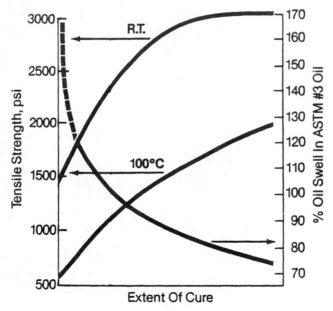

Figure 35.4   Physical properties as a function of the extent of cure.

rubber in the composition. These calculations can easily be performed if the composition of the TPV and the solubility of the different components in solvents are known. It has been commonly accepted that if the cross-link density is higher than 7×10^{-5} moles/cm^3 and/or the elastomer is at least 97% cured, then full cure has been achieved. EPDM, which has a very wide molecular-weight distribution, due to the presence of very low-molecular-weight ends, does not show significant improvement in tensile strength, because the low-molecular-weight ends do not effectively yield a three dimensional network [14].

C. Type of Curatives

For vulcanization of rubber, researchers have a wide choice of vulcanization chemicals [15] and accelerators or retarders to choose from, depending on the type of rubber used [16]. The most extensively studied vulcanization procedure is the sulfur system, because of its dominance in tire manufacturing. In dynamic vulcanization, the first curative used was zinc oxide with halobutyl rubber by Gessler [1]. Fisher [2] controlled the degree of vulcanization by limiting the amount of peroxide with EPDM. Coran et al. [3] used sulfur vulcanization for a majority of their experiments. The most obvious disadvantage of the latter type of curative is the development of an undesirable, sulfurous odor. The use of peroxide, on the other hand, with polyolefin as the plastic phase, leads to undesirable side reactions, because of the free radicals that are generated. In the case of polyethylene, the peroxy radicals lead to cross-linking of the polyethylene, yielding very viscous products that are difficult to process. In case of polypropylene, the peroxy free radicals abstract the hydrogen from the polyolefin chain, generating a more stable tertiary free radical, which undergoes chain scission.

The degradation of PP leads to a loss of properties. This effect can be minimized through the addition of other ingredients, e.g., PIB (polyisobutylene) [17], that preferentially degrade, thus protecting the plastic phase. Abdou-Sabet and Fath [4] demonstrated that this disadvantage can be overcome by using phenolic curatives to cross-link the EPDM phase. Not only were the problems caused by PP degradation eliminated, but also further significant improvement was achieved in terms of compression set, resistance to oil, and the processing characteristics of the material. (See Table 35.1.) This development has allowed the successful commercialization of TPV as a replacement for rubber. Surprisingly, the earlier work of Giller [18] on vulcanization of EPDM with phenolic curatives has been seriously doubted by Hofman [15b] in terms of its ability to achieve any commercial viability.

More importantly, significant improvement in processing was achieved over the sulfur cure, as can be seen in Table 35.2. This improvement was illustrated by extruding a tube under a slight air pressure to maintain the size of the tube, using barrel temperatures of 193°C to 232°C. The variable examined was the draw-down ratio, which is a measure of the integrity of the extrudate during its extension as the takeoff rate is increased. The draw-down ratio is the ratio of the annulus of the die to the cross-section of the tube at the failure point (i.e., the loss of integrity). It was found that dimethyloloctylphenol curatives allow the generation of very soft TPVs (as soft

Table 35.1   Effect of Cross-Linking Agent on Properties of 60/40 EPDM Rubber–PP Compositions

Property	Cross-Linking Agent			
	None	Sulfur	Dimethyl alkyl phenol	Peroxide
Hardness, Shore D	36	43	44	39
Tensile Strength, Wa	4.96	24.3	25.6	15.9
M100, XTa	4.83	8.00	9.72	8.07
Ultimate Elongation, %	190	530	350	450
ASTM #3 Oil Swell, %	disintegrated	194	109	225
Compression Set, % (22 hrs @100°C)	91	43	24	32
Odor	+++	—	++	++
Fabricability	++	—	+++	+++

Table 35.2   Extrusion Characteristics of Phenolic-Cured TPV

Sample Type[a]	Barrel Temperature 193°C		Barrel Temperature 232°C	
	Phenolic Curative	Sulfur Curative	Phenolic Curative	Sulfur Curative
Extrudate temperature °C	197	203	232	234
Output rate, g/min	199.5	201.8	182.5	171.2
Tube dimensions @ 381 cm/min takeoff rate				
O.D., mm	12.85	12.9	12.7	12.17
I.D., mm	9.91	9.91	9.91	9.53
Appearance	smooth	rough	smooth	smooth
Tube dimensions @ maximum draw down				
O.D., mm	7.7	9.68	5.46	7.11
I.D., mm	5.94	7.54	4.37	5.56
Area ratio	3.0	1.9	6.8[b]	3.7
Appearance	smooth	very rough fractured	smooth	very rough
Odor	mild	H_2S/S odor	mild	offensive

[a] 36/64 EPDM–PP blend, 45 Shore D Composition.
[b] Machine limit, no break.

as 35 Shore A) with a compression set approaching that of a thermoset rubber, while maintaining excellent thermoplastic processing.

The poor processing characteristics of the sulfur-cured TPVs are due to phase growth of the dispersed rubber particles. The polysulfidic bonds generated during vulcanization undergo a sulfur-exchange reaction, leading to significant coalescence of the rubber particles; see Fig. 35.5. Phase growth of the dispersed rubber phase leads to poor and variable processing characteristics.

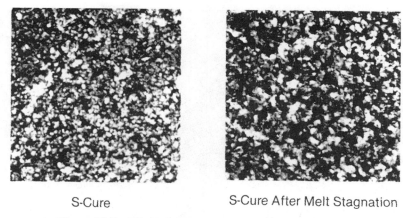

S-Cure   S-Cure After Melt Stagnation

Figure 35.5   Effect of the type of cross-link on morphology.

Finally, it should be noted that a conventional silicone rubber curative, namely, multifunctional organosilicone hydride [19], was satisfactorily applied in the partial vulcanization of elastomers containing carbon–carbon double bonds in a saturated plastic matrix.

D. Morphology

The preferred morphology for TPV is that of an elastomeric dispersed phase and a thermoplastic continuous phase. Such morphology should allow for flow in a shear field. Figure 35.1 illustrates the morphology of commercial products in scanning electron micrographs. In these micrographs, the rubber particles appear white. The micrographs are presented in terms of increasing rubber content, thus illustrating the achievement of reduced modulus and hardness. To validate the dispersed-phase morphology and to show that the electron micrographs are not that of two co-continuous phases wherein the microtoming of the fracture surface is always occuring at an angle perpendicular to the laminar structure, energy-dispersive microanalysis [20] was combined with electron microscopy. Inorganic particulates, such as aluminum silicate, were first incorporated into the EPDM phase, followed by melt mixing with PP and subsequent dynamic vulcanization. X-ray elemental analysis indicates the concentration of the aluminum in the EPDM phase, which correlates with the discrete, particulate nature of the EPDM phase.

The final morphology of the TPV is affected by the morphology of the blend at the outset of dynamic vulcanization. For a given polymer pair, melt blending is most efficient when the viscosities of the phases are matched [21]. Other parameters that affect the morphology and the properties of the TPV are the shear rates of the mixing process, the polymer ratio, the surface energies of the polymer pairs, cross-link density, the type of cross-link, the molecular-weight distribution of the rubber, and the presence of compounding additives (e.g., fillers, plasticizers, etc.). The most widely used combinations of polymer blends are combinations of olefinic

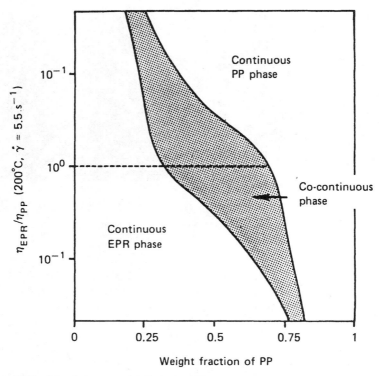

Figure 35.6 Morphology model for EPR–PP blends: viscosity ratio vs. weight fraction.

elastomers and olefinic thermoplastics. This type of combination is best exemplified by blends of EPDM and PPs which have similar solubility parameters. Romanini et al. [22] studied the effect of molecular weight (i.e., the viscosity ratio) on the phase morphology of simple blends prior to dynamic vulcanization. Figure 35.6 illustrates their findings, which shown that co-continuous two-phase morphological blends can be obtained over a wide range of polymer ratios by adjusting the viscosity ratio (e.g., from 80/20 to 20/80 for an EPDM–PP blend).

For an 80/20 EPDM–iPP [13] composition, PP is the minor component and is the dispersed phase in an EPDM matrix. During dynamic vulcanization of such a blend, the EPDM and the PP have to undergo a phase inversion to maintain the thermoplasticity of the blend. In the initial stages of dynamic curing, two co-continuous phases are generated, and as the degree of cross-linking advances during mixing, the continuous rubber phase becomes elongated further and then breaks up into polymer droplets. As these droplets form, the PP becomes the continuous phase. This process can easily be seen in the scanning electron photomicrograph (shown in Fig. 35.7), wherein the EPDM phase appears white and the PP phase black. The middle picture shows both phases to be co-continuous.

For TPV blends of natural rubber and iPP, it has been claimed that the stage at which the rubber phase is substantially vulcanized is co-continuous with the

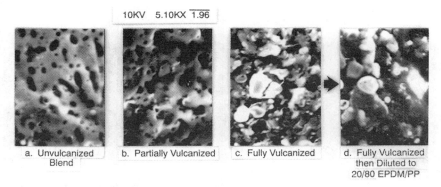

Figure 35.7   Morphology development for an 80/20 EPDM–PP TPV phase inversion.

polypropylene phase [23], yet the mechanism by which thermoplasticity is maintained has yet to be adequately explained. Much effort continues to be devoted to understanding the morphology and cross-linking of this blend. A recent paper by Ellul, Patel, and Tinker [24] describes the progress that these authors made in this area by using combinations of techniques, including scanning transmission electron microscopy (STEM) of sections and network visualization by TEM imaging of TPV.

E. Requirements for Stable Morphology

As a class of material, TPVs are preferred over simple blends, not only because of their significantly improved properties, but also because of their stable morphology. Prior to cross-linking, elastomers can be described as very viscous liquids. When blended into a thermoplastic matrix, under certain conditions the elastomer phase can undergo phase growth through coalescence which leads to changes in the performance of the product. This coalescence can be somewhat moderated through the incorporation of compatibilizers or graft copolymers [25] to stop or slow down the phase growth; see Fig. 35.8. The dynamic vulcanization of these rubber particles renders them into particles and eliminates their tendency to coalesce.

The choice of cure systems, however, can lead to the same phenomenon, that is demonstrated in Figs. 35.3 and 35.5. Cure systems that have a tendency to exchange, such as sulfidic, diisocyanate, and transesterification systems, are less stable than a C–C or a C–Si bond. It is also important to choose an appropriate process, in order to ensure that the preferred morphology is generated. This morphology includes the size, the shape, and the distribution of the rubber phase.

F. Fillers, Plasticizers, and Additives

Fillers and plasticizers can be added to TPVs to obtain the desired change in properties. For example, carbon black is widely used as a reinforcing agent. In TPVs, the thermoplastic phase provides the reinforcement to the matrix. Thus, there is little need to add carbon black above the amount that is needed to provide a black-colored

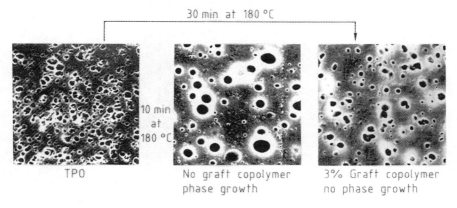

Figure 35.8   Stabilization of EPDM–PP morphology.

Table 35.3   Effect of Extender Oil and Carbon Black on EPDM–PP TPV[a]

Extender oil (phr)	Carbon black (phr)	Shore hardness	Ultimate tensile strength (MPa)	Ultimate elongation %	Stress at 100% strain (MPa)	Young's modulus (MPa)	Tension set (%)
0	0	48D	27.5	560	11.3	162	31
80	0	29D	15.2	550	6.4	47	19
0	80	51D	31.0	410	14.3	120	30
80	80	33D	23.0	530	7.2	23	16
160	80	74A	15.2	490	4.8	11.5	13

[a] TPV composition, phr: EPDM, 100; PP, 122; sulphur, 2.0.

product, if desired. The addition of plasticizers, on the other hand, allows the preparation of softer compositions that lead to significant improvement in processability and elastic recovery. The effects of carbon black and process oil can be seen in Table 35.3. In this table, paraffinic process oil is used as plasticizer in a polyolefinic polymer system, i.e., an EPDM–PP blend. In the melt, the oil partitions between the phases. This interphase transfer of the oil lowers the viscosity of the PP, which allows for enhanced flow. On crystallization of the polypropylene phase through cooling, most of the oil is expelled and is believed to be predominantly in the rubber phase. Some of the oil resides in the amorphous region of the plastic phase, thus improving the elasticity of the TPV [26].

The use of paraffinic oil has a moderate effect on the T_g of both the EPDM and the PP, such that a depression of 5°C and 15°C is obtained. More recently, Ellul demonstrated significant improvement of the low-temperature properties of TPVs through the use of ester plasticizers, instead of the usual process oil [27]. The use of iso-octylphathalate surprisingly gave a significant improvement by lowering the T_g of the EPDM by 34°C and of the isotactic PP by 36°C; see Table 35.4.

Additives can be used in the compositions, based on the application desired. For example, antioxidants, UV stabilizers, and flame-retardant additives can be included

Table 35.4   Effect of Plasticiser Type on T_g and T_m of EPDM–PP TPV^a

Plasticizer	Rubber T_g, °C	Plastic T_g, °C	T_m, °C
None	−41	+10	165
Polyurea grease	−47	−1159	
Paraffinic oil (Sunpar® 150)	−46	−5	158
Naphthenic oil (Cyclolube® 213)	−48	−5	158
Naphthenic oil (Cyclolube® 4053)	−50	−5	157
Naphthenic oil (Cyclolube® 410)	−51	−5	158
Amoco® 9012 polypropene	−45	−10	159
Alkylalkylether diester glutarate	−55	−11	159
Diisooctyldodecanedioate	−56	−14	158
Dioctyl sebacate	−60	−18	158
Dioctyl azelate	−60	−22	...
Diisooctylnonyl adipate	−64	−24	...
Butoxyethyl oleate	−66	−20	158
n−Butyl. oleate	−71	−24	155
n−Butyl tallate	−70	−24	155
Isooctyl tallate	−75	−26	155

^{a}Glass transition temperature T_g measured from tan δ peak at 10 Hz; T_m = melting point; TPV = 30/70 EPDM–PP, plus 40% plasticizer.

in compositions. Sometimes, coagents are also added, to achieve modification of the cure characteristics for ultimate property enhancement [28].

G. Choice of Rubbers and Plastics

In principle, a very large number of rubber–plastic blends—and consequently, their thermoplastic vulcanizates—can be made. There are at least 14 classes of rubbers and 22 classes of plastics [29], and these numbers are increasing every day with the introduction of new polymers (for example, copolymers of isobutylene and p-methyl styrene and derivatives thereof [30], as well as the new polymers based on metallocene catalysis, e.g., ethylene–styrene copolymers and terpolymers [31] and syndiotactic polystyrene [32]). The practical scope of TPV preparation from an elastomer–thermoplastic combination was largely explored by Coran and Patel, [33] who prepared the predominant TPVs from the 11 most common rubbers and 9 most common plastics, as shown in Table 35.5.

A prime criterion of elastomeric properties is ultimate elongation. Table 35.6 shows the percent elongation at break for each of the elastomer–thermoplastic compositions prepared. To classify the material as an elastomer, the ultimate elongation is required to be greater than 100%, according to ASTM D1566. A second criterion for rubberlike properties is a tension set of 50% or less after an extension of 100%, according to ASTM D-42. These criteria are shown in Table 35.6 for the blends in Table 35.5. From these two criteria, attractive compositions can be defined for further exploration. All rubber–PP blends, except for acrylate and CPE, did meet these criteria. The acrylate–PP TPV was later improved through the incorporation of a compatibilizer [34], which enhanced the TPV's classification to that of an elastomer.

Table 35.5   Elastomers and Thermoplastics in Compositions

Elastomers	Thermoplastics
Polyacrylate (ACM)	Acrylonitrile–butadiene–styrene copolymer (ABS)
Polybutadiene (BR)	Polyamide (nylon) (PA)
Chlorinated polyethylene (CPE)	Polycarbonate (PC)
Polychloroprene (CR)	Polyethylene (PE)
Ethylene–propylene–diene monomer (EPDM)	Poly(methyl methacrylate) (PMMA)
Ethylene–vinyl acetate copolymer (EVA)	Polypropylene (PP)
Butyl rubber (IIR)	Poly(tetramethylene terephthalate) (PTMT)
Nitrile rubber (NBR)	Polystyrene (PS)
Natural rubber (NR)	Styrene–acrylonitrile copolymer (SAN)
Poly trans-pentenamer (PTPR)	
Styrene–butadiene rubber (SBR)	

Table 35.6   Ultimate Elongation (%) of Elastomer–Thermoplastic (60/40) Compositions[a]

	Thermoplastics								
Elastomer	ABS	PA	PC	PE	PMMA	PP	PS	PTMT	SAN
ACM	144	163	140	20	21	18	20	135	18
BR	64	121	5	229	5	258	73	52	12
CPE	197	160	135	221	146	314	140	159	151
CR	96	6	91	390	5	151	67	65	7
EPDM	18	30	66	612	6	530	69	102	5
EVA	102	160	81	349	59	319	166	126	106
IIR	18	34	161	312	6	380	3	156	7
NBR	164	320	130	190	56	201	20	350	196
NR	56	42	21	360	58	390	85	62	14
PTPR	15	60	5	280	10	210	35	47	10
SBR	70	201	19	240	15	128	89	102	12
ACM	9.4	16.1	5.2	4.2	6.2	4.0	11.4	14.6	7.7
BR	9.9	16.3	2.1	19.3	3.5	20.8	11.6	12.8	8.3
CPE	13.7	17.3	20.8	10.5	17.0	12.3	14.0	13.0	17.9
CR	12.8	3.2	14.7	13.8	8.9	13.0	15.5	13.5	12.5
EPDM	3.2	7.7	15.7	16.4	6.0	24.3	7.9	12.2	5.6
EVA	9.6	10.9	9.6	18.9	9.3	17.8	12.7	3.4	12.9
IIR	1.7	4.0	1.3	14.9	5.4	21.6	0.9	1.4	4.3
NBR	13.6	21.5	18.2	17.6	10.8	17.0	7.7	19.3	25.8
NR	5.8	5.7	6.7	18.2	1.2	26.4	6.2	10.9	8.4
PTPR	11.0	10.8	2.5	12.1	4.7	22.7	6.9	12.1	13.4
SBR	10.8	14.6	7.3	17.1	5.7	21.7	15.8	21.7	8.1

[a] Spelled-out names for the elastomers and thermoplastics can be found in Table 35.5.

Another important property of these compositions is tensile strength. In general, the ultimate tensile strength (UTS) parallels the elongation values in Table 35.6. It is not surprising that those compositions having a high elongation also give high UTSs.

Table 35.7   Tension Set of 60/40 Rubber–Plastic Thermoplastic Vulcanizates

Rubber–Plastic	PP	PE	PS	ABS	SAN	PMMA	PTMT	PA	PC
IIR	23	28							
EPDM	16								
PTPR	20	27							
NR	24								
BR	27								
SBR	30								
EVA	36	36	70					26	
ACM							41	56	17
CPE	55	58		65	91	82	40	59	85
CR	33	37							
NBR	31				55		25	44	

The data show PP and PE to be the best thermoplastic candidates for blending and dynamic vulcanization with the elastomers, in accord with their hydrocarbon nature. Thus, the nonpolar elastomers interact best with nonpolar thermoplastics. Similarly, the polar rubbers interact best with the polar plastics, which correlate with a match of the solubility parameter δ [35] or the critical surface tension γ_c [36]. The smaller the difference between the solubility parameter of the two molten polymers, the smaller the droplet size of one polymer into the other during mixing. The smaller difference in surface tension for wetting is a fair estimate of the interfacial tension between the two phases, which in turn affects the droplet size during mixing. (Table 35.7.)

Another parameter [37] is the percent crystallinity in the plastic phase, W_c, which influences the elastic recovery and mechanical properties. The third polymer characteristic is the critical entanglement spacing N_c [37], which corresponds to the molecular weight that is sufficiently large for entanglements to occur in the neat rubber. A polymer that has a low N_c would have a high entanglement density and a low value for the molecular weight between entanglements. Low N_c values lead to high ultimate elongation in the TPV.

H. TPV's Relation to Thermosets

The utility of novel materials such as TPVs, requires a consideration of their performance and cost relative to other available materials. Since thermoplastic elastomers are intended primarily to replace thermoset rubbers, in order to conserve energy and achieve recyclability, comparing their performance to that of thermoset rubbers, in accordance with ASTM D2000 and SAE J 200, might be the first step foward using these products in a practical manner.

Figure 35.9 compares the performance of the various thermoset rubbers with that of TPV with regard to upper service temperature in air and the percent swell in ASTM oil #3 (now IRM 903) at that upper service temperature. Performance requirements, however, are not limited to thermal and oil resistance since other characteristics—namely, elastomeric recovery, impermeability, dynamic properties, UV resistance, and mechanical properties, among other parameters—

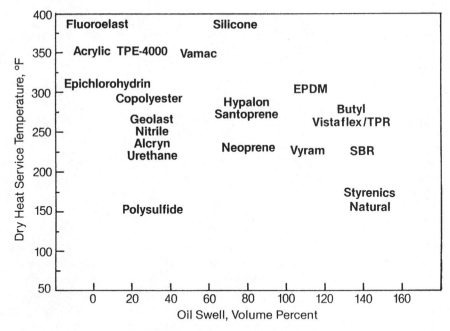

Figure 35.9 Elastomer performance chart.

also play a significant role. Nevertheless, Fig. 35.9 represents a good initial comparison.

Nonpolar elastomers in a matrix of a nonpolar thermoplastic materials, one class of TPVs, represent the most popular product of commercial and academic interest. This class of TPVs encompasses the products of at least seven hydrocarbon rubbers with polyolefin plastics using multiple cure systems. Hydrocarbon rubbers and polyolefin plastics are similar in terms of molecular structure, polarity, and the virtual absence of hydrogen bonding. Thus, it is not surprising that their interfacial tensions are similar. Consequently, the phase separation in this type of TPV is not controlled by strong thermodynamic forces. The most common product of this kind is a blend of EPDM and iPP, which is qualitatively better than thermoset rubbers.

Another outstanding characteristic of TPVs is their dynamic properties. Comparison of TPVs with thermoset rubbers by the Monsanto fatigue-to-failure test has repeatedly shown that EPDM–PP TPVs exhibit better fatigue performance than do conventional rubber [38], polychloroprene, hypalon, and EPDM thermosets. (See Fig. 35.10.) The dynamic mechanical properties of an EPDM–PP blend are shown in Fig. 35.11. The elastic modulus E', the loss modulus E'', and tan δ are normally used to predict performance in environments with vibration and for designs requiring resilience.

100% Extension, 100 Cycles/Min.

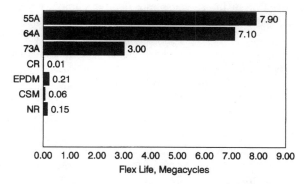

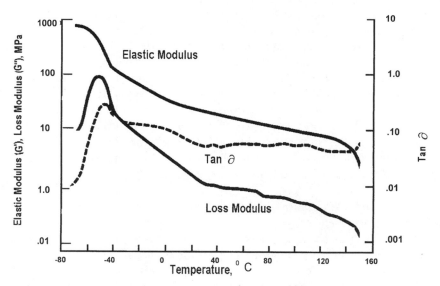

Figure 35.10   Dynamic properties and flex fatigue life of TPVs and elastomers.

Figure 35.11   Dynamic mechnical properties of EPDM–PP TPVs with 73 Shore A hardness.

III. TYPES OF TPVS

A. Nonpolar Rubber with Nonpolar Thermoplastic

1. *EPDM Elastomer with iPP Thermoplastic*   TPR, from Uniroyal, was the first commercial thermoplastic olefin elastomer based on dynamic vulcanization and was introduced in 1972. Shortly thereafter, EPDM producers introduced their own versions of mechanical blends of their polymers and polypropylene, commonly referred to as TPOs [38]. Thermoset replacement by TPE in general and TPV in particular, however, did not occur until 1981, with the introduction of Santoprene® thermoplastic rubber by Monsanto Company [39]. In Santoprene thermoplastic rubber, the complete vulcanization of the rubber phase [3], the choice of preferred vulcanization systems [4], and the generation of preferred morphology of elongated rubber particles led to a proper balance of good mechanical properties, good elastomeric resistance, moderate oil resistance, and excellent processing characteristics.

EPDM–PP TPVs can be prepared with hardnesses ranging from 35 Shore A to 50 Shore D. As the ratio of EPDM to PP is increased, the hardness decreases and the properties become more rubberlike and less rigid. Figure 35.12 illustrates the stress–strain curves for different hardnesses of this TPV at 25°C. The effects of temperature on the stress–strain properties can be seen in Fig. 35.13. As the temperature is increased, the material behaves in a more linear, or rubberlike, fashion. With decreasing temperature, the "knee" in the curve becomes more prominent, especially with the hard compositions.

Soft blends of EPDM–PP TPV—73 Shore A and below—have a compression set (see ASTM D395, Method B) that is very low. The level of this set is competitive with that of specially compounded thermoset rubber compounds. Figure 35.13 compares the variations in compression sets with time for three soft EPDM–PP TPVs with those of specially compounded neoprene and EPDM rubber compounds, showing that the TPVs to provide excellent compression sets. Compression-set measurement of TPE, unlike that of thermoset rubber, is subject to more variation, because the recovery of the TPV is tied to the crystallinity in the plastic matrix. Exposure to high temperatures for extended periods of time under compression followed by cooling after relaxation of compressive strain could lead to uncontrolled change in crystallinity and, thus, in recovery (Fig. 35.14). Stress relaxation provides a better measurement of elastomeric recovery [40]. Fig. 35.15 illustrates such a comparison between TPVs and conventional thermoset elastomers.

2. Natural-Rubber Elastomer with PE Thermoplastic   Natural rubber (NR) has an inherent affinity to blend with polyolefin plastics and further be dynamically vulcanized to form TPVs that have the preferred dispersed-phase morphology. This process occurs without the need for compatibilizers. However, natural rubber is a highly unsaturated rubber and suffers from poor oxidative stability. Its use as a vulcanizable component in a TPV has been hindered by the requirement that it withstand repeated exposure to high temperatures—higher than the melting point of the plastic used, i.e., higher than 165°C, and preferably 180–200°C in the case of PP—to accommodate mixing, fabrication, and recycling. Also, natural rubber is a naturally occurring material that contains thermally degradable proteins that generate undesirable, odorous products upon decomposition. Although the unpleasant effects associated with the degradation of TPVs based on natural rubber on exposure to heat are less dramatic than the effects of degradation of the vulcanized elastomer, they are still significantly more problematic than the effects of degradation of an EPDM-based TPV. These facts require consideration with regard to the selection of the cure system, the plastic phase, and a stabilizer package, in order to formulate around this instability.

Natural rubber and polyethylene can be dynamically vulcanized to form a TPV by using a peroxide or an efficient sulfur vulcanization system for NR. It is expected that

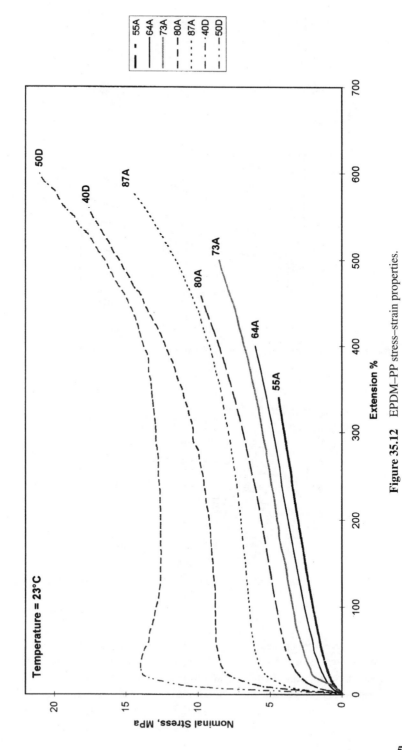

Figure 35.12 EPDM–PP stress–strain properties.

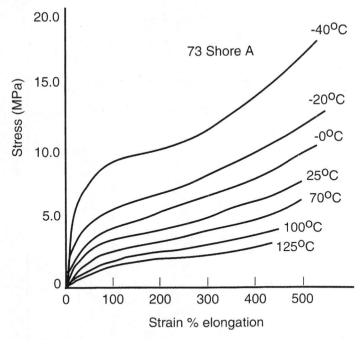

Figure 35.13 Effect of temperature on stress–strain properties for TPVs with 73 Shore A hardness.

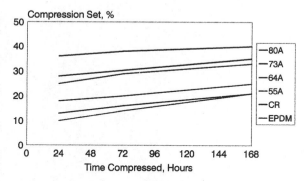

Figure 35.14 Compression-set resistance for TPVs and elastomers. Time dependence at 100°C.

the peroxide curatives partially cross-link the polyethylene. The sulfur curatives are based on low levels of elemental sulfur in order to minimize reversion of the NR cure. In a laboratory preparation, mixing at 150°C for a period of 3–4 minutes is adequate. The temperature of mixing must be above the melting point of the polyethylene (i.e., 125°C). Longer mixing times and higher levels of sulphur and accelerator in the mix cause the material to lose its thermoplasticity and mechanical properties

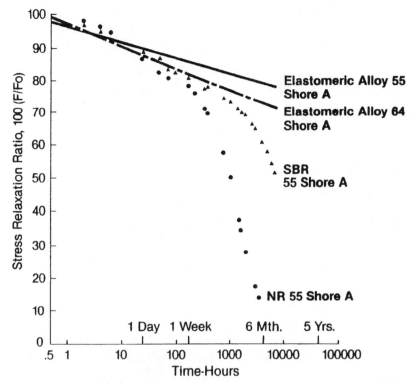

Figure 35.15   Stress relaxation for TPVs and elastomers. Time dependence at 70°C.

[41], due to degradation of the plastic phase and thermal oxidation of the NR. However, better dispersion of the components in the blends is only achieved for longer mixing times. An example of a formulation of an NR–PE TPV and its corresponding physical properties is shown in Table 35.8.

Just as with similar blends and their corresponding TPVs, the formation of the vulcanizate for NR–PE blends improves the tear strength and hardness while decreasing the tension set. Table 35.9 shows the effect of other additives on mechanical properties. On addition of carbon black to the TPV, there is an increase in tensile strength and the 100% modulus. With increasing PE content, the effect of carbon black is reduced. Other fillers have also been used to stiffen the TPV to reduce cost by increasing volume. The use of plasticizer along with filler is always necessary in these PE-based TPVs.

The properties of TPVs can be changed significantly through the selection of the cross-linking agent. This practice is more prevalent with NR–PE blends that have a high rubber content. Several cross-linking agents, such as sulphur, sulphur donors, peroxides, dimaleimides, diurethane, and phenolic resin, may be used. The degree of cross-linking and the related properties are dependent on the nature of the curative used. NR–PE TPVs have better physical properties [42] for sulfur-based cure systems, compared to peroxide curatives.

Table 35.8   Formulation and Physical Properties of NR–PE Blends and TPVs

	PE	A	Ac	B	Bc	C	Cb	Cc	Cbc	NR
Composition										
HDPE		70	70	50	50	30	30	30	30	
NR		30	30	50	50	70	70	70	70	
HAF black		–	–	–	–	–	40	–	40	
Dicumyl peroxide (DCP)		0	1	0	1	0	0	1	1	
Properties										
Young's modulus (MPa)	29.4	13.4	14.5	5.8	6.1	1.4	2.4	2.1	4.5	0.4
Modulus 100% (MPa)	–	–	–	7.2	8.7	2.2	4.6	4.1	5.7	0.8
Tensile strength (MPa)	32.1	13.5	14.6	10.9	11.3	9.5	13.0	13.7	12.7	2.3
Elongation at break (%)	30	65	24	420	430	460	S00	470	300	450
Tear strength ($kN \cdot m^{-1}$)	118.0	72.5	74.4	47.2	55.7	28.1	40.8	43.6	36.4	4.35
Tension set (%)	–	–	–	92	>100	72	64	70	68	>100

Table 35.9   Effect of Changes in Formulation on NR–PE TPVs

Composition	A	B	C	D	E	F	G	H	I	J
NR	70	70	70				70	70	70	70
CPE			20							
ENR							20			
EPDM				70	70				20	
NBR						70				
LDPE	30	30	30	30			27	27	30	30
PE–MA							3	3		
PP					30	30				
DCP	0.5		0.5					0.5		
ZnO		5		5	5	5			5	5
Stearic Acid		1		1	1	1			1	1
S		2.5		2	2	2			2.5	2.5
CBS		0.8							0.8	0.8
TMTD				1	1					
MBT				0.5	0.5	0.5				
Properties										
Modulus (MPa)	30	54	42	24	60	21	35	30	45	50
Tensile strength (MPa)	6.4	15.5	8.4	1.5	12.3	7.9	5.2	5.3	9.5	11.5
Tear Shear (kJ/m^{-2})	19.0		28.2	12.5	27.0		18.0	19.0	54.2	64.0
Tear Trouser (kJ/m^{-2})	17.0	120	12	15	28	0.6	19.8	15		

The mechanical properties of NR–PE TPVs can be enhanced severalfold by using diluents for the thermoset and plastic phases [43]. Diluents for the NR include EPDM, chlorinated polyethylene (CPE), and chlorosulphonated polyethylene (CSPE), and diluents for the PE include a combination of polyethylene modified by maleic anhydridc (PE–MA) and epoxidized natural rubber (ENR). It was observed that 10% of PE–MA and ENR diluents in their respective phases could improve the tensile strength by 45% over that of NR–LDPE blends. This improvement is due to the formation of graft compatibilizers from the reaction of the the anhydride group and the epoxy group of the ENR.

In comparison with vulcanized rubber, NR–PE blends are resistant to heat ageing [44]. This was determined by ageing NR–LDPE thermoplastic elastomers in air at various temperatures for varying duration. In general, for all blends including dynamically vulcanized and modified systems, at a particular time in the ageing process, the strength and modulus decrease with increasing temperature. However, the durations change in properties in severely oxidizing environments is substantially slower for the TPV than for the NR alone [45].

3. Natural-Rubber Elastomer with Polypropylene Thermoplastic   The driving force for attempts to develop analogues of the Santoprene® thermoplastic rubber that use NR as the elastomer is the belief that such analogues are more economical and have a high green strength due to strain-induced crystallization. These advantages would be in addition to the high use temperature of iPP. However, in practice, these TPVs have been unable to replicate the stability and the processability of EPDM–PP TPVs.

NR–iPP TPVs are made by intensive melt mixing of a mixture of NR and iPP after the NR is masticated to reduce the viscosity in the presence of a curative. The mixing temperature must exceed 165°C in order to melt the iPP. The mixing condition should have a maximum temperature of about 185°C. In laboratory conditions, mixing times of about five minutes are used. Higher temperatures are undesirable, because some degradation of the NR is likely to occur above 200°C. On the other hand, if the temperature is too low, the required degree of reactive mixing is not achieved. NR polymers that are to be used for the preparation of TPV should not contain gel or have an initial viscosity that is too high. These conditions lead to poor dispersion of the NR in the iPP matrix.

It is believed, without experimental verification, that the cross-linking reaction of NR in the iPP matrix leads to the formation of block polymers by grafting of iPP to the NR. It is known that mastication of NR [46] produces free radicals that may attack iPP. This mechanism is more plausible in the presence of cross-linking agents that generate free radicals.

Coran and Patel [47] cite NR–PP compositions cured with sulfur, urethane, and phenolic resin curatives. Tinker, Icenogle, and Whittle [48] describe thermoplastic NR–iPP blends that are peroxide cured. Peroxide cross-linking of NR–iPP TPVs is accomplished with about 0.7 phr peroxide in a blend of 85% NR and 15% iPP [49]. Larger amounts of peroxide lead to nonelastic TPVs, because of degradation in the molecular weight of iPP and oxidation of NR. A disadvantage of peroxides

Table 35.10   Typical Mean Properties of NR–iPP TPVs

Shore A hardness	50	60	70	80	90
M100 (MPa)	3.1	4.1	5.3	6.3	8.2
Tensile strength (MPa)	6.5	8.8	11.2	13.2	15.3
Elongation at break (%)	285	315	340	370	405
Tear strength					
Die C (kN·m^{-1})	22	31	38	43	51
Tension set (%)	9	10	14	18	23
Compression set (%)					
72 hours for 23°C	26	28	31	34	39
24 hours for 70°C	30	35	36	41	55
22 hours for 100°C	38	42	43	47	57
Volume swell (7 days)					
in ASTM No. 1 oil at 23°C	14	9	9	7	
in ASTM No. oil at 23°C	19	13	12	9	
in ASTM No. 3 oil at 23°C	71	53	47	35	
in ASTM No. 1 oil at 100°C	101	80	67	61	
in ASTM No. 2 oil at 100°C	151	123	108	82	
in ASTM No. 3 oil at 100°C	190	164	139	116	

is that they initiate oxidation of the NR. This effect is manifested as a deterioration in the resistance of NR to heat ageing. Other cross-linking systems based on elemental sulphur or sulphur donors, such as thiuram disulphides and dithiobismorpholine, are more suitable for the formation of NR–iPP blends. The cure system needs to be reversion resistant in order to obtain stable phase morphology [47]. Diurethane cross-linking reagents [50] have been used to cross-link the NR phase in these TPVs. Cross-linking of alkyl phenol-formaldehyde resins with NR has been studied [51] and has been utilized for the dynamic cross-linking of NR–iPP TPVs. N, N'-metaphenylene dimaleimide is a cross-linking agent for elastomers, including NR. The dimaleimide cross-links in NR are very stable to heating. This property makes dimaleimide an attractive material for the dynamic cross-linking of NR–iPP TPVs. Note that a free-radical activator, such as an organic peroxide or MBTS, is required for this function. A curative level of 7.5 phr with 1.5 phr of MBTS was used to dynamically cross-link NR–iPP blends containing 50–60% NR [46].

The hardness of the TPV can be adjusted by varying the ratio of NR to iPP. Other variables that can affect the hardness are plasticization with compatible oils and the addition of fillers. Some typical dynamic properties are given in Table 35.10 for grades of NR–iPP TPV ranging from 50 to 90 Shore A hardness [48]. The tensile strengths shown in Table 35.10 are considerably lower than the maximum strengths obtainable with vulcanized NR, although they are roughly comparable to those of typical nonblack SBR and BR vulcanizates. However NR–iPP TPVs do have compression sets comparable to that of a thermoset elastomer at 70°C. Just as with Santoprene® thermoplastic rubber, these TPVs are resistant to solvents, since the values for swelling at 100°C are only about two thirds of those observed for some NR vulcanizates at equivalent hardnesses.

Table 35.11   Typical Permeability Performance of Butyl–PP TPVs

Standard Test	Media	Trefsin® 3201–60	Trefsin® 3101–65 W302	Polyurethane	Butyl Thermoset (Bladder)
ASTM D-1434	Air[1]	2.1	0.3	0.7	0.3
ASTM D-3985	Oxygen transmission rate[2]	650	123	160	77–127
ASTM F-1249	Water-vapor transmission rate[3]	1.19	0.57	46	0.37–0.62

® Registered trademark of Advanced Elastomer Systems.
[1] Unit $= 10^9 \times m^3 \times m$ (at STP)/$m^2 \times$ day $\times$ KPa at 23°C.
[2] Unit $= cm^3 \times mm/m^2 \times$ day at 23°C (50–75% relative humidity).
[3] Unit $= gr \times mm/m^2 \times$ day at 23°C (90% relative humidity).

Paintability and bondability are attributes that have been cited for NR compositions, due to the unsaturation in the NR phase. Ellul and Hazelton [52] described a use of the unsaturation in the NR as permiting surface halogenation, in order to improve bondabiltiy. Riddiford and Tinker [53] described a TPV based on modified NR in which a portion of the double bonds have been epoxidized to produce an oil-resistant material.

4. Butyl Rubber Elastomer with iPP Thermoplastic   Butyl rubber thermoset compounds are used in the market wherein impermeability to gases, namely, O_2 and H_2O, is desired. TPVs that comprise butyl rubber [54] are capable of duplicating the inherent low gas permeability (see Table 35.11) of thermoset butyl rubber. These TPVs contain butyl, halogenated butyl rubber, or any of the newer halogenated *p*-methyl styrene–isobutylene copolymers [55]. TPVs using copolymers of isobutylene and the component of the elastomer phase are capable of outstanding long-term heat ageing, which is maximized with highly stable cure systems [56]. Such products can withstand prolonged exposures at temperatures up to 150°C. The advantage of thermal stability for the polyisobutylene-containing rubber phase is believed to be due, at least in part, to the fact that most elastomers tend to continue to cross-link, with consequent loss of tensile elongation. Butyl polymers, however, tend to chain scission, thus losing tensile strength but retaining elongation and softness.

B. Polar Rubber with Nonpolar Plastic

1. NBR Elastomer with iPP Thermoplastic   TPVs based on (Nitrile Butadiene Rubber) NBR as the elastomer and iPP as the thermoplastic should have the ideal combination of elastomeric properties and resistance to solvents. However, these TPVs are difficult to form, a problem that is due primarily to the difficulty in attaining, during the dynamic vulcanization process, the particle size of vulcanized rubber that is sufficiently small for forming good TPVs. Therefore, the preparation of a TPV from the dissimilar polymers NBR and iPP requires a compatibilizer in order to improve the interfacial adhesion between the two phases [57], to lower the differences in surface energy, and to enable the formation of a fine dispersion prior

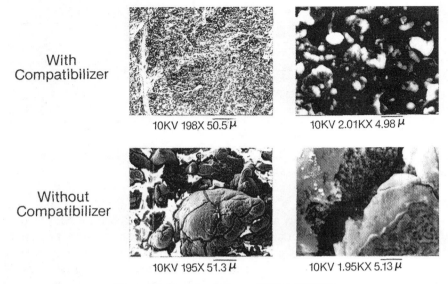

With
Compatibilizer

10KV 198X 50.5 μ   10KV 2.01KX 4.98 μ

Without
Compatibilizer

10KV 195X 51.3 μ   10KV 1.95KX 5.13 μ

Figure 35.16   Morphology of NBR–PP TPVs.

to vulcanization. A secondary effect of using a compatibilizer is a stronger interaction between the two phases, leading to increase in the tensile strength of the TPV. The function of this compatibilizer is to provide greater, but not complete, thermodynamic compatibility between the two polymers.

Two means of compatibilizing NBR with iPP are available: (1) the use of a block copolymer, and (2) the formation of polymer–polymer grafts between the NBR and PP. Compatibilization is reflected in improvement in the physical properties of the composition, accompanied by changes in morphology (see Fig. 35.16) and fracture surfaces. Also, PP can be functionalized with phenolic resin [58]. This functionality can subsequently react with the NBR, creating a polymer–polymer graft:

The phenolic modification of the PP gives improved properties of the resulting blend. (See Composition 2 in Table 35.12.) These properties are enhanced by dynamic vulcanization of the NBR to the TPV. (See Composition 3 in Table 35.12.) Maleic-anhydride-modified PP can be used to react with amine-terminated NBR

Table 35.12   Phenolic-Modified NBR–PP Blends

	Sample Number		
	1	2	3
Composition			
iPP	50	0	0
Phenolic-modified PP[a]	0	50	50
NBR	50	50	50
Phenolic resin cross-linker	0	0	1.7
Properties			
UTS (MPa)	7.2	10.1	10.5
Stress at 100% strain (MPa)	–	–	10.2
Young's modulus (MPa)	182	170	157
Ultimate elongation (%)	24	66	170
True stress at break (MPa)	8.9	16.8	28.4

[a]Prepared by melt mixing PP (100 parts) with dimethylolphenol resin (2 parts) and tin(II) dichloride dihydrate (0.4 parts).

Table 35.13   Data on Composition and Physical Properties for Compatibilized NBR–PP TPVs

	Composition Number					
	1	2	3	4	5	6
Recipe[a]						
PP	50	45	45	45	45	45
Maleic-modified PP	0	5	5	5	5	5
NBR	50	50	49.92	49.37	48.75	45
ATBN	0	0	0.08	0.63	1.25	5
Properties						
Ultimate tensile strength (MPa)	8.8	12.0	12.4	22.0	25.7	26.7
Ultimate elongation (%)	19	110	170	400	430	540
Stress at 100% strain (MPa)	–	12.0	12.1	12.3	12.5	12.9
Young's modulus (MPa)	209	200	212	185	184	237
True stress at break (MPa)	10	25	33	110	136	171
Tension set (%)	–	–	45	40	42	45
Breaking energy (J/cm^3)	1.2	11.0	18.4	54.9	61.7	86.5
Improvement in breaking energy (%)	–	–	67	399	461	686

[a]All parts by weight. All compositions contain 3.75 parts phenolic resin and 0.5 part tin(II) dichloride dihydrate as NBR vulcanizing system. Maleic acid (5 parts) melt mixed with PP (100 parts); 2,5-bis (t-butylperoxy)hexane (0.87 parts) then added.

(ATBN) to give a block copolymer. This block copolymer can be made independently or generated *in situ* during the mixing process. The resulting improvement is illustrated by the data in Table 35.13, which shows that 10% of the PP was replaced by maleic-anhydride-modified PP and that the percentage of amine-terminated NBR increased progressively. The tensile properties improved progressively with the amount

Table 35.14   Comparison of Resistance of EPDM–PP and NBR–PP TPVs to Fluid Media[a]

		Percent Retention		
Fluids and Conditions	Weight change (%)	Elongation	Ultimate tensile	100% modulus
ASTM No. 1 oil, 100°C	17.3/ − 13.2	87/103	82/120	94/113
ASTM No. 2 oil, 100°C	33.4/ − 4.2	81/104	78/118	95/108
ASTM No. 3 oil, 100°C	48.4/ − 0.3	70/85	71/90	93/94
Isooctane, 23°C	18.9/ − 10.6	83/104	77/99	84/88
Isooctane/toluene (70/30), 23°C	33.6/17.1	77/67	74/67	78/81
Isooctane/toluene (50/50), 23°C	18.3/17.7	75/67	76/67	79/81
Turpentine, 23°C	50.5/6.2	77/85	70/77	82/89

[a]Both the TPV of EPDM–PP and of NBR–iPP have 80 Shore A hardness. One week, total immersion, ASTM D 471. Data is arranged in the form (EPDM–iPP)/(NBR/iPP).

of ATBN present. In fact, amounts as small as 0.08 parts of ATBN caused a significant improvement. (See Composition 3.)

The compatibilization of PP and NBR to form alloys with improved properties is of particular use in generating TPV that complement the properties of EPDM–iPP TPVs. NBR–iPP TPVs have much lower swell on exposure to hydrocarbon-based fluids. Table 35.14 compares the resistance of these two TPVs (80 Shore A hardness) to a variety of fluids. In nonpolar hydrocarbons, the EPDM TPVs retain their tensile properties reasonably well; however, their swell is much greater than that of the NBR TPVs. This is a logical result of the greater polarity of NBR compared with that of EPDM rubber.

The physical properties of NBR–PP alloys are generally comparable to those of EPDM–PP alloys of the same hardness [59]. Both their tensile and their compression-strain properties are essentially the same. NBR–PP alloys have low compression sets compared to those of EPDM–PP alloys and are suitable for sealing applications in hydrocarbon media, whereas EPDM–PP alloys might not be. The useful temperature range for NBR–PP alloys is slightly narrower than that for EPDM–PP TPVs. The low-temperature limit is approximately −40°C for NBR–PP, versus −60°C for EPDM–PP, due to the higher glass transition temperature of NBR. The upper temperature limit is marginally lower (5–15°C) for NBR–PP than that of EPDM–PP. This is because the upper temperature limit is determined by the resistance to air oxidation: NBR has olefinic unsaturation in the polymer backbone, rendering it less resistant to air oxidation, while EPDM does not. NBR–PP TPVs have been commercialized by Advanced Elastomer Systems L.P. under the trademark Geolast® thermoplastic rubber [10].

2. Acrylate Rubber with iPP Thermoplastic   The acrylate rubber used in this kind of TPV specifically consists of ethylene, methylacrylate, and a monomer with a carboxylic-acid side group. This acrylate rubber is commonly known as Vamac®

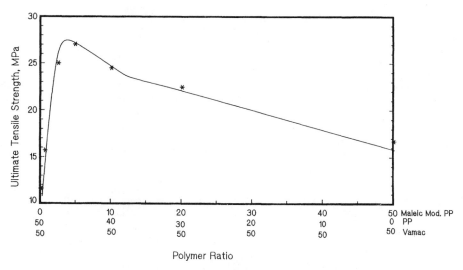

Figure 35.17   Effect of compatibilizer on acrylate rubber–iPP TPVs.

and can be cross-linked with multivalent amines. This particular acrylate rubber is of interest because of its properties at low temperatures and its superior heat-resistant properties. A 50:50 blend of Vamac® and polypropylene was dynamically vulcanized [60] with 4,4-methylene dianiline (MDA) and 4,4-methylene bis (cyclohexylamine) carbonate. To affect compatibilization, some PP was proportionally replaced with maleated PP. The results of this study can be seen in Fig. 35.17. Optimum tensile properties were achieved with less than 5 wt% of compatibilizer. The addition of extra maleated PP caused a significant drop in properties, due to at least two factors—the replacement of high-molecular-weight polypropylene with lower molecular-weight maleated PP and the increased level of competing reaction between the diamines with the carboxylic functionality of Vamac and succinic anhydride moeity of the maleated PP. Furthermore, the generation of primary chemical bonds between the two phases could have led to an interpenetrating network morphology that is expected to provide poor thermoplastic processing after dynamic vulcanization.

C. Nonpolar Rubber with Polar Thermoplastic

1. *EPDM Rubber with PA-6 Thermoplastic*   Blending of polyamides with elastomers has been studied extensively as a remedy to improve the impact strength of polyamides. However, due to the immiscible nature of the polymers, these blends have poor mechanical properties. Compatibilization is needed to convert such immiscible blends into commercially useful products. For polyamides containing primary amine groups, the most commonly used reactive functional groups are based on maleic anhydride and epoxide. Examples of maleic-anhydride-functionalized polymers are EPR-g-MA and EPDM-g-MA. Epstein [61] capitalized on this reactivity and contributed to the development of "supertough" nylons. Olivier [62] has ex-

Table 35.15   EPDM–Nylon TPVs*

Compounds	I	II	III	IV
EPDM	25	25	25	25
EPM	25			
EPM-g-MA		25	25	25
Nylon-11	50	50	50	50
n-butyl-4,4-bis (t-butyl peroxy) valerate	–	–	1.25	
Tensile strength, MPa	4.6	6.5	18.8	19
Ultimate elongation, %	25	50	200	245

*Used in all case 2.5p ZnO, 0.5p methyl tuads, 0.25p MBTS, 1p S.

Table 35.16   Tensile and Set Properties of EPDM–PBT TPVs

Rubber Phase	Tensile Strength (psi)	Elongation %	Tension Set (%)
EPDM	2450	110	fails
EPDM-g-glycidyl methacrylate	3300	285	15
EPDM-g-butyl methacrylate	2000	75	fails
EPDM-g-glycidyl acrylate	3041	200	18

tended this reactivity to the preparation of TPVs wherein the functionalized EPDM reacts with the amine group of the polyamide. The functionalized EPDM has either succinic anhydride or epoxide functionalities. Properties of such TPVs are shown in Table 35.15.

2. EPDM Rubber with PBT Thermoplastic   The need for TPVs that adhere to engineering thermoplastics has led to the use of PBT as the continuous thermoplastic phase of the TPVs. Moffett et al. and, later, Campbell et al. [63] demonstrated the formation of TPVs of this composition by using EPDM grafted with 3% acrylate monomers (e.g., butyl acrylate or glycidyl acrylate). The grafted elastomers were required in order to reduce the interfacial tension for EPDM–PBT TPVs. The elastomer phase was cured with peroxide to form TPVs with excellent tensile and set properties. (See Table 35.16.) Grafting with acrylate reduces the mean size of the dispersion of the EPDM and improves the mechanical properties of the TPV.

3. EPDM Rubber with iPP with PA-6 Thermoplastic   The high-temperature-use characteristics of TPV are governed by the loss of mechanical properties of the thermoplastic phase. Higher service temperatures for TPVs of EPDM rubber and iPP thermoplastic and of butyl rubber and iPP thermoplastic can be obtained using a compatibilized blend with PA-6 [64]. In these TPVs, maleated PP is the compatibilizer and the blend of iPP and PA-6 is the thermoplastic phase. Examples of the formulation and mechanical properties of such TPVs are shown in Table 35.17. These compatibilized systems have the characteristic tensile properties of TPVs. The removal of the compatibilizer leads to loss of tensile strength and failure of the speci-

Table 35.17 TPV Based on Compatibilized Blends of PA-6 and iPP

Composition	I	II
PA-6	10	10
80% PA-6 + 20% PP–MA	10	10
EPDM–iPP TPV	40	
Butyl–iPP TPV		40
Properties		
Tensile strength (psi)	2940	1920
Elongation (%)	240	78
100% modulus (psi)	2180	1790
Compression set (%)	54	77
Oil swell (%)	70	72

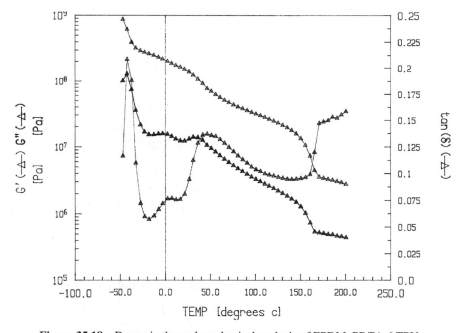

Figure 35.18 Dynamic thermal mechanical analysis of EPDM–PP/PA-6 TPVs.

men on tension or compression. The elevation in the elastic properties of the blend on addition of PA-6, shown by the improvement of the dynamic modulus of the blend, in comparison to the properties of an EPDM–iPP TPV, is shown in Fig. 35.18. In addition to improved use-temperature properties, these blends have substantial resistance to hydrocarbon solvents at elevated temperatures, since a part of the hydrocarbon polymer has been replaced with PA-6.

D. Polar Rubber with Polar Thermoplastic

1. *Acrylate Elastomer with Polyester Thermoplastic*   The need for resistance to hydrocarbon solvents has led to the development of TPVs that contain polar elastomer and thermoplastic components. Patel and, later, Venkataswamy [65] have shown TPVs of such compositions to exhibit low absorption of solvent and minimal degradation of properties on exposure to liquid hydrocarbons. The thermoplastic phase could be PET, PBT, or PC. The flexibility in the choice of the acrylic rubber phase leads to TPVs with very low T_g's. Compatibilization of the blend is achieved by having either a carboxylic-acid or an epoxy functionality on the elastomer.

IV. APPLICATION TECHNOLOGY

A. Rheology

Improving the rheological characteristic of TPV without sacrificing the elastomeric properties of the material is an area of current interest [66]. TPVs have extremely non-Newtonian flow properties [67]. Dynamic cross-linking of the rubber phase causes an increase in the viscosity of the TPV. The effect is apparent (see Fig. 35.19) at low degrees of cross-linking, i.e., about 50–60%; higher degrees of cross-linking do not further increase the viscosity [68]. The rheological behavior of these materials is similar to that of filled thermoplastics [67]. At low shear rates, the rheology is anamolous, indicating contact of the rubber particles, although no direct morphological evidence of this contact exists. These particle associations are more prevalent in soft compositions at low shear rates and higher temperatures. Similar interparticle forces have been observed in mineral suspensions [69] and have been characterized as long-lived entanglements between cross-linked elastomer particles [70]. It is believed that at the die and the exit hole, the rubber particles and the matrix are aligned to give the appearance of a pseudo co-continuous rubber phase. These materials show relatively low die swell (see Fig. 35.20), which makes it easy to control the dimensions of the extrusion profile. The effect of temperature on the viscosity is illustrated in Fig. 35.21, and generally these materials have viscosities highly sensitive to shear rate (shear thinning) for process control, may undergo wall slip, and do not obey the Cox–Merz correlation.

B. Processing

The processing parameters of a polymer are highly influenced by the flow properties of the material. This factor has been touched upon in the previous section. These materials have the same advantages of fast thermoplastic processing techniques and, as such, provide a cost advantage over compression molding of thermoset rubber, provided that the performance is comparable. Besides having these processing advantages, these materials, are recyclable, whereas thermoset scrap or used thermoset parts can only be discarded in landfills or possibly burned, to recover the fuel value. One of the disadvantages of these materials is a drying step prior to processing.

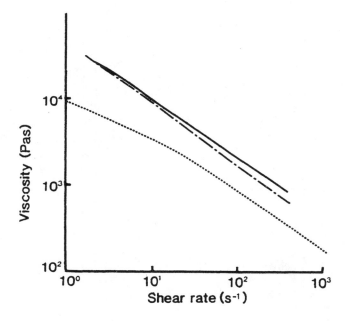

.... Blend   _ _ _ _ _ _ Lightly crosslinked   _____ Highly crosslinked

Figure 35.19   Effect of dynamic vulcanization on viscosity.

However, advances in technology are allowing the availability of materials that do not require drying [71].

C. Blow Molding

This method of fabrication of hollow parts provides the best cost savings compared with the fabrication of such hollow parts from thermosets, for which a solid core is needed. With a TPV, a parison is first extruded or injection molded, and then air is introduced and the molten material is blown against the mold cavity. The blow ratio is usually 2.5:1 or 3:1. Blown air should be dry and free from dust and oil. The mold should be vented. The high melt viscosity of the TPV enables the parison to retain its shape with minimum sag. Hollow TPV parts are depicted in Fig. 35.22.

D. Injection Molding

Injection molding is the most popular means for fabrication in a broad variety of reciprocating-screw injection-molding machines. Clamping tonnage should be three to five tons per square inch of projected shot area. The molding process should ex-

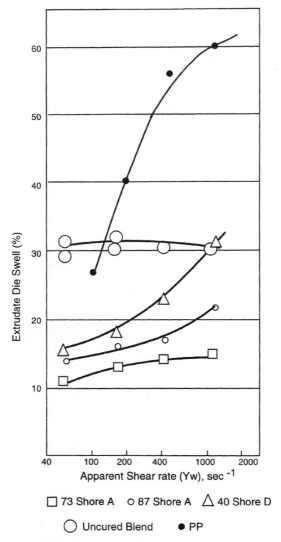

Figure 35.20   Effect of composition on die swell.

ploit the non-Newtonian characteristics of the material by employing high injection pressures and rapid ejection from the mold. Mold shrinkage is in the 1.5–2.5% range and is minimized by adequate mold packing and melt temperature.

E. Extrusion

TPVs can be extruded through complex dies to provide sheeting, tubing, and other profiles. By crosshead extrusion, TPV insulation cover can be applied to hose and to

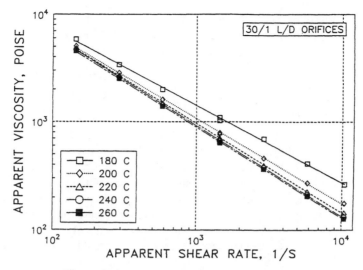

Figure 35.21   Effect of temperature on viscosity.

Figure 35.22   Hollow TPV articles: rack and pinion automotive parts.

electrical wire and cable. Extruders ($L/D \geq 20$) should be used with TPV for good melt homogeneity. A variety of screw designs are suitable, with a compression ratio between 2.0:1 and 4.0:1. The melt temperature should not exceed 250°C. Proper dimensional control of the extrudate requires that the the small amount of die swell be considered. Factors increasing the die swell are decreasing melt temperature, increasing shear rate at the die, and increasing thermoplastic-to-elastomer ratio.

V. SUMMARY

The history of dynamic vulcanization and the effects of complete vulcanization of the rubber component of the plastic–rubber blend has been presented, with emphasis on the commercial significance. The materials produced by this process are called "thermoplastic vulcanizates" and provide a route to producing thermoplastic elastomers other than the conventional reactor technology for synthesizing elastomeric block polymers. The parameters and molecular requirements to stabilize the dispersed-phase morphology for thermoplastic vulcanizates are discussed, and a wide range of different thermoplastic vulcanizates have been compared to thermoset elastomers.

VI. REFERENCES

1. A. M. Gessler and W. H. Haslett, U.S. Patent 3,037,954 (1962).

2. W. K. Fischer, U.S. Patent 3,758,643 (1973); U.S. Patent 3,835,201 (1974); U.S. Patent 3,862,106 (1975).

3. (a) A. Y. Coran, B. Das and R. P. Patel, U.S. Patent 4,130,535 (1978).
   (b) A. Y. Coran and R. P. Patel, *Rubber Chem. Technol.*, **54**, 91 (1981).
   (c) A. Y. Coran and R. P. Patel, *Rubber Chem. Technol.*, **54**, 982 (1981).
   (d) A. Y. Coran and R. P. Patel, *Rubber Chem. Technol.*, **55**, 116 (1982).
   (e) A. Y. Coran and R. P. Patel, U.S. Patent 4,355,139 (1982).
   (f) A. Y. Coran, R. P. Patel, and D. Williams, *Rubber Chem. Technol.*, **54**, 1063 (1982).
   (g) A. Y. Coran and R. P. Patel, *Rubber Chem. Technol.*, **56**, 210 (1983).
   (h) A. Y. Coran and R. P. Patel, *Rubber Chem. Technol.*, **56**, 1045 (1983).
   (i) A. Y. Coran, R. P. Patel, and D. Williams-Headd, *Rubber Chem. Technol.*, **58**, 1014 (1985).

4. S. Abdou-Sabet and M. A. Fath, U.S. Patent 4,311,628 (1982).

5. M.T. Payne and C.P. Rader, in *Elastomer Technology Handbook*, N. P. Cheremisinoff (ed.), CRC Press, Boca Raton, FL, 1993, p. 568.

6. A.Y. Coran, and R.P. Patel, *Rubber Chem. Technol.*, **53**, 141 (1980).

7. D. R. Paul and J. W. Barlow, in *Multiphase Polymers*, S. Cooper and G. M. Estes (eds.), American Chemical Society, Washington, DC, 1979, p. 315.

8. D.J. Lohse, *Ann. Tech. Conf. Soc. Plast. Eng.*, **301**, 1985.

9. (a) A. Y. Coran and R. P. Patel, *Rubber Chem. Technol.*, **56**, 1045 (1983).
   (b) O. Olabisi, L.M. Robeson, and M.T. Shaw, *Polymer–Polymer Misability*, Academic Press, New York, 1979, pp. 277 and 321.
   (c) D. R. Paul, in *Polymer Blends*, D. R. Paul and S. Newman (eds.), Academic Press, New York, 1978, Ch. 12.

10. (a) S. Abdou-Sabet, Y. L. Wang, and E. F. Chu, *Rubber and Plastics News*, November 4, 1985.

    (b) S. Abdou-Sabet, Y. L. Wang, and E. F. Chu, Paper No. 19, presented at the 128th National Meeting of the Rubber Division, American Chemical Society, Cleveland, Ohio (October 3, 1985).

11. J. G. Wallace, in *Handbook of Thermoplastic Elastomers*, B. M. Walker and C. P. Rader (eds.), Van Nostrand Reinhold Co., New York, 2nd Ed., 1988, Ch. 5.

12. (a) P. J. Flory, *Principles of Polymer Chemistry*, Cornell University Press, Ithaca, New York, 1953, p. 576.

    (b) G. Kraus, *Rubber Chem Technol.*, **30**, 928 (1957).

13. S. Abdou-Sabet, R. C. Puydak, and C. P. Rader, *Rubber Chem. and Technol.*, **69**, 476 (1996).

14. S. Abdou-Sabet and R. Patel, *Rubber Chem. and Technol.*, **64**, 769 (1991).

15. (a) W. Hofmann, *Vulcanizaton and Vulcanizing Agents*, Maclaren and Sons Ltd., London, 1967, p. 212.

    (b) W. Hofmann, *Vulcanizaton and Vulcanizing Agents*, Maclaren and Sons Ltd., London, 1967, p. 306.

16. *Kirk-Othmer Encyclopedia of Chemical Technology*, Vol. 21, John Wiley & Sons, New York, 1997, p. 460.

17. (a) K. Youckura and S. Shimono, U.S. Patent 4,785,045 (1988).

    (b) A. Madsuda, S. Shimizu, and S. Abe, U.S. Patent 4,247,652 (1981).

18. A. Giller, *Kant. Gummi Kuntst.*, **19**, 188 (1966).

19. J. D. Umpleby, U.S. Patent 4,803,244 (1989).

20. J. A. Chandler, E. Audrey, and M. Glauert, *X-Ray Microanalysis with the Electron Microscope*, Elsevier Publishing, Amsterdam, 1977, Ch. 5.

21. (a) G. W. Avgeropoulos, F. C. Weissert, G. G. A. Bohm, and P. H. Biddison, *Rubber Chem. Technol.*, **49**, 93 (1976).

    (b) J. Yin, E. Zhou, G. Jinn, F. Ji, and D. Huang, *Scientia Sinica*, **29**, 1233 (1986).

    (c) J. Karger-Kocsis, A. Kallo, and U. N. Kuliznev, *Polymer*, **25**, 279 (1984).

22. D. Romanini, E. Garagnani, and E. Marchetti. Paper presented at the International Symposium on New Polymeric Materials, organized by the European Physical Society (Macromolecular Section), Naples, Italy, June 9–13, 1986.

23. D. J. Elliot and A. J. Tinker, "Blends of Natural Rubber and Thermoplastics," in *Natural Rubber Science and Technology*, A. D. Roberts (ed.), Oxford University Press, Oxford, 1988.

24. M. D. Ellul, J. Patel, and A. J. Tinker, *Rubber Chem. Technol.*, **68**, 573, 1995.

25. (a) S. Datta and D. J. Lohse, *Macromolecules*, **28**, 5465 (1993).

    (b) D. J. Lohse, S. Datta, and E. N. Kresge, *Macromolecules*, **24**, 561 (1991).

26. S. Abdou-Sabet, Unpublished work at Advanced Elastomer Systems.

27. M. D. Ellul, U.S. Patent 5,290,886 (1994).

28. H. Schepres et. al., WO 97/39059, International Patent Application 1997.

29. *Handbook of Plastics, Elastomers and Composites*, C. A. Harper (ed.), 2nd Ed., McGraw-Hill, 1992.

30. (a) K. W. Powers and H. C. Wang, U.S. Patent 5,162,445 (1992).

    (b) K. W. Powers and H. C. Wang, *Elastomerics*, Jan. 1992.

    (c) H. C. Wang and K. W. Powers, *Elastomerics*, Feb. 1992.

31. (a) K. W. McKay, F. J. Timmers, and E. R. Feig, WO 96/07681 (PCT application 1996).
    (b) Y. W. Cheung and M. J. Guest, "Structure, Thermal Transitions and Mechanical Properties of Ethylene/Styrene Copolymers," in *Proceedings of Antec '96*, New Orlean, 1996.

32. N. Tshihara, *Macromol. Symp.*, **89**, 553–562 (1995).

33. A. Y. Coran, R. Patel, and D. Williams, *Rubber Chem. Technol.*, **55**, 116 (1982).

34. R. Patel, U.S. Patent 4,555,546 (1985).

35. (a) S. Wu, *J. Phys. Chem.*, **72**, 3332 (1968).
    (b) J. L. Gardon, *J.Phys. Chem.*, **67**, 1935 (1965).
    (c) G. I. Crocker, *Rubber Chem. Technol.*, **42**, 30 (1969).

36. W. A. Zisman, *Adv. Chem. Ser.*, **43**, 1 (1964).

37. J. D. Ferry, *Viscoelastic Properties of Polymers*, 3rd Ed., John Wiley and Sons, New York, 1980.

38. D. J. Synnott, D. F. Sheridan, and E. G. Kontos, in *Thermoplastic Elastomers from Rubber–Plastic Blends*, S. K. De and A. K. Bhowmick (eds.), Ellis Harwood, New York, 1990, Ch. 5.

39. G. E. O'Connor and M. A. Fath, *Rubber World*, **25** (December 1991); *ibid.*, **26** (January 1982).

40. T. Burton, J. L. Delanaye, and C. P. Rader, *Rubber & Plastics News*, December 12, 1988, p. 22; ibid December 16, 1988, p. 16.

41. D. J. Elliott, in *Thermoplastic Elastomers from Rubber–Plastic Blends*, S. K. De and A. K. Bhowmick (eds.), Ellis Harwood, New York, 1990, p. 71.

42. N. Roy Choudhury and A. K. Bhowmick, *J. Mat. Sci.*, **25**, 161 (1990).

43. (a) N. Roy Choudhury and A. K. Bhowmick, *J. Appl. Polym. Sci.*, **38**, 1091 (1989).
    (b) N. Roy Choudhury and A. K. Bhowmick, *J. Mat. Sci.*, **23**, 2187 (1988).

44. (a) D. J. Elliott, in *Development in Rubber Technology*, Vol. 3, A. J. Whelan and K. S. Lee (eds.), Applied Science, London, 1982.
    (b) N. Roy Choudhury and A. K. Bhowmick, *Polym. Degrad. Stab.*, **25**, 39 (1989).

45. (a) S. Akhtar, P. P. De, and S. K. De, *Mater. Chem. and Phys.*, **12**, 235 (1985).
    (b) S. Akhtar, D. Setua, P. P. De, and S. K. De, *Polym. Degrad. and Stab.*, **10**, 299 (1985).

46. D. J. Angier and W. F. Watson, *Trans. Instn. Rubb. Ind.*, **33**, 22 (1957).

47. A. Y. Coran and R. Patel, U.S. Patent 4,271,049, 1981.

48. A. J. Tinker, R. D. Icenogle, and I. Whittle, *Rubber World*, **199(6)**, 25 (1989).

49. (a) D. J. Elliott, "Natural Rubber Systems," in *Developments in Rubber Technology 3*, A. Whelan and K. S. Lee (eds.), Applied Science, London and New York, 1982.
    (b) D. S. Campbell, D. J. Elliott, and M. A. Wheelans, *NR Technol.*, **9(2)**, 21 (1978).

50. C. S. L. Baker, D. Barnard, and M. Porter, *Kautsch. Gummi. Kunst.*, **26(12)**, 540 (1973).

51. D. J. Elliott and B. K. Tidd, *Prog. of Rubber Technol.*, **37**, 83 (1973/4).

52. M. D. Ellul and D. R. Hazelton, *Rubber Chem. Technol.*, **67**, 582 (1994).

53. C. L. Riddiford and A. J. Tinker, Paper No. 33 presented at a meeting of the Rubber Division, American Chemical Society, Washington, DC, October 9–12, 1990; abstract in *Rubber Chem. Technol.*, **64**, 124 (1991).

54. (a) H. C. Wang, K. W. Powers, R. C. Puydak, and N. R. Dharamarajan, U.S. Patent 5,013,793 (1991).
(b) T. Ouhadi and D. S. T. Wang, "Development of Low Gas Permeable Thermoplastic Elastomers and Their Applications," *The Europe '91 Fifth International Conference on Thermoplastic Elastomer Markets and Products*, October 22–23, 1991.
(c) A. Y. Coran and R. P. Patel, U.S. Patent 4,130,534 (1978).
(d) R. C. Puydak and D. R. Hazelton, *Plast. Eng.*, **37**, 9 (1988).

55. (a) K. W. Powers and H. C. Wang, U.S. Patent 5,162,445 (1992).
(b) K. W. Powers and H. C. Wang, *Elastomerics*, Jan. 1992.
(c) H. C. Wang and K. W. Powers, *Elastomerics*, Feb. 1992.

56. (a). R. C. Puydak, D. R. Hazelton, and T. Ouhadi, U.S. Patent 5,073,597 (1991).
(b) R. P. Patel and S. Abdou-Sabet, U.S. Patent 5,621,045 (1997).

57. (a) D. R. Paul, in *Polymer Blends*, Vol. 2, D. R. Paul and S. Newman (eds.), Academic Press, New York, 1978, Ch. 12.
(b) S. Wu, *Polymer Interface Adhesion*, Marcel Dekker, Inc., New York, 1982.
(c) A. Zigman, *Adv. Chem. Ser.*, **43**, 1 (1964).

58. A. Y. Coran, R. P. Patel, and D. Williams-Headd, *Rubber Chem. Technol.*, **58**, 1014 (1985).

59. C. P. Rader, in *Handbook of Thermoplastic Elastomers*, 2nd Ed., B. M. Walker and C. P. Rader (eds.), Van-Nostrand Reinhold Co., New York, 1988, Ch. 4.

60. R. P. Patel, U.S. Patent 4,654,402 (1987).

61. B. E. Epstein, U. S. Patent 4,174,358 (1979).

62. E. J. Olivier, U. S. Patents 5,003,003 (1991) and 5,525,668 (1996).

63. (a) A. J. Moffet and M. E. J. Dekkers, *Polym. Sci. Eng.*, **32**, 1 (1992).
(b) J. R. Campbell, F. F. Khouri, S. Y. Hobbs, T. J. Shea, and A. J. Moffett, *Polym. Prepts., American Chemical Society*, **34(2)**, 846 (1993).

64. K. Venkataswamy, U.S. Patent 5,574,105 (1996).

65. (a) R. Patel, U.S. Patent 5,300,573 (1994).
(b) K. Venkataswamy, U.S. Patent 5,523,530 (1996).

66. T. Ouhadi, K. S. Shen, and S. Abdou-Sabet, *Plast. Eng.*, 37 (August 1996).

67. L. A. Goettler, J. R. Richwine, and F. J. Wille, *Rubber Chem. Technol.*, **55**, 1448 (1982).

68. D. J. Elliott and A. J. Tinker, "Thermolastic Natural Rubber Blends," in *Proceedings of the International Rubber Conference 1985*, Kuala Lumpur, Vol. 11, J. C. Rajarao and L. L. Amin (eds.), RRIM, Kuala Lumpur, 1986.

69. (a) H. Tanaka and J. L. White, *Polym. Eng. Sci.*, **20**, 949 (1980).
(b) V. M. Lobe and J. L. White, *Polym. Eng. Sci.*, **19**, 617 (1979).

70. Z. Krulis and I. Fortelny, *Eur. Polym. J.*, **33**, 513 (1997).

71. R. E. Medsker, D. R. Hazelton. G. W. Gilbertson, and J. E. Pfeiffer, "New Non-Hygroscopic UV Resistant Thermoplastic Vulcanizates for Extrusion," *SPE-ANTEC 97*, Toronto, Canada, May 1997.

36 Recycling of Polymer Blends and Mixtures

THOMAS S. ELLIS*

Polymers Department
General Motors Research and Development Center
Warren, Michigan 48090

I. INTRODUCTION

There is a growing political and economic awareness of the need to institute recycling as part of a broader objective of practicing design and manufacture from an environmental perspective. The terms *recycling*, *recycled*, and *recyclable*, although commonly applied to the process of recovering waste materials for reuse, have assumed different meanings that depend upon the context in which they are applied.

---

*Now with Delphi Automotive Systems Research and Development.
Polymer Blends, Volume 2: Performance. Edited by D. R. Paul and C. B. Bucknall.
ISBN 0-471-35280-2.   © 2000 John Wiley & Sons, Inc.

For example, there are administrative definitions based upon a compositional criterion that permits a material to be termed *recycled* provided that it contains a minimum of 25% by weight of postconsumer scrap, with the remainder being made up of unprocessed, or *virgin*, material. Similarly, a *recyclable* material has been defined by the U.S. Federal Trade Commission as material that must conform to processes that are *currently practiced and widely available*; that is, actual recycling of the material must be an established industrial process. However, the reuse of polymer waste, usually called regrind, that is produced as a by-product of a molding or extrusion process is not classified as *recycling* when the regrind is returned directly into the feed material. Neither can this kind of material be classified as *recyclable*, due to what are described as good housekeeping practices.

Unfortunately, none of these more precise definitions addresses the materials science requirements for recyclability. These concerns regarding recycling are the main focus of this chapter, and henceforth, any reference to recyclability in this chapter implies material integrity that facilitates diversion of the material from the waste stream, i.e., landfills, back into practical applications. By addressing some of the issues that allow this diversion to occur, several recent books [1, 2] have provided excellent accounts of the various practices and successes of recycling plastic materials from both relatively pure and randomly mixed wastestreams of unknown content and composition, including materials generated as solid municipal refuse. However, this chapter is restricted to examination of the recycling of manufactured thermoplastic polymer blends and alloys and inseparable polymer mixtures that are produced by melt processing and assembly operations, i.e., made by intent or design.

It is not necessary for a recycled material to possess properties that are comparable to those of the virgin resin. The properties of recycled compositions need only to meet the requirements of the intended application. Unfortunately, random and uncontrolled mixtures of polymers usually possess inferior mechanical, optical, and surface properties compared to their pure constituents and are typically described as incompatible. Compatibility is an often poorly defined qualitative term that is frequently confused by different interpretations; however, it is used here to describe the ability of two or more polymers to be processed, using currently available technology, into a single mixed material whose property profile is considered acceptable for normal applications. Although the word *incompatibility* is normally reserved for describing immiscible mixtures, compatibility in this sense can occur in both miscible and immiscible mixtures.

Many of the thermodynamic and processing issues that have to be considered in the design of a blend or alloy apply equally to the recycling of polymer mixtures; however, there are additional concerns that must be taken into account with recycling. Without differentiating between end-of-life (EOL) recycling and recovery or reuse during manufacture, the most important issue that controls the recycling of thermoplastics and their mixtures is the robustness of the constituents and of any required morphology in resisting thermal, oxidative, and shear degradation during multiple processing operations. Additional factors, such as the effects of long-term environmental degradation during service and contamination by other materials, such as coatings, paints, and absorbed fluids, must also be considered; however, these concerns apply equally to the recyclability of single-component materials.

II. BLENDS AND ALLOYS

Many widely used thermoplastics are, in fact, blends or mixed-polymer systems of considerable complexity, and over 30% by weight of total polymer consumption has been reported to be in the form of an alloy or blend [3]. For example, polycarbonate and acrylonitrile-butadiene-styrene (PC–ABS) blends are composed of three different polymers. The growing demand for improved properties, such as stiffness, ductility, and impact resistance, has been encouraged by the considerable material advances that can be achieved by blending or alloying polymers. This may be accomplished by incorporation of chemical reactivity between constituents and/or by control of morphology and composition during blending. Unfortunately, the incorporation of multiple constituents and compatibilization systems can introduce added sensitivity to degradation (e.g., chemical reactivity between constituents) and instability of morphology upon exposure to the repetitive, and hence abusive, thermal treatments that may be encountered during recycling. These undesirable processes are, of course, in addition to the usual thermal degradation and reduction of molecular weight that might be anticipated during melt processing.

A. Stability of Impact Modifiers

Impact-modified compositions probably represent the most frequently encountered multicomponent systems. Resistance to cracking and enhanced toughness also constitute the most prized attributes of polymeric materials. Accordingly, the morphological and thermal stability of the rubber additive is a primary requirement for recyclability. Polybutadiene (PB) provides excellent low-temperature impact properties; however, the carbon–carbon double bonds are particularly susceptible to thermal and oxidative degradation [4]. In such cases, the rubber degrades and cross-links, which reduces its effectiveness as an impact modifier and can also result in increased viscosity and processing difficulties.

The tolerance of ABS thermoplastics to multiple processing, up to five cycles, is quite good, because of the relatively low processing temperatures involved (220°C); however, a small loss of impact strength, attributable to degradation of the polybutadiene, has been reported [5]. PC–ABS alloys, wherein PC forms the matrix, require a higher processing temperature than does ABS alone and therefore are more prone to thermal degradation. Subjecting a PC–ABS blend to four or five injection-molding cycles has also been found to promote a significant loss of mechanical properties, which are accompanied by an increase in the melt viscosity [6]. Once again, cross-linking and oxidation of the rubber phase were identified as the main reasons for the observed changes.

Environmental and photooxidative degradation of ABS can also have a similar effect on the PB. Materials exposed outdoors for up to 10 years exhibited a large increase in the T_g of the PB component when examined by thermal analysis [7]. The greatest changes occured at the surface of the moldings, and a semiquantitative method, applying differential scanning calorimetry (DSC) was proposed to measure the amount of degradation. Samples exposed to sunlight for over 6 years exhibited lower impact strength; however, grinding and melt processing of aged samples was not performed, so it was difficult to assess the effect of exposure on the recyclability of these materials. If degradation is limited to the surface layer, the overall effect of environmental and photooxidative degradation on bulk properties of the recycled ABS is probably minimal.

The combination of stiffness, especially on exposure to high temperature, and solvent resistance displayed by alloys of polyamides, e.g., PA-66, with poly (2,6,-dimethyl-1,4-phenylene oxide) (PPO) has encouraged large-scale commercial development of these materials. However, the thermodynamic incompatibility of polyamides and PPO has necessitated the inclusion of a graft copolymer containing the two polymer backbones in order to develop the appropriate morphology [8, 9] and sufficient adhesion between the two polymers. The inclusion of this graft copolymer can be accomplished by *in situ* reactive extrusion of a maleic-anhydride-modified PPO, which reacts with the terminal amine group of the polyamide. (See Chapter 17.) The addition of small quantities (i.e., 2–10 wt%) of styrene–maleic anhydride (SMA) copolymers containing 8 wt% MA can also perform a similar function [10].

The alloy is approximately 50:50 by weight of polyamide and PPO; however, the polyamide forms a continuous phase containing a distinct dispersed-PPO phase. Resistance to impact and cracking has been a major consideration, and the inclusion of an elastomeric modifier, usually in the form of a block copolymer of polystyrene (PS) and a rubbery hydrocarbon polymer (e.g., polybutadiene (PB)), has been found to be necessary [11]. The resultant alloy is a highly complex marriage of four different polymer backbones and attendant stabilizers. Transmission electron microscopy (TEM) of one such alloy is shown in Fig. 36.1. The miscibility of PS and PPO drives the SBS copolymer into the PPO phase, which also results in a small decrease (i.e., 20°C) in the T_g of the PPO component. Staining with osmium tetroxide reveals the rubber phase as a black "rash" morphology in the PPO particles, which are approximately 0.5–2.5 µm in diameter [13].

The relatively high temperatures ($\geq 280°C$) required to process these particular PB-modified alloys, in association with abusive or repetitive processing, has been found to produce a significant loss of impact strength that can be correlated with degradation of the PB [12, 13]. Abusive processing was applied by allowing the material to reside in an injection-molding barrel for prolonged periods of time (e.g., 20 minutes), and repetitive processing included multiple injection molding and extrusions, respectively. Impact strength in both the notched condition and a high-speed (i.e., 5 mph) driven-dart method were found to be significantly affected. Figure 36.2 summarizes some of these findings. Similar results were obtained when the material was subjected to multiple extrusions and a cycle delay time in an injection-molding machine, respectively. By contrast, abusive processing of a PA-66 homopolymer, which contained a saturated and hence more thermally stable impact modifier, did not produce the same precipitous loss of impact strength.

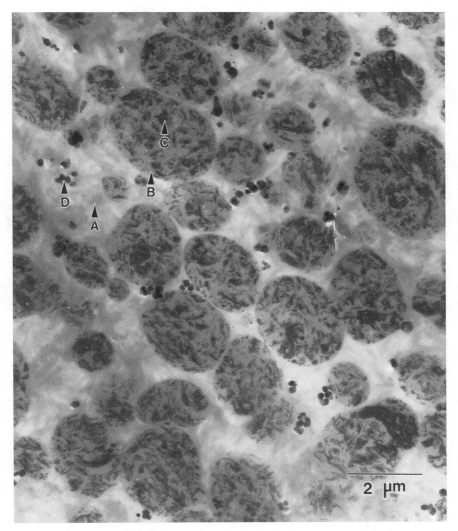

Figure 36.1   Transmission electron micrograph of a commercial PA-66–PPO alloy stained with osmium tetroxide. (A) PA-66 matrix, (B) PPO, (C) SBS elastomer, (D) carbon black–titanium dioxide pigment.

TEM failed to identify any large morphological changes in the rubber phase, and the correlation of loss of impact strength with PB degradation was made using NMR and thermal analysis. The NMR technique adopted provides a semiquantitative measure of molecular mobility of the elastomer. These results are also shown in Fig. 36.2. The solid-state NMR experiment applied a 90° Hahn-echo pulse sequence, which produced a double-exponential free-induction decay (FID) signal from the rigid (glassy) and mobile (rubbery) protons. By integrating the absolute-intensity

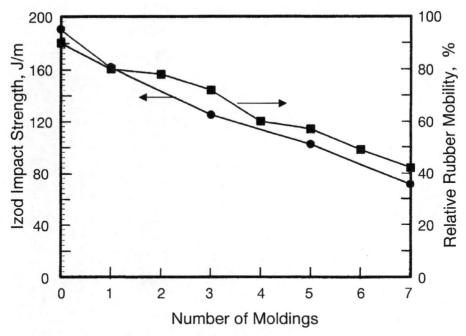

Figure 36.2   Effects of multiple injection molding on impact strength of a PA-66–PPO alloy, as measured by notched Izod impact test, and on the "mobility" of polybutadiene, as measured by NMR.

signal of the spectrum, a determination of the portion of the signal from the mobile protons was obtained. Calibration, provided by standards of known rubber content, allowed for an estimation of the content of mobile rubber in terms of wt%. The data obtained were semiquantitative; however, the technique is a simple, reliable method of assessing the state of a PB modifier in a material, and in principle, it can be applied to any blend or alloy.

Dynamic mechanical analysis (DMA) of materials subjected to the same multiple processing was also able to identify a shift to higher temperatures of the loss peak of T_g associated with the PB phase, as shown in Fig. 36.3. The β relaxation of the PA-66 obscured the relaxation maximum associated with the PB phase; however, after seven extrusions, the development of an additional loss peak at higher temperatures was quite visible. This observation is also in accord with the observation noted previously concerning the degradation of PB in ABS resins [7]. Evidently, the use of PB-based modifiers in alloys that require relatively high processing temperatures may have an adverse affect on the recyclability of the material. Replacement of the PB with a saturated rubber is a simple remedy, and PA-66–PPO alloys that are modified in this way are readily available.

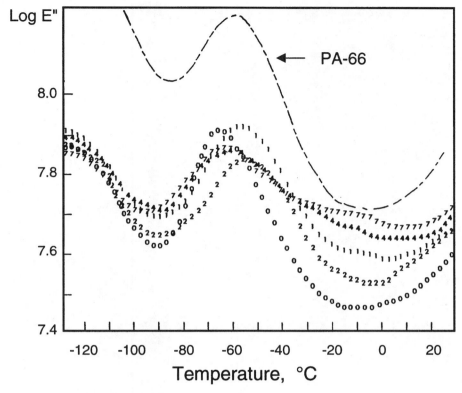

Figure 36.3   Dynamic mechanical loss $\log E''$, measured at 1 Hz, of a PA-66–PPO alloy subjected to zero, one, two, four, and seven extrusions, respectively. A PA-66 homopolymer is shown for comparison.

B. Stability of Morphology

Particle coalescence in a multiphase alloy, unless prevented or stabilized, compromises properties of the alloy. Chemical coupling across the interface or the addition of a compatibilizer helps to prevent this from happening; however, prolonged thermal exposure of the alloy may create instability, owing to thermal breakdown of the compatibilizing species. In the case of maleic-anhydride-based compatibilization, some of the species formed upon chemical reaction with an amine or hydroxyl functionality may also be susceptible to hydrolysis.

Extended thermal abuse of PA–PPO alloys has also been linked to a breakdown of the morphology, resulting in agglomeration of PPO particles, and is accompanied by transformation to brittle behavior [13]. The separation of the latter processes from the effects of degradation of the elastomeric modifier was demonstrated in compositions that contained a fully saturated rubber modifier that did not appear to degrade. The breakdown of morphology is seen most easily using TEM and scanning electron microscopy (SEM) techniques, an example of which is shown in Fig. 36.4.

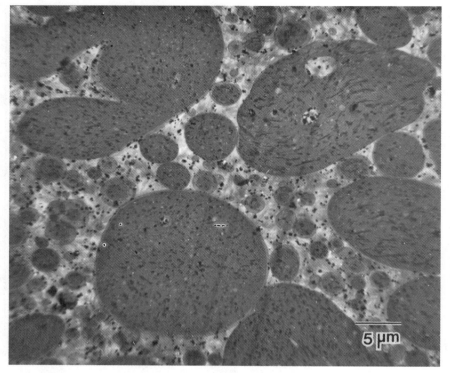

Figure 36.4 Transmission electron micrograph of a PA-66–PPO alloy after exposure to a processing temperature of 280°C in an injection-molding machine for 20 minutes.

In this micrograph, the alloy contains a thermally stable rubber modifier, and the breakdown of the compatibilizer appears to be responsible for the development of large PPO domains that are irregular in shape and up to 15–20 μm in diameter (c.f. Fig. 36.1). SEM of solvent-etched microtomed surfaces can also be used to provide a semiquantitative assessment of the increase of particle-size distribution caused by agglomeration [13].

The presence and state of the compatibilizing species remains among the least investigated and most difficult-to-measure characteristics of blends manufactured by reactive processing. The proprietary nature of alloying technology and blend formulation adds to the difficulty. The formation of covalent species between different polymer molecules is usually confirmed indirectly by microscopic or rheological measurements, as well as by measurements of mechanical properties. In the case of a PA-66–PPO alloy, sequential solvent extraction coupled with infrared spectroscopy has been shown to be effective in confirming the occurrence and chemistry of the reactive compatibilization [9].

The formation of the compatibilizer during reactive extrusion may also be inferred from a TEM micrograph, as shown in Fig. 36.5. This figure shows an ultra-microtomed sample after soaking in toluene to remove all soluble PPO. Although

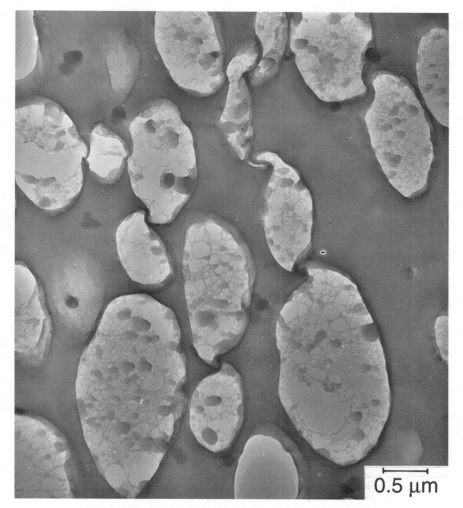

Figure 36.5 Transmission electron micrograph of an ultramicrotomed sample of a PA-66–PPO alloy treated with toluene to remove the PPO phase.

the morphology presented is not observed throughout the specimen and therefore is not representative of the material as a whole, the fibrous material suspended between the voids where the PPO resided is believed to be PA-66–PPO copolymer that cannot be removed by exposure to toluene. On the other hand, similar examination of an uncompatibilized alloy failed to show this kind of interfacial material and the blend exhibited well-defined interfacial regions. Generally, it is difficult to confirm the presence of a compatibilizer at the interface, and the example shown in Fig. 36.5 is probably caused by an aberrant locally high concentration of interfacial copolymer material.

C. Chemical Reactivity

Blends of two polyesters, e.g., PBT and PC, represent a significant proportion of commercially available blends. The partial miscibility of PBT with PC is a critical factor that allows the formation of compatible mixtures while retaining the beneficial crystalline character of the PBT. Unfortunately, this unique balance can be destroyed by reactivity, i.e., transesterification, between the components, leading to the formation of block copolymers and hence reduced melting temperature and crystallinity. Therefore, chemical stabilization to inhibit transreactions is essential. This subject has been studied extensively, because of the substantial and often adverse changes in properties that accompany transreaction (see Chapter 14) [14, 15].

There is little available information on the ability of these alloys to withstand prolonged thermal exposure in the form of repetitive processing; however, it is known that PBT–PC blends are susceptible to degradation in the absence of proper control of processing parameters. Several early studies [16, 17] addressed the robustness and recyclability of a commercial PBT–PC blend called Xenoy®, manufactured by the General Electric Co., and found increased viscosity and slight reductions in tensile properties and the low-temperature impact strength after the material had been extruded up to three times. The authors related these changes to reactivity between the constituents, which was evaluated by observing changes in thermal properties (e.g., melting temperature and heat of fusion) and solvent extractables. PBT–PC blends also contain PB-based elastomeric modifiers that may deteriorate upon multiple melt processing, as detailed previously.

Chemical degradation that leads to a reduction in molecular weight is a serious impediment to recycling, because such degradation makes it more difficult, if not impossible, to apply subsequent corrective action. For example, all polyesters and PCs are susceptible to hydrolysis, which results in a reduction in molecular weight. Drying prior to melt processing is essential. Long-term environmental hydrolytic degradation of PBT–PC blends, which are used in automotive bumpers, has been reported to lead to a significant reduction in impact strength [18, 19]. After five years of service in locations with high natural relative humidity, the impact strength of these blends was reduced to 500–600 J/m, compared with 790 J/m for freshly molded samples. The properties of the recycled materials, which included paint and adhesives, were reported to be even lower.

D. Rejuvenation

The prospect of rejuvenating a degraded or brittle material or of upgrading a material by blending or alloying it with another polymer or by adding a reinforcing filler is a viable option that should always be considered economically attractive because of the value that is added to a waste stream when it can be recycled. The addition of an appropriate impact modifier almost invariably improves the ductility and toughness of most polymers or their mixtures, but is usually accompanied by a decrease in the tensile modulus and tensile strength. Upgrading of a material in this way involves added cost, but may be preferable to consigning the material to a landfill.

Remedial impact modification was achieved by addition of 5–10 wt% SBS elastomer to a PA-66–PPO alloy that had been subjected to four processing cycles and had exhibited a loss of impact strength due to degradation of the PB. The elastomer boosted the notched Izod impact strength from 30 to 90 J/m [20]. Similar improvements were seen in driven-dart tests at both ambient temperature and −30°C. TEM and DMA indicated that at low levels (2–5 wt%), the additive was relatively well dispersed in the PPO phase, but at high levels (10 wt%), dispersion was very uneven, and large particles of elastomer could be seen.

Although rubber toughening of PA-66–PPO alloys is usually effected by adding modifiers that reside in the PPO phase, the presence of rubber in either phase has been reported to provide compositions with adequate impact resistance [21]. Accordingly, upgrading the alloy using an impact-modified PA-66 may also be an effective rejuvenation strategy. Similarly, an alloy that has been found to be deficient because of breakdown of the compatibilizer may be rejuvenated by additional compounding and compatibilization using, for example, styrene–maleic anhydride copolymers.

III. RECYCLING OF MANUFACTURED POLYMER MIXTURES

The recovery of pure polymers from waste streams is the most important prerequisite for economically successful recycling. Material-separation technologies continue to provide the highest value recycled material and should be pursued as the preferred recycling option. Considerable effort is currently being directed to separation and purification of mixed-material waste streams using sophisticated mechanical methods such as differential solvation, flotation, and electrostatic and air vortex (hydrocyclone) technologies [2]. Nevertheless, there are manufacturing designs and processes that generate intimately mixed residues that cannot be separated using these methods. Therefore, if the materials are to be recycled, the recycling must be done with the materials in the form of a blend. Processes producing inseparable mixture include:

  (i) coinjection molding,
 (ii) rotary table and multicomponent injection molding,
(iii) coextrusion and lamination,
(iv) welding (friction and thermal methods), and
 (v) adhesive bonding and mechanical fastening.

The first three of these processes above are essentially thermoforming processes. When it is not possible to manufacture multicomponent systems by melt-processing procedures alone, secondary assembly, involving welding, adhesives, and mechanical fasteners, can result in inseparable mixtures. Mechanical grinding, physical-separation processes, and occasionally manual separation, can be effective in producing reasonably pure recyclable materials from such mixtures. Nevertheless, successful separations still produce material streams that contain a finite level of contamination (0–5 wt%).

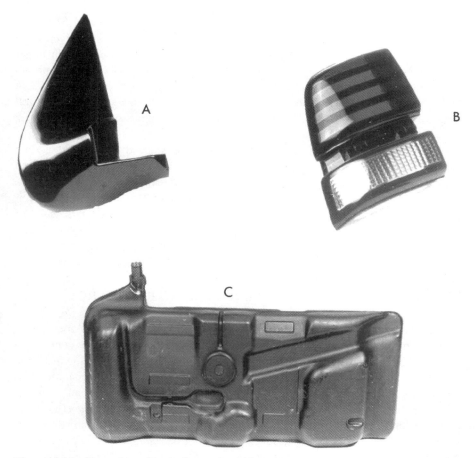

Figure 36.6 Examples of manufactured multicomponent automotive systems: (A) Co-injection-molded mirror housing (PBT–PC–ABS), (B) Rear-lens cover (PC–PMMA), (C) Co-extruded multilayer fuel tank (HDPE–maleated LLDPE–ethylene-vinyl alcohol copolymer).

The material and compositional content of mixtures produced by these processes are dictated by design criteria and not by the compatibilty of the respective polymers for recycling purposes. An advantageous feature of these types of scrap is the well-defined composition and content of the wastes generated; however, there may be local and long-range compositional heterogeneities that vary randomly and are extremely difficult to quantify. Occasionally, there is a need to identify correctly the polymers that are in a waste stream. Rapid computerized equipment, based upon infrared spectroscopy, has provided the most successful means of such identification [2]; however, there are still many difficulties associated with identifying blends.

The automotive waste stream represents a typical and significant source of mixed thermoplastic scrap, and some examples of the types of waste manufactured using previously listed the processes are shown in Fig. 36.6. The mirror housing mentioned in Fig. 36.6 was produced by coinjection molding of an impact-modified PBT–PC blend (as the outer skin) with an ABS (as the core). Consequently, when impact modifiers are included, this particular component contains five different inseparable polymers. Conjection molding is often proposed as a method of recycling wherein the recycled material forms the core of a molding. The rearlens cover in Fig. 36.6 is a rotary-table injection-molding assembly of PC and PMMA. In this single operation, the PC frame is molded first; then the mold is rotated, and the PMMA lenses are injected to form an overlay. The particular fuel tank shown is a coextruded blow molding of six layers comprising high-density polyethylene (HDPE), a maleated linear low-density polyethylene (LLDPE), and an ethylene–vinyl alcohol copolymer (EvOH), which usually contains approximately 32 mol% ethylene.

A. Compatibility Charts

Design and manufacturing engineers are beginning to recognize the need to specify compatible materials in designs and, whenever possible, to minimize the number of different polymers that are incorporated into a product. A favored strategy for assessing the recyclability of a mixed-polymer system, intended for implementation at the design stage, involves consulting one of the many compatibility charts that are now available. A representative example of a compatibility chart is presented as Fig. 36.7 and provides a relative measure of compatibility for a matrix of polymer pairings. Most of the pairs listed are classified as incompatible, and often there is little differentiation between members of a class of materials, e.g., polyamides and ABS. Use of these simple charts represents a good first step toward efficient recycling that allows nonspecialists to become aware of the problems associated with polymer mixtures. However, the brevity a one-page summary of a complex subject can lead to deficiencies in implementation that depend upon the nature of the individual components involved. Some recycling studies of PC–ABS mixtures and HDPE fuel containers, respectively, can be used to illustrate this point.

PC and ABS are generally considered to be compatible, for reasons that are discussed in the next section, and accordingly they are usually represented as such on compatibility charts. Several studies [22, 23] have indicated a satisfactory outcome from recycling PC–ABS mixtures that is independent of either the composition of the blend or the chemical composition of the styrene–acrylonitrile copolymer (SAN) in the ABS. Some ABS materials, however, contain additives that can degrade the PC during processing.

Figure 36.8 documents an unacceptable loss of impact strength, measured by driven-dart tests, in a particular recycled PC–ABS (ABS 2) blend. This loss occurred because it contained an additive, identified as a fatty-acid lubricant, that degraded the PC, causing evolution of volatiles during melt processing [22]. Notched Izod impact testing indicated a similar catastrophic loss of performance. Although decreases in molecular weight were also detected, the loss of impact strength was due mainly to void formation caused by gaseous degradation products. Figure 36.6 also shows that the impact properties of mixtures containing ABS 1 (31 wt% AN) and ABS 3 (24 wt% AN) displayed an almost linear dependence upon composition even though

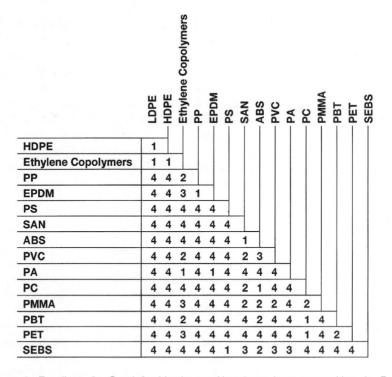

	LDPE	HDPE	Ethylene Copolymers	PP	EPDM	PS	SAN	ABS	PVC	PA	PC	PMMA	PBT	PET	SEBS
HDPE	1														
Ethylene Copolymers	1	1													
PP	4	4	2												
EPDM	4	4	3	1											
PS	4	4	4	4	4										
SAN	4	4	4	4	4	4									
ABS	4	4	4	4	4	4	1								
PVC	4	4	2	4	4	4	2	3							
PA	4	4	1	4	1	4	4	4	4						
PC	4	4	4	4	4	4	2	1	4	4					
PMMA	4	4	3	4	4	4	2	2	2	4	2				
PBT	4	4	2	4	4	4	4	2	4	4	1	4			
PET	4	4	3	4	4	4	4	4	4	4	1	4	2		
SEBS	4	4	4	4	4	1	3	2	3	3	4	4	4	4	

1 = Excellent   **2** = Good   **3** = May be good but depends on composition   **4** = Poor

Figure 36.7   Example of a compatibility chart available in the trade and product literature.

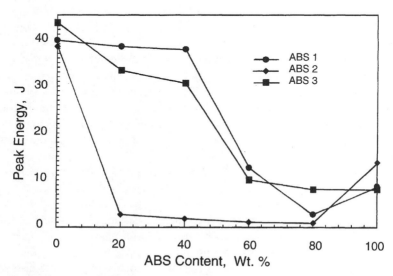

Figure 36.8   Impact strength, measured by a driven-dart impact tester, of PC–ABS blends containing different compositions of ABS.

there was a significant difference in their respective AN contents. The presence of flame retardants in ABS has also been reported to present an impediment to the recycling of mixed PC–ABS compositions [23]. Evidently, the general perception that PC–ABS mixtures are compatible is not universally correct.

Coextruded, multilayer films and containers contain polymers that are carefully chosen for their chemical resistance and ability to act as barriers to permeation by liquids and gases. They are often encountered in food-packaging applications; however, the example of these blends that is shown in Fig. 36.9 is a representative construction of a high-barrier-performance automotive thermoplastic fuel tank [24]. The sandwich structure consists mainly of HDPE (95% by weight) and two other polymers, a maleated LLDPE and a barrier polymer, which can be a polyamide or an ethylene–vinyl alcohol copolymer (EvOH). The maleated LLDPE, which provides adhesion by chemically reacting with the hydroxyl groups of the barrier layer, is usually referred to as the "tie" layer, and is believed to assist compatibilization between the HDPE and EvOH in the regrind layer.

Mixtures of polyolefins with EvOH copolymers are rarely included in compatibility charts; however, inspection of the chemical content of the two polymers suggests very poor compatibility. A TEM micrograph, reproduced in Fig. 36.10, of the recycled layer of the fuel container shown in Fig. 36.9 illustrates that the

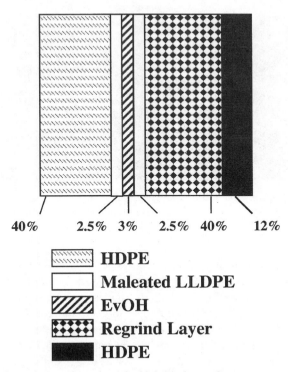

Figure 36.9 Schematic representation of a high-barrier-performance coextruded multilayer fuel tank. Compositions (wt%) of the polymers are indicated.

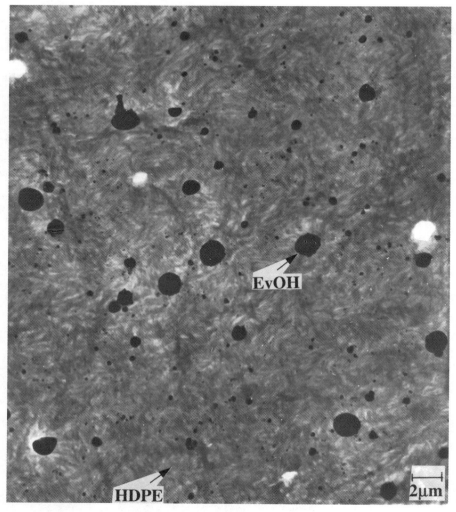

Figure 36.10 Transmission electron micrograph, obtained by staining with ruthenium tetroxide, of the morphology of the regrind layer of the fuel tank shown in Fig. 36.9.

minor component is well dispersed. The presence of voids where EvOH particles have been ripped out of the matrix during sample preparation indicates relatively poor adhesion with the HDPE matrix. Nevertheless, the tensile properties and impact strength of the HDPE were found to be largely unaffected by the presence of the EvOH [24]. In this instance, the high ductility of HDPE imparts a considerable tolerance towards the inclusion of small quantities (e.g., 5 wt%) of what is considered to be an incompatible component. By contrast, studies of polypropylene–EvOH blends [25], which contain significantly greater quantities of EvOH (i.e., 15–30 wt%), indicated a more serious decrease in impact properties when the EvOH content was 25 wt% or greater. This was attributed to the inherent incompatibility of the polymers and the production of a laminar morphology during processing.

B. Thermodynamic Features

The compatibility of PC–ABS compositions is founded upon a moderately small, positive thermodynamic interaction [26] between PC and the SAN copolymers. Although the magnitude of the interaction depends upon the AN content of the copolymer, a minimal interaction occurs when the AN content of the copolymer is approximately 22–24% by weight. Accordingly, when the composition of the SAN matrix is within this range, blends of PC and ABS do not require any additional compatibilization, have good impact resistance, and constitute a high-volume commercial blend.

Recycling of mixtures of PC and PMMA has attracted much attention because of the use of PC and PMMA in automotive lighting components. This application alone accounts for more than 18,000 tons of material each year in North America that is potentially available for recycling from industrial waste and EOL vehicles. PC and PMMA are also close to miscibility [27], signifying the prospect of compatible mixtures. With suitable impact modification they produce alloys with a good balance of properties [22, 28]. DMA of PC–PMMA blends, as shown in Fig. 36.11, shows a shift of the maxima in tan δ for both polymers towards each other in the blends, consistent with some degree of mutual solubility. Included in this figure are the results obtained for an actual sample of automotive lens scrap (R) in which PMMA is the major component (55 wt%). This sample also contains 5 wt% of a core–shell impact modifier, which produces a decrease in stiffness, as shown in the figure. The decrease in the storage modulus log E' as the PC content increases is in accord with measurements of the tensile modulus. The desirable stiffness of PC at temperatures up to 140°C, characterized by log E' in Fig. 36.11, can be compromised by the presence of relatively small quantities (e.g., 25 wt%) of PMMA.

The inclusion of only small quantities of PMMA in PC can have a disastrous effect on the notched impact resistance of PC [22], as shown in Fig. 36.12. The introduction of flaws, in the form of small inclusions of PMMA, imparts notch sensitivity to the blend. PC is known to exhibit a much lower impact strength when a very sharp notch is present [28]. By contrast, the toughness of the blends, measured by a driven-dart impact test, is reduced significantly only when the PMMA is present at elevated (i.e., 40–50 wt%) levels. These results are also shown in Fig. 36.12, which emphasizes the benefits of applying more than one test protocol for evaluating the impact resistance of recycled mixtures. Application of infrared spectroscopy and DSC has shown that PC and PMMA undergo transesterification above 200°C [29]; however, there is no information on the consequences of these effects in the context of multiple melt processing.

As with pure materials, the tensile modulus and strength of recycled mixtures are also of great importance. Figure 36.13 summarizes the behavior of PC–PMMA mixtures. The tensile moduli of these mixtures are linear with composition, whereas

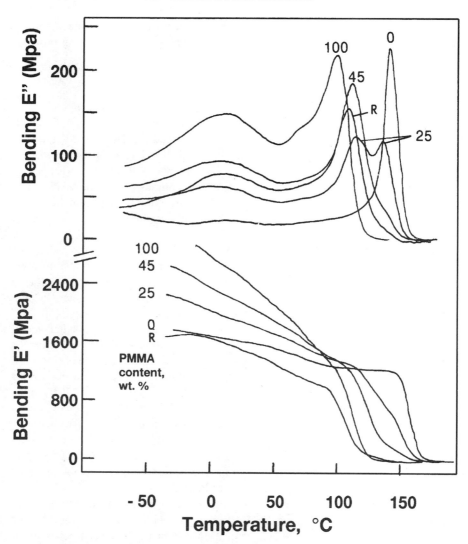

Figure 36.11   Dynamic mechanical storage log E' and loss log E'' moduli recorded at 1 Hz for blends of PC and PMMA, as a function of PMMA content.

the tensile strengths suffer a negative deviation from linearity. These results typify the behavior of most incompatible binary blends; however, in some instances, the tensile modulus can be enhanced to values greater than dictated by linear additivity, because of orientation effects. Addition of either an acrylic core–shell or a SAN-grafted EPDM impact modifier, at approximately 10–15 wt%, can improve the notched Izod impact strength considerably (e.g., by 200–500 J/m for compositions containing up to 55 wt% PC) [22, 28], leading to material properties normally associated with commercial PC–ABS blends. This modification also invariably produces

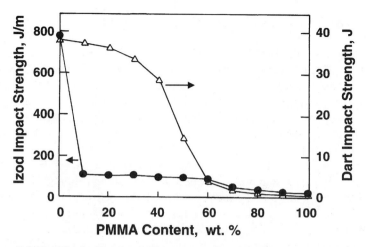

Figure 36.12   Notched Izod and driven-dart impact strength of PC–PMMA blends.

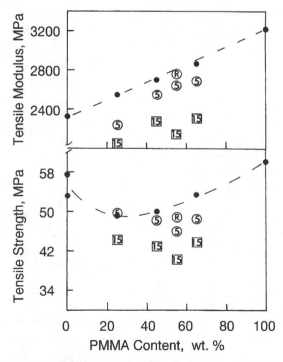

Figure 36.13   Tensile properties of PC–PMMA mixtures (●). Blends containing a SAN-grafted EPDM elastomeric impact modifier at 5 (⑤) and 15 ([15]) wt%, respectively, are also shown. Data points (Ⓡ) were obtained from a recycled mixture, derived from automotive lens scrap, containing 55 wt% PMMA and 5 wt% of a core–shell impact modifier.

a decrease in the modulus and the strength, as illustrated by the data points shown in Fig. 36.13.

Binary combinations of ABS and PMMA have also been explored in the context of recycling [22]. The composition of the SAN copolymer in the ABS imparts an additional complexity in that complete miscibility occurs with PMMA when the AN content is between approximately 9 and 28 wt% AN [30]. Some ABS resins contain copolymer of approximately 24 wt% AN and therefore form homogeneous mixtures, whereas other ABS resins may contain more than 30 wt% AN and thus form heterogeneous blends [22].

Miscible and immiscible mixtures of ABS and PMMA have exhibited significant differences in impact strength as a function of blend composition [31]. The immiscible mixtures exhibited a pronounced negative deviation from linearity with a 50% reduction in the impact strength of the ABS at just 10 wt% PMMA. Miscible mixtures exhibited an almost linear dependence of impact strength with composition. These results were obtained using ABS materials that possessed widely different AN contents (i.e., 21 and 35 wt% AN, respectively). In a separate study [22] the differences based on phase behavior were reduced considerably, as shown in Fig. 36.14. One of the immiscible blends exhibited a 50% drop in impact strength on addition of 20 wt% PMMA. Similar results were obtained from driven-dart impact tests, signifying low notch sensitivity in these mixtures. The presence of pigments in the ABS of the miscible blend and a reduced difference in the AN content of blends in this latter study may have been responsible for the lack of variation between the different ABS blends.

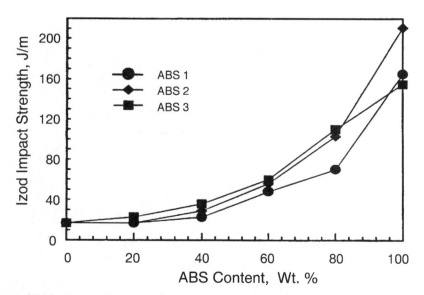

Figure 36.14 Notched Izod impact strength of ABS–PMMA blends with ABS of different AN contents (wt%). The immiscible blends are ABS 1 and 2 (31% and 29% AN, respectively). The miscible blend is ABS 3 (24% AN).

C. Composition and Morphological Features

The importance of morphology, such as dispersion, particle size, and the formation of cocontinuous phases, in determining the properties of binary systems is well recognized; however, with blends of three or more polymers, a situation that arises frequently in recycling, the influence of morphology on properties is less understood. Numerous complex morphologies may be found that are dependent upon compositional as well as thermodynamic factors. Although multiple dispersions are more prevalent, encapsulation of one phase by another in a matrix of a third polymer has also been seen quite frequently [32–35]. An example of encapsulation is shown in Fig. 36.15 for a mixture of PC, ABS, and PMMA, wherein the ABS forms the matrix and the unstained PMMA encapsulates the PC phase.

Similar observations have been noted in ternary blends of PC, SAN copolymers, and PBT. Transmission electron microscopy revealed that in a ternary blend of PS and PC in a PBT matrix, the PS phase was encapsulated by the PC [32]; however, as styrene was replaced by acrylonitrile in the form of SAN copolymers, two separate dispersed phases were produced. The underlying reason for the kind of morphology developed was explained in terms of the Harkins spreading coefficient: In a three-component blend, where polymer 2 is the matrix, polymer 3 will encapsulate polymer 1 if the Harkins spreading coefficient.

$$\lambda_{31} = \gamma_{12} - \gamma_{32} - \gamma_{13}, \tag{36.1}$$

where the parameter γ_{ij} is the interfacial tension between polymers i and j, is positive. This situation is depicted in Fig. 36.16.

This reasoning was successful for describing morphologies in mixtures of PC, PBT, PS, and SAN copolymers; however, it does not account for effects attributable to changes in the composition of the mixture. A more recent analysis [36, 37], founded on calculations that minimize the total interfacial free energy $\Sigma A_i \gamma_{ij}$ of the system, is probably more appropriate and was also successful in describing morphologies of mixtures comprising up to four polymers. The interfacial area A_i is composition dependent, whereas the interfacial tension γ_{ij} depends in part on thermodynamic interactions. True measurements of γ_{ij} are uncommon and difficult to obtain; however, estimations can be made using values of liquid–solid surface tension for each polymer [32–35].

A concise discussion [38] of the theoretical developments describing polymer–polymer interfaces suggests that the relationship $\gamma_{ij} \approx \chi_{ij}^{0.5}$ (or B_{ij}) should be observed. (See Chapter 3.) Therefore, in a simple approximation, it might be possible to predict morphologies using the thermodynamic interaction parameter alone. For example, the morphology seen in Fig. 36.15 can be predicted using values of the thermodynamic interaction energy density B between the constituents, where $B_{\text{PC–SAN (30 wt\% AN)}} = 0.2 \, \text{cal cm}^{-3}$ [26]; $B_{\text{PMMA–SAN (30 wt\% AN)}} = 0.01 \, \text{cal cm}^{-3}$ [39], and $B_{\text{PC–PMMA (30 wt\% AN)}} = 0.06 \, \text{cal cm}^{-3}$. Similarly, when the PMMA content is increased and PMMA becomes the matrix, the spreading coefficient is negative. Consequently, the PC and ABS form a binary dispersion, as shown in Fig. 36.17.

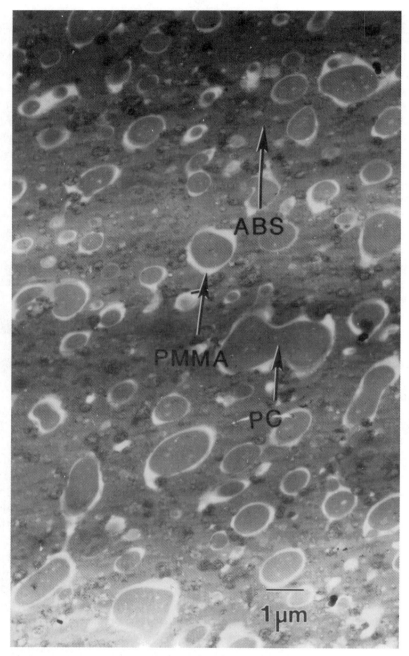

Figure 36.15 Transmission electron micrograph, obtained by staining with ruthenium tetroxide, of an 18/14/58/10 PC–PMMA–ABS–elastomer mixture.

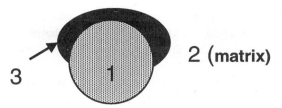

Figure 36.16 Schematic representation of the encapsulation of polymer phase 1 by component 3 in a ternary blend. (Polymer 2 is the matrix.)

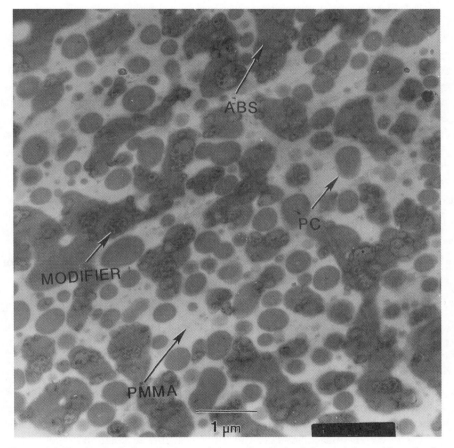

Figure 36.17 Transmission electron micrograph, obtained by staining with ruthenium tetroxide, of a 17/41/27/15 PC–PMMA–ABS–elastomer mixture.

Ternary compositions of PC, PMMA, and ABS that contain an ABS with a high AN content (e.g., 34 wt%), which is not particularly compatible with PC, have been found to display enhanced impact strength when small quantities of PMMA are present [40]. This apparently beneficial influence disappears in binary and ternary mixtures of PC, PMMA, and ABS when PMMA is the major component, and the blends exhibit poor notched Izod impact strength. The composition shown in Fig. 36.17 has a notched Izod impact strength of only 40 J/m. When PC forms the continuous phase, this value increases to almost 200 J/m, even without addition of further impact modifiers. These examples serve to illustrate that if the overall material composition of a particular mixed waste is unsuitable for recycling, then supplementation of a minor component may be necessary to upgrade the mixture to the desired composition.

IV. FUTURE TRENDS AND CHALLENGES

Many industries worldwide are beginning to struggle with the issues concerning the recycling and recyclability of their products. The automotive industry, in particular, is a major consumer of polymer blends and alloys and a producer of components comprising mixtures of polymers. Presently, approximately 75% by weight of all vehicles are recycled; however, most of the residual matter that is not recycled is polymeric. Current projected European legislation and commitments to environmentally responsible manufacturing by Asian and North-American automotive companies call for vehicle designs that are to be 85–90% recyclable by the year 2005. These developments will present great challenges to materials manufacturers and consumers to formulate recyclable blends and alloys, promote methods to recycle polymers and their inseparable mixtures, and help develop markets and applications for recycled materials.

It would be unfortunate if the hard-won success of high-performance polymer blends and alloys as premier engineering materials was compromised by the recycling of inferior materials that fail to meet specified requirements because of a lack of appropriate safeguards or failure to screen properties during use or repetitive processing. Perhaps the greatest challenge to prevent this problem from occurring is the determination of the factors that facilitate the recycling of a material and maintain the confidence that it will be able to perform to a required specification.

Design of materials and products for recyclability must continue to be an integral part of manufacturing operations; however, the development of alloys and blends and the use of multiple polymers in a particular application should be viewed in the wider context of life-cycle analysis (LCA) [41–43]. This tool, although difficult to apply with complete accuracy because of the difficulty in obtaining reliable data, assesses the total economic and environmental impact of products and processes. Sometimes referred to as a "cradle to grave" assessment, LCA assesses the life cycle of a product from the production of its raw materials through to its energy recovery or its final placement in a landfill. Recycling is only one element of LCA, but it is often used to emphasize the economic and environmental effectiveness of recycling materials. Once the economic benefits of material or component recyclability are identified, cost of materials can become a secondary consideration.

Recycling directly into applications can be achieved only a limited number of times. Material degradation, chemical and physical, must eventually render recycling impractical to the extent that energy recovery or tertiary recycling, such as depolymerization to feedstock molecules, are the only options available. The latter option should not be viewed negatively, however. Provided that appropriate, well-managed facilities are in place, energy recovery should be viewed as an opportunity to obtain the maximum value from a material.

V. ACKNOWLEDGMENT

I would like to thank the following people for their contributions to the content of this chapter: Mike Balogh, Bob Bullach, Sooho Kim, John Laverty, Michele Lesinski, Tim McMinn, Mark Myers, John O'Gara, Kevin Sheehan, and Curt Wong.

VI. REFERENCES

1. R. J. Ehrig (ed.), *Plastics Recycling: Products and Processes*, Hanser, Munich, 1992.
2. J. Brandrup, M. Bittner, W. Michaeli, and G. Menges (eds.), *Recycling and Recovery of Plastics*, Hanser, Munich, 1995.
3. L. A. Utracki (ed.), *Encyclopaedic Dictionary of Commercial Polymer Blends*, Chemtec Publishing, Toronto, 1994.
4. O. Chiantore, M. P. Luda di Cortemiglia, and M. Guaita, *Macromol. Chem.*, **190**, 3143 (1989).
5. J. K. Kim and C. K. Kang, *Polym.-Plast. Technol. Eng.*, **34**, 875 (1995).
6. J. I. Eguiazabal and J. Nazabal, *Polym. Eng. Sci.*, **30**, 527 (1990).
7. H. E. Bair, D. J. Boyle, and P. G. Kelleher, *Polym. Eng. Sci.*, **20**, 995 (1980).
8. S. Y. Hobbs, M. E. Dekkers, and V. H. Watkins, *J. Mater. Sci.*, **24**, 2025 (1989).
9. J. R. Campbell, S. Y. Hobbs, T. J. Shea, and V. H. Watkins, *Polym. Eng. Sci.*, **30**, 1056 (1990).
10. H.-J. Sue and A. F. Yee, *J. Mater. Sci.*, **24**, 1447 (1989).
11. C.-R. Chiang and F.-C. Chang, *Polymer*, **38**, 4807 (1997).
12. J. J. Laverty, *Polym. Eng. Sci.*, **28**, 360 (1988).
13. J. J. Laverty, T. S. Ellis, J. O'Gara, and S. Kim, *Polym. Eng. Sci.*, **36**, 347 (1996).
14. R. S. Porter and L. Wang, *Polymer*, **33**, 2019 (1992).
15. D. G. Hamilton and R. R. Gallucci, *J. Appl. Polym. Sci.*, **48**, 2249 (1993).
16. A. W. Birley and X. Y. Chen, *Brit. Polym. J.*, **16**, 77 (1984).
17. A. W. Birley and X. Y. Chen, *Brit. Polym. J.*, **17**, 297 (1985).
18. W. H. Tao, A. Golovoy, M. Zinbo, and D. R. Bauer, *Soc. Plast. Eng. ANTEC 93*, **39**, 3427 (1993).
19. A. Golovoy, M. F. Cheung, and H. van Oene, *Polym. Eng. Sci.*, **28**, 200 (1988).
20. J. J. Laverty and T. S. Ellis. unpublished work.
21. C. Koning and L. V. D. Vondervoort, *Soc. Plast. Eng. ANTEC 92*, **38**, 1435 (1992).

22. J. J. Laverty, R. L. Bullach, T. S. Ellis, and T. E. McMinn, *Polymer Recycling*, **2**, 159 (1996).

23. H. Larsson and H. Bertilsson, *Polymer Recycling*, **1**, 243 (1995).

24. T. S. Ellis, J. J. Laverty, and M. M. Lesinski, Paper 980096, *Proceedings of SAE International Congress and Exposition*, Detroit, 1998.

25. N. Walling and M. R. Kamal, *Adv. Polym. Technol.*, **5**, 269 (1996).

26. T. A. Callaghan, K. Takakuwa, D. R. Paul, and A. R. Padwa, *Polymer*, **34**, 3796 (1993).

27. T. A. Callaghan and D. R. Paul, *J. Polym. Sci.: Part B: Polym. Phys.*, **32**, 1813 (1994).

28. T. W. Cheng, H. Keskkula, and D. R. Paul, *Polymer*, **33**, 1606 (1992).

29. M. Rabeony, D. T. Hseih, R. T. Garner, and D. G. Peiffer, *J. Chem. Phys.*, **97**, 4505 (1992).

30. M. E. Fowler, J. W. Barlow, and D. R. Paul, *Polymer*, **28**, 1177 (1987).

31. B. Y. Kim, G. S. Shin, Y. J. Kim, and T. S. Park, *J. Appl. Polym. Sci.*, **47**, 1581 (1993).

32. S. Y. Hobbs, M. E. J. Dekkers, and V. H. Watkins, *Polymer*, **29**, 1598 (1988).

33. N. Nemirovski, A. Siegmann, and M. Narkis, *J. Macromol. Sci.-Phys.*, **B34**, 459 (1995).

34. B. K. Kim, C. H. Choi, and X. M. Xie, *J. Macromol. Sci.-Phys.*, **B35**, 829 (1996).

35. B. K. Kim, L. K. Yoon, and X. M. Xie, *J. Appl. Polym. Sci.*, **66**, 1531 (1997).

36. H. F. Guo, S. Packirisamy, N. V. Gvozdic, and D. J. Meier, *Polymer*, **38**, 785 (1997).

37. H. F. Guo, N. V. Gvozdic, and D. J. Meier, *Polymer*, **38**, 4915 (1997).

38. J. T. Koberstein, "Interfacial Properties," in *Encyclopeadia of Polymer Science and Engineering*, Vol. 8, John Wiley and Sons, New York, 1987, pp. 237–279.

39. M. Nishimoto, H. Keskkula, and D. R. Paul, *Polymer*, **30**, 1279 (1989).

40. K. J. Choi, G. H. Lee, S. J. Ahn, K. H. Shon, and H. M. Jeong, *J. Appl. Polym. Sci.*, **59**, 557 (1996).

41. R. A. Lee, M. H. Prokopyshen, and S. D. Farrington, Paper 971158, *Proceedings of SAE International Congress and Exposition*, Detroit, 1997.

42. G. Lambert, *Automobile Life Cycle Tools and Recycling Technologies*, SP-966, Society of Automotive Engineers (1993).

43. G.A. Keoleian, S. Spatari, R. Beal, R. L. Williams, and R. D. Stephens, *Int. J. Life Cycle Assess.* **3** (1), 18 (1998).

Index

Acrylate rubber, with iPP thermoplastic, 544–545
Additive localized shear banding, plastic void growth, and rubber bridging model, 245–246
Adhesion, core–shell toughening, 156–157
Aging, effects on impact strength, 73–74
Amorphous blends, 339–342
Amorphous polymers, blends with crystalline polymers, 344–346
Analytical methods, quasielastic mechanical properties, 2–4

Barrier materials
  blending, 359–393
  blends, 383–385
  control of blend morphology, 374–385
  water vapor properties, 382–383
Bifurcation, of cracks by thermoplastic particles, 261
Bisphenol A polycarbonate, toughening, 160
Blending, barrier materials, 359–393
Blend morphology
  barrier materials through control of, 374–385
  effects in rubber-toughened polymer blends, 290–291
Blow molding, thermoplastic vulcanizates, 549
Bound rubber, GC analysis, immiscible elastomer blends, 500
Bowing, of cracks by thermoplastic particles, 261
Bridging model, rubber particles, 243–244
Brittle fractures, 69–72

Cavitated particles, stresses in, 96–97
Cavitation
  core–shell particles, 167–169
  rubber particle, 87–101
Charpy impact test, 68, 203–204
  limitations of, 75–76
Chemical interactions, microfibrillar reinforced composites, 464–468
Chemical reactivity, recycling of blends and alloys, 566
Chemistry, core–shell polymers, 140–143

Classical mechanics, 2–3
Compatibility, thermoplastic vulcanizates, 521–522
Compatibility charts, manufactured polymer mixtures, 569–573
Compatibilization, 349–352
  immiscible elastomer blends, 501–502
Compatibilizers
  definition, 464
  in reinforced polymer blends, 410
  role in development of morphologies, 375–378
Composition, manufactured polymer mixtures, 577–580
Compositional gradient copolymers, 484–488
Compression, processing of microfibrillar reinforced composites, 468–472
Continuous-fiber composites, from polymer blends, 456–457
Core–shell impact modifiers, 137–175
  characterization, 145–147
  toughening polymeric matrices with, 155–166
Core–shell particles
  cavitation of, 167–169
  semicrystalline thermoplastics, 183
Core–shell structure
  formation, 143–144
  manufacturing, 144–145
Core–shell tougheners, effects on other physical properties, 166–167
Core–shell toughening
  adhesion, 156–157
  glass transition temperature, 156–157
  microstructural variants, 157
  particle size, 155–156
Crack bifurcation, deflection and, by rubber particles, 243
Crack deflection, by thermoplastic particles, 261
Crack deflection models, thermoplastic particle toughening, 262
Crack growth
  behavior, modeling of, 275–285
  under cyclic loading, 274–275
  under monotonic loading, 270–271

583